GW00383774

ENCYCLOPEDIA OF

UNDERWATER LIFE

ENCYCLOPEDIA OF

UNDERWATER LIFE

AQUATIC INVERTEBRATES AND FISHES

EDITED BY
ANDREW CAMPBELL
AND JOHN DAWES

OXFORD
UNIVERSITY PRESS

This edition published by
OXFORD
UNIVERSITY PRESS

Great Clarendon Street, Oxford OX2 6DP

Oxford University Press is a department of the University of Oxford. It furthers the University's objective of excellence in research, scholarship, and education, by publishing worldwide in:

Oxford New York

Auckland Bangkok Buenos Aires Cape Town Chennai Dar es Salaam Delhi Hong Kong Istanbul Karachi Kolkota Kuala Lumpur Madrid Melbourne Mexico City Mumbai Nairobi São Paulo Shanghai Singapore Taipei Tokyo Toronto

British Library Cataloguing-in-Publication Data
Data available

ISBN 0-19-280674-2
ISBN 978-0-19-280674-1
This edition first published in 2005

Printed in China

10 9 8 7 6 5 4 3 2 1

The Brown Reference Group plc
(incorporating Andromeda Oxford Limited)
8 Chapel Place
Rivington Street
London EC2A 3DQ

Advisory Editors

Dr. W. Nigel Bonner
British Antarctic Survey,
Cambridge, England

Professor Fu-Shiang Chia
University of Alberta,
Edmonton, Canada

Dr. Richard Connor
University of Massachusetts
at Dartmouth,
North Dartmouth,
Massachusetts

Dr. John Harwood
Gatty Marine Laboratory,
University of St. Andrews,
Scotland

Dr. John E. McCosker
Steinhart Aquarium,
California Academy of Sciences,
San Francisco, California

Dr. R. M. McDowall
Ministry of Agriculture
and Fisheries,
Christchurch,
New Zealand

Dr. Bernd Würsig
Texas A&M University
College Station,
Texas

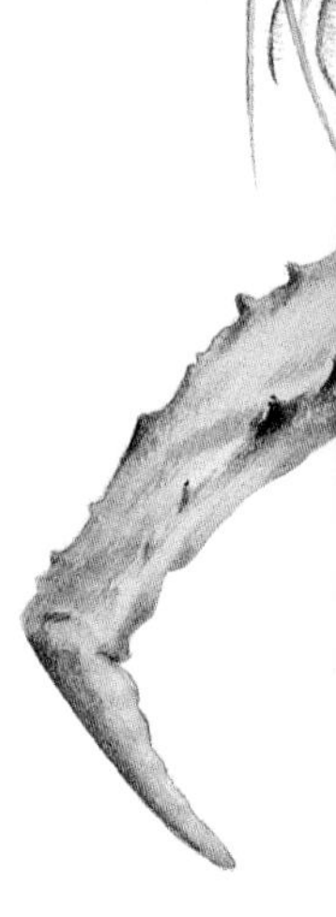

Artwork Panels

Mick Loates
Denys Ovenden
Colin Newman
Priscilla Barrett
S. S. Driver
Roger Gorringe
Richard Lewington
Kevin Maddison
Malcolm McGregor
Norman Weaver

Contributors

SSA	Sheila S. Anderson, British Antarctic Survey, England
RGB	Roland G. Bailey, University of London, England
GJB	Gerald J. Bakus, University of Southern California, Los Angeles
CCB	Carole C. Baldwin, National Museum of Natural History, Washington, D.C.
KEB	Keith E. Banister, (formerly) British Museum, London, England
RB	Robin Best, Instituto Nacional de Pequisas de Amazonia, Brazil
RCB	Robin C. Brace, University of Nottingham, England
BB	Bernice Brewster, British Museum, London, England
AC	Andrew Campbell, Queen Mary College, University of London, England
JEC	June E. Chatfield, Gilbert White Museum, Selborne, Hampshire, England
JD	John Dawes, Manilva, Málaga, Spain
GD	Gordon Dickerson, (formerly) Wellcome Research Laboratory, Beckenham, England
GDi	Guido Dingerkus, American Museum of Natural History, New York
AWE	Albert W. Erickson, University of Seattle, Seattle, Washington
SAF	Svein A. Fosså, Akvariekonsulenten, Grimstad, Norway
PRG	Peter R. Garwood, University of Newcastle upon Tyne, England
JCG	John Craighead George, North Slope Borough Dept. of Wildlife Management, Barrow, Alaska
GJH	Gordon J. Howes, British Museum, London, England
JJ	Jack Jackson, Woking, Surrey, England
JL-P	Johanna Laybourn-Parry, University of Lancaster, England
CL	Christina Lockyer, British Antarctic Survey, England
JMcC	John E. McCosker, California Academy of Sciences, San Francisco, California
RMcD	Bob McDowall, Ministry of Agriculture and Fisheries, Christchurch, New Zealand
LP	Lynne R. Parenti, Smithsonian Institution, Washington, D.C.
TP	Theodore W. Pietsch, University of Washington, Seattle, Washington State
PSR	Philip S. Rainbow, Queen Mary College, University of London, England
IW	Ian J. Winfield, Centre for Ecology and Hydrology, Cumbria, England

Crustaceans
see page 62

CONTENTS

AQUATIC INVERTEBRATES

FISHES

IUCN CATEGORIES

Ex Extinct, when there is no reasonable doubt that the last individual of a taxon has died.

EW Extinct in the Wild, when a taxon is known to survive only in captivity or as a naturalized population well outside the past range.

Cr Critically Endangered, when a taxon is facing an extremely high risk of extinction in the wild in the immediate future.

En Endangered, when a taxon faces a very high risk of extinction in the wild in the near future.

Vu Vulnerable, when a taxon faces a high risk of extinction in the wild in the medium-term future.

LR Lower Risk, when a taxon has been evaluated and does not satisfy the criteria for CR, EN or VU.

Note: The Lower Risk (LR) category is further divided into three subcategories: Conservation Dependent (cd) – taxa that are the focus of a continuing taxon-specific or habitat-specific conservation program targeted toward the taxon, the cessation of which would result in the taxon qualifying for one of the threatened categories within a period of five years; Near Threatened (nt) – taxa that do not qualify for Conservation Dependent but which are close to qualifying for VU; and Least Concern (lc) – taxa that do not qualify for the two previous categories.

PREFACE

ALL LIFE ON EARTH ORIGINATED IN THE PRIMEVAL seas some 4,000 million years ago. After eons of evolution, the waters that cover over two-thirds of planet Earth are home to a bewildering array of creatures, from tiny single-celled animals to giant squid and monstrous sharks, and from beautiful and delicate sea anemones, corals, and sponges to grotesque angler fishes and other denizens of the deep.

The aim of the *Encyclopedia of Underwater Life* is – to hazard an obvious pun – to give the reader an "in-depth" insight into this largely hidden underwater world and reveal the secrets of its diverse inhabitants. The *Encyclopedia of Underwater Life* proceeds from the microscopic to the gargantuan: from the tiniest invertebrates, such as the amoebas and ciliates, or vertebrates such as the pygmy goby and its relatives, to the giant squid, the whale shark, and the basking shark. The common denominator of all the taxonomic groupings described is that they lead an entirely aquatic lifestyle.

Aquatic invertebrates are invertebrate animals that live in the sea, in freshwater, or in moist terrestrial habitats. The term *invertebrates* refers to the fact that none of these animals has a bony or cartilaginous backbone. Although they are not animals, and hence cannot be invertebrates, **Protists** make up an important group of largely aquatic nonvertebrate organisms; they include many parasites whose aquatic environment is that of the bodies of their hosts.

Some invertebrate phyla, while overwhelmingly aquatic in habits, contain groups that have terrestrial forms. For example, although segmented worms are mainly marine, earthworms live in damp soil; also, slugs and snails are terrestrial variations of the mainly aquatic mollusks. For completeness, such terrestrial forms are also considered here. The diversities of form and biology are immense – a salmon and an elephant have more in common with one another than do many apparently related members of invertebrate phyla.

Thanks to the prominent part they play in human lives, birds and mammals are hugely popular subjects of study and amateur interest. By contrast, aquatic invertebrates and protists are seen by some people as the poor relations of the animal kingdom. This is far from the case. For sheer beauty, the microscopic architecture of diatoms and sea anemones is not easily surpassed. For their capacity to cause devastating disease in humans, the malaria and bilharzia parasites are without equal. In addition, for complexity of structure and intelligence, squids and octopuses rival fishes in their mastery of the water and in their fascinating behaviors.

Molecular analysis has resulted in a major revision of invertebrate systematics, and this text endeavors to reflect the latest findings. However, some chapter groupings remain mere "flags of convenience" and should not be taken as indicating taxonomic affinities between the different groups described.

With over 24,000 species known to science, **Fishes** live almost everywhere: from the cold, lightless waters of the deepest oceans to lakes high in the Andes mountains, on land, in mud, underground, in the air, even in trees. There are luminous fishes, transparent fishes, and electric fishes. The size range is colossal, from species that are fully grown at 9mm (0.4in) to species reaching 12.5m (41ft) in length. Some species have countless millions of individuals, while others survive precariously with just a handful of individuals.

Fish classification continues to be a matter of great debate among ichthyologists; as with the invertebrates, DNA studies have radically altered the understanding of relationships. While the present survey broadly adopts the scheme proposed by Joseph S. Nelson's authoritative volume *Fishes of the World* (3rd edn., 1994), the editors have given individual compilers free rein to follow their preferred classifications. The meticulously compiled web resource FishBase (www.fishbase.org) has also been of invaluable, not least by supplying a standard set of common English species names.

The bulk of the *Encyclopedia of Underwater Life* comprises general entries describing the biology, distribution, diet, breeding, and conservation status of particular groups; these groups are treated variously at the level of phyla (invertebrates), orders (fishes), or families (oysters, nurse sharks). Each entry incorporates a fact panel providing a digest of key data and often also includes outline drawings. For large groups these facts are consolidated in a separate table. Special features also focus in detail on particular subjects.

The invertebrate section includes five of these special features:
- the fascinating world of the drifters and wanderers of the world's seas – the plants and animals that make up the plankton;
- the life cycle of the malaria parasite, whose aquatic environment consists of the body fluids of its host and victims;
- the origins and organization of coral reefs;
- the economic importance of shrimps, crabs, and their relatives;
- the life cycles and importance of disease-causing (pathogenic) parasites, including flukes and the bilharzia parasite.

The fish section contains other special features, including:
- how some species of fishes have evolved to survive out of water;
- the amazing ability that sockeye salmon possess to return to breed in the very same river in which they were born;
- the various methods employed by fishes to generate light that attracts prey and mates, confuses predators, and acts as camouflage;
- the legendary coelacanth, first described in 1938.

As well as the outline drawings, there are many color illustrations by highly gifted wildlife artists that vividly bring their subjects to life (because fishes vary widely in size, it is not possible to show them to scale in the plates). Throughout, color photographs from diverse sources complement and enhance the text and artwork.

We are much indebted to the contributors who updated the text of the original 1985 edition by Banister and Campbell; sadly, Keith Banister, who died in 1999, could not be among them. Thanks are also due to the design, editorial, and production team who have seen this book through to publication. For all involved in this undertaking, a major reward is in knowing that the work will help raise awareness about the fragility of life in the oceans, lakes, and rivers, and the pressing need to conserve it at all costs.

ANDREW CAMPBELL
QUEEN MARY COLLEGE, UNIVERSITY OF LONDON

JOHN DAWES
MANILVA, SPAIN

What is an Aquatic Invertebrate?

WHAT DO CRABS, SEA URCHINS, EARTH-worms, and corals have in common? Until recently, it was believed that these very diverse groups of animals had very few shared characteristics, apart from the fact that they all lack a backbone. Of the 1,300,000 or so known species of animals, about 1,288,550 – more than 98 percent – have no backbone. Invertebrates, as they are termed, make up the vast bulk of animals, measured both in terms of numbers of species recognized and numbers of individuals. Some invertebrates, such as garden snails and earthworms, are conspicuous and familiar animals, while others, although very abundant, pass unnoticed by most people.

However, recent research in the field of genetics has shown that in fact invertebrates, like the fruit fly *Drosophila*, share much genetic material with the higher vertebrates like man. Therefore, in terms of genes rather than conspicuous bodily structures, all members of the animal kingdom have a lot in common.

Although they are not animals in the strictest taxonomic sense, protists make up an important group of largely aquatic nonvertebrate organisms. Heterotrophic protists, which feed on other organisms and are mobile, are animals in the broadest sense. Protists are single-celled life forms. Unlike other single-celled organisms (bacteria), protists have a membrane-bound nucleus that contains their genetic material, as do animal cells. In the past, animal-like protists were considered to be invertebrates and were grouped within the animal kingdom.

Invertebrate and protist body forms range in size from microscopic *Amoeba*, which may be just one micrometer in diameter, to the Giant squid 18m (59ft) in length, a ratio of 1:18,000,000. They include life forms as diverse as the Desert locust and the sea anemones. They inhabit all regions of the globe, and all habitats, from the ocean abyss to the air. Life almost certainly originated in the seas, and virtually all the major invertebrate groups, or phyla, listed (right) have marine representatives. Somewhat fewer (almost 14 phyla) have conquered freshwater. Fewer still (about 5 phyla) live on land and of these only the

jointed-limbed groups (arthropods) have mastered the air and really dry places. Most numerous among arthropods are the uniramians (e.g. insects, millipedes, centipedes) and the chelicerates (e.g. scorpions, spiders, ticks, and mites).

Many invertebrates, like slugs (whether of the garden or sea), are free living; others, such as barnacles, are attached to the substrate throughout their adult life: yet others live as parasites in or on the bodies of plants or other animals. Some invertebrates are of great commercial significance, either as direct food for man (e.g. prawns and oysters), or as food for man's exploitable reserves (e.g. the planktonic copepods on which herring feed). Others (e.g. earthworms) are much appreciated because they improve the soil for agriculture.

Many protists live as parasites, either inside the human body or in the bodies of domestic animals and plants. Because they can cause considerable damage, parasitic protists are of great medical and economic importance. Diatoms and other phytoplanktonic forms are hugely important as the basis of virtually all aquatic food chains.

This great diversity of form and lifestyle in all organisms has led zoologists to classify them according to type and evolutionary connections. To be certain that they are speaking of the same life forms, biologists give each species a unique scientific name. These two-part names are known as Linnean binomials, from the 18th-century Swedish botanist Carolus Linnaeus, who devised this system of classification. Thus, the Common earthworm is *Lumbricus terrestris*. Every species is classified into one of the major groups, or phyla. A phylum comprises all those organisms that are thought to have a common evolutionary origin.

Left and below *Among the most well-known aquatic invertebrates are cephalopods like the Day octopus* (Octopus cyanea), *while at the other end of the scale, both in size and familiarity, are inconspicuous jelly animals such as hydras* (Chlorohydra viridissima).

KINGDOM PROTISTA

Subkingdom Protozoa

FLAGELLATES Phylum Euglenida p12
1,000 species

TRYPANOSOMES & ALLIES Phylum Kinetoplastida p12
600 species

CILIATES Phylum Ciliophora p12
8,000 species

MALARIAL PARASITES Phylum Apicomplexa p12
5,000 species

DINOFLAGELLATES Phylum Dinoflagellata p12
4,000 species

DIATOMS, SLIME NETS, & ALLIES
Phylum Stramenopila p12
9,000 species

AMOEBAS & ALLIES Phylum Rhizopoda p12
200 species

RADIOLARIANS & ALLIES Phylum Actinopoda p12
4,240 species

FORAMINIFERA Phylum Granuloreticulosa p12
40,000 species

DIPLOMONADS Phylum Diplomonadida p12
100 species

PARABASILIDS Phylum Parabasilida p12
300 species

CRYPTOMONADS Phylum Cryptomonada p12
200 species

MICROSPORANS Phylum Microspora p12
200 species

ASCETOSPORANS Phylum Ascetospora p12
c.30 species

CHOANOFLAGELLATES Phylum Choanoflagellata p12
400 species

GREEN MICRO-ALGAE Phylum Chlorophyta p12
7,000–9,000 species

OPALINIDS Phylum Opalinida p12
c.400 species

KINGDOM ANIMALIA

Subkingdom Parazoa

SPONGES Phylum Porifera p20
5,000 species

Subkingdom Mesozoa
Phyla Placozoa, Monoblastozoa, Rhombozoa, and Orthonectida pp5&7

Subkingdom Metazoa

Radiata (radially symmetrical animals)

SEA ANEMONES, CORALS, & JELLYFISHES
Phylum Cnidaria p24
c.9,900 species

COMB JELLIES Phylum Ctenophora p36
100 species

Bilateria (bilaterally symmetrical animals)
DEVELOPMENT TYPE: PROTOSTOMES
Subgroup 1: (neither ecdysozoans nor lophotrochozoans)

GNATHOSTOMULIDS
Phylum Gnathostomulida p38
80 species

ACOELOMORPHS
Phylum Acoelomorpha p38
c.280 species

GASTROTRICHS Phylum Gastrotricha p38
c.450 species

ARROW WORMS Phylum Chaetognatha p38
c.100 species

Subgroup 2: Ecdysozoans (animals that molt)

WATER BEARS Phylum Tardigrada p40
Over 800 species

VELVET WORMS Order Onychophora p41
c.110 species

CRUSTACEANS
Phylum Athropoda; Subphylum Crustacea p42
c.39,000 species

HORSESHOE CRABS
Phylum Athropoda; Subphylum Chelicerata p64
4 species

SEA SPIDERS
Phylum Athropoda; Subphylum Pycnogonida p65
c.1,000 species

PRIAPULANS Phylum Priapula p66
16 species

KINORHYNCHS Phylum Kinorhyncha p66
c.150 species

LORICIFERANS Phylum Loricifera p66
10 species

HORSEHAIR WORMS Phylum Nematomorpha p66
c.320 species

ROUNDWORMS Phylum Nematoda p68
c.25,000 species

Subgroup 3: Lophotrochozoans (animals that have either a lophophore or a trocophore larvae):

FLATWORMS Phylum Platyhelminthes p72
c.18,500 species

RIBBON WORMS Phylum Nemertea p72
c.900 species

MOLLUSKS Phylum Mollusca p78
c.100,000 species

SIPUNCULANS Phylum Sipuncula p102
c.250 species

ECHIURANS Phylum Echiura p103
c.140 species

SEGMENTED WORMS Phylum Annelida p104
c.16,500 species

SPINY-HEADED WORMS Phylum Acanthocephala p113
c.1,100 species

ROTIFERS Phylum Rotifera p112
c.1,800 species

ENDOPROCTS Phylum Endoprocta p114
c.150 species

HORSESHOE WORMS Phylum Phoronida p116
11 species

MOSS ANIMALS Phylum Bryozoa p118
4,000 species

LAMPSHELLS Phylum Brachiopoda p120
c.335 species

DEVELOPMENT TYPE: DEUTEROSTOMES

SPINY-SKINNED ANIMALS
Phylum Echinodermata p122
c.6,000 species

ACORN WORMS & ALLIES
Phylum Hemichordata p136
c.90 species

SEA SQUIRTS & LANCELETS Phylum Chordata p138
Chordates include invertebrates such as lancelets and sea squirts as well as all vertebrates.

With one exception, animal phyla are made up exclusively of invertebrates. The phylum Chordata includes all animals with a hollow dorsal nerve cord. Nearly all the chordates – including fishes, amphibians, reptiles, and birds – have a backbone, but some are invertebrate, such as the sea squirts and lancelets. As new discoveries are made, and classification reassessed through new techniques of molecular analysis, the exact number of phyla is in a constant state of flux.

The technical classification that zoologists traditionally employ leads from the most primitiveand simple life forms to the most complex and advanced. In order to achieve some form of system, various levels of organization are recognized, which give clear distinctions between phyla. The most fundamental of these organizational levels concerns the number of cells in the body. A cell is the smallest functional unit of an animal, governed by its own nucleus, which contains genetic material known as DNA (deoxyribonucleic acid). Other features are employed, such as the way the embryos develop, the number of layers of cells, the shape and symmetry of the larvae and adults, and the way the various internal cavities are formed. However, recent research into the structure of DNA, and particularly the way in which individual genes are arranged in the genetic code of organisms, has revealed that most animals have much more in common with each other from a genetic standpoint than had previously been imagined. Also this study has provided some rather surprising information on the affinities of various groups.

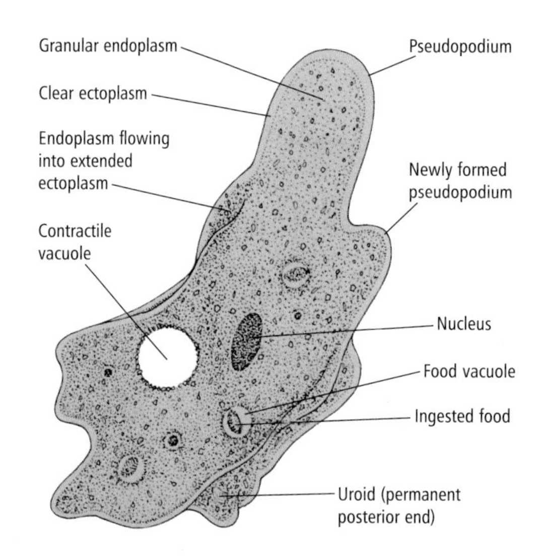

__Right__ The single protozoan cell may be quite a complex structure, with specialized parts called organelles responsible for different functions such s feeding and locomotion.

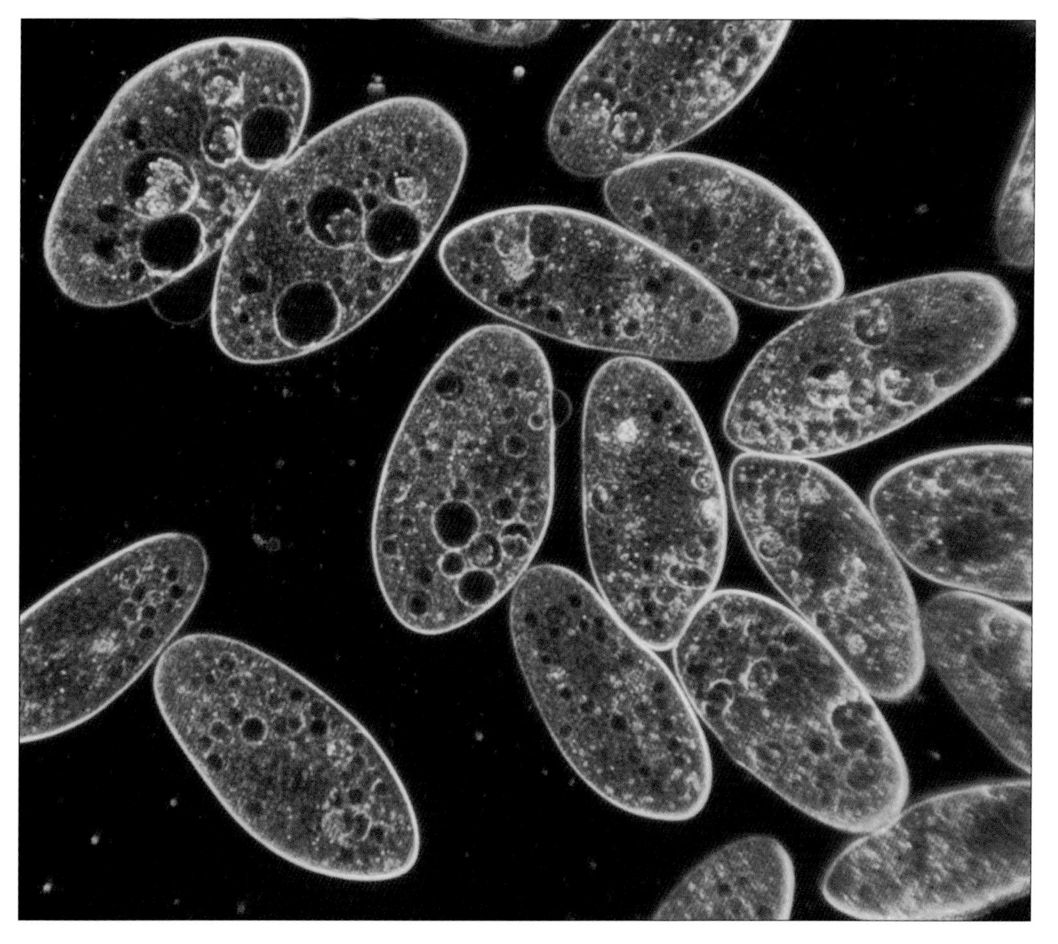

__Below__ Paramecia belong to a group of aquatic protozoans known as ciliates. These single-celled microorganisms reproduce by binary fission – dividing in half to form duplicates of themselves.

The Simplest Organisms?

Unicellular Organisms

Life forms with bodies made up of a single cell represent a separate level of organization from all the rest, for their body processes are performed by the one cell, which therefore itself cannot be specialized. In multicellular animals the responsibility for different life processes is shared out, different cells being specialized for different tasks, e.g. receiving stimuli (sensory cells), communication (nerve cells), movement (muscle cells) etc. Currently some 80,000 species of protists are known, but it is likely that many more remain to be discovered.

Single-celled life forms are sometimes assumed to be low on the scale of evolutionary sophistication. In reality their single cell is often large and complex. The essential life processes are carried

out by special regions (organelles) within the cell – nucleus for government, bubblelike food vacuoles for energy acquisition, vacuoles that contract and expand to regulate water levels within the cell and so on.

Protists can be loosely separated into plantlike and animal-like types. Animal-like protists are often termed protozoans. Like animals, protozoans are heterotrophic – they obtain their nourishment by breaking down organic materials. Plantlike protists (e.g. micro alga *Volvox*) are autotrophs (self-feeders). They are generally referred to as algae. Plantlike protists have organelles called chloroplasts, which are characteristic of plant cells. The chloroplast enables the cell to synthesize organic materials, such as sugars, from mineral salts, carbon dioxide, and water in the presence of sunlight (photosynthesis). The difficulty is that while some protists are clearly plantlike and others clearly animal-like, some, like the green flagellate *Euglena*, can feed by both autotrophy and heterotrophy. For reasons such as these, all protists have been classified in their own kingdom (Protista) for several decades.

Origins of Life

EVOLUTIONARY BEGINNINGS

Protozoans are important because, as the "simplest animals" they are likely to provide keys to two fundamental questions. These concern the origin of life, and the origin of multicellular metazoan animals. The origin of life is shrouded in scientific speculation. The biblical account of the origin of life as presented in Genesis is an historic attempt to answer one of man's most fundamental questions. Belief in the idea of a Divine creation is a matter of faith that cannot be tested by science. Nor can science yet tell us how life first began, although it has also been shown that primitive ideas such as spontaneous generation of life are completely wrong.

It is thought that the earth is not quite 5,000 million years old, and realistic estimates suggest that life began in its simplest form 4,000 million years ago. The first sedimentary rocks, not quite 4,000 million years old, contain fossils of simple cells that resemble those of present-day bacteria, that is, they lacked a distinct nucleus (i.e.they were prokaryotes). These cells lived in a primeval atmosphere devoid of oxygen. The appearance of oxygen on the earth, 1,800 million years ago, brought with it many new evolutionary possibilities. The protozoans, green algae, higher plants, and animals appeared a lot later, the earliest known fossils having been taken from rocks 1,000 million years old. All these organisms are made up of cells with distinct, membrane-enclosed nuclei (i.e. they are eukaryotes). Thus the startling fact emerges that for three-quarters of the period for which life has existed on earth the only cells were prokaryotes, resembling bacteria. It was not until the Cambrian

Above *The beautiful Christmas-tree worm* (Spirobranchus giganteus) *is a member of the phylum Annelida – segmented worms – an ancient group that has existed at least since the Cambrian 500 million years ago.*

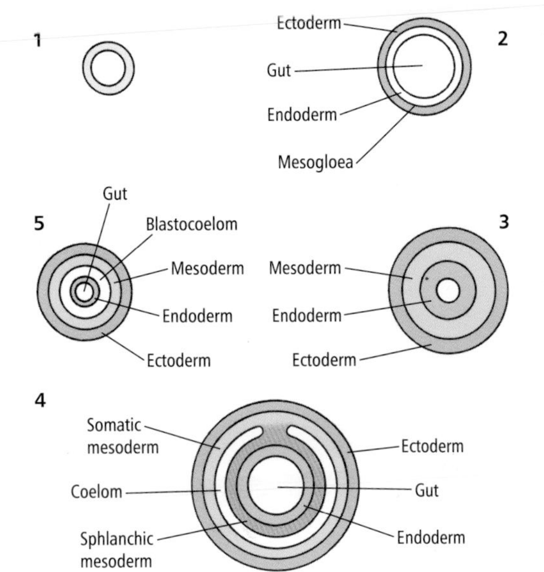

Right *The number and arrangement of cell layers distinguish various degrees of complexity in many-celled animals, from the single layer of the monoblastic sponges and mesozoans **1** through the diploblastic jellyfishes and comb jellies **2** to the three layers of most animals. The bulky middle layer (mesoderm) of triploblastic animals may be solid, as for example in flatworms and ribbon worms **3**, or divided into inner and outer parts separated by a cavity or coelom, **4**. Some groups, such as nematode worms, have a body cavity that is not formed within the mesoderm and is called a blastocoelom **5**.*

period 545–490 million years ago that invertebrates such as mollusks, trilobites, lampshells and echinoderms became established. Invertebrates with soft delicate bodies, such as segmented worms, flatworms, and sea squirts, have left a scant fossil record.

For life to have appeared, many conditions had to be fulfilled. It seems quite possible that the physical conditions prevailing on the surface of the early earth could have generated simple organic molecules such as amino acids, and then proteins, from inorganic molecules. The big unanswered question is how such substances could form themselves into organized living systems capable of reproducing their own kind.

Multicellular Animals

GROWING COMPLEXITY

The origins of multicellular animals are also speculative, but rather more can usefully be said about their possible early history. Because so many of the early animals had soft bodies, they left very little fossil record. Therefore, all theories about the early evolution of animals rely mainly on the study of similarities between developing embryos and adults of animals in different groups, which allows inferences to be drawn about common ancestry. Two chief theories have been put forward.

According to one theory, protozoan animals gave rise to multicellular ones by colony formation. A number of types of colonial protists are known to exist, such as *Volvox*.

The famous 19th-century German biologist Ernst Haeckel proposed that a hollow *Volvox*-like ancestor could have developed into a two-layered organism. Views differ as to whether or not this was a planktonic or a bottom-dwelling organism, but it may have somewhat resembled the planula larvae of the sea anemones and jellyfishes and could have given rise to bottom-dwelling animals such as adult hydroids and sea anemones.

Although most animals are composed of three layers of cells, often in a highly modified form, there is evidence that the evolution from two-layered animals did occur in evolutionary history.

A different theory proposes that multicellular animals arose from single-celled animals containing many nuclei by the growth of cell walls between each nucleus. A number of protozoans, for example *Opalina*, are like this. According to this theory, the primitive multicellular animal would lack a gut, as in gutless flatworms. However, the latter have three layers of cells, which raises the question – what is the origin of the two-layered animals? Because of this and other criticisms, this theory is now generally discarded in favor of the colonial one.

While it is not certain how multicellular animals evolved from protozoan ancestors, it is possible to distinguish groups of metazoans on

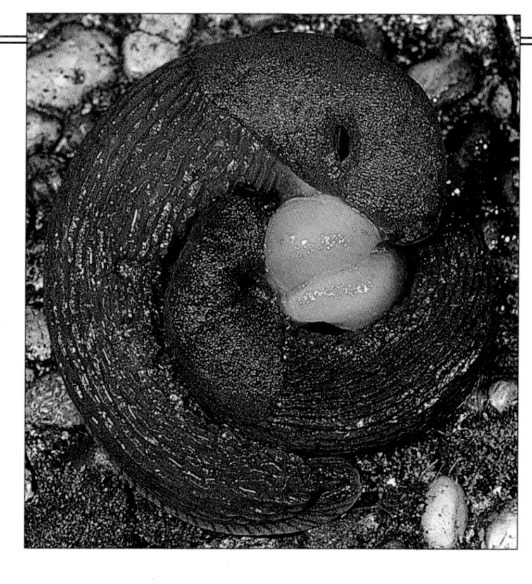

Above *Red slugs* (Arion rufus) *mating. As triploblastic animals, mollusks such as slugs and snails have a coelom – a fluid-filled body cavity that allows the body wall and gut muscles to move independently.*

Below *Sea anemones may well have been some of the earliest multicellular animals. Here, the tentacles of a group of sea anemones ensnare a* Mastagias sp. *jellyfish, another member of the phylum Cnidaria.*

the basis of relative simplicity or complexity of structure. (Some biologists divide invertebrates into protostomes and deuterostomes, see below.)

In metazoans there are three categories that can be used to determine level of complexity: how the cells are organized; how many layers of cells are to be found within the body; and whether or not a body cavity is present.

There are few types of cells in the most lowly metazoans and in sponges and these cells are never arranged into groups of similar cells (tissues). Such animals are said to have a cellular grade of organization. In the next step, as found in jellyfishes and allies and comb jellies, cells with similar functions are arranged together into tissues, each tissue having its own function or series of functions – these animals have a tissue grade of organization. In all animals apart from those just mentioned, requirements for functional specialization increase such that specific organs (often comprising a series of tissues) have evolved. Thus all animals from flatworms to man are said to have an organ grade of organization.

The second way to divide multicellular animals is to look at the number of layers of cells that make up the animal's body. Those groups that are of the mesozoan type as well as the sponges consist of just one layer of cells. However, in the jellyfishes and comb jellies, two layers appear (ectoderm outside and endoderm inside). This "diploblastic" condition contrasts markedly with the single layer of cells seen in the protozoan colony *Volvox*, which is described as monoblastic. The two layers develop from the egg and remain throughout adult life, separated from each other by a sheet of jellylike mesogloea.

All animals "above" the jellyfishes and comb jellies are equipped with three layers of cells and are known as triploblastic. Here the ectoderm and endoderm are separated by a third cell layer, the mesoderm. The mesoderm forms the most bulky part of many animals, contributing the musculature of the body wall as well as that of the gut. In some of these triloblastic animals, e.g. flatworms (Platyhelminthes), the mesoderm is solid and not itself divided into two layers by a body cavity

Above *In common with many aquatic invertebrates, nudibranchs (shell-less sea slugs) are hermaphroditic. One of these colorful* Tambja verconis *has laid a roselike egg mass (center).*

(coelom), so they are described as acoelomate. Without a body cavity these animals are at a disadvantage because the body movements affect the movements of the gut and vice versa. To be able to move the gut independently of the body wall is a great advantage, as it enables sophisticated digestive activities. Such a condition is reached only in coelomate animals, in which the mesoderm is divided into an outer section forming the body wall and an inner section forming the muscles of the gut, and the two are separated largely by a body cavity (coelom) within the mesoderm. The possession of a fluid-filled body cavity is the hallmark of all the more advanced invertebrates and the major phyla, including mollusks, annelids, the different arthropods, echinoderms and chordates; all have a coelom (are coelomates). Even so, improved techniques of research have changed the understanding of the origin of body cavities.

In some groups there is a body cavity between the body wall and the gut. This cavity used to be called a pseudocoelom, not a true coelom, for it is not formed inside the mesoderm. Generally it arises from the blastocoel, an embryonic cavity that appears in the early stages of development, and contains fluid. In animals such as the nematode worms, it performs an important function as an incompressible, fluid skeleton. This cavity is not "false" (i.e. pseudo), and many authorities now describe animals such as these as blastocoelomate.

Some evolutionary biologists suggest a different way of grouping invertebrates, believing that two main evolutionary lines have emerged in the animal world, the protostomes (first mouth) and deuterostomes (second mouth). In the early embryos of protostomes such as annelid worms and insects, the mouth is formed at or near the site of the blastopore. In deuterostomes such as the echinoderms (starfishes and sea urchins) and the chordates, the anus forms at the blastopore. Other distinctions can be seen by comparing the development of the two types. In the annelids the nerve cord is on the underside, a double, solid structure reminiscent of the nerve cord of insects. In the chordates it is a single hollow structure along the upper side of the animal. The annelid coelom is formed by the splitting of the mesoderm, but in the echinoderms and the chordates it develops from two pouches of the primitive gut. The sea-dwelling representatives of these two "lines" have characteristically different larvae.

Recent DNA sequence analysis of the different members of the animal kingdom has shown that the relationships believed to exist among different phyla, based on their structural organization, are now not likely to be correct, and different phyla have been repositioned within the scheme of an evolutionary tree accordingly. The sequence of the multicellular phyla, set out at the start of this section, reflects this new thinking. Here the genetic affinities between invertebrates that periodically molt their outer cuticle, and those that do not are particularly reflected.

These distinctions are not definite evidence of links between phyla, but do indicate possible evolutionary affinities. Certainly the chordates are likely to have arisen from a deuterostome type of ancestor, and the form of development shown by annelids and mollusks, as well as their DNA sequences, places them far from the echinoderms and chordates in any phylogeny or "family tree."

Symmetry

Body Plans

One of the most obvious differences separating the phyla of the animal kingdom is the overall appearance of the animals. The majority of animals, including the vertebrates, worms, and jointed-limbed forms, are bilaterally symmetrical: complementary right and left halves are mirror images. Their front and hind ends are however dissimilar; in these animals there has been some specialization at the head end. In the lowly flatworms the head is only feebly developed, but it is identifiable. In the jointed-limbed animals, such as the insects, the head is clearly defined. The evolution of a head end (cephalization) has come about in direct response to the development of forward movement. Clearly it is an advantage to have specialized sensory equipment (e.g. eyes and smelling and tasting receptors at the front of the body) to deal efficiently with environmental stimuli as they occur in the direction of travel. In this way, prey and predators can be identified quickly.

The position of sense receptors is often associated with the mouth opening. These factors have led to the development of aggregations of nervous tissue to integrate the messages coming from the receptors and to initiate a coordinated response (e.g. attack or flight) in the muscle systems of the body. This brain, be it ever so simple, came to lie at the front end of the body, often near the mouth, and the typical head arrangement was formed. In addition to a distinct head, other features of bilaterally symmetrical animals are thoracic regions (modified for respiration) and abdominal regions (modified for digestion, absorption, and reproduction). The function of locomotion may be undertaken by either or both of the regions. In the mollusks the principles of bilateral symmetry are disguised somewhat by the unique style of body architecture, but if these animals are reduced to their simplest form, as in the aplacophorans and chitons, a basic bilateral symmetry is visible.

However, some animals have managed without a head. Despite their relatively high position in the table of phyla, the echinoderms (e.g. starfishes, sea urchins) are headless. This is all the more surprising when one realizes that the earliest echinoderms and their likely ancestors were bilateral animals. Still, they managed without a head, and in the present-day echinoderms the nervous tissue is spread fairly evenly throughout the body and all sensory structures are very simple and widely distributed. Modern adult echinoderms are fundamentally radially symmetrical; the body parts are equally arranged around a median vertical axis which passes through the mouth. There is no clear front end, nor left and right sides, in most of the species. Radial symmetry is best for a stationary lifestyle, in which food is collected by nets or fans of tentacles. This was almost certainly the lifestyle of the earliest echinoderms, as it is of most of present-day Cnidaria (jellyfishes and allies), the other phylum to display a clear radial symmetry. The radial symmetry in the Cnidaria is evident, whereas in the echinoderms a unique five-sided body form has been imposed on it, as seen in present-day starfishes and brittle stars.

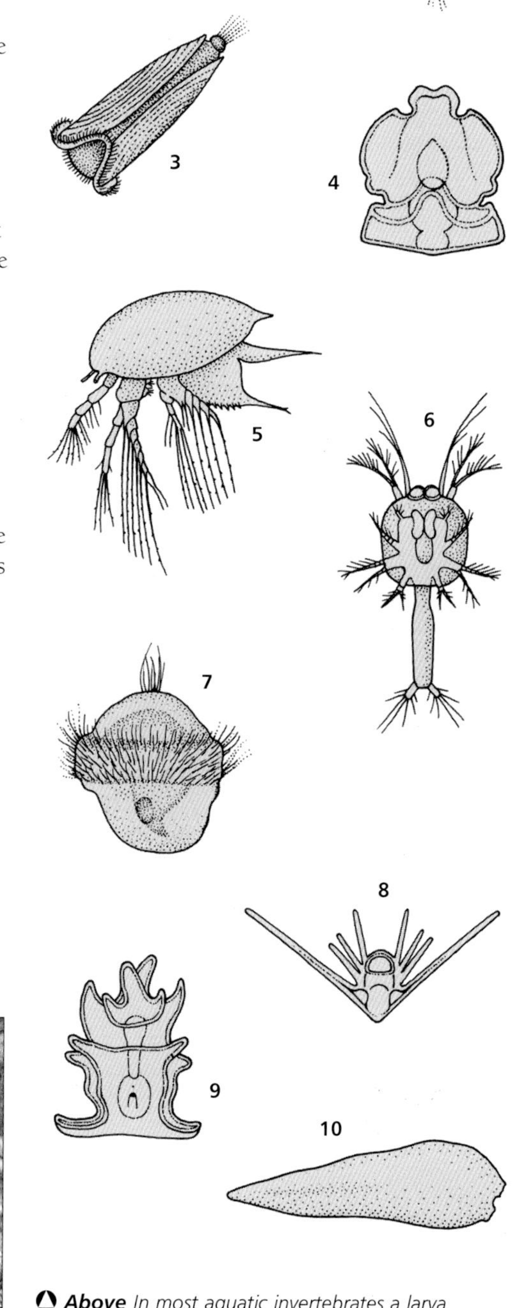

Above *In most aquatic invertebrates a larva hatches from the egg, floats free, and metamorphoses into adult form. In some, an adult-like juvenile emerges. Adults with simple body architecture usually have larvae that are simple in form (e.g. the planula of sponges and jellyfishes). The tadpole-like sea squirt larva may have provided the evolutionary springboard from which vertebrates developed:* ***1*** *planula (sponge, jellyfishes, and other cnidarians);* ***2*** *trochophore (many worms);* ***3*** *cyphonautes (sea mat);* ***4*** *tornaria (hemichordates, including acorn worm);* ***5*** *nauplius (first stage of many crustaceans);* ***6*** *zoea (crabs and other decapods);* ***7*** *trochophore (mollusk);* ***8*** *pluteus (brittle star, sea urchin);* ***9*** *auricularia (sea cucumber);* ***10*** *"tadpole" larva of a sea squirt.*

Above and left *Most aquatic invertebrates have bilateral symmetry in common, however diverse their general appearance. For example, crabs and other crustaceans show a straightforward symmetry around a midline. Minor asymmetries may occur within this basic pattern, such as in the male Fiddler crab (*Uca pugnax*; left), which has evolved one claw far larger than the other. This outsized limb is used to attract the attention of females for mating, and to wrestle with other males during courtship struggles.*

Yet two major invertebrate groups show a different, radial symmetry without a head – the jellyfishes and their relatives, and the echinoderms, such as this Fromia *sp. starfish (above). Five-pointed symmetry around the axis of the mouth is called pentamerism.*

Finally, there are those animals which are essentially without regular form or are asymmetrical. These are the sponges, which grow in a variety of fashions, including encrusting, upright or plantlike, or even boring into rocks. While each species has a characteristic development of the canal system inside the body, its precise external appearance depends very much on prevailing local conditions, such as exposure, currents and form of substrate. The simple body form of the sponges of course rules out the development of nerves and muscles, so a coordinating nerve system has not evolved.

The bilaterally symmetrical bodies of coelomate metazoans may be unsegmented (americ), divided into a few segments (oligomeric), or many units (metameric). The bodies of many types (e.g. ectoprocts or echinoderms) show division into a few segments during their development, but these may be masked in the adult. In the annelids and arthropods the repetition of structural units along the body (metamerism) allows for the modification of the segmental appendages to fulfil various functions. In the arthropods these appendages or limbs, often jointed, carry out a wide range of activities, from locomotion to copulation. AC

DRIFTERS AND WANDERERS

The ecology of zooplankton

THE WORD PLANKTON MEANS "DRIFTER" OR "wanderer". It refers to those plants and animals that are swept along by water currents rather than by their own swimming ability. (Animals that swim and determine their own direction are called nekton.) The greatest diversity of plankton exists in the world's oceans, but lakes and some rivers also have their own plankton communities. Here, examples from the sea will be used. Plants of the plankton are known as phytoplankton and animals of the plankton as zooplankton. While many zooplankton are minute – less than 5mm (0.2in) long – a few are large, for example some jellyfishes, whose tentacles may be 15m (49ft) in length. Some zooplankton can swim, but not sufficiently well to prevent them from being swept along by currents in the water. However, their swimming ability may be sufficient to allow them to regulate their vertical position in the water, which can be very important as the position or depth of their food can vary around the daytime/nighttime cycle (for example, phytoplankton rises by day and sinks by night, whereas zooplankton does the reverse).

Seawater contains many nutrients important for plant growth, notably nitrogen, phosphorus, and potassium. Their presence means that phytoplankton can photosynthesize and grow while they drift in the illuminated layers of the sea. Two forms of phytoplankton, dinoflagellates and diatoms, are particularly important as founders in the planktonic food webs, for upon them most of the animal life of the oceans and shallow seas ultimately depends. By their photosynthetic activity, the dinoflagellates and diatoms harness the sun's energy and lock it into organic compounds such as sugars and starch, which provide an energy source for the grazers that feed on the phytoplankton.

Zooplankton comprises a wide range of animals. Virtually every known phylum is represented in the sea, and many examples of marine animals have planktonic larvae. Such organisms may be referred to as meroplankton or temporary plankton. Good examples are the developing larvae of bottom-dwellers such as mussels, clams, whelks, polychaete worms, crabs, lobsters, and starfish. These larvae ascend into the surface waters and live and feed in a way totally different from that of their adults. Thus the offspring do not compete with the adults for food or living space, and the important task of dispersal is achieved by ocean currents. At the end of their planktonic lives, the temporary plankton must settle on the seabed and change into adult forms. If the correct substrate is missing then they fail to mature. Often complex physiological and behavioral processes occur before satisfactory settlement is achieved, and many settling larvae have elaborate mechanisms for detecting textures and chemicals in substrates.

Right *Radiolarians have geometrically shaped skeletons made of silica. These minute protozoans form part of the vast oceanic plankton community.*

Below Oncea *sp. are tiny holoplanktonic crustaceans; copepods such as these are the main source of protein in the oceans. This group is mating.*

In addition to the temporary plankton there is a holoplankton: organisms whose entire lives are spent drifting in the sea. Of these, the most conspicuous element (around 70 percent) are crustaceans. The most abundant class of planktonic crustaceans are the copepods, efficient grazers of phytoplankton especially in temperate seas. The euphausids make up another very important group of crustaceans, and in some regions, for example, the southern oceans, they can occur in enormous numbers as "krill," providing the staple diet of the great whales. All these crustaceans have mechanisms for straining the seawater to extract the fine plant cells from it. Other noncrustacean holoplanktonic forms that sieve water for food are the planktonic relatives of the sea squirts. Some rotifers live as herbivores in the surface waters of the sea, but they are a much more important component of the plankton of lakes and rivers. Along with many invertebrate larvae, these herbivorous holoplanktonic forms are important in harvesting the energy contained in the planktonic and pass it on to the carnivorous zooplankton by way of the food webs of the sea's surface.

There are many types of carnivorous zooplankton in the oceans and members of many phyla are involved. Protozoans feeding on bacteria or other protozoans occur. Some, like the foraminiferans and radiolarians, form conspicuous deposits on the seabed after they die, thanks to their durable mineralized shells or tests. Cnidarians provide a range of temporary and permanent planktonic carnivores. Many hydroid medusae spend only part of the life cycle of the hydroids in the plankton, while others like the Portuguese man-of-war are permanent plankton dwellers, often taking food as large as fishes. The ctenophores, such as *Pleurobrachia* and *Beroë*, are efficient predators of copepods, often outcompeting fishes (for example, herrings) for them. Thus they are of economic significance as competitors of commercial fish stocks. Other carnivores include pelagic gastropods, polychaetes, and arrow worms.

Oceanographers regard the occurrence of certain species in surface waters as an indication of the origins of water currents. Thus, in Northwest Europe, plankton containing the arrow worm *Sagitta elegans* has been demonstrated to come from the clean open Atlantic, whereas water containing *S. setosa* is known to have a coastal origin. Different chaetognaths also appear at different depths in the ocean and are indicative of different animal communities. AC

Protozoans

PROTOZOANS LIVE IN AN UNSEEN WORLD. They are invisible to the naked eye but occur all around us, beneath us, and even within us – they are ubiquitous. Their overriding requirement is for free water, so they are found mainly in floating (planktonic) and bottom-dwelling communities of the sea, estuaries, and freshwater environments. Some live in the water films around soil particles and in bogs, while others are parasites of other animals, notably causing malaria and sleeping sickness in humans.

Most protozoans are microscopic single-celled organisms living a solitary existence. Some, however, are colonial. Colony structure varies; in the green chlorophyte *Volvox* numerous individuals are embedded in a mucilaginous spherical matrix, while in other flagellates, for example *Diplosiga*, a group of individuals occurs at the end of a stalk. Branched stalked colonies are characteristic of some ciliates, such as *Carchesium*.

Basic Life Forms

FORM, FUNCTION, AND DIET

Many protozoans have evolved a parasitic mode of life involving one or two hosts. Some species live in the guts and urogenital tracts of their hosts. Others have invaded the body fluids and cells of the host, for example the malarial parasite *Plasmodium*, which lives for part of its life cycle in the blood and liver cells of mammals, birds, and reptiles and the other part in mosquitoes (see Malaria and Sleeping Sickness). Five of the seven protozoan phyla are exclusively parasitic, but among the mainly freeliving flagellates, amoebae, and ciliates there are some species that have opted for parasitism. Notable among the parasitic flagellates are the haemoflagellates causing various forms of trypanosomiasis, including sleeping sickness. The opalinid gut parasites of frogs and toads are another example. In most multicellular animals the essential life processes are carried out by specialized tissues and organs. In protozoans all life processes occur in the single cell, and the building blocks for these specialized functions are small tubular structures termed microtubules. These microtubules have reached their most complex organization in the ciliates, the most advanced of all protozoans. Apart from a few flagellate groups, the ciliates are the only protozoans to possess a true cell mouth, or cytostome. In addition, they typically have two types of nuclei (dimorphism), each performing a different role. The macronucleus – which is often large and may be round, horseshoe-shaped, elongated, or resemble a string of beads – controls normal physiological functioning in the cell, while the micronucleus is concerned with the replication of genetic material during reproduction. It is quite common for a ciliate to possess several micronuclei. Other protozoans have nuclei of one type only, although some species may have several. The exceptions are the foraminiferans, which show nuclear dimorphism at some stages in their life cycle. The most widely known and researched ciliates are species of the genera *Paramecium* and *Tetrahymena*, but these represent only a minute fraction of the 8,000 or so species so far described by science.

The majority of protozoans feed on bacteria, algae, other protozoans, microscopic animals, and

Right *Slender filaments of a microscopic foraminiferan* (Globigerinoides *sp.*) *catch the light. The filaments, or pseudopodia, radiate out from the rest of the cell inside a hard shell or test. They are used to trap food and for movement.*

Below *Protozoan forms:* **1** Actinophrys *(heliozoan);* **2** Opalina *(opalinid);* **3** Acineta *(suctorian);* **4** Euglena *(phytoflagellate);* **5** Elphidium *(foraminiferan);* **6** Trypanosoma *(hemoflagellate);* **7** Hexacontium *(radiolarian).*

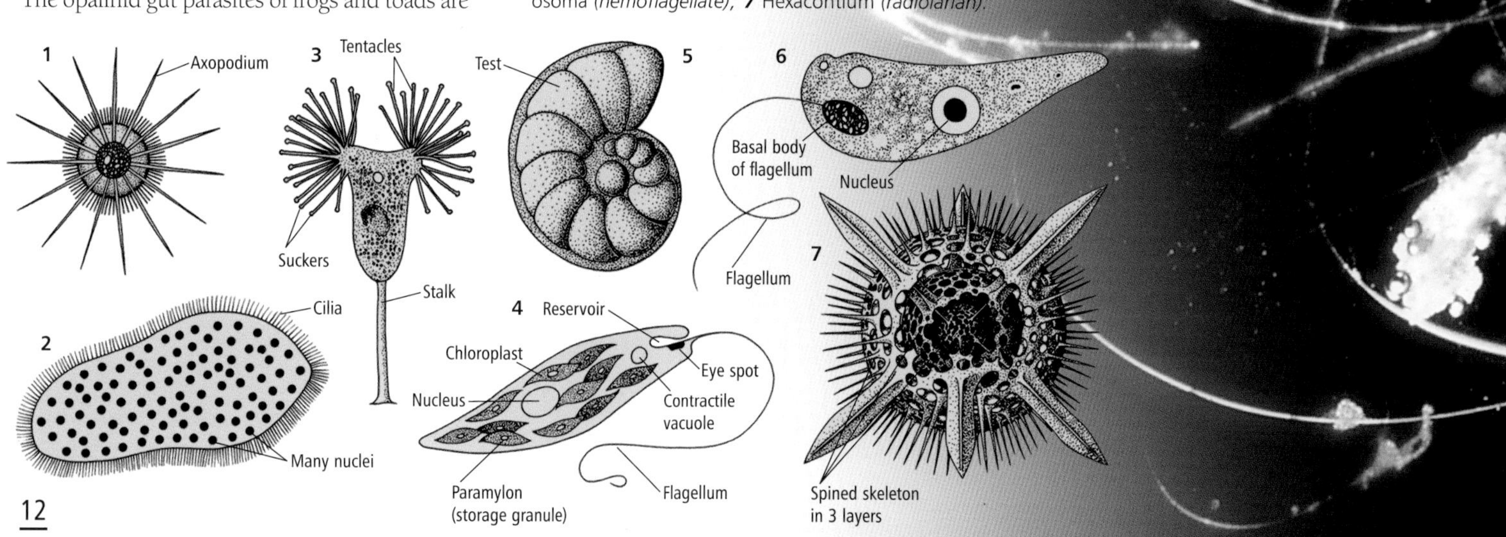

FACTFILE

Protozoans

Kingdom: Protista

About 80,000 species in 17 phyla.

Distribution Worldwide in aquatic and soil environments, parasitic in invertebrates and vertebrates.

Fossil record Earliest record of amoebalike organisms Precambrian, possibly up to 1,000 million years ago.

Size Microscopic, 1μ–5mm (most 5–250μm).

Features Unicellular, freeliving or parasitic; mostly solitary, some ciliates and flagellates colonial; move by means of pseudopodia, flagella or cilia; some amoebae with tests or shells; ciliates possess a mouth pore for ingestion (cystostome) and nuclei of two sizes; reproduction mainly asexual; sexual reproduction in some groups.

Flagellates Phylum Euglenida
1,000 species, including *Euglena*.

Trypanosomes and Allies
Phylum Kinetoplastida
600 species, including *Trypanosoma*, *Leishmania*.

Ciliates Phylum Ciliophora
8,000 species, including *Paramecium*, *Tetrahymena*, *Vorticella, Balantidium, Acineta, Didinium, Coldpidium*.

Malarial parasites Phylum Apicomplexa
5,000 species, including *Plasmodium*, *Coccidia*, the Gregarines and their allies.

Dinoflagellates Phylum Dinoflagellata
4,000 species, including *Ceratium*, *Noctiluca*.

Diatoms, Slime nets, and Allies
Phylum Stramenopila
9,000 species, including *Diatoma*, *Pinnularia*.

Amoebas and Allies Phylum Rhizopoda
200 species, including *Amoeba*, *Arcella, Difflugia, Naegleria*.

Radiolarians, Heliozoans, and Allies
Phylum Actinopoda
4,240 species, including *Actinophys, Actinosphaerium, Actipylina*.

Foraminifera Phylum Granuloreticulosa
40,000 species, including *Elphidium*, *Globigerina*, *Pilulina*.

Diplomonads Phylum Diplomonadida
100 species, including *Enteromonas*, *Giardia*.

Parabasilids Phylum Parabasilida
300 species, including *Trichonympha*.

Cryptomonads Phylum Cryptomonada
200 species, including *Chilomonas*.

Microsporans Phylum Microspora
800 species, including *Encephalitozoon*.

Ascetosporans Phylum Ascetospora
c.30 species, including *Paramyxa, Marteilia*.

Choanoflagellates Phylum Choanoflagellata
c.400 species, including *Monosiga*.

Green micro-algae Phylum Chlorophyta
7,000–9,000 species, including *Chlamydomonas*, *Eudorina*, *Volvox*.

Opalinids Phylum Opalinida
c.400 species, including *Opalina, Cepedea*.

in the case of parasites on host tissue, fluids, and gut contents. Their diet incorporates complex organic compounds of nitrogen and hydrogen, and they are said to be heterotrophic. Some flagellates and green algae, such as *Euglena* (Phylum Euglenida) and *Volvox* (Phylum Chlorophyta) however, possess photosynthetic pigments in chloroplasts. These protozoans are capable of harnessing the sun's radiant energy in the chemical process of photosynthesis to construct complex organic compounds from simple molecules – they are said to be autotrophic. A number of such protozoans must, however, combine autotrophy with heterotrophy, in varying degrees. Such organisms lie on the boundary between animal and plantlike nutrition.

Many protozoans are simply bound by the cell wall, but skeletal structures in the form of secreted shells or tests are common among the members of the phyla Rhizopoda, Actinopoda, and Granuloreticulosa and usually have a single chamber. The exclusively marine foraminiferans (phylum Granuloreticulosa), however, are exceptional in having shells with numerous chambers. Shells and tests may be formed of calcium carbonate or silica, or from organic substances such as cellulose or chitin.

Most freeliving and some parasitic species need to move around their environment to feed and to move toward and away from favorable and unfavorable conditions; in some cases special movement is required in reproductive processes. The various protozoan groups achieve movement using different structures.

Members of the phylum Rhizopoda (including the genus *Amoeba*) produce so-called pseudopodia – flowing extensions of the cell. They may be extended only one at a time, as in *Naegleria*, or several at a time, as in *Amoeba proteus*, *Arcella,* and *Difflugia*. Heliozoan sarcodines (phylum Actinopoda), which resemble a stylized sun, possess long slender pseudopodia, called axopodia, which radiate from a central cell mass. Each axopodium is supported by a large number of microtubules arranged in a parallel fashion along the longitudinal axis. Heliozoans move slowly, rolling along by repeatedly shortening and lengthening the axopodia. A well-known example of these so-called sun organisms is *Actinosphaerium*. The foraminiferans, for example *Elphidium*, which bear complex chambered shells, have a complicated network of pseudopodial strands that branch and fuse with each other to produce a linking complex of structures called reticulopodia. Like the

Main picture *A colony of* Volvox, *a genus of microscopic unicellular organisms widely distributed in freshwater environments. Each colony is a hollow sphere composed of thousands of cells, and each cell has two flagella pointing outwards, which enable the colony to swim.*

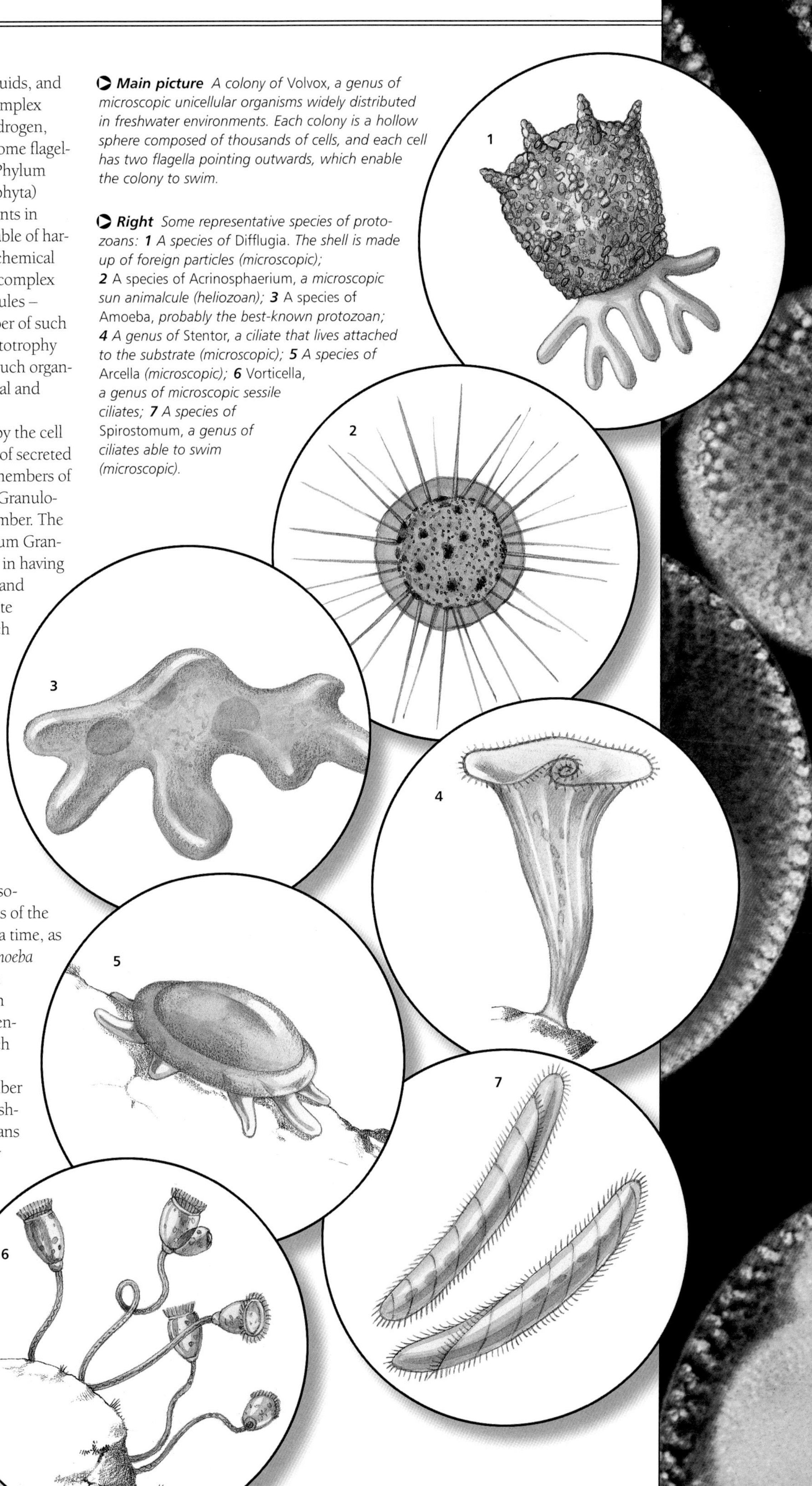

Right *Some representative species of protozoans:* ***1*** *A species of* Difflugia. *The shell is made up of foreign particles (microscopic);* **2** *A species of* Acrinosphaerium, *a microscopic sun animalcule (heliozoan);* **3** *A species of* Amoeba, *probably the best-known protozoan;* **4** *A genus of* Stentor, *a ciliate that lives attached to the substrate (microscopic);* **5** *A species of* Arcella *(microscopic);* **6** Vorticella, *a genus of microscopic sessile ciliates;* **7** *A species of* Spirostomum, *a genus of ciliates able to swim (microscopic).*

axopodia of heliozoans, the reticulopodia of foraminiferans are supported by microtubules.

The other means of movement is by the beating action of the filamentous cilia and flagella, which are permanent outgrowths of the cell rather than, like pseudopodia, its temporary pseudopodial extensions.

Cilia and flagella are structurally similar, but cilia are shorter. Normally flagellates carry only one or two flagella, while in the ciliates the cilia are numerous and usually arranged in ordered rows, each called kinety. The number of kinety is constant in each species and is used as an aid in identification. In some cases cilia may fuse to form cirri, which resemble short thick hairs, or structures which are sail-like. Each cilium and flagellum is about 0.15–0.3μm in diameter and is supported by a core (axoneme) made up of two centrally positioned microtubules surrounded and joined by cross-bridges to nine double microtubules. This 9+2 arrangement of microtubules is common in cilia and flagella throughout the living world – from amoebas to invaders of human lung linings. Movement in cilia and flagella involves the passage of waves along them from one axis to the other. Most flagella move in two-dimensional waves, while cilia move in three-dimensional patterns coordinated into waves that result from fluid forces (hydrodynamic forces) acting on the automatic beating of each cilium.

A Diversity of Strategies

REPRODUCTION

Reproduction in the protozoans does not usually involve sex or sexual organelles – it mainly is asexual. In most free-living species, asexual reproduction occurs by a process called binary fission, whereby each reproductive effort results in two

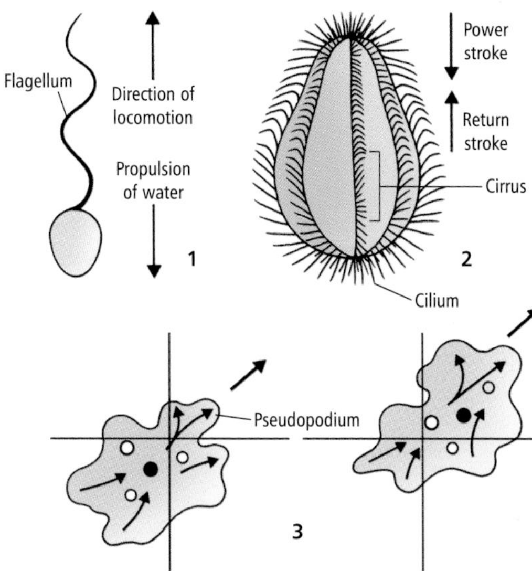

Above *Protozoans move in three ways:* ***1*** *By a single flagellum that pulls the cell through the water;* ***2*** *By rows of cilia coordinated to act like the oars of a rowboat, having a power stroke and a recovery stroke;* ***3*** *By amoeboid movement, whereby pseudopodia are extended and the rest of the body flows into them.*

identical daughter cells by the division of a parent cell. In the flagellates, including the parasitic species, the plane of division is longitudinal, while in the ciliates it is normally transverse, and prior to division of the cytoplasm the mouth is replicated. Members of the phyla Rhizopoda, Actinopoda, and Granuloreticulosa do not normally have a fixed plane for division. In shelled and testate species the process is complicated by the need to replicate skeletal structures. In testate species of amoeba – for example *Difflugia* – cytoplasm destined to become the daughter is extruded from the aperture of the parent test. Preformed scales in the cytoplasm then form a test around the extruded cytoplasm. When the process is complete the two amoebae separate.

Most freeliving species normally reproduce asexually providing conditions are favorable. Sexual reproduction is usually resorted to only in adversity, such as drying up of the aquatic medium when the normal cells would not survive. The ability to undergo a sexual phase is not widespread in amoebae and flagellates and is restricted to a limited number of groups. Some species may never have reproduced sexually in their evolutionary history, while others may have lost sexual competence. Both isogamous (reproductive cells or gametes alike) and the more advanced anisogamous (reproductive cells or gametes dissimilar) forms of sexual reproduction may occur in protozoans.

The foraminiferans are unusual among freeliving species in having alternation of asexual and sexual generations. Here each organism reproduces asexually to produce many amoeba-like organisms that secrete shells around themselves. When mature, these organisms produce many identical gametes, which are usually liberated into the sea. Here, they fuse in pairs to produce individuals that in turn secrete a shell, grow to maturity, and repeat the cycle.

Almost all of the ciliates are capable of sexual reproduction by a process called conjugation, which does not result in an immediate increase in numbers. The function of conjugation is to facilitate an exchange of genetic material between

Below Asexual reproduction in protozoans is by simple division (binary fission). In flagellates **1** (e.g. Euglena) division is longitudinal. In ciliates **2** (e.g. Tetrahymena) division is complicated by the replication of the oral apparatus. In shelled amoebae **3** (e.g. Euglypha) the daughter is extruded from the parent. Amoebae **4** have no fixed plane of division.

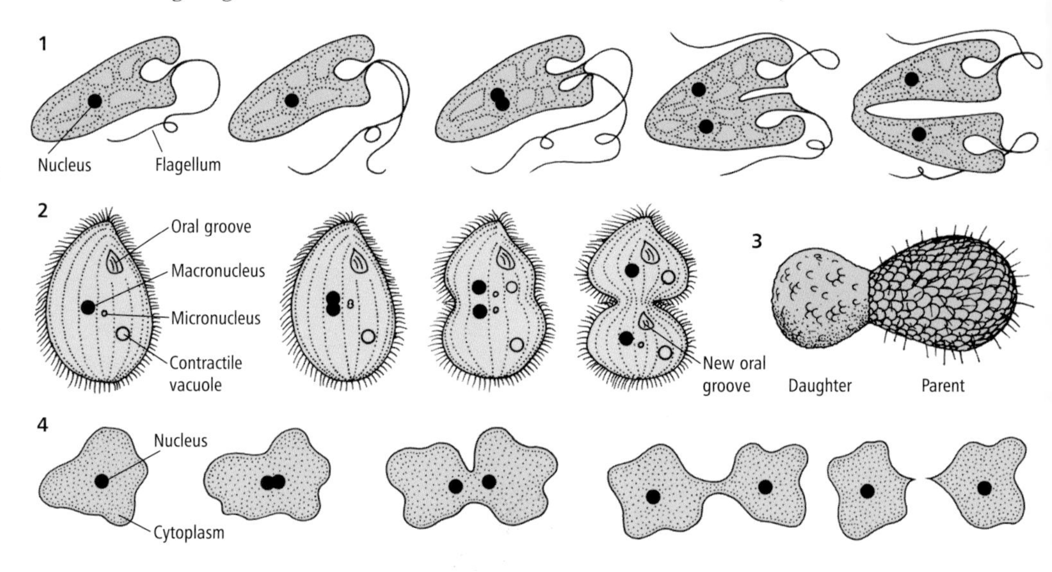

Prey Capture in Carnivorous Protozoans

Carnivorous protozoans prey on other protozoans, rotifers, members of the Gastrotricha, and small crustaceans. The mode of capture and ingestion is often spectacular and frequently the prey are larger than the predators.

Among the ciliates, the sedentary species of the subphylum Suctoria have lost their cilia, which have been replaced by tentacles, each of which functions as a mouth. When other ciliates, such as *Colpidium*, collide with a tentacle, they stick to it. Other tentacles move toward the prey and also attach. The cell wall of the prey is perforated at the sites of attachment and the prey cell contents are moved up the tentacle by microtubular elements within the tentacle. A single species such as *Podophrya* can feed simultaneously on four or five prey. *Didinium nasutum* is a ciliate that feeds exclusively on *Paramecium*, which it apprehends using extrudable structures called pexicysts and toxicysts. The former hold the prey while the latter penetrate deeply into it, releasing poisons. *Didinium* consumes the immobilized prey whole, its body becoming distended by the ingested *Paramecium*.

The heliozoan *Actinophrys* also feeds on ciliates captured, on contact, by radiating axopodia. Once attached, the prey is progressively engulfed by a large funnel-shaped pseudopodium produced by the cell body. Occasionally when an individual *Actinophrys* has captured a large prey, other *Actinophrys* may fuse with the feeding individual to share the meal. In such instances, after digesting their ciliate victim, the heliozoan predators separate again.

The foraminiferan *Pilulina* has evolved into a living pitfall trap. This bottom-dwelling species builds a bowl-shaped shell or test with mud, camouflaging the pseudopodia across the entrance. When copepod crustaceans come into contact with the pseudopodia, they get stuck and are drawn down into the animal. The radiolarians, which possess a silica-rich internal skeleton, deal with copepod prey by extending the wave flow along the axopodia to the broad surfaces of the prey's exoskeleton, and rupturing the prey by force. The axopodia then penetrate and prise off pieces of flesh, which are directed down the axopodia to the main cell body for ingestion. JL-P/AC

Left As it maneuvers a Paramecium into a position where it is easier to devour, a Didinium nasutum expands its snout and gullet.

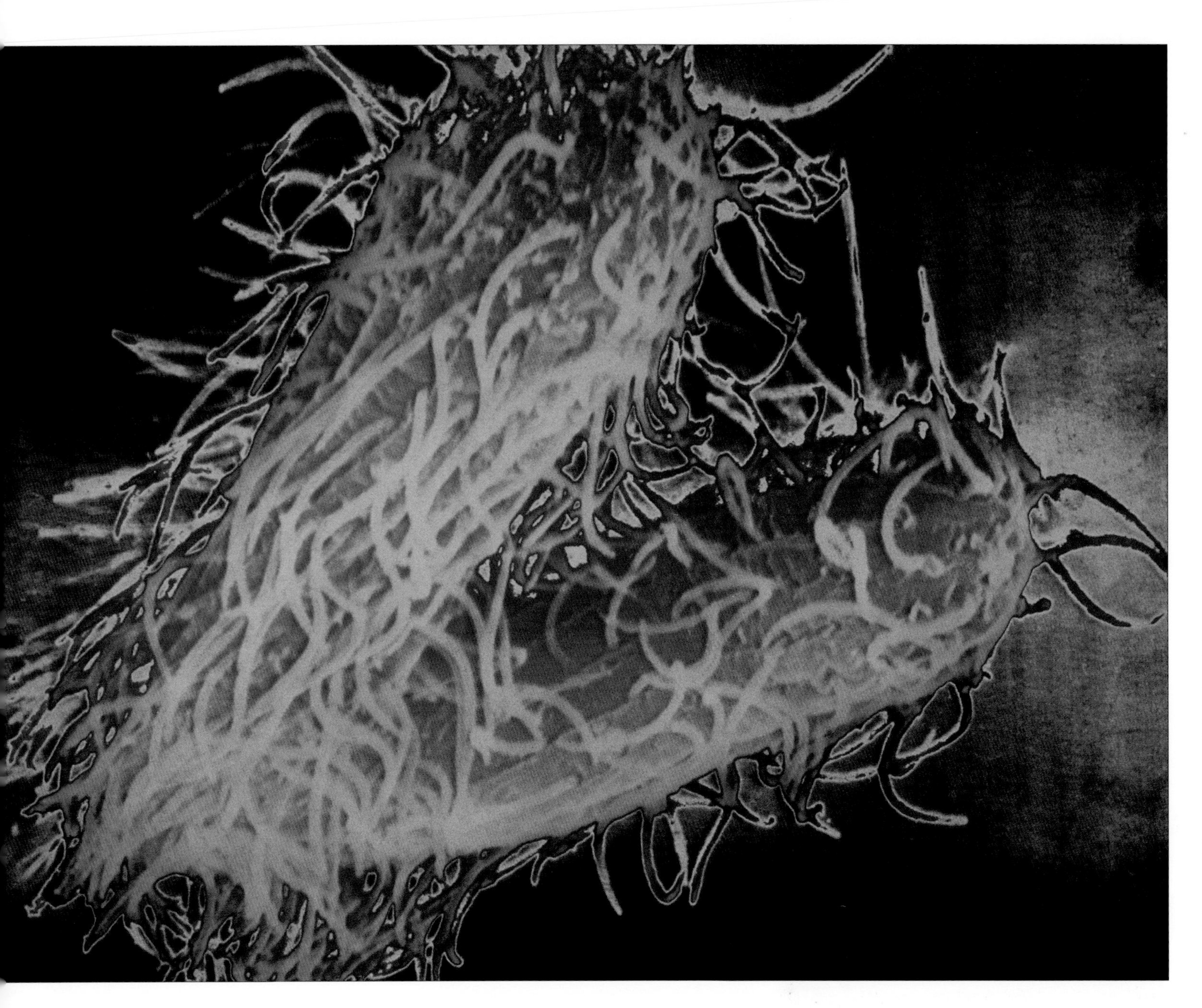

individuals. During this process, two ciliates come together side by side and are joined by a bridge of cell contents (cytoplasm). A complex series of divisions of the micronucleus occurs, including a halving of the pairs of chromosomes (or meiosis).

In the final stages of conjugation, a micronucleus passes from each individual into the other. Essentially the micronuclei are gametes. Each received micronucleus fuses with an existing micronucleus in the recipient. The ciliates separate and, after further nuclear divisions, eventually undergo binary fission.

All members of the parasitic phyla, except for some groups in the Apicomplexa, produce spores at some stages in their life cycles. The Apicomplexa contains a number of parasites of medical and veterinary importance, including the malarial parasites *Plasmodium* and the *Coccidia* responsible for coccidiosis in poultry. Some species, like *Plasmodium*, have complex life cycles involving two hosts with an alternation of sexual and asexual phases. In *Plasmodium* the sexual phase is initiated in humans and is completed in the mosquito; following this phase many thousands of motile spores (sporozoites) are reproduced that are infective to humans, and which are transmitted when the mosquito feeds. In the human body, repeated phases of multiple asexual division take place within the red blood cells and liver cells (see Malaria and Sleeping Sickness). The phyla Microspora and Ascetospora are parasites of a wide range of vertebrates and invertebrates, while other forms parasitize algae.

Recycling Agents

ECOLOGY

The ecology of protozoans is very complex, as one would expect in a group of ubiquitous organisms. They are found in the waters and soils of the world's polar regions. Some have adapted to warm springs and there are records of protozoans living in waters as warm as 60°C (134°F). Protozoans occur commonly in planktonic communities in marine, brackish, and freshwater habitats, and also in the complex bottom-dwelling (benthic) communities of these environments. Little is known about protozoans in the marine deeps, but there is a record of foraminiferans living at 4,000m (13,000ft) in the Atlantic. Ciliates, flagellates, and various types of amoebae are also common in soils and boggy habitats.

Since many protozoans exploit bacteria as a food source, they form part of the decomposer food web in nature. Recent research indicates that protozoans may stimulate the rate of decomposition by bacteria and thus enhance the recycling of minerals such as phosphorus and nitrogen. The exact mechanism is not entirely clear, but protozoans grazing on bacteria may maintain the bacterial community in a state of physiological youth and hence at the optimum level of efficiency. There is also evidence to suggest that some protozoans secrete a substance that promotes the growth of bacteria. JL-P/AC

Above *Ciliate protozoans reproducing by conjugation. In protozoans, this process is bidirectional, with each of the pair of organisms exchanging genetic material with the other.*

MALARIA AND SLEEPING SICKNESS

Protozoan diseases of humans

BETWEEN ONE AND THREE MILLION PEOPLE DIE annually from malaria, and another 500 million contract the disease – statistics that exemplify the virulence of protozoan diseases. The most serious pathogens are *Plasmodium*, which produces malaria, and various trypanosome species responsible for diseases broadly called trypanosomiasis, or sleeping sickness.

Malaria is caused by four species of the genus *Plasmodium*. The life cycle is similar in each species but there are differences in disease pathology. *Plasmodium falciparum* causes malignant tertian malaria and accounts for about 50 percent of all malarial cases. It attacks all red blood cells (erythrocytes) indiscriminately so that as many as 25 percent of the erythrocytes may be infected. In this species stages not involving the erythrocytes do not persist in the liver, so that relapses do not occur. *Plasmodium vivax* produces benign tertian malaria, which invades only immature erythrocytes; thus, the level of cells infected is low. Here, however, other stages remain in the liver, causing relapses. Benign tertian malaria is responsible for approximately 45 percent of malarial infections. The other two species are relatively rare. *Plasmodium malariae*, causing quartan malaria, attacks mature red blood cells and has persistent stages outside the blood cells. Little is known about *P. ovale* because of its rarity.

The diseases are named after the fevers that the parasites cause, tertian fevers occurring every three days or 48 hours and quartan fevers every four days or 72 hours. The naming practice is based on the Roman system of calling the first day one, whereas we would call the first day nought.

Once inside an erythrocyte, the parasite feeds on the red blood cell contents and grows. When mature it undergoes multiple asexual fission to produce many individuals called merozoites, which, by an unknown mechanism, rupture the erythrocyte and escape into the blood plasma. Each released merozoite then infects another erythrocyte. The asexual division cycle in the red blood cells is well synchronized so that many erythrocytes rupture together – a phenomenon responsible for the characteristic fever that accompanies malaria. The exact mechanism producing the fever is not fully understood, but it is believed to be caused by substances (or a substance), possibly derived from the parasite, that induce the release of a fever-producing agent from white blood cells, which fight the disease. When the parasite has undergone a series of asexual erythrocytic cycles, some individuals produce the male and female gametocytes, which are the stages infective to the mosquito host. The stimulus for gametocyte production is unknown.

Malaria is still one of the greatest causes of death in humans. Tens of millions of cases are reported each year and many are fatal. Successful control measures are available, and in countries such as the USA, Israel, and Cyprus the disease has been eradicated. In developing countries, however, control measures have little impact on malaria. Broadly, eradication programs involve the use of drugs to treat the disease in humans, and a series of measures aimed at breaking the parasite's life cycle by destroying the intermediate mosquito host.

Like many insects, the mosquito has an aquatic larval stage. The draining of swamps and lakes deprives the mosquito of an environment for breeding and its larval development, but residual populations continue to breed in irrigation canals, ditches, and paddy fields. Spraying oil on the water surface asphixiates the larvae, which have to come to the water surface periodically to breathe. The poison Paris Green can effectively kill larvae when added to the water. Biological control using fish predators of mosquito larvae, such as the guppy, aid in reducing larval populations.

Adult mosquitoes can be killed by spraying houses with various insecticides, such as hexachlorocyclohexane and dieldrin. In the past DDT was very successful but its toxicity to more highly developed animals now precludes its use. Biological control measures involve releasing sterile male mosquitos into the population, thereby decreasing reproduction rates, and the introduction of bacterial, fungal, and protozoan pathogens of the mosquito. Chemotherapy in humans involves four broad categories of treatment. First, there are prophylactic drugs, such as proguanil, which prevent recurring erythrocytic infections when taken on a regular basis. Second, there are drugs, such as chloroquine, that destroy the blood stages of the parasite. Third, there are drugs that destroy gametocytes. Lastly, there are drugs that, when taken up by the mosquito during feeding on humans, prevent further development of the parasite in the insect. Drug resistance by

Below *Life cycle of sleeping sickness:* ***1*** *Trypanosomes in human blood are taken up by the tsetse fly when feeding;* ***2*** *They enter the midgut, where division occurs before migrating forward after 48 hours to the foregut (proventriculus);* ***3*** *Trypanosomes remain in the proventriculus for 10–15 days before migrating to the salivary glands;* ***4*** *Over a 30–50-day period in the salivary gland, the trypanosomes change into crithidial forms and multiply before becoming infective metacyclic forms;* ***5*** *When the tsetse bites another human, metacyclic trypanosomes enter the victim's body and multiply at the site of infection;* ***6*** *Trypanosomes then invade the blood stream and reproduce by binary fission. They then may follow two courses (7 or 8);* ***7*** *They enter the tissue space of various organisms and lymph nodes and then invade the central nervous system to cause typical sleeping sickness symptoms. The trypanosomes do not enter cells, but remain between them;* ***8*** *They are taken up by further tsetse flies to continue the cycle.*

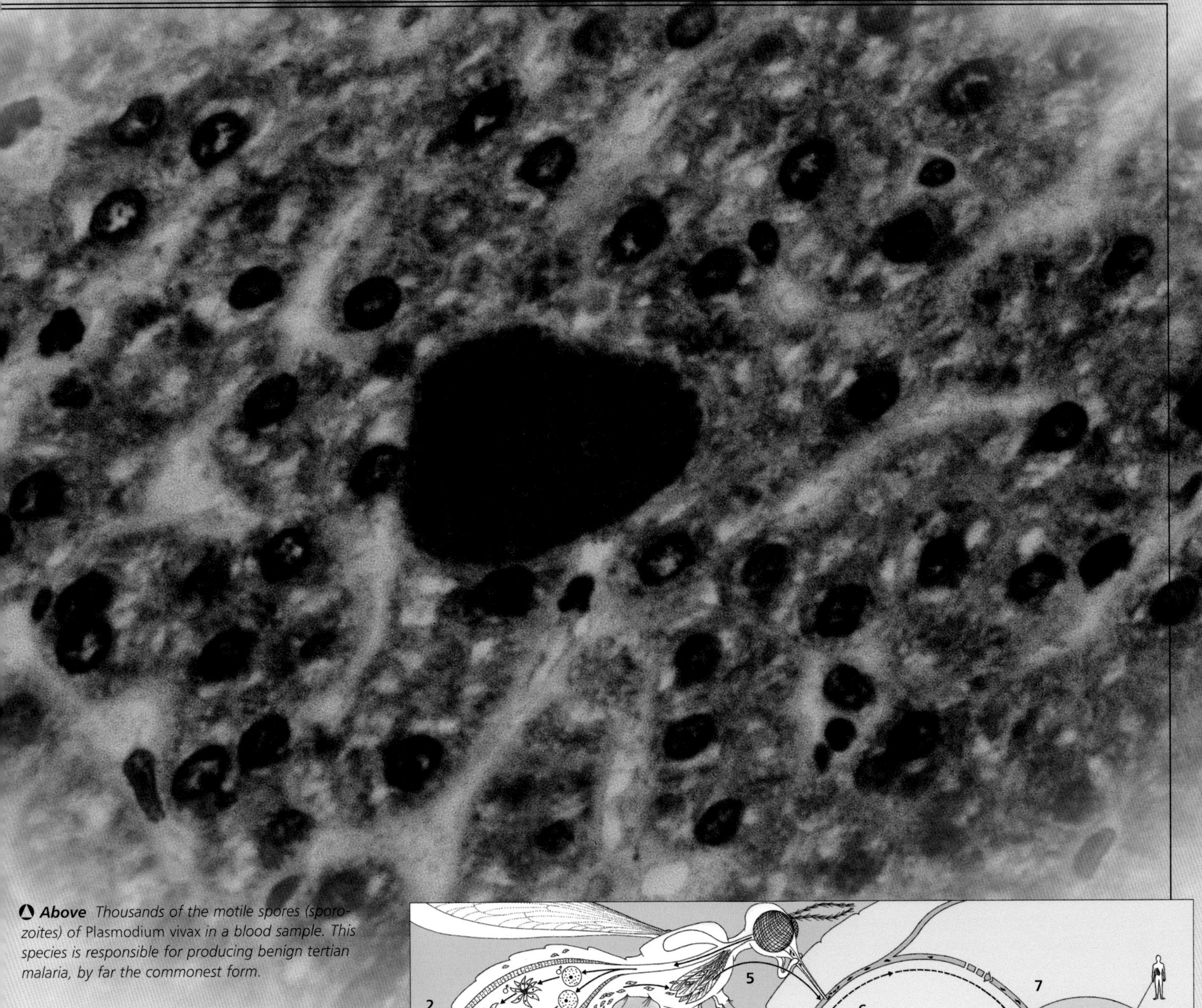

Above *Thousands of the motile spores (sporozoites) of* Plasmodium vivax *in a blood sample. This species is responsible for producing benign tertian malaria, by far the commonest form.*

Plasmodium is increasing; *P. falciparum* has become resistant to chloroquine in some parts of Africa and South America, and has to be treated by a combination of quinine and sulfonamides.

The flagellates *Trypanosoma rhodesiense* and *T. gambiense* cause African trypanosomiasis or sleeping sickness. The two-host life cycle involves a tsetse fly (genus *Glossina*) and humans. In humans, trypanosomes live in the blood plasma and lymph glands, progressing later to the cerebrospinal fluid and the brain. The disease is typified by mental and physical apathy and a desire to sleep. The disease is fatal if untreated, *T. rhodesiense* running a more acute course than *T. gambiense*. Control measures include insecticide use, introduction of sterile males, clearing vegetation in which *Glossina* spends the whole of its life cycle, and the use of drugs in humans. Control is complicated by the fact that *T. rhodesiense* also infects game animals, so that a reservoir population of the parasite persists. JL-P/AC

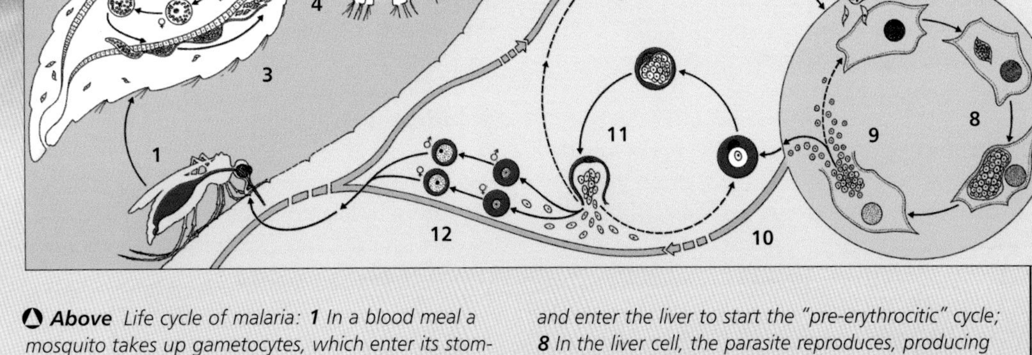

Above *Life cycle of malaria:* **1** *In a blood meal a mosquito takes up gametocytes, which enter its stomach;* **2** *The gametocytes mature to produce either thin motile male gametes or larger female gametes. The male gametes fertilize the female gametes to produce a zygote;* **3** *The zygote penetrates the stomach wall, where it develops into an oocyst;* **4** *The oocyst ruptures, releasing many sporozoites into the mosquito's body cavity;* **5** *The sporozoites migrate to the mosquito's salivary glands;* **6** *When the insect next feeds, it releases sporozoites into the human bloodstream;* **7** *The sporozoites migrate through the bloodstream and enter the liver to start the "pre-erythrocitic" cycle;* **8** *In the liver cell, the parasite reproduces, producing masses of merozoites;* **9** *In some species, released merozoites may reinfect other liver cells, which act as a reservoir of the parasite;* **10** *Merozoites in the bloodstream enter red blood cells;* **11** *Within the red blood cell division occurs to produce masses of merozoites that reinfect other red blood cells. Cell rupture is synchronized – a characteristic fever develops, repeated on each release every 48 or 72 hours, depending on species;* **12** *Some merozoites form gametocytes, taken up by mosquitos to renew the infection cycle.*

Sponges

SINCE ANCIENT TIMES, THE HUMBLE BATH sponge has been harvested and used by people, particularly in the Mediterranean region. Bath sponge species are the best known of a group of animals whose relationship to other organisms is still a matter of debate.

Until the early 19th century sponges were regarded as plants, but they are now generally considered to be a group (phylum Porifera) of animals placed within their own subkingdom, the Parazoa. They probably originated either from flagellate protozoans or from related primitive metazoans.

A Simple Structure

FORM, FUNCTION, AND DIET

Sponges range in size from the microscopic to 2m (6.6ft). They often form a thin incrustation on hard substrates to which they are attached, but others are massive, tubular, branching, amorphous, or urn-, cup- or fan-shaped. They may be drab or brightly colored, the colors derived from mostly yellow to red carotenoid pigments.

All sponges are similar in structure. They have a simple body wall containing surface (epithelial) and linking (connective) tissues, and an array of cell types, including cells (amoebocytes) that move by means of the flow of protoplasm (amoeboid locomotion). Amoebocytes wander through the inner tissues, for example, secreting and enlarging the skeletal spicules and laying down spongin threads. Sponges are not totally immovable but the main body may show very limited movement through the action of cells called myocytes, but they often remain anchored to the same spot.

Although sponges are soft-bodied, many are firm to touch. This solidity is due to the internal skeleton comprised of hard rod- or star-shaped calcareous or siliceous spicules and/or of a meshwork of protein fibers called spongin, as in the bath sponge. Spicules may penetrate the sponge surface of some species and cause skin irritation when handled.

Above *A freshwater sponge* (Ephydatia fluviatilis)*; its green color is the result of action of micro-algae such as* Chlorella.

Right *Clearly visible on this group of Yellow tube sponges* (Aplysina fistularis) *is the large exhalant opening, or osculum.*

Sponges are filter feeders, straining off bacteria and fine detritus from the water. Oxygen and dissolved organic matter are also absorbed and waste materials carried away. Water enters canals in the sponge through minute pores in their surface and moves to chambers lined by flagellate cells called choanocytes or collar cells. The choanocytes ingest food particles, which are passed to the amoebocytes for passage to other cells. Eventually the water is expelled from the sponge surface, often through volcano-like oscules at the surface. Water is driven through the sponge mainly by the waving action of flagella borne by the choanocytes.

Different Strategies

SEXUAL REPRODUCTION

Sponges reproduce asexually by budding off new individuals, by fragmentation of parts that grow into new sponges, and, particularly in the case of freshwater sponges, by the production of special gemmules. These gemmules remain within the body of the sponge until it disintegrates, when they are released. In freshwater sponges, which die back in winter in colder latitudes, the gemmules are very resistant to adverse conditions, such as extreme cold. Indeed, they will not hatch unless they have undergone a period of cold.

In sexual reproduction, eggs originate from amoebocytes and sperms from amoebocytes or transformed choanocytes, usually at different times within the same individual. The sperms are shed into the water, the eggs often being retained within the parent, where they are fertilized. Either solid (parenchymula) or hollow (amphiblastula) larvae may be produced; many swim for up to several days, settle, and metamorphose into individuals or colonies that feed and grow. Others creep on the substrate before metamorphosis. Some mature Antarctic sponges have not grown over a period of 10 years.

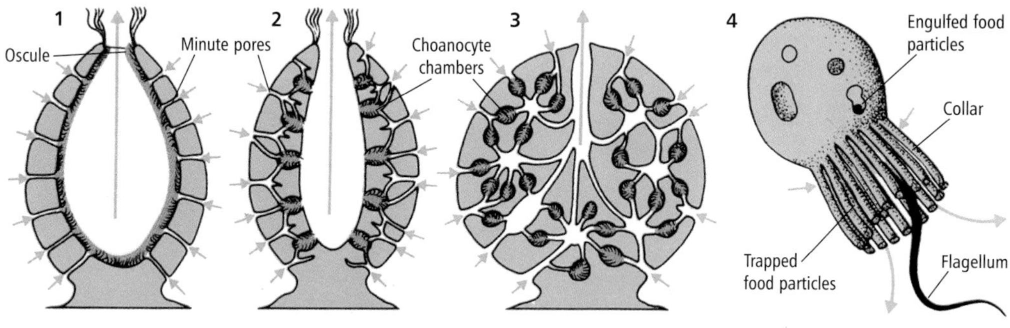

Left *1, 2, and 3: Water enters canals in the sponge through minute pores in their surface and moves to chambers lined by flagellate cells called choanocytes or collar cells* ***4****. The choanocytes ingest food particles, which are passed to the amoebocytes for passage to other cells. Eventually the water is expelled from the sponge surface. Water is driven through the sponge mainly by the beating action of the flagellae borne on the choanocytes. Choanocytes may line the body cavity* ***1****, or the wall is folded so these cells line pouches.*

FACTFILE

SPONGES

Subkingdom: Parazoa

Phylum: Porifera

About 5,000 species in 790 genera and 80 families.

Distribution Worldwide, freshwater and marine, intertidal to deep sea.

Fossil record Originated in Cambrian 570–500 million years ago; 390 genera identified from Cretaceous (135–65 m.y.a.).

Size From microscopic to 2m (6.6ft); the largest sponges occur in the Antarctic and the Caribbean.

Features Form variable; solitary or colonial; mostly porous, filter-feeding organisms mostly attached direct to substrate, without "stem"; lack organs and have little in way of definite tissues, but with complex array of cell types; skeleton lacking or of siliceous or calcareous spicules, or of organic spongin fibers; generally hermaphrodite; sexual and asexual reproduction.

GLASS OR SILICEOUS SPONGES
Class Hexactinellida (Hyalospongiae)
About 600 species. Marine, below tidal levels but more common in deeper waters. Skeleton of complex silica spicules, with basic pattern of 6 rays. Genera and species include: *Aphrocallistes*, **Venus' flower basket** (*Euplectella aspergillum*), *Holascus*, *Pheronema*.

CALCAREOUS SPONGES Class Calcarea
About 400 species. Marine. Skeleton of calcareous spicules, which are needlelike or 3- or 4-rayed. Genera include: *Acyssa, Clathrina, Leucilla, Leucosolenia, Scypha*.

TYPICAL SPONGES Class Demospongiae
About 4,000 species. Marine and freshwater. Skeleton lacking or of silica spicules, spongin fibers or both. Spicules when present not 6 rayed. Genera and species include: *Aplysina, Cliona,* **Caribbean sponge** (*Cribochalina vasculum*), *Ephydatia, Haliclona,* **Bath sponge** (*Hippospongia communis*), **Caribbean fire sponge** (*Neofibularia nolitangere*), *Siphonodictyon*, **Bath sponge** (*Spongia officinalis*), *Spongilla*.

CORALLINE SPONGES Class Sclerospongiae
About 15 species. Marine, in tropical, shallow, subtidal caves or underneath corals. Skeleton with calcareous base and entrapped silica spicules and organic fibers. Sponge forms thin layer over calcareous base. Genera include: *Ceratoporella, Stromatospongia*.

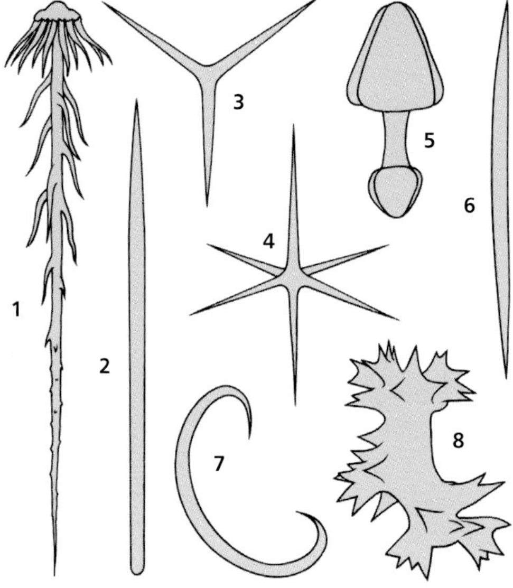

Abundant on the Continental Shelf

ENVIRONMENT AND CONSERVATION

Sponges live in large numbers in all the seas of the world. They occur in greatest abundance on firm substrates, relatively few being adapted to life on unstable sand or mud. Their vertical range includes the lowest part of the shore subject to tidal effects and extends downwards as far as the abyssal depths of 8,600m (27,000ft). One family of siliceous sponges, the Spongillidae, has invaded freshwater lakes and rivers throughout the world.

Sponges living between tide marks are typically confined to parts of the shore that are seldom exposed to the air for more than a very short period. Some sponges live a little higher up the shore, but they are found only in shaded situations or on rocks facing away from the sun.

Some sponges are killed by even a relatively

Above *A species of* Axinellida *sponge (Demospongiae) growing on the trunk of a fan coral. Large examples of sponges of this genus are commonly known as "Elephant's ear sponges."*

Right *Finger sponges* (Suberites *sp.*) *off the Australian coast. Even within a single genus, sponges may take a huge variety of forms.*

Left *Sponge spicules, which are made of calcium carbonate or silica, help support the body of the sponge. Their various shapes may characterize particular types of sponges and can be important in identifying them.* ***1*** *Monaxon spicule with barbs* (Farrea beringiana); ***2*** *Monaxon spicule* (Mycale topsenti); ***3*** *Triaxon spicule* (Leucoria heathi); ***4*** *Hexaxon spicule* (Auloraccus fissuratus); ***5*** *Monaxon spicule with terminal processes* (Mycale topsenti); ***6*** *Monaxon spicule* (Raspaigella dendyi); ***7*** *Monaxon spicule with recurved ends* (Sigmaxinella massalis); ***8*** *Polyaxon spicule (genus* Streptaster).

Right *Boring sponges (here,* Siphonodictyon *sp.) are widespread within tropical stony corals, to which they can cause considerable damage.*

short exposure to air, and it is in the shallow waters of the continental shelf that sponges achieve their greatest abundance in terms both of species and individuals.

Cavernous sponges are often inhabited by smaller animals, some of which cause no harm to the sponge, although others are parasites. Many sponges contain single-celled photosynthetic algae (zoochlorellae), blue-green algae, and symbiotic bacteria, which may provide nutrients for the sponge. Sponges are eaten by sea slugs (nudibranchs), chitons, sea stars (especially in the Antarctic), turtles, and some tropical fishes.

Usually more than half of the species of tropical sponges living exposed rather than under rocks are toxic to fish. This is believed to be an evolutionary response to high-intensity fish predation, nature having selected for noxious and toxic compounds that prevent fish from consuming sponges. Some toxic sponges are very large, such as the gigantic Caribbean sponge (*Cribochalina vasculum*), while others, such as the Caribbean fire sponge (*Neofibularia nolitangere*), are dangerous to the touch – in humans they cause a severe burning sensation lasting for several hours. Toxins probably play an important role in keeping the surface of the sponge clean, by preventing animal larvae and plant spores from settling on them. Some sponge toxins may prevent neighboring invertebrates from overgrowing and smothering them.

Sponge toxins have been used in studies on the transmission of nerve impulses. They show considerable potential as biodegradeable antifouling agents and possibly as shark repellants.

Bath sponges owe their usefulness to the water-absorbing and retaining qualities of a complex lattice of spongin fibers; the fibers are also elastic enough to allow water to be squeezed out of the sponge. A number of species are harvested (mainly off Florida and Greece), principally *Spongia officinalis*, with a fine-meshed skeleton, and *Hippospongia equina*, with a coarser skeleton. They grow on rocky bottoms from low-tide level down to great depths and may be collected either by using a grappling hook from a boat or by divers. Curing sponges simply involves leaving them to dry in the sun, allowing the soft tissues to rot, pounding and washing them, leaving only the spongin skeleton.

Cultivation of sponges from cuttings has been used successfully, although such projects are probably less viable than making synthetic products.

Sponges contain various antibiotic substances, pigments, unique chemicals such as sterols, toxins, and even anti-inflammatory and antiarthritis compounds. Boring sponges of the family Clionidae may cause economic loss by weakening oyster shells. These sponges excavate chambers both chemically and mechanically. GJB/AC

ANIMALS OF THE MESOZOAN PLAN

Mesozoan animals are a taxonomic enigma. Here they are regarded as an assemblage of animals showing similar evolutionary attainments. Ever since they were first discovered in 1883, scientists have been changing their opinions about their evolutionary position and relations with other animals. Now they are thought to comprise four phyla (phylum Placozoa, phylum Monoblastozoa, phylum Rhombozoa, and phylum Orthonectida) containing about 90 species, which are free living or parasitic on marine invertebrates. None is larger than about 8mm (0.3in) in length. They are multicellular animals constructed from two layers of cells, and are therefore distinct from the protozoa, but their cell layers do not resemble the endoderm and ectoderm of the metazoans (see Aquatic Invertebrates). The features of these animals mean that they cannot be assigned to any other animal phylum. Some scientists believe that certain species may be degenerate flatworms, in other words they may have been previously more complex; the other more widely held opinion, however, is that they are simple multicellular organisms holding a position intermediate between Protozoa and Metazoa.

Dicyemida, members of the phylum Rhombozoa are all parasites in the kidneys of cephalopods (for example *Octopus*), while Orthonectida infect echinoderms (starfish, sea urchins etc), mollusks (snails, slugs etc.), Annelida (earthworms etc.), and ribbon worms (see Flatworms and Ribbon Worms). Despite their simple morphology, Dicyemids have evolved complex life cycles involving several generations. The first generation, called a nematogen, occurs in immature cephalopods. Repeated similar generations of nematogens are produced asexually by repeated divisions of special central (axial) cells that give rise to wormlike (vermiform) larvae **1**. When the host attains maturity, the parasite assumes the next generation, or rhombogen, which looks superficially similar to the nematogen but differs in its cellular makeup. The individuals are hermaphrodite and produce infusariiform larvae, which look superficially like ciliate protozoans. The fate of the larvae is uncertain, but it is believed that another intermediate host is involved in the life cycle. Genera included in the Phylum Rhombozoa are *Dicyema*, *Dicyemmerea,* and *Conocyema*.

Orthonectids **2** live in the tissues and tissue spaces of their marine invertebrate hosts, for example nemerteans, polychaetes, ophiuroids, and bivalves. The asexual phase looks like an amoeboid mass and is called a plasmodium because it resembles the protozoan *Plasmodium*. GJB/AC

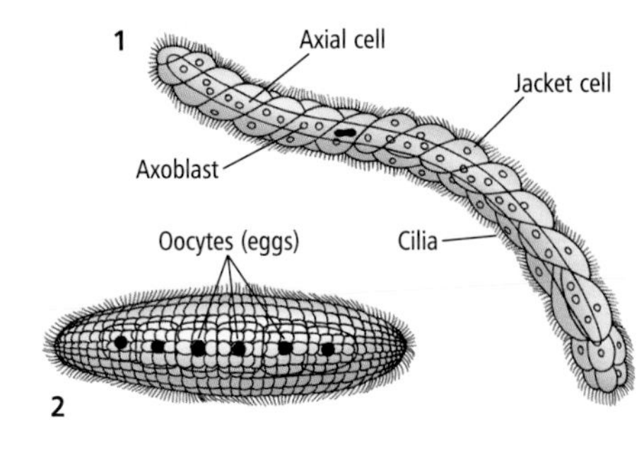

Sea Anemones and Jellyfishes

SEA ANEMONES, CORALS, AND JELLYFISHES ARE perhaps the most familiar members of the phylum Cnidaria. This diverse assemblage, whose name derives from the Greek cnidos, *meaning "stinging nettle," contains an enormous number of animals, many of which are characterized by their possession of stinging cells (cnidoblasts). Cnidarians are mainly marine animals; there are only a few freshwater species, of which the best known are the hydras.*

The cnidarians are multicellular animals and have a two-layered (diploblastic) construction, in which both the differences between cells and organ development are limited. These restrictions have, however, been partially offset in colonial types by the specialization of individuals (polymorphism).

Cnidarians

FORM AND FUNCTION

There are two life-history phases: polyp and medusa. The polyp is the sedentary phase and consists of three regions: a basal disk or pedal disk that anchors it; a middle region or column within which is the tubular digestive chamber (gastrovascular cavity); and an oral region that is ringed by tentacles. In colonial types a tubular stolon links adjacent polyps. The medusa is the mobile phase and is effectively an inverted polyp. By virtue of the fluid (water) it contains, the digestive cavity plays an important role in oxygen uptake and excretion. This fluid additionally acts as a hydrostatic skeleton through which body wall muscles can antagonize one another.

Since the medusa is the sexual phase, it can be argued that it is the original life form, with the predominantly bottom-living (benthic) polyp acting as an intermediate, multiplicative asexual stage. However, in the class Hydrozoa the medusa is frequently reduced or even lost, and in the class Anthozoa it is totally absent. Emphasis in the class Scyphozoa lies, to the contrary, with the medusa stage, as the evolution of the highly mobile and graceful jellyfish testifies; the polyp phase in jellyfish is a relatively inconspicuous component in the life cycle.

The outer (ectodermal) and inner (endodermal) cell layers of the body are cemented together by the jellylike mesogloea, which in the jellyfish forms the bulk of the animal. The mesogloea contains a matrix of elastic collagen fibers that aid both the change and maintenance of body shape. This change is particularly obvious in the pulsating swimming movements characteristic of jellyfish, during which contractions of the swimming bell brought about by radial and circular muscles are counteracted by vertically running, elastic fibers.

Muscle contraction results in an increase in bell depth and hence fiber stress; fiber shortening subsequently restores the bell to its original shape. In the medusae of hydrozoans, the resulting water jets are concentrated and directed by the shelflike velum projecting inwards from the rim of the bell, where there are tentacles, towards the mouth. Structural support in the relatively large anthozoan polyps is also provided by septa (mesentaries), which contain retractor muscles. When mobile, polyp locomotion may be brought about in a number of ways: by creeping upon the pedal disk, by looping, or, rarely, by swimming (for example, the anemones, *Stomphia*, *Boloceroides*).

Left *A delicate, featherlike colony of Ostrich-plumed stinging hydroids* (Aglaophenia cupressina). *All species of this genus have modified side branches (corbulae) that act as reproductive structures.*

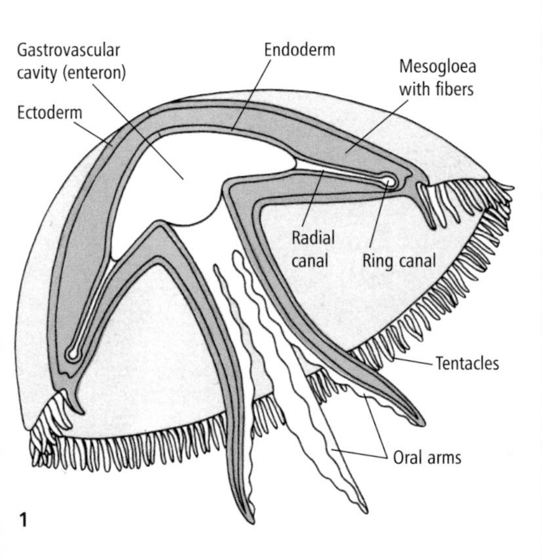

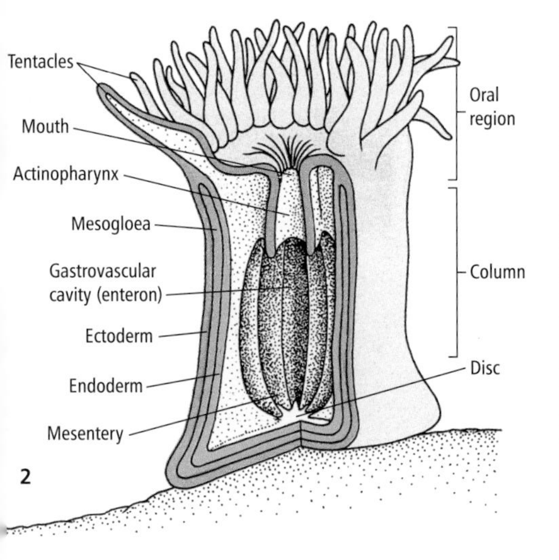

3
Oral disc
Pinnule
Stomodeum
Hollow tentacle
Mesogloea
Epidermis on colony surface
Mesentery
Spicules of endoskeleton
Mesentery filament
Tubular stolon between polyps

Above *The three main forms of cnidarians: **1** A medusa (jellyfish); **2** A solitary polyp (sea anemone); **3** A colonial polyp (soft coral). While medusae are free-swimming, polyps are usually sessile. The single exterior opening to the enteron acts as both mouth and anus.*

FACTFILE

SEA ANEMONES AND JELLYFISHES

Phylum: Cnidaria

Classes: Hydrozoa, Scyphozoa, Anthozoa

About 9,900 species in 3 classes.

Distribution Worldwide, mainly marine free-swimming and bottom-dwelling.

Fossil record Precambrian (about 600 million years ago) to present.

Size Width microscopic to several meters.

Features Radially symmetrical animals with cells arranged in tissues (tissue grade); possess tentacles and stinging cells (cnidoblasts); body wall of two cell layers (outer ectoderm and inner endoderm) cemented together by a primitively noncellular jellylike mesogloea and enclosing a digestive (gastrovascular) cavity not having an anus; there are two distinct life-history phases: free-swimming medusa and sedentary polyp.

HYDRAS AND ALLIES Class Hydrozoa
About 3,200 species in about 6 orders. **Fossil record:** some hydroids: many hydrocorals. **Features:** of 4- (tetramerous) or many- (polymerous) fold symmetry; solitary or colonial; life cycle can include polyp and medusa or exclusively one or other; mesogloea without cells; digestive (gastrovascular) system lacks a stomodeum (gullet); stinging cells (nematocysts) and internal septa absent; sexes separate or individuals bisexual; gametes mature in the ectodermis, which frequently secretes a chitinous or calcareous external skeleton; medusa has shelf-to-bell rim (velum); tentacles generally solid. Orders: Actinulida; Hydroida (hydroids, sea firs): Milleporina; Siphonophora; Stylasterina; Trachylina.

JELLYFISHES Class Scyphozoa
About 200 species in 5 orders (4 if Cubomedusae are elevated to class level, see main text). **Fossil record:** minimal. **Features:** dominant medusa form with 4-fold (tetramerous) symmetry; polyp phase produces medusae by transverse fission; solitary (either swimming or attached to substrate by stalk); mesogloea partly cellular; digestive (gastrovascular) system has gastric tentacles (no stomodeum) and is usually subdivided by partitions (septa); sexes usually separate; gonads in endodermis; complex marginal sense organs; skeleton absent; tentacles generally solid; exclusively marine.
Orders: Coronatae; Cubomedusae; Rhizostomae; Semaeostomae; Stauromedusae.

SEA ANEMONES, CORALS Class Anthozoa
About 6,500 species in probably 14 orders in 2–3 subclasses. **Fossil record:** several thousand species known. **Features:** exclusively polyps; predominantly with 6-fold (hexamerous) or 8-fold (octomerous) symmetry; pronounced additional tendency to bilateral symmetry; solitary or colonial; have flattened mouth (oral) disk with an inturned stomodeum; cellular mesogloea; sexes separate or hermaphrodite; gonads in endodermis; digestive (gastrovascular) cavity divided by partitions (septa) bearing gastric filaments; skeleton (when present) is either a calcareous external skeleton or a mesogloeal internal skeleton of either calcareous or horny construction; some forms specialized for brackish water; tentacles generally hollow.

SUBCLASS ALCYONARIA (OR OCTOCORALLIA)
Orders include: **soft corals** (Alcyonacea); **blue coral** (Coenothecalia); **horny corals** (Gorgonacea); **sea pens** (Pennatulacea); Stolonifera; Telestacea. One species, the **Broad sea fan** (*Eunicella verrucosa*; order Gorgonacea), is listed as Vulnerable.

SUBCLASS ZOANTHARIA (OR HEXACORALLIA)
Orders include: **anemones** (Actinaria); **thorny corals** (Antipatharia); Ceriantharia; Corallimorpharia; **hard or stony corals** (Madreporaria); Zoanthidea. The **Starlet sea anemone** (*Nematostella vectensis*; order Actinaria) is classed as Vulnerable by the IUCN Red List.

Hydras and Allies

CLASS HYDROZOA

The hydras and their allies (class Hydrozoa) are considered to be the group that exhibits the most primitive medley of features. The class contains a plethora of medusa and polyp forms that are, for the most part, relatively small. One can plausibly imagine the early hydrozoan life cycle as being similar to that of the hydrozoan order Trachylina. Here the medusae have a relatively simple form and the typical cnidarian larva, the planula, gives rise in turn to a hydralike stage that buds off the next generation of medusae. Significantly, this stage is predominantly free-swimming (pelagic), but in other hydrozoan orders subsequent polyp elaboration has resulted in the interpolation of bottom-living, hydralike colonies. Further evolution in specialized niches where dispersal is not at a premium has led in turn to the secondary reduction of medusae; indeed, most hydroids lack or almost totally lack a medusoid phase.

Early hydroids were probably solitary inhabitants of soft substrates. Subsequent evolution produced types living in sand (Actinulida) and freshwater (Hydridae – hydras). Most colonial types, however, occur on hard surfaces, anchored

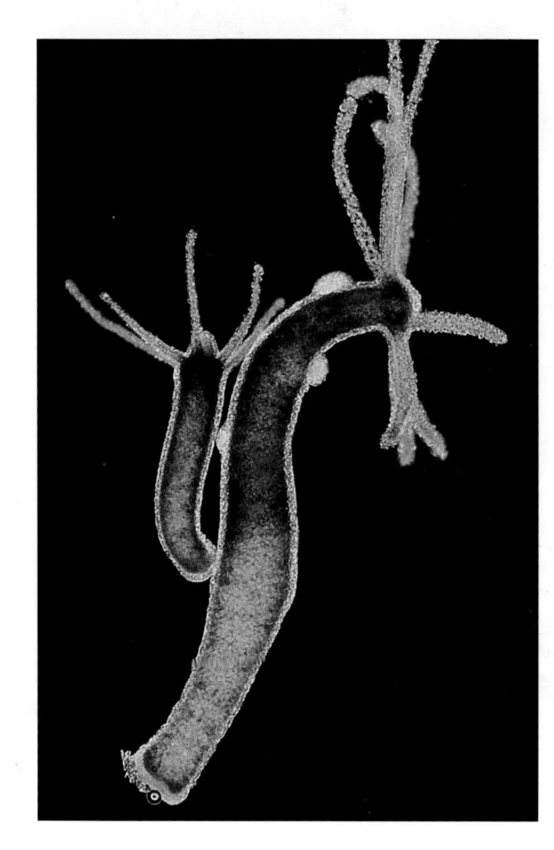

Right Green hydra (*Hydra viridissima*), *a freshwater polyp, showing asexual budding. Unlike those of the colonial hydroids, these buds detach themselves from the parent to form new organisms.*

by rooting structures. The interconnecting stems (stolons) are protected and supported by a chitinous casing (perisarc), which may or may not enclose the polyp heads. The functional interconnection of members of these colonies permits a division of labor between polyps and an associated variety in form (polymorphism). While one form (gastrozooid) retains both tentacles and a digestive cavity, the form that defends the colony (dactylozooid) has lost the cavity. Another form, the gonozooid, is dedicated solely to the budding-off of medusae or, in species lacking medusae, to producing gametes for reproduction. The delicate branching of hydroid colonies is extremely variable, but universally serves to space out member polyps and to raise them well above the substrate, thereby reducing the chance of clogging by silt and sediments.

The evolution of various forms in the class Hydrozoa has culminated in the formation of the complex floating siphonophore colonies (oceanic hydrozoans), each colony composed of a diverse array of both medusae and polyps; they are characteristic of warmer waters. Essentially each individual within the colony is interlinked by a central stolon. In addition to the three polyp forms found in hydroids, there can be up to four forms of medusae: **1** muscular swimming bells that propel the entire colony (for example, *Muggiaea*, *Nectalia*); **2** gas-filled flotation bells (for example, the Portuguese man-of-war – *Physalia physalis*); **3** bracts, which play either a supportive or protective role, or both; **4** medusa buds. Freed from the substrate, these colonies are able to reach large sizes with, for example, the trailing colonial stemwork of the Portuguese man-of-war often extending for several meters below the apical float. Such colonies are capable of paralyzing and ingesting relatively large prey items, such as fish, and can even deliver a potentially dangerous sting to humans. Recent research indicates that some species (for example, those of the genus *Agalma*) may attract large prey by moving tentacle-like structures, which are replete with stinging cells (cnidoblasts) and which bear a remarkable resemblance to small zooplankton (copepods).

Right *Physalia physalis, the Portuguese man-of-war, Atlantic (diameter of float 30cm/12in).*

At one time it was thought that the pinnacle of this evolutionary line was illustrated by animals such as the by-the-wind-sailor (*Vellela vellela*), which has a disklike, apical float bearing a rigid "sail" that catches the wind, thus facilitating drifting. It is now thought, however, that these organisms consist simply of one massive polyp floating upside-down, and that they are related to gigantic bottom-living hydroids. The large size of these bottom-living giants – up to 10cm (4in) in *Corymorpha* and 3m (10ft) in *Branchiocerianthus* – has been permitted by their adoption of a deposit-feeding lifestyle, often at great depth in still water.

Finally, two groups of hydrozoans produce a calcareous external skeleton resembling that of corals: they are the tropical milleporine and stylasterine hydrocorals.

Left *Sea anemones and jellyfishes:* **1** Cyanea lamarckii, *a jellyfish, N Atlantic (diameter 20cm/8in);* **2** Aurelia aurita, *the common jellyfish, medusa phase, Mediterranean and N Atlantic (diameter 25cm/10in);* **3** Actinia equina, *the beadlet anemone, Mediterranean and N Atlantic in intertidal zone (height 7cm/2.8in);* **4** Metridium senile, *the Plumose anemone (height 8cm/3in);* **5** Obelia geniculata, *shallow rocky habitats of NW Europe (colony height 4cm/1.6in);* **6** Sertularia operculata, *a hydroid, a colony of polyps (height of colony 45cm/17.7in);* **7** Eunicella verrucosa, *a sea fan, Mediterranean and N Atlantic (height 30cm/12in);* **8** Peachia hastata, *a "sit-and-wait" burrowing anemone, Mediterranean and N Atlantic (height 10cm/4in);* **9** Corynactis viridis, *an anemone-like animal, N Atlantic (diameter 5cm/2in);* **10** Alcyonium digitatum, *or "dead man's fingers" a colony of polyps, Mediterranean and N Atlantic (colony height 20cm/7.9in).*

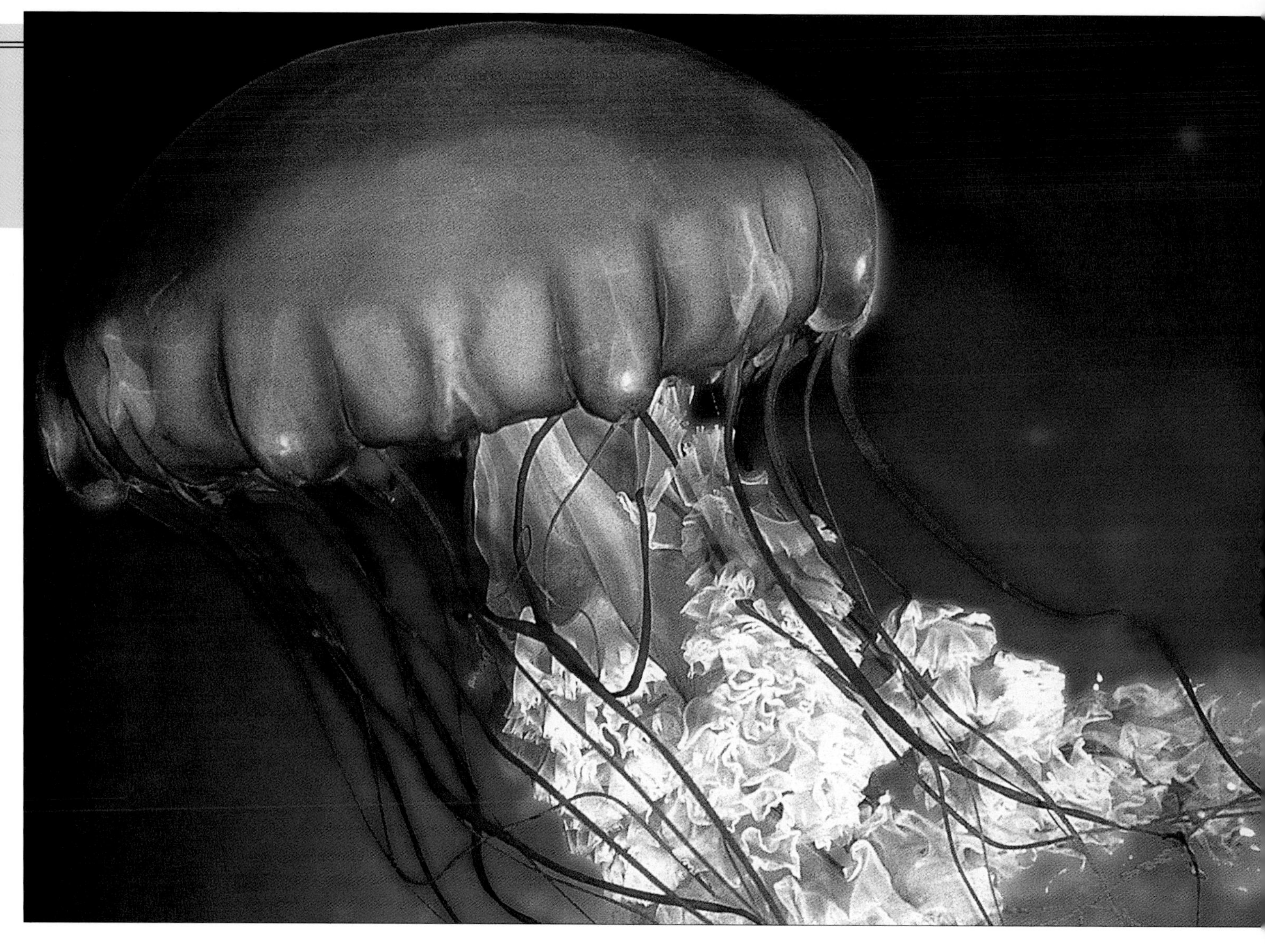

Above *The Sea nettle* (Chrysaora fuscescens) *is named for the powerful sting that its tentacles can deliver. Dense swarms of this species may gather off the coasts of California and Oregon in fall and winter.*

Below *Jellyfish polyps attached to the bed of a landlocked marine lake on Kakaban Island off Borneo. This lake is renowned for its abundance of jellyfishes, of four different stingless species.*

Jellyfishes

CLASS SCYPHOZOA

Among cnidarians, the jellyfishes have most fully exploited the free-swimming mode of life, although the members of one scyphozoan order (the Stauromedusae) are bottom-living, with an attached, polyplike existence. Jellyfish medusae have a similar though more complex structure than the medusae of hydroids, with the disk around the mouth prolonged into four arms, a digestive system comprising a complex set of radiating canals linking the central portion (stomach) to a peripheral ring element, and a relatively more voluminous mesogloea. The mesogloea in some genera (for example *Aequoria*, *Pelagia*) helps buoyancy by selectively expelling heavy chemical particles (anions) (such as sulfate ions), which are replaced by lighter ones (such as chloride ions). A wide size range of prey organisms are taken, although many species, including the common Atlantic semaeostomes of the genus *Aurelia*, are feeders on floating particles and thus concentrate on small items. The arms of *Aurelia* periodically sweep around the rim of the bell, gathering up particles that accumulate there after deposition on the animal's upper surface. In contrast, the arms around the main mouth of Rhizostomae have become branched and have numerous sucking mouths, each capable of ingesting small planktonic organisms such as copepods. Within this group are the essentially bottom-living, suspension-feeding forms of the genus *Cassiopeia*, which lie upside down on sandy bottoms, their frilly arms acting as strainers. The bell shapes of members of two orders are distinctive: coronate medusae have bells with a deep groove and cubomedusae have bells that are cuboid in shape.

The gametes of jellyfish are produced in gonads that lie on the floor of the digestive cavity and are initially discharged into it. Fertilization normally occurs after discharge. Many species, however, have brood pouches located on the undersurface where the larvae are retained. After release, larvae settle and give rise to polyps, which produce additional polyps by budding. These polyps also produce medusae by transverse division (fission), a process that results in the formation of stacks (strobilae) of ephyra larvae. When released, the ephyra larvae feed mainly on protozoans and grow and change into the typical jellyfish.

To trap prey, cnidarians normally employ stinging cells (cnidoblasts). The discharge of these cells is now thought to be under nervous control. Discharge involves a collagenous thread being rapidly shot out, uncoiling and turning inside-out, sometimes to expose lateral barbs. Hollow stinging cells often contain a toxin that can enter the body of the prey. The released toxins can be very potent; especially dangerous are those of the cubomedusan sea wasps (e.g., genus *Chironex*), which have been responsible for killing several humans, particularly off Australian coasts. Victims usually succumb rapidly to respiratory paralysis. Cnidoblasts may be pirated by sea slugs and used for their own protection. Some authorities treat these jellyfish, with their box-shaped medusae and powerful stinging cells, as a separate class, the Cubomedusae.

Corals and Sea Anemones

CLASS ANTHOZOA

Corals and sea anemones (class Anthozoa) exist only as polyps. Sea anemones (order Actiniaria) always bear more than eight tentacles and usually have both tentacles and internal partitions (mesentaries) arrayed in multiples of six.

Many anemone species, especially the more primitive ones, are burrowers in mud and sand but most dwell on hard substrates, cemented there (permanently or temporarily) by secretions from a well-differentiated disk. The disk around the mouth (oral disk) is equipped with two grooves (siphonoglyphs) richly endowed with cilia, which serve to maintain a water flow through the relatively extensive, digestive cavity. The oral disk extends inward to produce a tubular gullet, or stomodeum, which acts as a valve, closing in response to increases in internal pressure. In common with jellyfish, some anemones feed on particles suspended in the seawater for which leaflike tentacles, prodigious mucus production, and abundant food tracts lined with cilia are required; a good example is the common plumose anemone, genus *Metridium*. Asexual reproduction occurs by budding, breaking up, or fission; sexual reproduction may involve either internal or external fertilization of gametes. Some species brood young, either internally or externally at the base of the column.

Members of two other orders are also anemone-like: the cerianthids have greatly elongated bodies adapted for burrowing into sand, but have only one oral groove (siphonoglyph). Zoanthids lack a pedal disk, are frequently colonial and often live attached to other organisms (epizoic).

Also included in the subclass Zoantharia are the hard (stony) corals (order Madreporaria) whose polyps are encased in a rigid, calcium carbonate skeleton. The great majority of hard corals live in colonies, which are composed of vast numbers of small polyps (about 5mm, 0.2in), but the less abundant solitary forms may be large (*Fungia* up to 50cm, 20in, across); most are tropical or subtropical in distribution. In colonial forms the polyps are interconnected laterally; they form a superficial living sheet overlying the skeleton, which is itself secreted from the lower outer (ectodermal) layer.

Overleaf *Sandalled anemones* (Actinothoe sphyrodeta) *customarily attach themselves to rocks or other hard substrates.*

ASSOCIATIONS AND INTERDEPENDENCE IN ANEMONES, CRABS, AND FISHES

Cnidarians are involved in a variety of associations with other animals, ranging from obtaining food or other benefits from another animal (commensalism) to being interdependent (symbiosis).

An example of commensalism occurs between the hydroid genus *Hydractinia* and hermit crabs, particularly in regions deficient in suitable polyp attachment sites. This is understandable since the shells inhabited by such crabs provide substitute sites and, moreover, the relationship provides *Hydractinia* with the opportunity for scavenging food morsels. Whether the crab benefits is unclear, although the development of defensive dactylozooids in *H. echinata* specifically in response to a chemical stimulus emanating from crabs suggests that it does and that there is therefore a mutually beneficial coexistence (mutualism). The association between the cloak anemone *Adamsia palliata* and the crab *Pagurus prideauxi* is, to the contrary, far more intimate. These species normally form a partnership when small. Over time the crab outgrows its shelly refuge, but by secreting a horny foot (pedal) membrane the anemone progressively enlarges the shell lip, thus obviating any need for the crab to change shells. A crab lacking an anemone will, on contact with *Adamsia*, recognize it and attempt to transfer it to its shell.

A range of intermediate degrees of association is provided by the anemone *Calliactis parasitica*, which associates with hermit crabs but also frequently lives independently. An interesting, one-sided association is that of the Hawaiian crab genus *Melia*, which has the remarkable habit of carrying an anemone in each of its two chaelae, thereby enhancing its aggressive armament; the crab even raids their food.

Clown fishes (*Amphiprion* spp.) live in the tentacles of sea anemones. Their hosts protect these fishes and the fishes protect the anemones from would-be predators (chiefly other fishes, deterred by the territorial behavior of the clown fish). Clown fishes also apparently act as anemone cleaners. Although it has been suggested that inhibitory substances are secreted by these fishes to reduce the discharge of stinging cells, recent work has failed to confirm this. It is more likely that stinging is avoided by the secretion of a particularly thick mucous coat, which during acclimatization of the fish to its host anemone may become modified, with the levels of certain excitatory (acidic) components being reduced. RB/AC

Above *Sea pens (order Pennatulacea) live on soft seabeds. The large, stemlike, primary polyp houses a skeletal rod that becomes embedded in the substrate as a result of waves of contractions. Secondary polyps are arranged laterally on this stem.*

Right *Coral bleaching and death, perhaps caused by rising sea temperatures and lower salinity, is a growing problem. This coral is* Acropora *sp.*

Below *The octocorallian Mushroom soft coral* (Anthomastus ritteri) *is a deep-sea species found off the US west coast and Baja California. When threatened, with all its tentacles retracted, it resembles a fungus.*

Corals exhibit a great diversity of growth forms, ranging from delicately branching species to those whose massive skeletal deposits form the building blocks of coral reefs. An interesting growth variant is shown by *Meandrina* and its relatives in which polyps are arranged continuously in rows, resulting in a skeleton with longitudinal fissures – hence its popular name, the brain coral.

Closely related to the hard corals are the members of the order Corallimorpharia, which lack a skeleton. They include the jewel anemone (genus *Corynactis*), so named because of its vivid and highly variable coloration. Since it reproduces asexually, rock faces can become covered by a multicolored quiltwork of anemones. The black or thorny corals (order Antipatharia) form slender, plantlike colonies bearing polyps arranged around a horny axial skeleton; they have numerous thorns.

Octocorallian corals comprise a varied assemblage of forms, but all possess eight featherlike (pinnate) tentacles. The polyps project above and are linked together by a mass of skeletal tissue called coenenchyme, which consists of mesogloea permeated by digestive tubules. Thus, in contrast to hard corals, the octocorallians have an internal skeleton. This assemblage includes the familiar gorgonian (horny) corals, sea whips and fans, and the precious red coral, genus *Corallium*. Most have a central rod composed of organic material (gorgonin) around which is draped the coenenchyme and polyps, the former frequently containing spicules, which may impart a vivid coloration. Such is the case with *Corallium*, whose central axis consists of a fused mass of deep red calcareous spicules; this material is used in jewelry. The tropical organ pipe coral, genus *Tubipora* (order Stolonifera), produces tubes or tubules of fused spicules that are crossconnected by a regular series of transverse bars. In contrast, the soft corals (order Alcyonacea) contain only discrete spicules within the coenenchyme (for example, dead man's fingers, genus *Alcyonium*). The order Coenothecalia is represented solely by the Indo-Pacific blue coral, genus *Heliopora*, which has a massive skeleton composed of crystalline aragonite fibers fused into plates (lamellae): its blue color is imparted by bile salts. Many species in most of these groups have several forms (especially gastrozooids, dactylozooids, and gonozooids). Many polyps (siphonozooids) act as pumps, promoting water circulation through the colonial digestive system. Familiar examples are the sea pansy (genus *Renilla*) and the sea pen (genus *Veretillum*), both of which emit waves of phosphorescence when disturbed. These waves are controlled by the nervous system and are inhibited by light. Their role is unclear, although it is likely that they are a response to intrusion by would-be predators.

The cnidarian nervous system shows a certain amount of organization and local specialization. This is especially evident in anemones where nerve tracts accompany the retractor muscles responsible for protective withdrawals. The marginal ganglia of scyphomedusae and the circumferential tracts of hydromedusae have been found to contain pacemaker cells that are responsible for initiating and maintaining swimming rhythms. In *Polyorchis* it has been found that the giant nerves controlling movement are all coupled together electrically, ensuring that they function collectively as a giant ring nerve fiber capable of initiating synchronously muscle contraction from all parts of the bell.

Similarly, the behaviors of individual polyps in hydroid and coral colonies are integrated by the activities of colonial nerve nets. Additional powers for integrating control are provided by conduction pathways created by sheets of electrically coupled epithelial cells. For example, the shell-climbing behavior of anemones of the genus *Calliactis* seems to depend on the interplay of activities between two epithelial systems – one on the outside (ectodermal), the other inside (endodermal) – and the nerve net, although conclusive evidence as to the exact cellular locations of these additional systems has been difficult to obtain. RB/AC

AGGRESSION IN ANEMONES

A number of anemones display a well-defined aggressive sequence, which, for the most part, is used in confrontation with other anemones. These anemones all possess discrete structures located at the top of the column that contain densely packed batteries of stinging cells (cnidoblasts): they are called acrorhagi. They can be inflated and directed at opponents.

The common intertidal beadlet anemone (*Actinia equina*), whose distribution encompasses the Atlantic seaboards of Europe and Africa and which also occurs in the Mediterranean, is an example upon which attention has been focused. Although this species can vary considerably in color (red to green), the acrorhagi are always conspicuous thanks to their intense bluish hue. Aggression is triggered by the contact of tentacles. One individual usually displays column extension and bends so that some of its simultaneously enlarging acrorhagi make contact with the opponent (after 5–10 minutes). There follows a discharge of the stinging cells (cnidoblasts) which normally results in a rapid withdrawal by the victim.

Experiments suggest that in common with more-advanced animals, contest behavior is ritualized, but that in these less-advanced forms it depends on simple physiological rules rather than on complex behavioral ones. The "rules" apparently decree that larger anemones should act aggressively more rapidly than smaller ones, and so win contests.

The North American anemone *Anthopleura elegantissima* reproduces asexually by fission. In consequence, intertidal rocks can become entirely covered by a patchwork of asexually produced anemones. Close inspection reveals that each densely packed clone – the mass of asexually produced offspring – is separated from its neighbors by anemone-free strips, and that they are maintained by aggressive interactions involving acrorhagi. It is clear, therefore, that the aggressive behavior of individuals constituting the boundary of a clone serves to provide territorial defense for the entire clone, the central members of which, significantly, are more concerned with reproduction than aggression. Thus there is, as in hydroid colonies, a functional division of labor, despite the lack of physical interconnection between the clonal units. Individuals at the interclonal border have more and larger acrorhagi than centrally placed members, a difference apparently dependent solely on the former experiencing aggressive contact with non-clonemates. Such a dichotomy is thought to indicate the presence of a sophisticated self/nonself recognition system. RB/AC

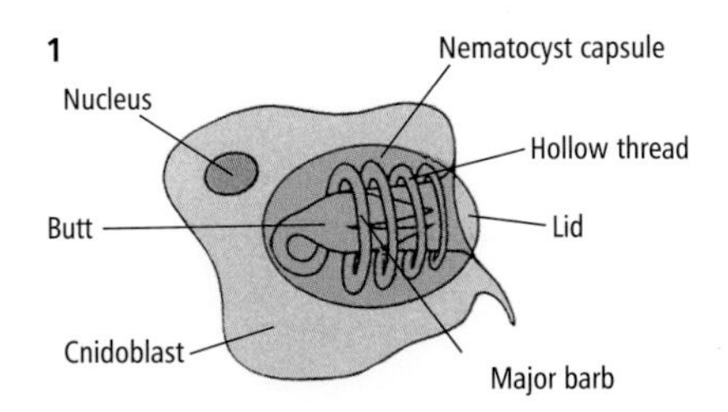

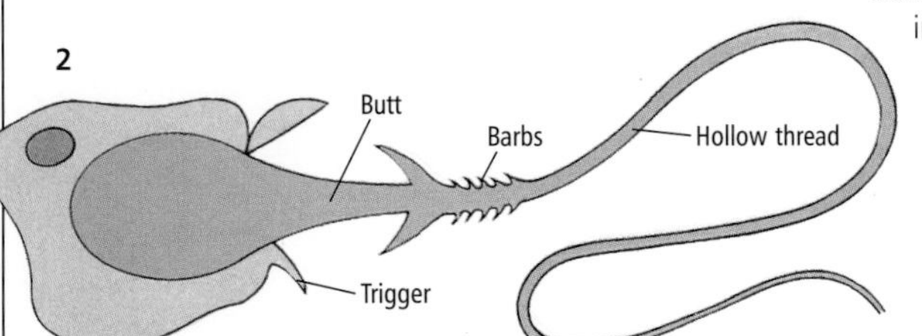

Left *The stinging cell (cnidoblast) of a cnidarian:* **1** *before discharge of the nematocyst and* **2** *after discharge.*

Below *A corallimorph* (Pseudocorynactis *sp.*) *attacking a Sea star* (Linckia laevigata).

SPECIAL FEATURE

THE LIVING AND THE DEAD

The origins and biological organization of coral reefs

CORAL REEFS ARE EXTRAORDINARY OASES IN the midst of oceanic deserts, for they support immensely rich and diverse faunas and floras but occur primarily in the tropics where the marked clarity of the water indicates a relative dearth of planktonic organisms and other nutrients. "Coral" consists of the skeletons of hard or stony coral.

The success of reef-building (hermatypic) corals in tropical waters, where high light intensities prevail throughout the year, is strictly dependent on the nutrient-manufacturing (autotrophic) activities of interdependent (symbiotic) algae (zooxanthallae) that live within each polyp. Such dependence also necessarily restricts the algae to these waters. Moreover, since these algae flourish best at temperatures higher than 20°C (68°F), reef development is further limited to depths of less than 70m (230ft), where light intensities are greatest. Corals do, however, survive both at higher latitudes and in deeper waters, but where they do their capacity to secrete limestone for reef building is found to be severely curtailed as a result of the reduced metabolic support provided by the algae. Finally, restrictions on the distribution of corals are also imposed by the deposition of silt, freshwater runoff from land, and cold, deepwater upwelling. The two former factors, for example, restrict reef development in the Indo-Pacific Ocean towards offshore island sites, while the last hinders coral growth off the west equatorial coast of Africa, where the Guinea current surfaces. It should not be forgotten, though, that in the development of most reefs, encrusting (calcareous) algae (for example, *Lithophyllum*, *Lithothamnion*) normally

Above *Polyps of the Sun coral* Tubastraea faulkneri. *These corals, which are found throughout Pacific tropical waters, including the Great Barrier Reef, extend their tentacles at night in order to catch small zooplankton.*

Left *Colonies of sea fans may be several meters high, as with these* Melithea *sp. fans off Fiji in the Southern Pacific. Sea fans can fall prey to several diseases, including aspergillosis and red-band disease.*

play an extremely important part. In many cases the limestone they produce acts as a cement.

In 1842 Charles Darwin distinguished three main geomorphological categories of reef that are still in use: fringing reefs, barrier reefs, and atolls. As the name suggests, the first are formed close to shore, on rocky coastlines. Barrier reefs are, on the contrary, separated from land by lagoons or channels, which have usually been produced as a result of subsidence. (The best known and largest barrier reef is the Great Barrier Reef off the northeast coast of Australia, although the name is somewhat misleading as along its length (1,900km/1,200 miles) occur a host of different reef configurations – more than 2,500. Finally, atolls are found around subsiding volcanoes.

The continuation of coral growth is heavily dependent upon changes in water level. At present the world is in a period between glaciations, and the rising sea level permits vertical growth to continue at about 0.3–1.5cm (0.1–0.6in) per year, a rate that apparently has been maintained over the last 100,000 years. Core drillings taken at Eniwetok atoll in the Pacific have extended downwards for up to 1.6km (1mile) before hitting bed rock: from analyses of both the fauna and flora in cores obtained, it has proved possible to reconstruct past fluctuations in sea level. The majority of the world's coral reefs started development during the Cenozoic era (not later than 65 million years ago) and consist predominantly of corals of the order Madreporaria.

All coral reefs have a similar biological organization, with the reef plants and animals, as on rocky shores, lying in zones in accordance with their tolerances to physical factors. This zonation is most evident on the exposed, windward faces of those reefs subject to continuous wave crash, where especially prolific growths of both corals and algae develop. However, a reef can grow outwards only if debris accumulates on the reef slope; with increasing water depths, such material tends to slide down the slope and thus becomes unavailable. Below 30–50m (100–165ft), hermatypic corals are replaced by nonhermatypic ones, and by fragile, branching alcyonarians (gorgonians, etc.). Above the slope is the reef crest, which, in the most exposed situations, is dominated by encrusting calcareous algae (for example, the genus *Porolithon*). These algae form ridges that are full of cavities, thereby providing numerous recesses which are colonized by a multitude of invertebrates, including zoanthids, sea urchins, and vermetid gastropods. Where wave action is not too severe, windward reef crests are usually dominated by a relatively small number of coral species, notably stout *Acropora* and hydrocoralline *Millepora* species.

Zonation is far less marked on the leeward side of the reef crest, where a different set of problems has to be faced, of which sediment accumulation from land runoff is perhaps the most acute. Nevertheless, the relatively sheltered nature of this habitat permits the rapid proliferation of branched corals. In common with the coral faunas of windward slopes, those of leeward faces display dramatic changes of coral form with increasing depth, changes that can be attributable either to the replacement of species by others or to changes in species forms. For example, on Caribbean reefs, dominant species such as *Monastrea annularis* display both stout (shallow-water) and branching (deepwater) growth forms. Recent work on *M. cavernosa* has indicated that forms found at equivalent depths are distinctive: the polyps of the shallow-water form are open continuously, while those of the deepwater form (which house far fewer zooxanthellae) are open only at night. RB

Comb Jellies

SMALL, TRANSLUCENT, GLOBULAR ANIMALS, comb jellies float through the open seas, like ghosts, capturing prey with their whiplike tentacles. The order name Ctenophora was derived from the Latin for "comb bearers" to reflect their most distinct feature – the rows of comb plates.

The comb jelly's body consists of three zones – a voluminous mesogloea sandwiched, as in sea anemones and jellyfishes, between thin ectodermal (outer) and endodermal (inner) cell layers. Most noticeable, however, are the eight rows of plates of cilia (comb rows), whose activities serve to propel the animal while it is searching for zooplanktonic prey.

Combing the Sea

MOVEMENT

The waves of movement of the comb plates responsible for swimming are initiated and synchronized by impulses arising within the apical sense organ, which functions primarily as a statocyst, detecting tilting, although it is also sensitive to light. It contains a sensory epithelium bearing four groups of elongate sensory "balancer" cilia, upon which a calcareous ball (statolith) perches.

The orientation of the animal is controlled very simply: irrespective of whether the comb jelly is swimming upward or downward, deflection of the balancer cilia away from an upright position brought about by the statolith results in a change in the frequency of beating of the cilia, which is generated spontaneously. The overall effect exerted through the differential activitation of the four groups of "balancer cilia," and subsequently the four pairs of comb rows to which they are electrically connected, is to ensure that the animal swims vertically rather than obliquely. Since the power stroke of the individual comb plates is opposite to direction of waves of activity passing along the comb rows, the animal normally travels mouth forward. However, this activity is also under control exerted by the nerve net, which upon receipt of a mechanical stimulus can cause either a cessation or reversal of beating. There is also some evidence to suggest that the synchronization of beating of the comb plates within any one row might be dependent partially upon direct mechanical coupling.

Angling with Tentacles

FORM, FUNCTION, AND DIET

Most comb jellies resemble species in the common genus *Pleurobrachia* (order Cydippida), which lives in the colder waters of both the Atlantic and Pacific oceans, and is often found stranded in tidal pools. Its globular body is up to 4cm (1.5in) in diameter, and has two pits into which the tentacles can be retracted. These tentacles function as drift nets, catching passing food items while the animal hovers motionless. When extended, these appendages may be up to 50cm (20in) in length. They have lateral filaments and bear numerous adhesive cells (colloblasts), each of which has a hemispherical head fastened to the core of the tentacle by a straight connective fiber and by a contractile spiral one, the latter acting as a lasso. Once caught prey is held by the colloblasts, which produce a sticky secretion, until transferred to the central portion (stomach) of the digestive system following the wiping of the tentacles over the mouth. When feeding upon pipefishes of a similar size to itself, *Pleurobrachia* will play them in much the same way as an angler tires out a hooked fish.

The stomach, in which digestion commences, leads to a complex array of canals where there is further digestion and intake of small food particles which are subsequently broken down by intracellular digestion. These canals are especially routed alongside those body regions having high energy-consumption levels, notably the eight comb rows. The gonads lie in association with the lining of the gastric system; gametes pass out through the mouth. Following external fertilization a larva – a miniature version of the adult – is produced.

Right While most comb jellies feed on zooplankton, the large-mouthed Beroë cucumis *feeds exclusively on other jellies* (Bolinopsis infundilbulum). *It has been referred to as the "Jaws of the midwater jelly*

The common comb jellies of temperate waters are all like *Pleurobrachia*, which has the most primitive body form; shape in the other orders departs from this. The elongate lobate comb jellies are laterally compressed and have six lobes projecting from their narrow mid region, four of which are delicate and two stout. They serve to capture food: the tentacles are small and lack sheaths. *Mnemiopsis* (order Lobata), which is about 3cm (1.2in) in size and occurs in immense swarms, has, in association with the production of these lobes, four long and four short comb rows. Elongation and compression have been carried much further in the Cestida, resulting in organisms resembling thin gelatinous bands. Species in the two genera concerned (*Cestum*, *Velamen*) are referred to collectively as Venus's girdle, and are found in tropical waters and the Mediterranean, only occasionally straying into northern waters. The graceful swimming of these forms is dependent mostly upon undulations of the whole body, brought about by muscle fibers embedded in the mesogloea. They feed entirely by means of tentacles set in grooves running along the oral edge.

In the order Beroidea, the thimble-shape body is similarly laterally compressed, but occupied mainly by the greatly enlarged stomodeum rather than mesogloea. Species in this group are up to 20cm (8in) in length and often have a pinkish color. There are no tentacles; instead food is caught by lips, which curl outward to reveal a glandular and ciliated (macrocilia) area. Relatively large food items are taken in rapidly by a combination of suction pressure from the contraction of radial muscles in the mesogloea and ciliary action. The common North Sea species *Beroë gracilis* feeds exclusively upon *Pleurobrachia pileus*.

The very small order Ganeshida includes one genus *Ganesha* with two pelagic species somewhat intermediate in form between members of Cydippida and Lobata. The order Platyctenida is a curious group, which, contrary to other flattened ctenophores, are compressed from top to bottom and have, for the most part, assumed a benthic

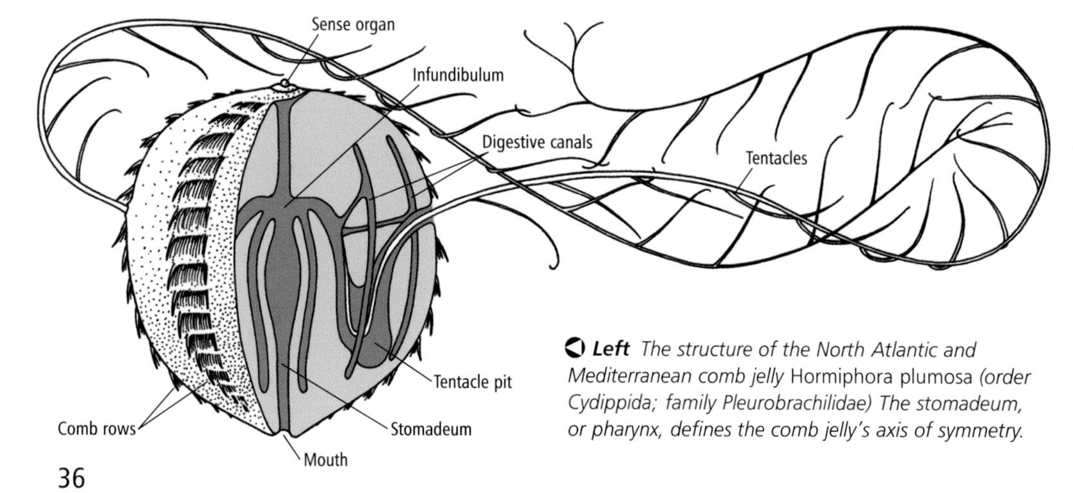

Left The structure of the North Atlantic and Mediterranean comb jelly Hormiphora plumosa *(order Cydippida; family Pleurobrachilidae) The stomadeum, or pharynx, defines the comb jelly's axis of symmetry.*

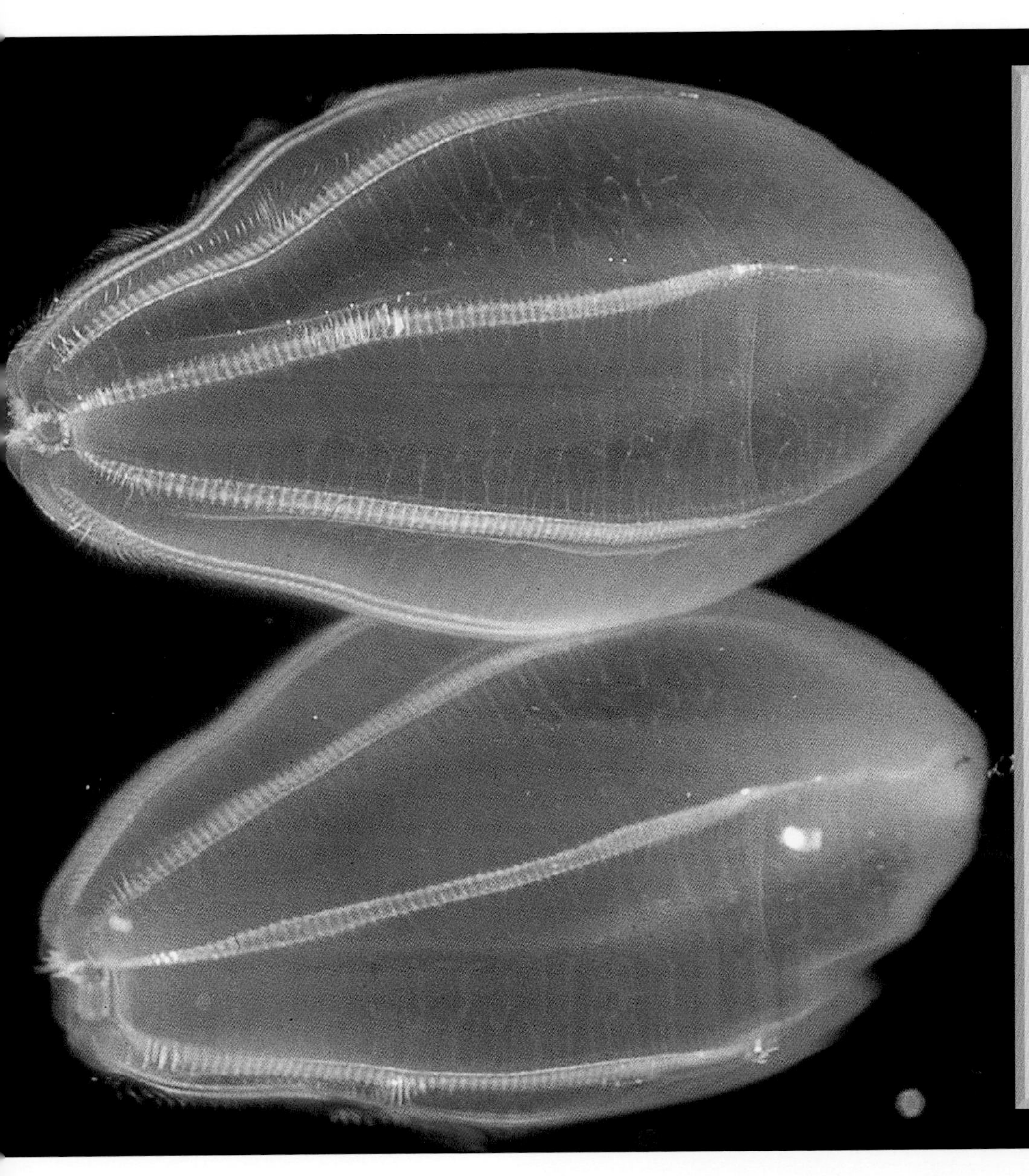

FACTFILE

COMB JELLIES

Phylum: Ctenophora

Orders: Beroida, Cestida, Cydippida, Ganeshida, Lobata, Platytecnida, Thalassocalycida

About 100 species in 7 orders.

Distribution Worldwide, marine

Fossil record None

Size Length from very small about 0.4cm (0.15in) to over 1m (3.3ft).

Features Basically radially symmetrical but masked by superimposed bilateral symmetry; body wall 2-layered (diploblastic) with a thick jellylike mesogloea and nerve net (these features making them similar to sea anemones and jellyfishes); 8 rows of plates of fused cilia (comb plates) upon whose activity locomotion predominantly depends; tentacles when present help in the capturing of zooplankton; digestive/gastrovascular system with a stomodeum (gullet), stomach, and a complex array of canals; one phase in life cycle (not equivalent to polyp or medusa); hermaphroditic.

ORDER BEROIDA (class Nuda)
Includes the species *Beroë gracilis.*

ORDER CESTIDA
Includes: **Venus's girdle** (genera *Cestum*, *Velamen*).

ORDER CYDIPPIDA
Includes: *Pleurobrachia, Dryodora, Ctenella.*

ORDER GANESHIDA
One genus *Ganesha.*

ORDER LOBATA
Includes: *Mnemiopsis, Deiopea, Ocyropsis.*

ORDER PLATYCTENIDA
Includes: *Coeloplana, Ctenoplana, Tjalfiella, Gastrodes.*

ORDER THALASSOCALYCIDA
One species *Thalassocalyce inconstans.*

creeping lifestyle. Their ciliated undersurface is derived from part of the stomodeal wall. Of the four genera involved, *Coeloplana* and *Tjalfiella* are flattened creeping forms, the latter being practically sessile – attached to the substrate. In contrast, *Ctenoplana* is partially planktonic, having become adapted to creeping on the water's surface also. The most specialized, however, is *Gastrodes*, which is a parasite of the free-swimming sea squirt genus *Salpa*. The flattened adult stage is a free-living, benthic form, but the planula-type larva bores into the tunicate test and develops into a bowl-shaped, intermediate parasitic stage. The final, small-order Thalassocalycida contains comb jellies with extremely fragile bodies and an umbrella-shaped expansion at one end. *Thalassocalyce inconstans* is the sole species of the order. RB/AC

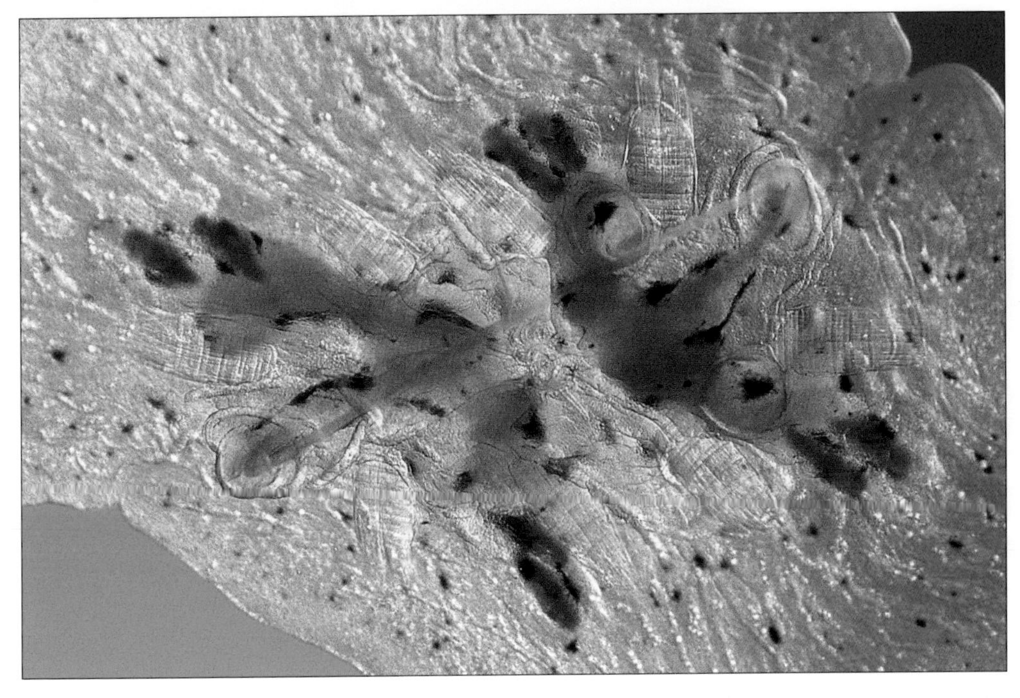

Right *Usually found on soft corals in warm waters, this* Ctenoplana *species (order Platyctenida) is in its larval stage. A highly modified ctenophore species, it is sometimes mistaken for a nudibranch or a flatworm.*

Gnathostomulids, Acoelomorphs, Gastrotrichs, and Arrow Worms

This disparate grouping of organisms embraces a huge variety of marine worms, ranging in size from the small arrow worms to the infinitesimally tiny gnathostomulids, which can live in the gaps between sand grains.

Gnathostomulids

PHYLUM GNATHOSTOMULIDA

These microscopic acoelomate marine worms are commonly found in shallow saltwater to a depth of several hundred meters. Members of the order Filospermoidea have a lengthy tapering rostrum (front end lies in front of the mouth), and the adult males lack a penis and the adult females lack a vagina. In the Bursovaginoidea, members have less elongated bodies lacking the slender rostrum, and the adults have a penis, bursa, and vagina. These animals were first described from meiofaunal samples in 1956 and accorded phylum status in 1969. While they are undoubtedly acoelomate, their relationships with other animal groups are far from clear. Their outer surface bears cilia (one cilium per epithelial cell) and they glide through the sediment by a combination of ciliary action and flexure of the body brought about by muscular activity. They feed on bacteria and the microscopic jaws aid in the ingestion of food. On account of their minuscule size, diffusion plays an important part in the absorption of oxygen, the removal of carbon dioxide and the distribution of nutrients around the body. Sensory structures, often pitlike, on the outer surface and associated with cilia detect environmental stimuli and connect with the superficial nervous system. Many aspects of the biology of these animals are not well understood and the identification and classification of species is a specialist task.

Above *Acoelomorphs of the order Acoela. Formerly grouped with the platyhelminthes (flatworms), new molecular and morphological evidence has caused them to be reclassified.*

Acoelomorphs

PHYLUM ACOELOMORPHA

Scientists are constantly reviewing the taxonomic positions and evolutionary relationships of various animal groups. Formerly thought to be basal among the Protostomes (flatworms), acoelomorphs, on the basis of molecular evidence, are now believed to belong to the Lophotrochozoa, except for the orders Acoela and Nemertodermatida, which occupy a unique position as the most basal extant group of bilaterians (bilaterally symmetrical animals).

For many years the lowly gutless flatworms have been classified as an order, the Acoela, within the class Turbellaria of the problematic flatworms. In light of the molecular evidence, there is now a suggestion that these apparently simple yet remarkable animals, which have a mouth but no gut, and which frequently associate with symbiotic algae, may constitute a new phylum. This phylum also includes the Acoela's sister order Nemertodermatida, where the similarities between the two involve their ciliation, body-wall structure, reproductive organs, and the relation of the statocyst to the nervous tissue. The phylum has been called the Acoelomorpha. Evidence for this view is consolidated by the fact that the acoelomorphs display some structural details and specialized characteristics that set them apart from the other flatworms. Were the acoelomorphs to be removed from the group, those animals remaining would show a clear affinity with each other and would appear to have been evolved from a single evolutionary line. Thus, creating a separate phylum for the acoelomorphs is as much a recognition of the integral similarities of the other flatworms as it is of the differences shown by the acoelomorphs. (See also Flatworms and Ribbon Worms.)

Gastrotrichs

PHYLUM GASTROTRICHA

Gastrotrichs are minute animals often found in habitats shared with rotifers. They are "wormlike" and live among detrital particles in both fresh and seawater. Gastrotrichs show bilateral symmetry but have a pseudocoel and not a true coelom.

The genus *Chaetonotus* has a rounded head attached to the trunk by an elongated neck; the tail is usually forked and the outer surfaces of the body are covered with sticky knobs (papillae). The underside bears hairs (cilia) arranged in bands whose coordinated beating causes the animal to glide over the substrate. The upper surface of the animal is usually armed with spines and scales.

The gastrotrich pharynx generates a sucking action, which is employed to draw food into the gut. It takes single-celled algae, bacteria, and protozoans. The reproductive strategies vary according to the groups of gastrotrichs. Males are unknown in the class Chaetonotoidea, where reproduction is by parthenogenesis among females. Here two types of eggs are laid. In one, hatching takes place very quickly and the young mature in about three days. In the other case, eggs can lie dormant, surviving desiccation, and will hatch when more favorable circumstances prevail. The members of the class Macrodasyida are hermaphrodites. Gastrotrichs are of little ecological or economic importance.

Arrow Worms

PHYLUM CHAETOGNATHA

Arrow worms are small, inconspicuous marine animals that were unknown before 1829. Their torpedo-shaped bodies, combined with their

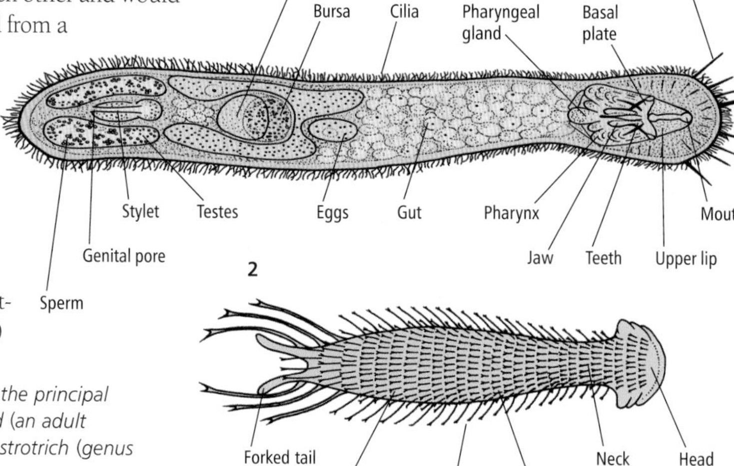

Right *Body plans showing the principal features of* **1** *a gnathostomulid (an adult* Problognathia minima*),* **2** *a gastrotrich (genus* Chaetonotus*), and* **3** *an arrow worm.*

◐ *Left* Arrow worms in marine plankton (of which they form a significant part). This group of animals was not known until the early 19th century, when fine-meshed towed nets capable of sieving out plankton from sea water were developed.

ability to dart rapidly forward through the water for short distances, explain their common name.

Arrow worms are transparent and almost all are colorless. All known species are similar in appearance and all live in the plankton, except *Spadella*, which lives in the sea floor. Their evolutionary position remains unclear despite various attempts to show relationships with other phyla. They are now generally supposed to be without close affinities, lying at the base of the protostome line.

The long, narrow body carries a head at the front which bears curved hooks of chitin. These grasp prey and maneuver it into the mouth. Small planktonic animals form the diet, especially crustaceans such as copepods, but occasionally larger food such as other arrow worms and fish larvae may be taken. The mouth carries small teeth that assist in the act of swallowing. A pair of small, simple eyes is also borne on the head. These eyes may or may not be pigmented and serve mainly as light detectors rather than as organs that actually form an image.

The trunk carries two pairs of lateral fins and the body terminates in a fishlike tail segment. The gut runs straight through from mouth to anus and has two small pouches, or diverticula, at the front.

Arrow worms are hermaphroditic and their sex organs are relatively large for the overall body size. The ovaries lie in the trunk region and the testes and seminal vesicles lie toward the tail, separated by a partition. The form of the seminal vesicles, where sperm is stored, is used to identify the various species. The breeding cycle varies in length according to distribution. It is completed rapidly in tropical waters, in about six weeks in temperate latitudes, and it may take up to two years in polar regions. Fertilization takes place internally and fertilized eggs are released into the sea where they develop as larvae before maturing into adults. The larvae resemble small adults. In the genus *Eukrohnia* the larvae are brooded in special pouches.

The outside of the body is equipped with small mechanoreceptors that respond to vibrations in the water such as are generated by certain prey species. This facility allows arrow worms to detect prey and engulf them. If they themselves are damaged, for example in an attack by a predator, lost parts, including the head, can regrow. AC

3

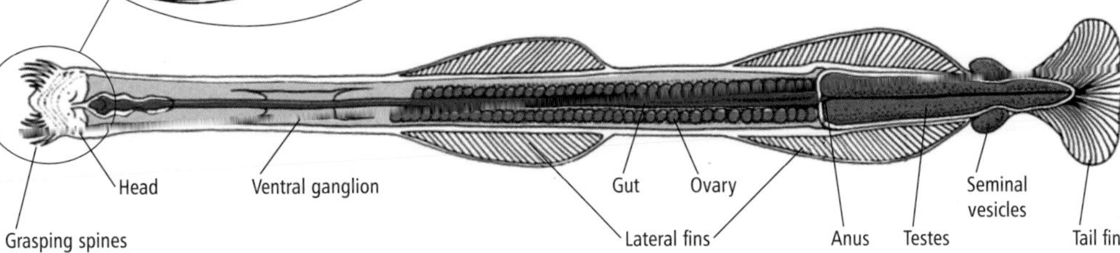

FACTFILE

GNATHOSTOMULIDS, ETC.

Phyla: Gnathostomulida, Acoelomorpha, Gastrotricha, Chaetognatha.

GNATHOSTOMULIDS Phylum Gnathostomulida
About 80 species in 2 orders and 25 genera. **Distribution:** worldwide, exclusively marine, inhabiting sediments like anoxic sands and muds from coastal regions down to several hundred meters. **Fossil record:** apparently none. **Size:** up to 2mm (0.1in) long. **Features:** adults: freeliving, acoelomate, bilaterally symmetrical, unsegmented wormlike microscopic animals, with bodies divisible into head, trunk, and tail regions. There is a unique pharyngeal region armed with jaws (hence the phylum name meaning "jaw mouthed"). The gnathostomulids are either protandric hermaphrodites (starting adult life as males and becoming females later) or they are simultaneous hermaphrodites, being male and female at the same time. The fertilized eggs develop in the protostome fashion and the young develop directly into adults without a larval phase. Order Filospermoidea; Order Bursovaginoidea.

ACOELOMORPHS Phylum Acoelomorpha
Controversial phylum with over 280 species in 2 orders – Acoela and Nemertodermatida. Genera and species include: *Convoluta convoluta* (order Acoela), *Meara* (order Nemertodermatida).

GASTROTRICHS Phylum Gastrotricha
About 450 species. **Distribution:** widespread, marine and freshwater. **Fossil record:** none. **Size:** microscopic, less than 3 mm (0.12in). **Features:** free living, usually with some external areas covered with cilia; body unsegmented, blastocoelomate but body cavity filled with cells; body of three layers (triploblastic) often with one or more pairs of adhesive organs; excretory protonephridia when present consist of a single flame cell; cuticle covered with spines, scales, or bristles.

CLASS CHAETONOTOIDA
Mainly fresh water. Front end of body usually distinct from trunk; adhesive tubes at rear; protonephridia present; mainly parthenogenetic (asexually reproducing) females. Genera include: *Chaetonotus*.

CLASS MACRODASYIDA
Marine, chiefly reported from Europe. Bodies straight with adhesive tubes on front, rear, and sides of the body; no protonephridia; hermaphrodite. Genera include: *Macrodasys*.

ARROW WORMS Phylum Chaetognatha
About 100 living species in 7 genera. **Distribution:** all oceans, planktonic apart from 1 benthic genus. **Fossil record:** one dubious record. **Size:** length between 0.4cm (0.16in) and 10cm (4in), most less than 2cm (0.8in). **Features:** small bilaterally symmetrical freeliving animals, formerly believed to have deuterostome development, but new research suggests they are basal protostomes, lack circulatory and excretory systems; body torpedo-shaped with paired lateral fins and tail fin; mouth at front and armed with strong grasping spines. Genera: *Bathyspadella, Eukrohnia, Heterokrohnia, Krohnitta, Pterosagitta, Sagitta, Spadella*.

Water Bears

TARDIGRADES (LITERALLY "SLOW WALKERS") are minute fat-bodied animals, measuring some 0.1 to 1.0 mm in length. Because of the perceived similarity of their body shape to that of bears, they are commonly referred to as "water bears" (a term first coined by the 18th-century German naturalist Johann August Goeze). Their customary habitat is in ditches, lakes, and shallow coastal waters. They can often be found in the water film that clings to damp clumps of moss or lichen – hence their alternative common name of "moss bears" – and also live in soil, freshwater, and marine sediments.

Resilient Survivors

LOCOMOTION AND BIOLOGY

Water bears move by slow crawling, attaching themselves with claws at the end of each leg. In addition to their four pairs of short, unjointed legs (which are known as lobopodial limbs, and are also found in velvet worms), there are further signs of segmentation in their muscles and nervous system. In *Echiniscus* species the cuticle is arranged in segmental plates. Most water bears feed by piercing plant cells with two sharp stylets and sucking out the contents via a muscular pharynx. Some water bears feed on detritus, while others are predatory. There are two, probably salivary, glands leading into the mouth. Defecation may be associated with molting, as in *Echiniscus,* which leaves its feces in its cast cuticle.

The three "Malpighian" glands leading into the gut are believed to serve an excretory function. Water bears have no respiratory organs, blood vessels, or heart because diffusion is sufficient for transport of essential foods in such small animals. The nerve cord along the underside is well organized with ganglia, and the sense organs include two eyespots and sensory bristles.

Water bears have separate sexes, the females usually being the more numerous. However, certain species are hermaphrodite, while males have yet to be discovered in other species. In some species the females breed asexually (namely, by parthenogenesis). In general, reproduction involves copulation with subsequent internal fertilization; some females store transferred sperm in seminal receptacles. Between one and 30 eggs are laid at a time. They may be thin-shelled and hatch soon after laying or thick-shelled to resist hazardous environmental conditions.

The newly hatched young resemble adults and grow by increasing the size but not the apparently fixed number of their constituent cells – a feature that may be associated with their very small size. Legs increase from two pairs to four before adulthood. Maturity is reached in about 14 days and they live between three and 30 months, passing through up to 12 molts.

Because the environments they inhabit are subject to extreme temperature changes, water bears have developed great resilience to severe conditions. They can survive desiccation and, experimentally at the other extreme, immersion into chemicals or liquid helium at –227°C. Their body covering, or cuticle, is not made of chitin like the exoskeleton of arthropods, but rather is composed of a water-permeable protein compound that is capable of shrinkage and swelling.

Thus, when drought occurs, tardigrades contract their bodies into a "barrel" (or "tun") state of dehydrated suspended animation, known as cryptobiosis, that can last for several years. When conditions improve, they swell up and return to life once more.

PSR

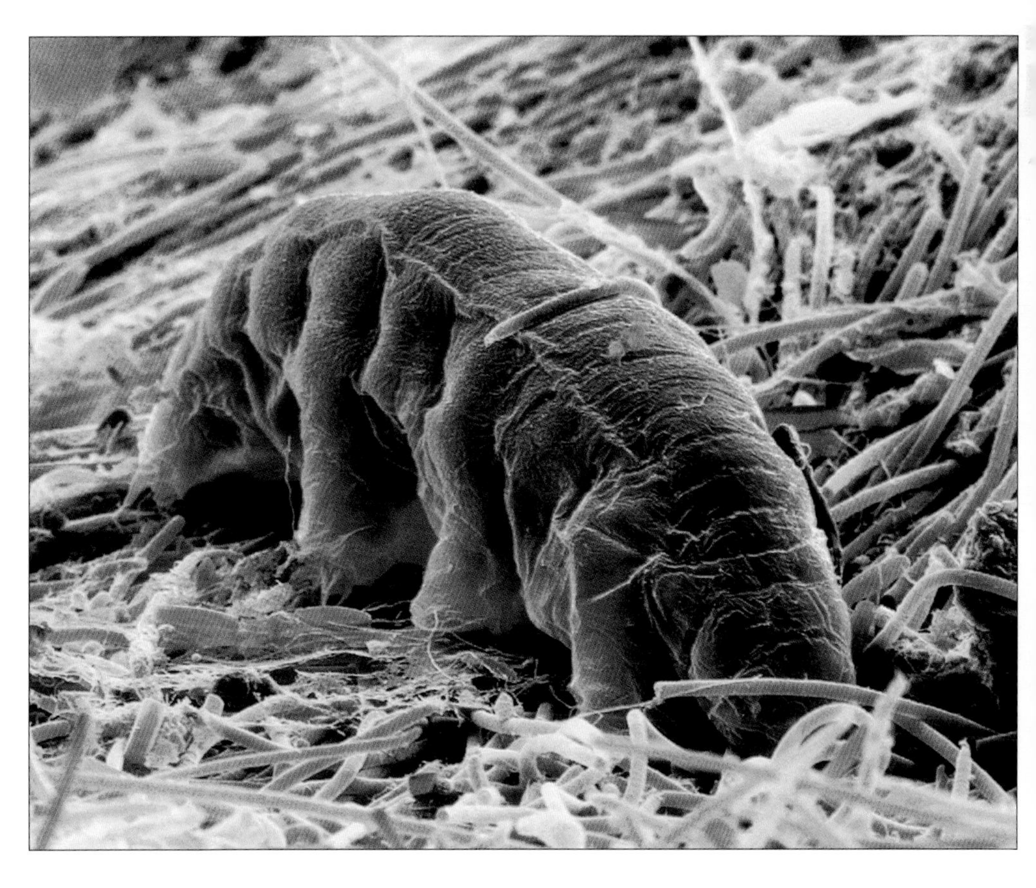

Above *An electron micrograph image of a water bear of the genus* Macrobiotus *shows clearly the characteristic segmented, cylindrical body form of the marine tardigrades.*

FACTFILE

WATER BEARS

Phylum: Tardigrada

Over 800 species in 2 classes and 15 families.

DISTRIBUTION Worldwide, particularly on damp moss, also in soil, freshwater, and marine sediments. Fossil record: species from the Cretaceous period, 136–6 5 million years ago.

SIZE Microscopic to minute – 50μm to 1 .2mm (under 0.05in).

FEATURES Fat-bodied, with a soft, chitinous cuticle, periodically molted; 4 pairs of stout, unjointed, clawed legs; sucking mouthparts with paired stylets; simple gut; segmented muscles; nerve cord on underside, with swellings (ganglia); no circulatory or respiratory organs; 3 "Malpighian" glands are possibly excretory; probably a fixed number of cells; only sense organs are 2 eyespots; remarkable suspended animation (anabiosis) during desiccation; sexes distinct, internal fertilization (some asexual reproduction).

CLASS EUTARDIGRADA

Species include: *Macrobiotus* species (family Macrobiotidae).

CLASS HETEROTARDIGRADA

Genera and species include: *Batillipes* species (family Batillipedidae); genus *Echiniscus* (family Echiniscidae).

Velvet Worms

THE SEGMENTED BODIES OF THE VELVET worms, or onycophorans, recall the body structure of annelid worms and arthropods (insects and crustacea), and a study of their features reveals that they lie somewhere between these two major invertebrate groups. For this reason, some biologists have tended to regard the onycophorans as "missing links" or "living fossils."

Voracious Predators

FORM AND FUNCTION

Unlike these annelids and arthropods, the onycophorans are now a small group with restricted distribution. They are almost all carnivores, feeding on smaller invertebrates such as insects like termites, annelid worms, and snails. They frequently capture prey by spraying it with the secretion of their slime glands, which discharge via the oral papillae. Jets of adhesive secretion can be discharged over remarkable distances, which harden on exposure to air and entangle the victims. The prey can then be dealt with at leisure, the jaws being effective in cutting it up prior to swallowing. The circulation system is like that of the arthropods, with blood bathing the organs as it passes through the various compartments of the haemocoel. The blood is pumped around the body by a dorsal heart. The transfer of gases to and from the tissues is the responsibility of many tubular tracheae, which open at the surface via small openings known as spiracles (as in the insects). The trachaea run inwards from the spiracles and pass down into the tissues. The transfer of oxygen into the body and of carbon dioxide out, is effected by diffusion and the blood is not involved directly in delivering respiratory gases. Although this system is very similar to that seen in the insects, it is thought to have evolved independently. Water balance and excretion are controlled by pairs of nephridia carried in most of the leg-bearing body segments. Onycophorans are dioecious (separate sexes). Mating between adults occurs but has rarely been observed. Some strange mechanisms exist for the acceptance of sperms by the female. For example, in one African species, a bundle of sperm is deposited on the external surface of the female, and the cuticle breaks down so the sperms can enter her body and pass into the blood stream and eventually pass into the ovaries where the eggs are fertilized. In some species, eggs with protective shells are laid and develop externally (ovipary). In others the embryo develops and is nurtured inside the mother's body, using food supplied by the mother, so that young onycophorans are born resembling miniature adults (vivapary). In yet others the eggs develop inside the mother's body depending on the yolk supply they contain (ovovivipary). The young then emerge as miniature adults. There are no larval phases in onycophorans. AC

FACTFILE

VELVET WORMS

Phylum: Onycophora

About 110 species in 2 families.

DISTRIBUTION Circumtropical (family Peripatidae) and temperate southern hemisphere (circumaustral) (family Peripatopsidae), freeliving, generally in leaf litter.

FOSSIL RECORD A number of fossil species date from the Lower Cambrian (530 mya), but unlike the modern species many of these species were marine.

SIZE About 5mm to 15 cm (0.4 – 6 inches).

FEATURES Caterpillarlike "worms" with poorly developed heads, and serially arranged lobelike walking legs. The head carries three pairs of ornaments or appendages, the well developed antennae, one pair of jaws, and a pair of oral or mouth papillae. The mouth opens by way of circular lips and there are small eyes at the base of the antennae. The outer surface of the body is covered with a thin cuticle formed from chitin, and appears segmented or annulated in many cases, as do the antennae and the legs. There is a true coelom, but it is restricted to the cavities of the gonads, as in the insects and crustacea, and the body organs, gut etc. are bathed in blood, which circulates through the spaces of the haemocoel.

FAMILY PERIPATIDAE
Genus *Peripatus.*

FAMILY PERIPATOPSIDAE
Genus *Peripatopsis.*

Above *Onycophorans occur in Africa, Southeast Asia, South America, and (most numerously) in Australia, where many new species are being identified. This is* Peripatus acacioi, *from Brazil.*

Below *A tropical velvet worm* (Macroperipatus torquatus) *from the island of Trinidad, West Indies, captures its prey by entangling it in a mass of sticky threads produced by its slime glands.*

Crabs, Lobsters, Shrimps, and Allies

WITH THEIR TOUGH EXOSKELETONS RICH in chitin – which require molting for growth to take place – and their paired jointed appendages, or limbs, arranged on segments down the body, crustaceans are clearly arthropods. They are primarily aquatic, mostly marine, but also include freshwater representatives.

This chapter deals with the more familiar – and by far the most speciose – order of crustaceans, the decapods, which comprise crabs, lobsters, and shrimps. In many cases, these are the larger crustaceans exploited worldwide by humans for food. However, also treated here are krill (order Euphausiacea) – the tiny organisms that, in their millions, constitute the diet of Earth's largest mammals, the baleen whales.

Segmented and Armored

ANATOMY AND PHYSIOLOGY

The numerous segments of crustaceans are usually grouped into three functional units (tagmata) – the head, thorax, and abdomen. The head is made up of five segments, while the thorax and abdomen (which often ends in a nonsegmental telson) vary in number of segments. Some of the smaller crustaceans contain up to 40 segments, whereas the decapods (shrimps, crabs, lobsters), the biggest crustaceans, have 19 – five in the head, eight in the thorax, and six in the abdomen. Some thoracic appendages may fuse to the back of the head, forming a cephalothorax, and a shield-like carapace may cover the head and part or all of the thorax: crustacean evolution is the story of reduction in the number of segments and their grouping into tagmata, with consequent specialization of the appendages for specialized roles. Although the fossil record goes back well over 500 million years, it reveals little about the origin of crustaceans.

The cuticle of crustaceans provides a tough, protective layer that also serves as an external skeleton (exoskeleton). The cuticle is rich in chitin, a polysaccharide similar in structure to the cellulose found in plants, and is secreted by the underlying single layer of living cells, the epidermis. The endocuticle of a decapod is comprised of an outer pigmented and inner calcified and uncalcified layers. The thin outermost epicuticle is secreted by tegumental glands.

The cuticle, which is initially soft and flexible, is progressively hardened inward from the epicuticle by the deposition of calcium salts and by sclerotization (tanning), involving the chemical cross-linking of cuticular proteins to form an impenetrable meshwork.

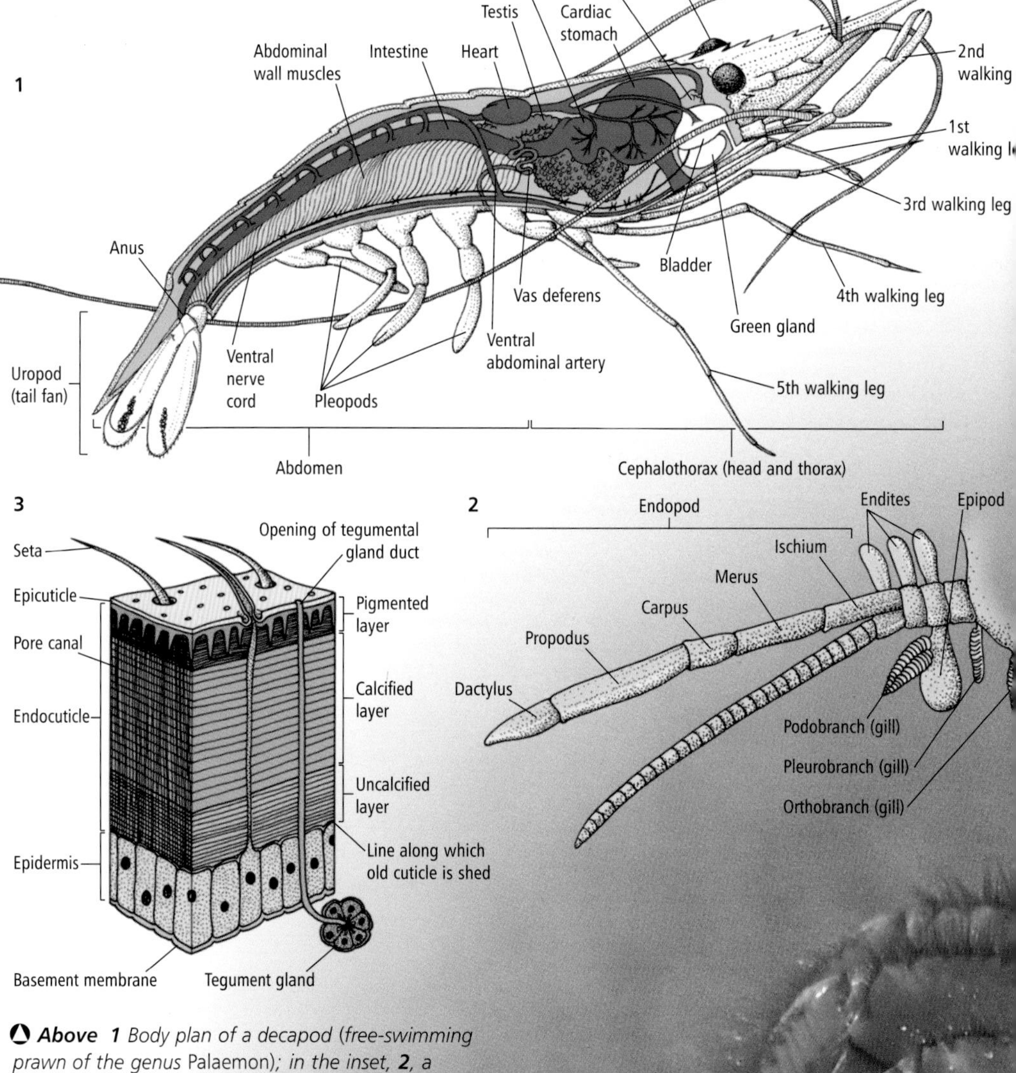

***Above** 1 Body plan of a decapod (free-swimming prawn of the genus* Palaemon*); in the inset, **2**, a typical thoracic appendage of this order of crustaceans is illustrated. The cross-sectional diagram **3** shows the different layers of the crustacean cuticle.*

***Right** The Ghost crab* (Ocypode quadrata) *inhabits sandy burrows on beaches along the east coast of the USA. Its large eyes are sensitive to changes in the intensity of light, and it is mainly nocturnal.*

A rigid exoskeleton prevents a gradual increase in size, and so a crustacean grows, stepwise, in a series of molts involving the secretion of a new cuticle by the epidermis and the shedding of the old. Before a molt, food reserves are accumulated and calcium is removed from the old cuticle and much of its organic material resorbed. It is then split along lines of weakness as the crustacean swells (usually by an intake of water) before hardening of the new cuticle. The rest of the old exoskeleton may then be eaten to reduce energy loss.

During and immediately after molting, the

temporarily defenseless crustacean hides to avoid predators. Some adult crabs cease molting, but many adult crustaceans molt throughout life.

Crustaceans' body organs lie in a bloodfilled space – the hemocoel, and the blood often contains hemocyanin, a copper-based respiratory pigment that bears oxygen and is equivalent to hemoglobin in vertebrates. The coelom, the major body cavity of many animals, is restricted to the inner cavity of the coxal glands – paired excretory and osmoregulatory organs each consisting of an end sac and a tubule that opens out at the base of the second antenna (antennal gland) or the second maxilla (maxillary gland).

The "typical" crustacean head bears five pairs of appendages. Each of the first two segments bears a pair of antennae – the first antennae (antennules) and the second antennae (antennae). All crustaceans have two pairs of antennae. The antennae have probably shifted in the course of evolution, since they now lie in front of the mouth, which may have been terminal in a pre-crustacean ancestor. They are typically sensory, although in some crustaceans they are employed in locomotion or even used by a male to clasp the female when mating.

Situated behind the mouth is the third segment bearing the mandibles, the main jaws developed from jawlike extensions (gnathobases) at the base of the appendages, while the remainder of the limb has been lost or reduced to a sensory palp. The fourth and fifth head segments bear the first maxillae (maxillules) and second, or true, maxillae, respectively; these accessory jaws assist mastication and are similarly derived by partial or complete reduction of the limb to a gnathobase.

The alimentary canal of crustaceans consists of a cuticle-lined foregut and hindgut, with an intermediate midgut giving rise to blind pockets (ceca), perhaps modified as a hepatopancreas combining the functions of digestion, absorption, and storage. The nervous system consists of a double ventral nerve cord, primitively with concentrations of the cord (ganglia) in each segment, but often further concentrated into a number of large ganglia.

FACTFILE

CRABS, LOBSTERS, SHRIMPS, ETC.

Superorder: Eucarida

About 8,600 species in 154 families.

Distribution Worldwide; marine, some in fresh water.

Fossil record Crustaceans appear in the Lower Cambrian, over 530 million years ago.

Size From tiny (0.5–5cm/0.2–2in) krill to lobsters 60cm (24in) long.

KRILL Order Euphausiacea
About 90 species in 2 families of marine pelagic filter feeders. Species include whale krill (*Euphausia superba*).

DECAPODS Order Decapoda
About 8,500 species in 151 families. Species include Common shrimp (*Crangon crangon*), cleaner shrimps (families Stenopodidae and Hippolytidae), American lobster (*Homarus americanus*), American spiny-cheeked crayfish (*Orconectes limosus*), Caribbean spiny lobster (*Panulirus argas*), Robber crab (*Birgus latro*), Christmas Island crab (*Geocarcoidea natalis*).

See Decapod Suborders and Infraorders ▷

From Swimming to Walking

EVOLUTION AND BREEDING BIOLOGY

The ancestors of crustaceans were probably small marine organisms living on the sea bottom but able to swim, with a series of similar appendages down the length of a body not divided into thorax and abdomen; all appendages would have been used in locomotion, feeding, and respiration. The limbs of early crustaceans were two-branched (biramous), with an endopod (inner limb) and exopod (outer limb) branching from the base (protopodite). The ancestral limb would have had two flat leaflike lobes, as in several living crustaceans of the order Branchiopoda (for example, fairy shrimps and brine shrimps). Such limbs are usually moved in rhythm to create swimming and feeding currents, food being passed along the underside of the body to the mouth – so promoting the evolution of jaws from limb gnathobases. Extensions from the outer side of the limb bases often act as respiratory surfaces.

One major line of crustacean evolution has produced large walking animals. The two-branched swimming and filtering leg has evolved into an apparently one-branched (uniramous) walking leg (stenopodium), strictly the endopod alone, the exopod being reduced and lost during larval development or previously in evolution. Such cylindrical legs have a reduced surface area and are not suitable as respiratory surfaces, which are particularly necessary in large crustaceans. Exites or extensions of the body wall at the base of the legs are therefore used as gills. The decapods (shrimps, crabs, lobsters) also show increased development of the head (cephalization). The first three pairs of thoracic limbs are adapted as accessory mouthparts (maxillipeds) and a carapace shields the cephalothorax.

Crustaceans typically have separate sexes and the fertilized egg as it divides shows a characteristic modified spiral cleavage pattern. Crustaceans usually develop through a series of larval stages of increasing size and numbers of segments with their associated limbs. The simplest larval stage is the nauplius larva, with three pairs of appendages – the first and second antennae and the mandibles. This larva occurs in many living crustaceans, often as a pelagic dispersal stage that swims and suspension-feeds with the three pairs of appendages. These limbs therefore have different functions in larva and adult, usually relinquishing their larval roles to other appendages that develop further down the body. Nauplius larvae are followed by a variety of larger larvae, according to the type of crustacean, although many crustaceans bypass the nauplius equivalent by developing within the egg.

The following account covers the more important crustacean groups. For a complete list of classes within the phylum Crustacea, see The 6 Classes of Crustaceans table on p.61.

Krill and Decapods – the Eucarids

SUPERORDER EUCARIDA; ORDER EUPHAUSIACEA

The planktonic krill and the decapods (shrimps, lobsters, crabs) are classified in the superorder Eucarida. They have a well-developed carapace, fused dorsally with all thoracic segments, and stalked eyes. Fertilized eggs are usually carried beneath the abdomen of the female and hatch as planktonic zoeae, with a large carapace, prominent eyes and well-developed thoracic appendages. Krill and primitive decapods, however, hatch as nauplius larvae.

Krill (order Euphausiacea) have primitive features of generalist members of the class Malacostraca. All are marine; the eggs hatch as nauplii. None of the thoracic appendages are adapted as

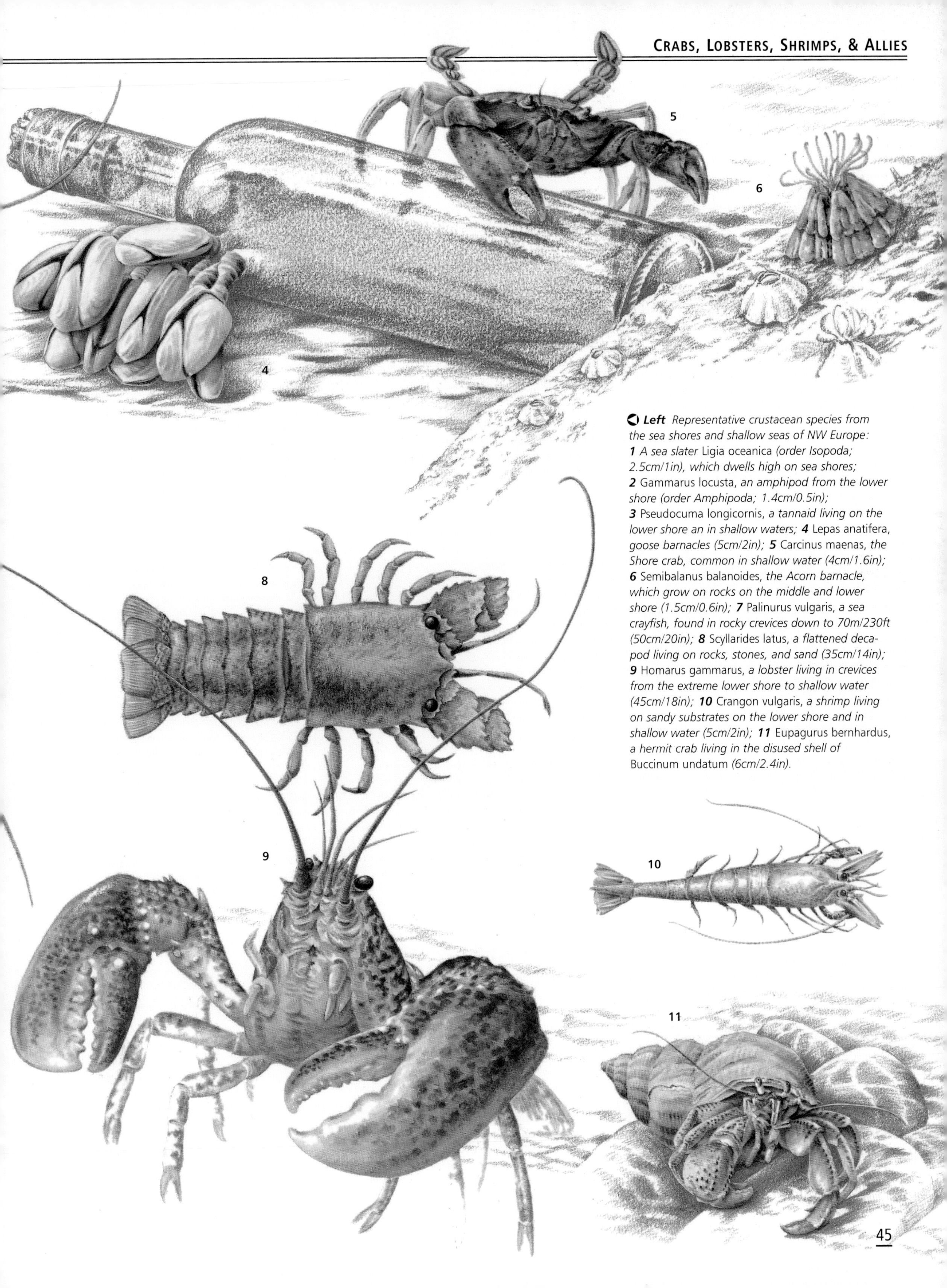

Left *Representative crustacean species from the sea shores and shallow seas of NW Europe:* **1** *A sea slater* Ligia oceanica *(order Isopoda; 2.5cm/1in), which dwells high on sea shores;* **2** Gammarus locusta, *an amphipod from the lower shore (order Amphipoda; 1.4cm/0.5in);* **3** Pseudocuma longicornis, *a tannaid living on the lower shore an in shallow waters;* **4** Lepas anatifera, *goose barnacles (5cm/2in);* **5** Carcinus maenas, *the Shore crab, common in shallow water (4cm/1.6in);* **6** Semibalanus balanoides, *the Acorn barnacle, which grow on rocks on the middle and lower shore (1.5cm/0.6in);* **7** Palinurus vulgaris, *a sea crayfish, found in rocky crevices down to 70m/230ft (50cm/20in);* **8** Scyllarides latus, *a flattened decapod living on rocks, stones, and sand (35cm/14in);* **9** Homarus gammarus, *a lobster living in crevices from the extreme lower shore to shallow water (45cm/18in);* **10** Crangon vulgaris, *a shrimp living on sandy substrates on the lower shore and in shallow water (5cm/2in);* **11** Eupagurus bernhardus, *a hermit crab living in the disused shell of* Buccinum undatum *(6cm/2.4in).*

Above *A Pederson cleaner shrimp* (Periclimenes pedersoni) *living among the tentacles of a sea anemone for protection. Cleaner shrimps rid the skin and mouths of fishes of parasites, which they eat.*

maxillipeds, and all have fully developed exopods. The thoracic appendages also bear epipodites, which take the form of external gills not covered by the carapace.

Most krill have luminescent organs, usually on the eyes, at the bases of the seventh thoracic limbs, and on the underside of the abdomen. They are probably used for communication in swarming and reproduction. Krill are pelagic and filter feed when phytoplankton conditions are suitable, otherwise preying on larger planktonic organisms. The first six pairs of thoracic appendages are adapted as a filter basket. Phytoplankton-rich seawater enters at the tips of the legs and is strained as it passes between the leg bases. Whale krill reach about 5cm (2in) long, dominate the zooplankton of the Antarctic Ocean, and are the chief food of many baleen whales (see Drifters and Wanderers).

In decapods, the first three pairs of thoracic appendages are adapted as auxiliary mouthparts (maxillipeds), theoretically leaving five pairs as legs (pereiopods) – hence deca poda "ten legs." (In fact, the first pair of pereiopods is often adapted as claws.) Historically decapods have been divided into swimmers (natantians) and crawlers (reptantians) – essentially the shrimps and prawns on the one hand and the lobsters, crayfish, and crabs on the other. More modern divisions rely on morphological characteristics, and decapods are now divided between the suborders Dendrobranchiata and Pleocyemata.

Members of the Dendrobranchiata are all shrimplike, characterized by their laterally compressed body and many-branched (dendrobranchiate) gills. Their eggs are planktonic and hatch as nauplius larvae. The Pleocyemata have gills that lack secondary branches, being platelike (lamellate) or threadlike (filamentous), and their eggs are carried on the pleopods of the female before hatching as zoeae.

Prawns and Shrimps

SUPERORDER EUCARIDA; ORDER DECAPODA

The terms shrimp and prawn have no exact zoological definition, and are often interchangeable.

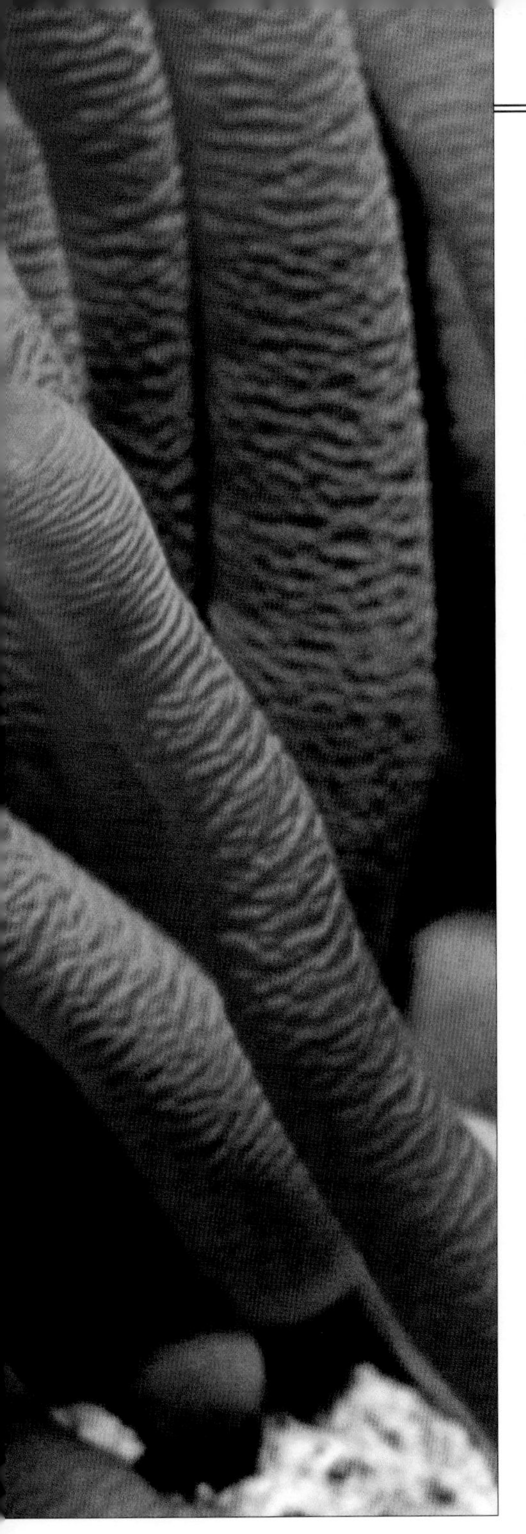

Above *The larger crustaceans are of great economic importance. Here, langoustines* (Nephrops norvegicus) *from the North Sea, a great seafood delicacy, await transport to market.*

Below *Tiny krill such as this (note the luminescent organs) make up a large proportion of the marine zooplankton that sustain filter-feeding whales. Krill themselves feed on diatoms, which they filter from the water.*

The most important "shrimp" families of the suborder Dendrobranchiata are the penaeids and sergestids, the Penaeidae including the most commercially important shrimps in the world (genus *Penaeus*), particularly dominating seas of tropical and subtropical latitudes.

Among the "shrimp" families in the much larger suborder Pleocyemata, the Stenopodidae comprise cleaner shrimps, which remove ectoparasites from the bodies of fish. Carideans (infraorder Caridea) are typified by possessing a second abdominal segment of which the lateral edges overlap the segments to either side. They are the dominant "shrimps" of northern latitudes although present throughout the world's oceans, from the intertidal zone down to the deep sea. Some shrimps (for example, *Macrobrachium* species) complete their entire life cycles in freshwater, but many shrimps living in rivers return to estuaries to breed and release their zoea larvae.

In addition to the totally pelagic species, many shrimps are essentially bottom-dwellers, only swimming intermittently. In adult shrimps the thoracic pereiopods are responsible for walking (and/or feeding) and the five pairs of abdominal pleopods for swimming. Flexion of the abdomen is used occasionally for rapid escape. Shrimp zoea larvae have two-branched thoracic appendages, the exopods being used for swimming; the pleopods take over the swimming function in postlarval and adult stages as the exopods are reduced. By the adult stage the "walking" pereiopods are single-branched.

Most pelagic shrimps are active predators feeding on crustaceans of the zooplankton, such as krill and copepods. Bottom-dwelling species are usually scavengers but range from catholic carnivores to specialist herbivores.

Shrimps usually have distinct sexes, although in some species, including *Pandalus borealis*, some of the females pass through an earlier male stage (protandrous hermaphroditism). Successful copulation usually requires molting by the female immediately before mating, when spermatophores are transferred from the male. Eggs are spawned

between two and 48 hours later and are fertilized by sperm from the spermatophore. The eggs of penaeids are shed directly into the water, but in most shrimps the eggs are attached to bristles on the inner branches (endopods) of pleopods 1–4 of the female. The incubation period usually lasts between one and four months, during which time the female does not molt. Most shrimp eggs hatch as zoeae and pass through several molts over a few weeks, to postlarval and adult stages. Most shrimps are mature by the end of their first year and typically live two to three years. (See also The Economic Importance of Crustaceans on page 62).

Lobsters and Freshwater Crayfish

ORDER DECAPODA; INFRAORDER ASTACIDEA

Lobsters and freshwater crayfish belong to a group known historically as the macrurans ("large-tails") but now divided into three infraorders – the Astacidea (lobsters, freshwater crayfish, scampi), the Palinura (spiny and Spanish lobsters), and the Thalassinidea (mud lobsters and mud shrimps). All have well-developed abdomens, compared with, for example, anomurans or brachyurans. For the group's importance as a source of food for humans, see Economic Importance of Crustaceans.

Lobsters and freshwater crayfish walk along the substratum on their four back pairs of single-branched thoracic legs (pereiopods). The first pair of pereiopods is adapted as a pair of formidable claws for both offense and defense. The abdomen bears pleopods, but they cannot move the heavy body and have become variously adapted for functions that include copulation or egg bearing.

Lobsters are marine carnivorous scavengers, usually living in holes on rocky bottoms. The commercially important American lobster reaches 60cm (2ft) in length and weighs up to 22kg (48lb). Lobsters are very long lived and they may survive 100 years.

MIGRATIONS ALONG THE SEABED

The spiny lobsters of the family Palinuridae (also sometimes called rock lobsters or sea crayfish) lack claws but have defensive spines on the carapace and antennae. Among the best known are two American species, *Panulirus argas*, found in the shallow seas off Florida and the Caribbean, and *P. interruptus*, which lives off California. (The European spiny lobster is *Palinurus elephas*, the generic name a curious anagram of that of the American species.) Spiny lobsters spend their days in rock or coral crevices, emerging by night to forage for invertebrate prey. They return from their wanderings to one of several dens within a feeding range of hundreds of meters, and after several weeks may move several kilometers to a new location. The homing capacity of *P. argas* is now known to involve the use of a magnetic map, for the lobsters can somehow derive enough positional information from Earth's magnetic field to determine the direction home. *Panulirus. argas* is the first invertebrate confirmed to use a magnetic map sense for navigation. Many spiny lobsters also take part in spectacular mass migrations. *Panulirus argas*, for example, will abandon its normal behavior in the fall, and as many as 100,000 individuals move south, by both day and night, in single files of up to 60 individuals (BELOW). They may cover 15km (9.3 miles) a day and travel for 50km (31 miles) at depths between 3 and 30m (10–100ft). The lobsters may be primed to migrate by annual changes in temperature and daylight length, but the immediate trigger is a sharp temperature drop associated with a fall storm – usually the first strong winter squall. The spiny lobsters maintain alignment in the queue by touch but may respond initially by sight, recognizing the rows of white spots along the abdomens of their companions. PSR

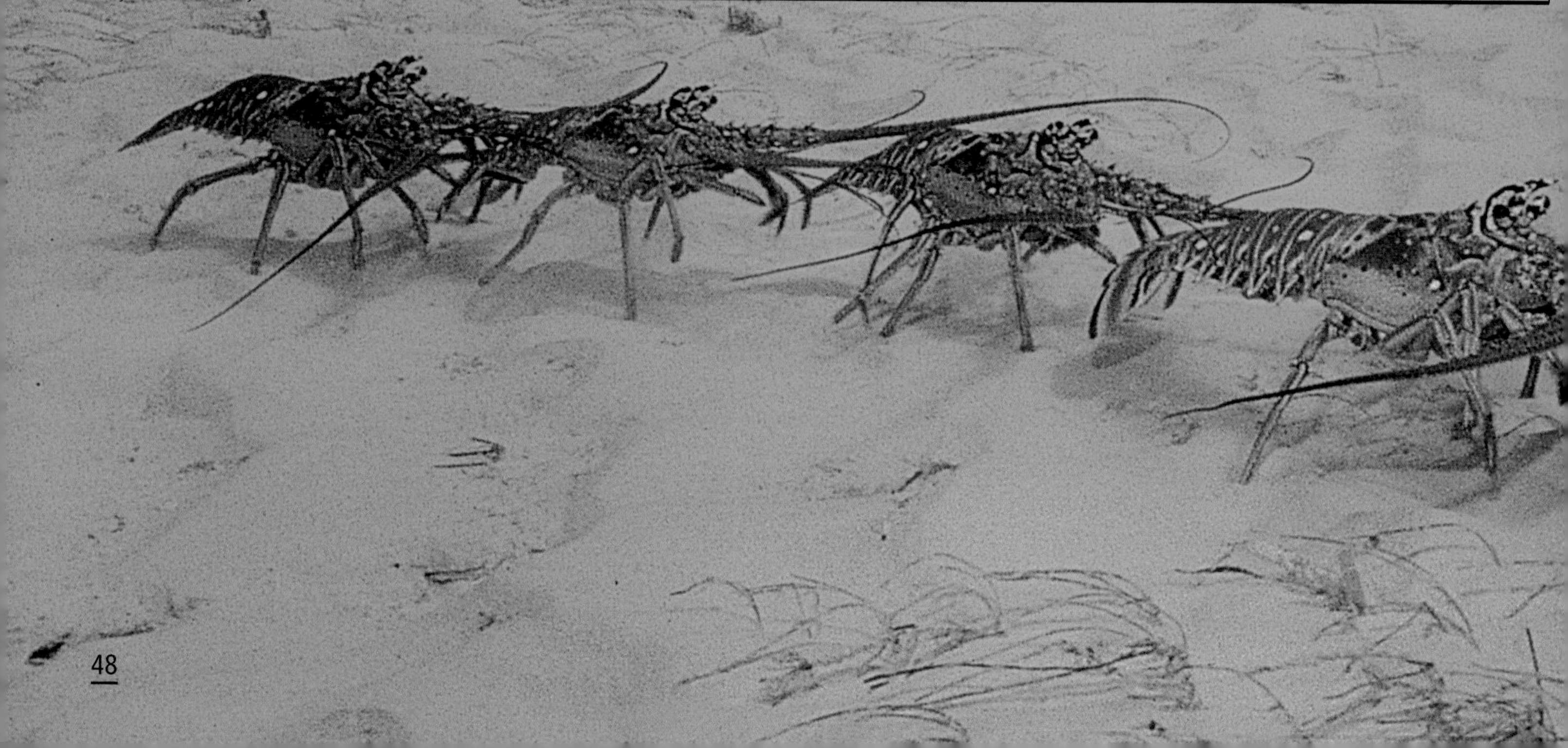

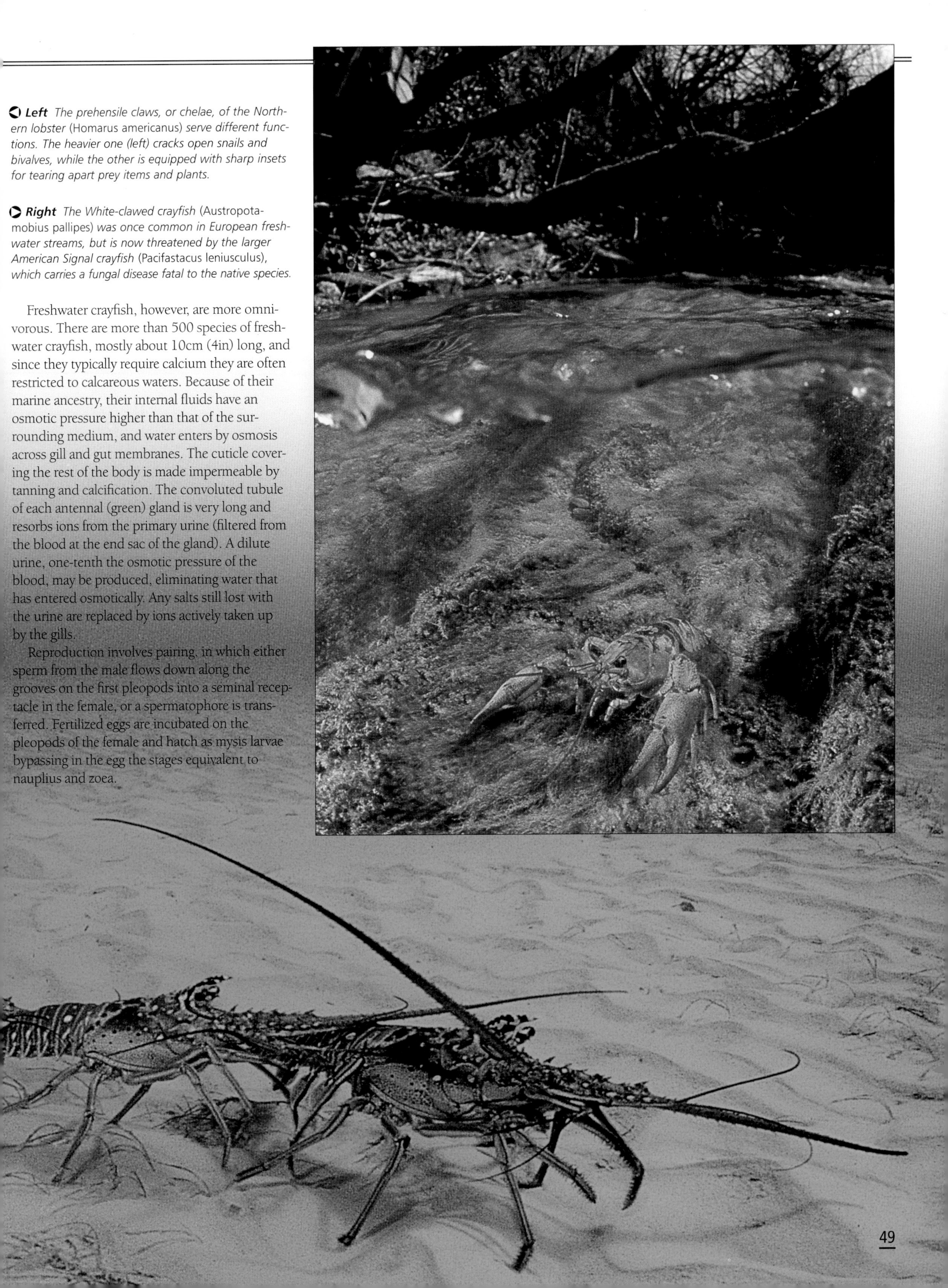

Left The prehensile claws, or chelae, of the Northern lobster (Homarus americanus) serve different functions. The heavier one (left) cracks open snails and bivalves, while the other is equipped with sharp insets for tearing apart prey items and plants.

Right The White-clawed crayfish (Austropotamobius pallipes) was once common in European freshwater streams, but is now threatened by the larger American Signal crayfish (Pacifastacus leniusculus), which carries a fungal disease fatal to the native species.

Freshwater crayfish, however, are more omnivorous. There are more than 500 species of freshwater crayfish, mostly about 10cm (4in) long, and since they typically require calcium they are often restricted to calcareous waters. Because of their marine ancestry, their internal fluids have an osmotic pressure higher than that of the surrounding medium, and water enters by osmosis across gill and gut membranes. The cuticle covering the rest of the body is made impermeable by tanning and calcification. The convoluted tubule of each antennal (green) gland is very long and resorbs ions from the primary urine (filtered from the blood at the end sac of the gland). A dilute urine, one-tenth the osmotic pressure of the blood, may be produced, eliminating water that has entered osmotically. Any salts still lost with the urine are replaced by ions actively taken up by the gills.

Reproduction involves pairing, in which either sperm from the male flows down along the grooves on the first pleopods into a seminal receptacle in the female, or a spermatophore is transferred. Fertilized eggs are incubated on the pleopods of the female and hatch as mysis larvae bypassing in the egg the stages equivalent to nauplius and zoea.

Squat Lobsters and Hermit Crabs

ORDER DECAPODA; INFRAORDER ANOMURA; SUPERFAMILY GALATHEOIDEA

Squat lobsters, hermit crabs, and mole crabs are part of a collection of probably unrelated groups intermediate in structure and habit between lobsters and crabs. The abdomen is variable in structure – often being asymmetrical or reduced and held in a flexed position (anom ura: odd-tailed – hence the scientific name Anomura given to these crustaceans). The fifth pair of pereiopods is turned up or reduced in size.

Hermit crabs are thought to have evolved from ancestors that regularly used crevices for protection, eventually specializing in the use of the discarded spiral shells of gastropod mollusks. The abdomen is adapted to occupy the typically righthanded spiral of such shells, although rarer lefthanded shells may also be used. The pleopods are lost at least from the short side of the asymmetric abdomen but those on the long side of females are adapted to carry fertilized eggs. At the tip of the abdomen, the uropods are modified to grip the interior of the shell posteriorly, and toward the front of the animal, thoracic legs may be used for purchase. One or both claws can block up the shell opening. The shell offers excellent protection and still affords mobility (see Househunting Hermits).

Not all species of hermit crabs live in gastropod shells. Some occupy tusk shells, coral, or holes in wood or stone. The hermit crab *Pagurus prideauxi* lives in association with a sea anemone, *Adamsia carciniopados*, whose horny base surrounds the crab's abdomen and greatly overflows the originally occupied shell. This hermit crab avoids the risk of being eaten when transferring from one shell to another, for the protecting anemone moves simultaneously with its associate. The crab is protected by the stinging tentacles of the anemone, which in turn profits from food particles which its crustacean partner releases into the water during feeding.

Hermit crabs live as carnivorous scavengers on sea bottoms ranging from the deep sea to the seashore. They have also taken up an essentially terrestrial existence in the tropics. Land hermit crabs range inland from the upper shore, often occupying the shells of land snails. The Coconut crab has abandoned the typical hermit crab form, and appears somewhat crablike, but with a flexed abdomen. It lives in burrows and holes in trees, which it can climb, feeding on carrion and vegetation, and can drink. Land hermit crabs have reduced the number of gills that would tend to dry out in the air and collapse under surface tension. The walls of the branchial chamber are richly supplied with blood, which enables them to act as lungs, and some species have accessory respiratory areas on the abdomen enclosed in the humid microclimate provided by the shell. Land hermit

HOUSEHUNTING HERMITS

Hermit crabs typically occupy the empty shells of dead sea snails (BELOW *Aniculus maximus* in a tun shell), so gaining protection while at the same time keeping their mobility. They can discriminate if offered a selection of shells, and will differentiate between shells of different sizes and species, to choose the one that fits their body most closely. Hermit crabs (family Paguridae) change shells as they grow, although in some marine environments there may be a shortage of available shells, and a hermit crab may be restricted to a shell smaller than is ideal. Some hermit crabs are aggressive and will fight fellows of their own species to effect a shell exchange: aggression is often increased if the shell is particularly inadequate.

Hermit crabs may encounter empty shells in the course of their day-to-day activity but the vacant shell is usually "spotted" by sight; the hermit crab's visual response increases with the size of an object and its contrast against the background. The hermit crab then takes hold of the shell with its walking legs and will climb onto it, monitoring its texture. Exploration may cease at any time, but if the shell is suitable the hermit crab will explore the shell's shape and texture by rolling it over between the walking legs and running its opened claws over the surface. Once the shell aperture has been located, the hermit crab will explore it by inserting its claws one at a time, occasionally also using its first walking legs. Any foreign material will be removed before the crab rises above the aperture, flexes its abdomen, and enters the shell backward. The shell interior is monitored by the abdomen as the crab repeatedly enters and withdraws. The crab will then emerge, turn the shell over, and re-enter finally. PSR

Left The Coconut crab (Birgus latro), *a hermit crab relative, is a terrestrial species from Pacific islands that is especially partial to coconuts. Scavenging widely, these large crabs are perfectly capable of scaling palms in search of their favorite food.*

Right Some 120 million Christmas Island crabs (Gecarcoidea natalis) *live on this small island in the Indian Ocean. At the beginning of each wet season, the adults migrate en masse from the forests to the coast, where they breed and release eggs into the sea.*

crabs are not fully terrestrial, for they have planktonic zoea larvae. The adults must return therefore to the sea for reproduction.

Squat lobsters are well named, for they have relatively large symmetrical abdomens flexed beneath the body. They typically retreat into crevices. Porcelain crabs are anomuran relatives of squat lobsters that look remarkably like true crabs. Mole crabs are anomurans that, by flexing the abdomen, burrow into the sand when waves break low on warm sandy shores. The first antennae form a siphon channeling a ventilation current to the gills and the setose (bristly) second antennae filter plankton.

True Crabs

ORDER DECAPODA; INFRAORDER BRACHYURA

There are 4,500 species of true crabs, all of which possess a much-reduced symmetrical abdomen held permanently flexed beneath the combined cephalothorax (brachyuran: short-tailed). The terminal uropods are lost in both sexes: four pairs of pleopods are retained on the female's abdomen to brood eggs and in the male only the front two pairs remain to act as copulatory organs. Crabs have a massive carapace extended at the sides, and the first of the five pairs of thoracic walking legs is adapted as large claws. Typically crabs are carnivorous walkers on the sea bottom.

The reduction of the abdomen has brought the center of gravity of the body directly over the walking legs, making locomotion very efficient and potentially rapid. The sideways gait assists to this end. The shape of a crab is therefore the ultimate shape in efficient crustacean walking. Mainly as a result, the brachyurans have enjoyed an explosive adaptive radiation since their origin in the Jurassic era (206–144 million years ago) and various anomurans (for example, mole crabs and porcelain crabs) approach or duplicate the crablike form.

Crabs live in the deep sea and extend up to and beyond the top of the shore. The blind crab *Bythograea thermydron* is a predator in the unique faunal communities surrounding deepsea hydrothermal vents, regions of activity in the earth's crust which emit hot sulfurous material 2.5km (1.6 miles) below the surface of the sea. On the other hand, crabs of the family Ocypodidae, such as the burrowing ghost and fiddler crabs, live at the top of tropical sandy and muddy

Above *One of the largest crabs on the Pacific coast of North America, the Puget Sound King crab* (Lopholithodes mandtii), *can grow up to 30cm (12in) across. It feeds on seastars and other echinoderms.*

Below *The true crab species* Caphyra laevis, *found around the Philippines, lives in a commensal relationship with the soft coral* Xenia elongata, *which it uses for camouflage.*

shores, and the distribution of the genus *Sesarma* extends well inland. Crabs have also invaded the rivers – the common British shore crab extends up estuaries – and in the tropics crabs of the family Potamidae and some Grapsidae are truly freshwater in habit.

Most crabs burrow to escape predators – descending back first into the sediment, and several typically remain burrowed for long periods. Of these, the Helmet crab has long second antennae, which interconnect with bristles to form a tube for the passage of a ventilation current down into the gill chamber of the buried crab. Some crabs have become specialist swimmers, with the last pair of thoracic legs adapted as paddles. The more terrestrial crabs, such as ghost crabs, are very rapid runners – their speed and night-time activity contributing to their common name. Other crabs, particularly spider crabs (see Japanese Spider Crabs), cover themselves with small plants and stationary animals (e.g. sponges, anemones, sea mats) for protective camouflage. Pea crabs may live in the mantle cavities of bivalve mollusks, feeding on food collected by the gills of the host, and female coral gall crabs become imprisoned by surrounding coral growth. The female crab is left with just a hole to allow entry of plankton for food and the tiny male for reproduction.

Decapod Suborders and Infraorders

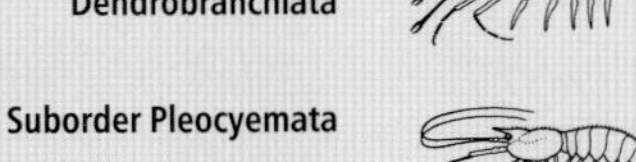

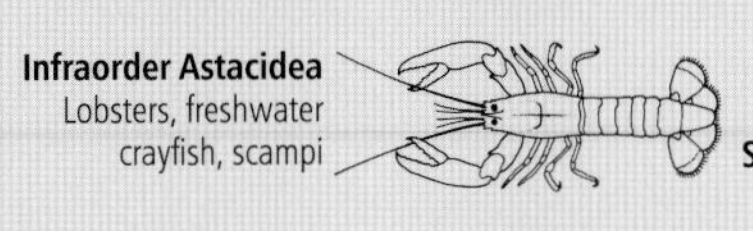

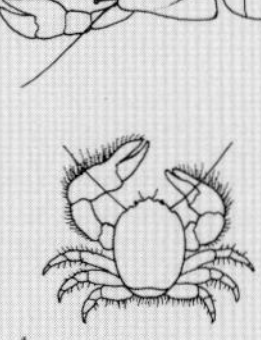

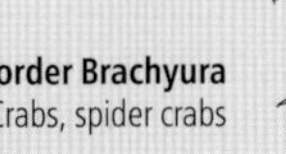

SUBORDER DENDROBRANCHIATA

7 families of free-swimmers with many-branched gills and free-floating eggs hatching as nauplius larvae; carnivorous, 0.5–20cm (0.2–8in) long. The family Penaeidae includes: Banana prawn (*Penaeus merguiensis*), Brown shrimp (*P. aztecus*), Giant tiger prawn (*P. monodon*), Green tiger prawn (*P. semisulcatus*), Indian prawn (*P. indicus*), Kuruma shrimp (*P. japonicus*), Pink shrimp (*P. duorarum*), White shrimp (*P. setiferus*), and Yellow prawn (*Metapenaeus brevicornis*).

SUBORDER PLEOCYEMATA

144 families; gills platelike (lamellate) or threadlike (filamentous), not many-branched; eggs carried by female before hatching as zoeae.

BANDED OR CLEANER SHRIMPS
INFRAORDER STENOPODIDEA

2 families of bottom-dwelling cleaners of fish, about 5cm (2in) long. Species include: *Stenopus* species.

SHRIMPS, PRAWNS, PISTOL SHRIMPS
INFRAORDER CARIDEA

36 families of marine, brackish, and fresh-water swimmers and walkers; predatory scavengers, dominant shrimps in N oceans; 0.5–20cm (0.2–8in) long. Species include: Brown shrimp (*Crangon crangon*), Pink shrimp (*Pandalus montagui*), Brown pistol shrimp (*Alpheus armatus*), Sand shrimp (*Crangon septemspinosa*), Pederson cleaner shrimp (*Periclimenes pedersoni*), Common shore shrimp (*Palaemonetes vulgaris*).

LOBSTERS, FRESHWATER CRAYFISH, SCAMPI
INFRAORDER ASTACIDEA

7 families of mostly marine, some fresh-water, bottom walkers, hole dwellers; predatory scavengers; up to 60cm (2ft) long. Species include: American lobster (*Homarus americanus*), European lobster (*H. gammarus*), and Dublin Bay prawn or Norway lobster (*Nephrops norvegicus*).

SPINY AND SPANISH LOBSTERS
INFRAORDER PALINURA

4 families, of marine bottom walkers or hole dwellers; often spiny but lack permanent rostrum (frontal spine) of Astacidea; predatory scavengers up to 60cm (2ft) long. Species include: American spiny lobsters (*Panulirus argus* and *P. interruptus*) and European spiny lobster (*P. elephas*).

MUD SHRIMPS, MUD LOBSTERS
INFRAORDER THALASSINIDEA

11 families; bear a carapace up to 9cm (3.5in) long. Live in shallow water in deep burrows in sand or mud. Species include: *Callianassa subterranea*.

ANOMURANS
INFRAORDER ANOMURA

13 families, including:

Hermit crabs
Superfamily Paguroidea
6 families; carapace to 19cm (7.5in); inhabiting marine and terrestrial gastropod shells; scavengers feeding on detritus. Species include: *Pagurus bernhardus*, stone crabs (*Lithodes* species), land hermit crabs and coconut crab (*Birgus latro*).

Squat lobsters, Porcelain crabs
Superfamily Galatheoidea
4 families with carapace to 4cm (1.6in); marine, in holes, under stones; scavenging detritivores. Species include: squat lobster (*Galathea* species), porcelain crab (*Porcellana platycheles*).

Mole crabs
Superfamily Hippoidea
1 family with carapace to 5cm (2in), marine sand burrowers, filter feeders. Species include: mole crab (*Emerita talpoida*).

CRABS, SPIDER CRABS
INFRAORDER BRACHYURA

71 families with carapace up to 45cm (18in); mostly bottom dwellers, walking on seabed, but some parasitic/commensal with fish, some burrowers, some swimmers; predaceous scavengers. Groups, families and species include: 7 families of spider crabs, including *Macrocheira kaempferi* (family Majidae), Edible crab (*Cancer pagurus*), and Dungeness crab (*C. magister*) (family Cancridae); helmet crabs (family Corystidae), including *Corystes cassivelaunus*; swimming crabs (family Portunidae), including Blue crab (*Callinectes sapidus*), Henslow's swimming crab (*Polybius henslowi*), and Shore or Green crab (*Carcinus maenas*); Chinese mitten crab (*Eriocheir sinensis*), *Sesarma* species (family Grapsidae); pea crabs (family Pinnotheridae), including *Pinnotheres* species; and ghost crabs (*Ocypode* species) and fiddler crabs (*Uca* species) (family Ocypodidae).

Most crabs are carnivorous scavengers, although the more terrestrial ones may eat plant matter. The fiddler crabs process sand or mud in their mouthparts, scraping off the nutritious microorganisms with specialized spoon- or bristle-shaped setae.

Reproduction involves copulation. The second pair of pleopods on the male's abdomen acts like pistons within the first pair to transfer sperm to the female for storage. The eggs are fertilized as they are laid and are held under the broad abdomen of the female. They hatch as prezoea larvae, molting immediately to the first of several planktonic zoea stages. Zoea larvae have a full complement of head appendages, but only the first two pairs of thoracic appendages (destined to become the first two or three pairs of adult maxillipeds), used by the zoea for swimming. After several molts the megalopa stage is reached, with abdominal as well as thoracic appendages. It settles on the sea bottom and metamorphoses into a crab. PSR

JAPANESE SPIDER CRABS

Spider crabs belong to a superfamily (7 families) of true (brachyuran) crabs with long legs and a superficial resemblance to spiders. Included in their number is the world's largest crustacean, the giant spider crab (*Macrocheira kaempferi*), whose Japanese name is *Takaashigani* – the "tall-leg crab".

This remarkable animal may measure up to 8m (26.5ft) between the tips of its legs when they are splayed out on either side of the body. Its claws may be 3m (10ft) apart when held in an offensive posture. The main body (cephalothorax) of the crab is, however, relatively small and would not usually exceed 45cm (18in) in length or 30cm (12in) in breadth.

Giant spider crabs have a restricted distribution off the southeast coast of Japan. They live on sandy or muddy bottoms between 30–50m (100–165ft) deep. They have poor balance and so live in still waters, hunting slow-moving prey that includes other crustaceans as well as echinoderms, worms, and mollusks. Little detail is known of their life history, but they probably pass through zoea and megalopa larval stages before they metamorphose into juvenile crabs. Large specimens are probably more than 20 years old.

The crabs are caught relatively infrequently, but they are used for food. Because of their size they command respect, and their nippers can inflict a nasty wound.

The existence of the giant crabs was first reported in Europe by Engelbert Kaempfer, a physician for the Dutch East India Company who visited Japan in 1690, in his *History of Japan* published in English in 1727 (although he himself died in 1716). The crabs were given their Latin name in honor of Kaempfer in the 19th century by C. I. Temminck, the director of the Leiden Museum, which received much of the natural history collections of the Dutch East India Company. PSR

Other Crustaceans

ASIDE FROM THE MORE FAMILIAR DECAPODS, the crustaceans also include a host of smaller, less commonplace arthropods, some of which are major components of plankton. Certain branchiopod genera, such as Daphnia *and* Artemia, *are well-known to aquarium hobbyists as commercial fish food.*

The small crustaceans are predominantly aquatic, and mostly marine, but include freshwater representatives. One group, the woodlice, has even successfully colonized land.

Branchiopods

CLASS BRANCHIOPODA

Fairy shrimps and water fleas are among the branchiopods, small, mostly freshwater, crustaceans. They have leaflike trunk appendages used for swimming and filter-feeding. Flattened extensions (epipodites) from the first segment (coxa) of these limbs act as gills, giving the class its scientific name of "gill legs" (Branchiopoda).

Fairy shrimps (order Anostraca) live in temporary freshwater pools and springs, (some, the brine shrimps, in salt lakes), typically in the absence of fish. Unlike other branchiopods they lack a carapace (hence an ostraca). They have elongated bodies with 20 or more trunk segments, many bearing appendages of the one type. They are usually about 1cm (0.4in) long, but some reach 10cm (4in). Fairy shrimps swim upside-down, beating the trunk appendages in rhythm, simultaneously filtering small particles with fine slender bristles (setae) on the legs. Collected food particles are then passed along a groove on the underside to the mouth.

When mating, a male fairy shrimp clasps the female with its large second antennae. The female lays her eggs into a brood sac. The eggs on release are extremely resistant to drying out (desiccation). Some eggs will hatch when wetted, but others require more than one inundation – a successful evolutionary strategy, ensuring that populations are not wiped out when insufficient rain falls to maintain the pool long enough to complete the fairy shrimp life cycle. Fairy shrimps hatch as nauplii and grow to maturity in as little as one week.

Brine shrimps are found in salt lakes, the brackish nature of which eliminates possible predators. The eggs similarly resist drying out and are sold as aquarium food. *Artemia salina* has a remarkable resistance to salt, surviving immersion in saturated salt solutions. The gills absorb or excrete ions as appropriate and this brine shrimp can produce a concentrated urine from the maxillary glands.

Water fleas (suborder Cladocera) are laterally compressed with a carapace enclosing the trunk but not the head, which projects on the underside as a beak. Overall length is just 1–5mm (0.04–0.2in). The powerful second antennae are used for swimming. The trunk is usually reduced to about five segments, of which two may bear filtering appendages. Most water fleas, including *Daphnia* species, live in freshwater and filter small particles, but some marine cladocerans are carnivorous.

Water fleas brood their eggs in a dorsal chamber and asexual reproduction (parthenogenesis) is common. Populations may consist only of females reproducing parthenogenetically for several generations until a temperature change or limitation of food supply induces the production of males. The brood chamber, which now encloses fertilized eggs, may be cast off at the next molt and can withstand drying, freezing, and even passage through the guts of vertebrates. Some *Daphnia* species show cyclomorphosis, cyclic seasonal changes in morphology, often involving a change in head shape.

In the clam shrimps (suborder Spinicaudata), most of which are some 1cm (0.4in) in length, the remarkably clamlike bivalve carapace encloses the whole body, but in tadpole shrimps (order Notostraca) part of the trunk extends behind the large shieldlike carapace and the overall body length may reach 5cm (2in). Either may be found with fairy shrimps in temporary rain pools.

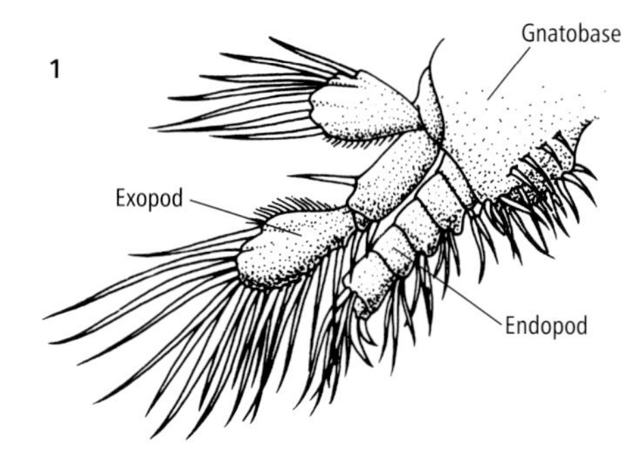

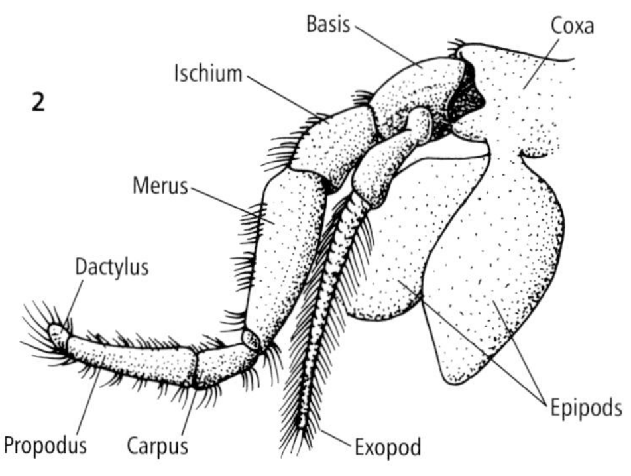

Remipedes

CLASS REMIPEDIA

The first extant remipede crustacean (*Speleonectes lucayensis*) was described only in 1981, after its discovery in an anchialine cave on Grand Bahama Island. Since then, another eleven species have also been found in such caves, which lack a surface connection to the sea and appear to act as a reservoir for relict crustacean taxa. Living remipedes are indeed very similar to a fossil arthropod (*Tesnusocaris*) from the Carboniferous period, and can therefore be considered as "living fossils."

Remipedes may therefore have a body form closest to that of the earliest crustaceans. They are small and elongate with 20 to 30 similar body segments, each with biramous appendages beating in metachronal rhythm. They are believed to be carnivorous, injecting digestive fluids into their prey, before sucking up the semifluid remains.

Cephalocarids

CLASS CEPHALOCARIDA

Cephalocarids were first described in 1955, and all those so far discovered are minute inhabitants of marine sediments, feeding on bottom detritus. The body, under 4mm (0.2in) in length, is not divided into thorax and abdomen, and the leaf-shaped body appendages are very similar to each other and indeed to the second maxillae.

Branchiuran fish lice and Tongue worms

CLASS MAXILLOPODA; SUBCLASS BRANCHIURA

Branchiurans are ectoparasites living on the skin or in the gill cavities of fish, feeding on the blood and mucus of the host, and are therefore commonly called fish lice. They have a large shieldlike carapace for protection, and have large claws on the antennules and large suckers (modified first maxillae) with which they hang on tight.

Tongue worms (pentastomids) are the most unlikely looking crustaceans and in the past have

Above *The Common water flea* (Daphnia pulex) *is widespread, and can be found in many eutrophic (oxygen-poor, nutrient-rich) environments such as stagnant ponds and ditches, feeding on bacteria and fine algae.*

Left *Crustacean limbs exhibit great diversity: these diagrams show the thoracic appendages of* **1** Hutchinsoniella *(class Cephalocarida) and* **2** *a malacostracan of the superorder Syncarida.*

been put into their own phylum. It is now believed, however, particularly from DNA evidence, that they are indeed crustaceans, probably related to branchiurans. Tongue worms are wormlike parasites in the respiratory tracts of carnivorous vertebrates, typically reptiles, although six species are found in mammals, including man, and two in birds. The life cycle usually involves a herbivorous vertebrate intermediate host for the developing larvae, such as a fish or a rabbit.

The front end of a tongue worm has five short protuberances bearing, respectively, the mouth and four hooks used by the animal to anchor itself and in feeding. (Once it was thought that there was a mouth on each of the five – hence the name pentastomid.) The mouth is modified for sucking

FACTFILE

CRUSTACEANS

Subphylum: Crustacea

About 39,000 species in 6 classes.

Distribution Worldwide, primarily aquatic, in seas, some in freshwater, woodlice on land.

Fossil record Crustaceans appear in the Lower Cambrian, over 530 million years ago.

Size From microscopic (0.15 mm/0.006in) parasites to goose barnacles 75cm (30in) and lobsters 60cm (24in) long.

Features 2 pairs of antennae; typically segmented body covered by plates of chitinous cuticle subject to molting for growth; body divided into head, thorax, and abdomen; head and front of thorax may fuse to form cephalothorax, often covered by shieldlike carapace; segments from 19 in decapods to 40 in some smaller species; compound eyes; paired appendages on each segment typically include the antennae, main and 2 accessory pairs of jaws (mandibles, maxillules, maxillae), and typically 2-branched limbs on thorax (called pereiopods) and abdomen (generally called pleopods) usually jointed, variously adapted for feeding, swimming. walking, burrowing, respiration, reproduction, defense (pincers); breathing typically by gills or through body surface; body organs in blood-filled hemocoel; double nerve cord on underside; foregut, midgut, hindgut; sexes typically separate; development of young via series of larval stages.

BRANCHIOPODS Class Branchiopoda
Comprises around 1,001 species in 23 families, 3 orders, and 2 subclasses. Mostly small freshwater filter feeders.

REMIPEDES Class Remipedia
2 species in 2 families and 1 order Nectiopoda. Blind swimmers in marine (anchialine) caves without surface connection to the sea. **Length** 5mm (0.4in).

CEPHALOCARIDS Class Cephalocarida
9 species in 4 genera and 1 family and 1 order. Bottom-dwellers feeding on deposited detritus.

MAXILLOPODS Class Maxillopoda
Over 9,705 species in 6 subclasses.

MUSSEL SHRIMPS AND SEED SHRIMPS
Class Ostracoda
5,650 species in 43 families and 4 orders. **Length** Small, 1–3mm 0.04–0.12in), mostly bottom dwelling, with two-valved carapace.

MALACOSTRACANS Class Malacostraca
23,000 species in 497 families and 16 orders.

See The 6 Classes of Crustaceans ▷

the host's blood. The worm shape is an adaptation to life in confined passages that must not be blocked, or the host will suffocate. The adults move little, and the nervous system is reduced, as befits a parasite. There are no excretory or respiratory organs, or heart. Fertilized eggs are released in the host's feces or by the host sneezing, and lie in vegetation before being ingested by a herbivore.

In the case of *Linguatula serrata*, the eggs are eaten by a rabbit or hare and the larvae hatch out under the action of digestive juices. The larva bores through the gut wall and is carried in the blood to the liver. Here it forms a cyst and grows through a series of molts to approach adult form. If the rabbit is eaten by a dog or wolf, the larva is released and passes up the dog's esophagus to the pharynx, and so to its adult location in the host's nasal cavity. *Porocephalus crotali* lives in snakes, with mice as intermediate hosts, and other tongue worms live in crocodiles with fish intermediate hosts. Ancestral pentastomids may have been parasitic on fish much as branchiurans now. It is a plausible evolutionary step for the parasite to have been stimulated to leave this host when eaten by a crocodile, to take up residence in the mouth of the predator.

Above *Exposed in the intertidal "barnacle zone" of the shore, these* Chthalamus entatus *barnacles keep their shell plates firmly closed to avoid dehydration.*

Barnacles

CLASS MAXILLOPODA; SUBCLASS THECOSTRACA; INFRACLASS CIRRIPEDIA

Barnacles are sedentary marine crustaceans, permanently attached to the substrate. For protection barnacles have carried calcification of the cuticle to an extreme and have a shell resembling that of a mollusk. The shell is a derivative of the cuticle of the barnacle head and encloses the rest of the body. Stalked goose barnacles, which commonly hang down from floating logs, are more primitive than the acorn barnacles that abut directly against the rock and dominate temperate shores. (For barnacles' economic impact, see Cemented Crustaceans).

Barnacles (infraclass Cirripedia) feed with six pairs of thoracic legs (cirri), which can protrude through the shell plates to filter food suspended in the seawater. Goose barnacles trap animal prey, but most acorn barnacles have also evolved the ability to filter fine material, including phytoplankton and even bacteria, with the anterior cirri.

Barnacles are hermaphrodites. They usually carry out cross-fertilization between neighbors. Fertilized egg masses are held in the shell until release as first-stage nauplii. Indeed it was J. Vaughan Thompson's observations (1829) of barnacle nauplius larvae of undoubted crustacean pedigree that removed lingering suspicions regarding the possible molluskan nature of barnacles. There are six nauplius stages of increasing size that swim and filter phytoplankton over a period of a month or so, before giving rise to a nonfeeding larva – the cypris larva, named for its similarity to the mussel shrimp genus *Cypris*. This settlement stage of the life cycle is able to drift and swim in the plankton before alighting and choosing a settlement site in response to environmental factors, which the larva detects by an array of sense organs.

Barnacles colonize a variety of substrates, including living animals such as crabs, turtles, sea snakes, and whales – with a moving host the barnacle does not need to use energy in beating its cirri. Some barnacles have evolved to become parasites, which bear little similarity to their free-living relatives, except as larvae. Members of the superorder Rhizocephala (literally "root-headed") parasitize decapod crustaceans.

Left *An exoparasitic fish louse attached to its host, a Variegated lizardfish* (Synodus variegatus). *Parasitic copepods grow far larger than their free-living relatives.*

Copepods

CLASS MAXILLOPODA; SUBCLASS COPEPODA

The class Copepoda includes 220 families of mostly minute sea- and freshwater inhabitants, which provide a major source of food for fish, mollusks, crustaceans, and other aquatic animals. As dominant members of the marine plankton, copepods of the order Calanoida may be among the most abundant animals in the world. On the other hand, members of the order Harpacticoida are common inhabitants of usually marine sediments, and species of the order Cyclopoida may be either planktonic or bottom-dwelling (benthic) in the sea or in freshwater. Some copepods are parasitic, infesting other invertebrates, fish (often as fish lice on the gills), and even whales. Some of them attain 30cm (12in) in length.

Free-living copepods are small and capable of rapid population turnover. The large first antennae may be used for a quick escape, but more usually they act as parachutes against sinking, while the thoracic limbs are the major swimming organs.

Many calanoid copepods can filter feed on phytoplankton, using the feathery second maxillae to sieve a current driven by the second antennae, which beat at about 1,000 strokes per minute. The setae on the second maxillae are adapted to trap a particular size of microscopic plant organisms (phytoplankton) and the first pair of thoracic limbs (maxillipeds) scrape off the filtered material.

Calanoid copepods cannot survive in the marine planktonic ecosystem by filtering alone. Filter feeding is an energy-sapping process that requires the driving of large volumes of water through a fine sieve. There must therefore be a minimum level of phytoplankton in the sea simply to repay the cost of filtering. Phytoplankton populations in temperate oceans reach such concentrations only for brief periods in the year – in the North Atlantic, during the spring bloom and perhaps again in the fall. For much of the year, therefore, the calanoid copepod feeds raptorially – seizing large prey items including other members of the zooplankton. During the rich spring "bloom," filtering becomes worthwhile and the copepods can pass through several generations, multiplying very quickly, before returning to low metabolic tickover, often using fat reserves for the rest of the year. The number of copepod generations per year is related to the time period of the phytoplankton bloom.

Copepods pair, rather than copulate directly, the male transferring a packet of sperm (spermatophore) to the female, perhaps in response to a "chemical message" (pheromone) from her.

CEMENTED CRUSTACEANS

Barnacles are the most successful marine fouling organisms, and more than 20 percent of all known species have been recorded living on artificial objects, including ships, oil rig legs (RIGHT), and power station outlets. Goose barnacles (*Lepas* sp.) once attached themselves in enormous clumps to slow-moving sailing ships. They have been supplanted now by acorn (or rock) barnacles (*Balanus* sp.) on motor-powered vessels, although stalked whale barnacles (*Conchoderma* sp., which grow, as their common name indicates, on whales' skins) are also found on the hulls of supertankers.

The minute planktonic (nauplius) larvae of barnacles, which feed on phytoplankton, are dispersed in the sea and build up fat reserves to support the non-feeding settlement (cypris) stage through further dispersal, site selection, and metamorphosis. The cypris larva, a motive pupa according to Charles Darwin in the 1850s, has a low metabolic rate and lasts as long as a month before alighting on a chosen substrate. The cypris then walks, using sticky secretions on the adhesive disks of the antennules, responding to current strength and direction, light direction, contour, surface roughness, and the presence of other barnacles, as it monitors the suitability of a site for future growth. Finally, cypris cement is secreted from paired cement glands down ducts to the adhesive disks so that they become embedded permanently, whereupon the cypris metamorphoses to the juvenile barnacle, which feeds and grows, developing its own cement system.

Barnacles on a ship's hull create turbulence and drag, slowing the vessel down and increasing its fuel consumption. To counter this, antifouling paint is used, which releases toxin in sufficient concentration to kill settling larvae or spores. Copper, the most common antifouling agent, is very toxic and must be released at a rate of 10 micrograms per sq cm per day to prevent barnacle settlement and growth.

Barnacles accumulate heavy metals, specifically zinc as detoxified granules of zinc phosphate. Barnacles are therefore suitable monitors of zinc availability in the marine environment, their high concentrations being easily measurable. Thames estuary barnacles, for example, have been found to contain an almost incredible zinc concentration of 15 percent of their dry weight. PSR

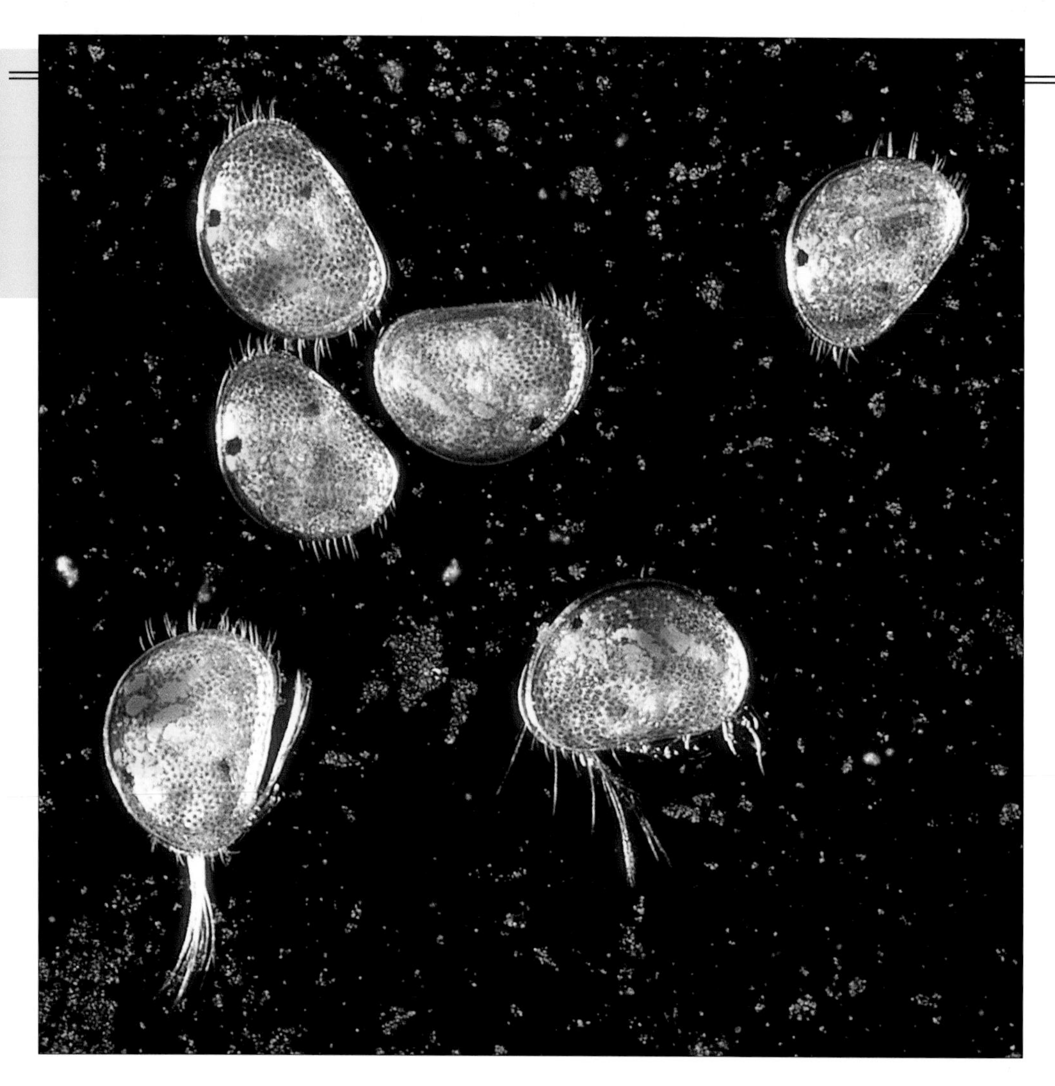

Left *The soft, vulnerable bodies of these freshwater ostracods (mussel shrimps) are protected by their hard, rounded carapaces.*

The eggs are fertilized as they are laid into egg sacs carried by the female. Eggs hatch as nauplii and pass through further nauplius and characteristic copepodite stages before adulthood.

The larval stages ensure the dispersal of parasitic copepods. Copepods that are parasitic on the exterior of their host (exoparasites) may show little anatomical modification, but endoparasites often consist of little more than an attachment organ and a grossly enlarged genital segment with large attendant egg sacs.

Mussel shrimps

CLASS OSTRACODA

Mussel shrimps or ostracods (class Ostracoda) are small bivalved crustaceans widespread in sea and fresh water. Some are planktonic, but most live near the bottom, plowing through the detritus on which they feed. Algae are another favored food.

The rounded valves of the carapace completely enclose the body, which consists of a large head and a reduced trunk, usually with only two pairs of thoracic limbs. The antennae are the main locomotory organs. Both pairs may be endowed with long bristles (setae) to aid propulsion when swimming, or the first antennae may be stout for digging or even hooked for climbing aquatic vegetation.

Malacostracans

CLASS MALACOSTRACA

Including more than half of all living crustacean species, the class Malacostraca is very important in marine ecology and has also successfully invaded freshwater and land habitats.

Malacostracans are modifications of a shrimp-like body plan. The thorax consists of eight segments bearing limbs, of which up to three of the front pairs are accessory mouthparts or maxillipeds. A carapace typically covers head and thorax, though this may have been lost in some peracarids (see below). Members of the primitive superorder Syncarida also lack a carapace but this absence may be of more ancient ancestry: the anaspidaceans are bottom-dwelling feeders on detritus in Southern Hemisphere fresh waters, and the bathynellaceans are minute, blind detritivores living in freshwater sediments.

Left *Peering out from its burrow under a coral head, a mantis shrimp* (Odontodactylus scyallarus) *waits to strike at its next prey.*

Most adult malacostracans are bottom-dwellers (benthic), and the single-branched (uniramous) thoracic legs are adapted for walking. Some malacostracans can swim using pleopods, the limbs of the first five abdominal segments. The appendages of the sixth abdominal segment are directed back as uropods to flank the terminal telson and form a tail fan.

The gills of malacostracans are usually situated at the base of the thoracic legs in a chamber formed by the carapace, and are aerated by a current driven forward by a paddle on the second maxillae. Malacostracans typically ingest food in relatively large pieces. The anterior part of the stomach is where the large food particles are masticated before they pass to the pyloric stomach, where small particles are filtered and diverted to the hepatopancreas for digestion and absorption. Remaining large particles pass down the midgut and hindgut to be voided.

The most primitive malacostracans are to be found in the order Leptostraca. They are marine feeders on detritus, with a bivalved carapace, leaflike thoracic limbs and eight abdominal segments. Mantis shrimps (order Stomatopoda) are marine carnivores that wait at their burrow entrances for unsuspecting prey, including fish, before striking rapidly with the enormous claws on the second thoracic appendages.

The vast majority of the remaining malacostracans are grouped within the two superorders Peracarida and Eucarida. The peracarids are a superorder of malacostracans containing nine orders, of which the isopods (woodlice or pill bugs and others) and amphipods (sand hoppers and beach fleas) contain about 10,000 species between them. Peracarids hold their fertilized eggs in a brood pouch formed on the underside of the body by extensions from the first segments (coxae) of the thoracic legs, the eggs hatching directly as miniature versions of their parents. The carapace, when present, is not fused to all the thoracic segments. The first thoracic segment is joined to the head.

The small sediment-dwelling peracarid crustaceans of the order Thermosbaenacea inhabit thermal springs, fresh and brackish lakes, and coastal ground water. Opossum shrimps (order Mysida), although typically marine, are common also in estuaries, where they may be found swimming in large swarms and may constitute the major food of many fish. They filter small food particles with their two-branched thoracic limbs, and on many species a balancing organ (statocyst) is clearly visible on each inner branch of the pair of uropods. Their common name comes from the large brood pouch on the underside of the female's thorax.

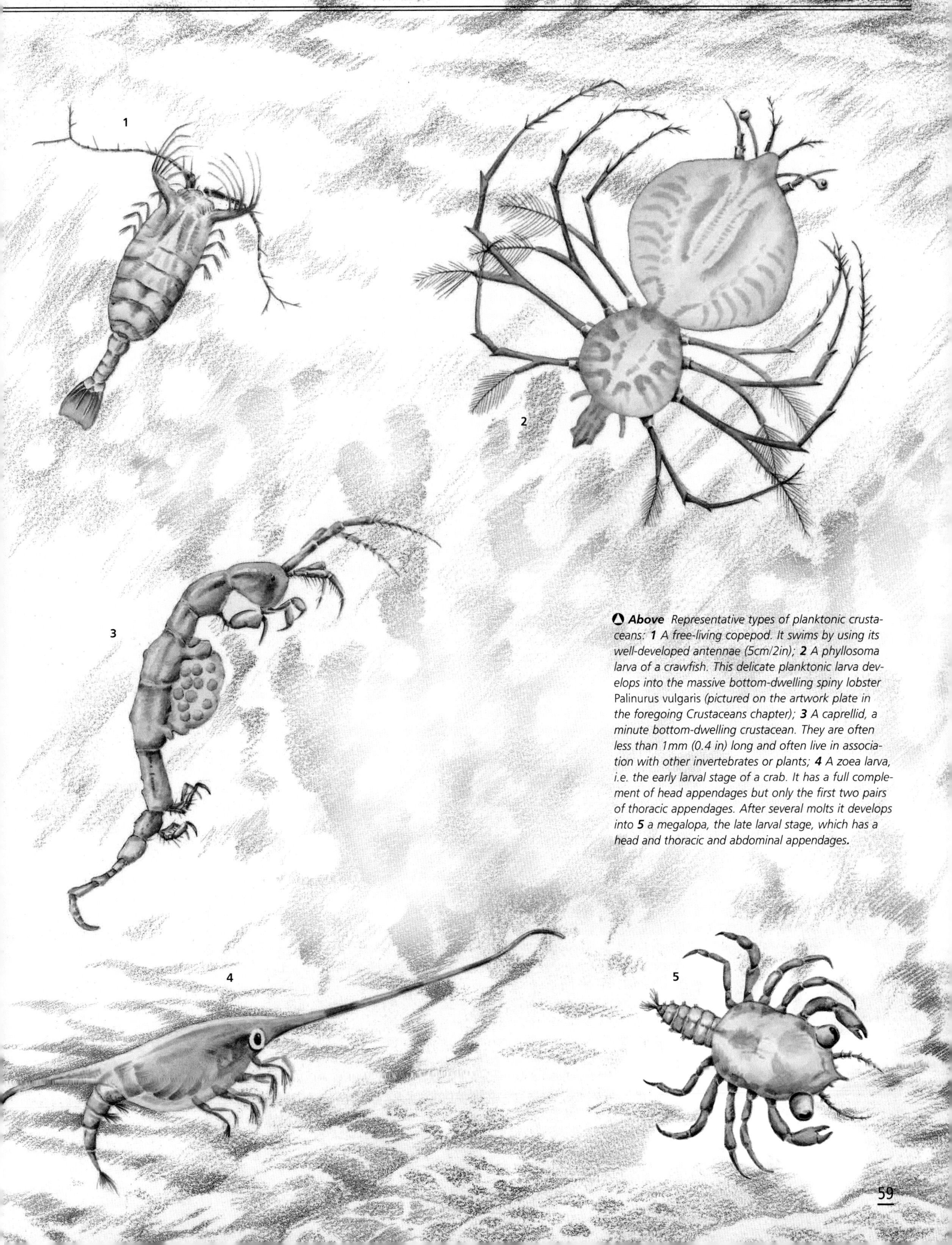

Above Representative types of planktonic crustaceans: **1** A free-living copepod. It swims by using its well-developed antennae (5cm/2in); **2** A phyllosoma larva of a crawfish. This delicate planktonic larva develops into the massive bottom-dwelling spiny lobster Palinurus vulgaris (pictured on the artwork plate in the foregoing Crustaceans chapter); **3** A caprellid, a minute bottom-dwelling crustacean. They are often less than 1mm (0.4 in) long and often live in association with other invertebrates or plants; **4** A zoea larva, i.e. the early larval stage of a crab. It has a full complement of head appendages but only the first two pairs of thoracic appendages. After several molts it develops into **5** a megalopa, the late larval stage, which has a head and thoracic and abdominal appendages.

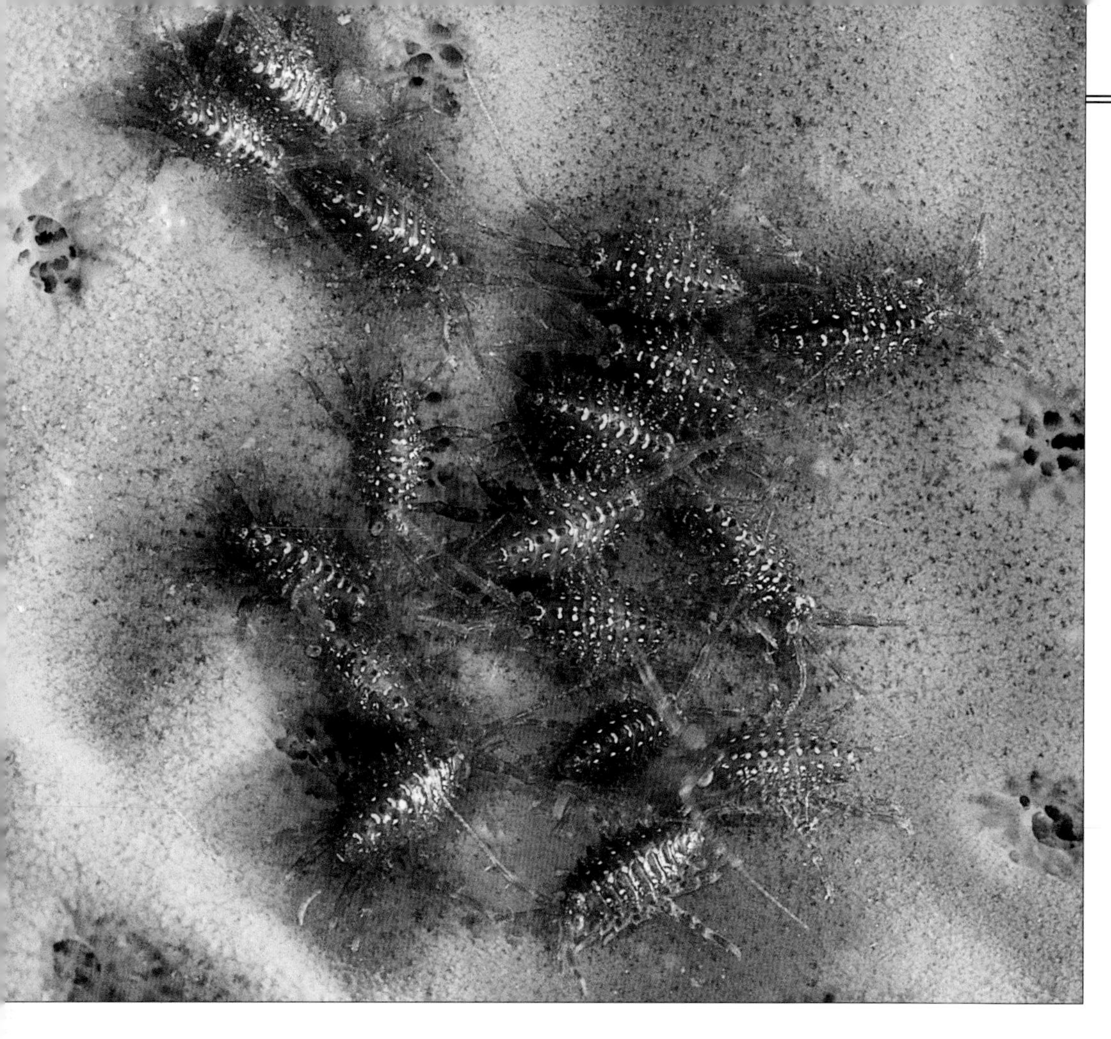

Above *The freshwater shrimp* Spinacanthus parasiticus *is endemic to Lake Baikal in Siberia, where it lives in symbiosis with the* Lubomiskiya *sponge.*

Below *A marine relative of the woodlouse is the Sea slater* (Ligia oceanica)*, which lives in seashore rock crevices, emerging at night to feed on detritus.*

Woodlice and Other Isopods

SUPERORDER PERACARIDA; ORDER ISOPODA

Woodlice (or pill bugs) are the crustacean success story on land. They are the most familiar members of the order Isopoda. Like other isopods, they are flattened top-to-bottom (dorsoventrally) and lack a carapace covering the segments. The first pair of thoracic limbs is adapted as a pair of maxillipeds, leaving seven pairs of single-branched thoracic walking legs (pereiopods). The five pairs of abdominal pleopods are adapted as respiratory surfaces. Isopods typically molt in two halves, the exoskeleton being shed in separate front and rear portions.

Most isopods walk on the sea bed, but some swim, using the pleopods, and others may burrow. The woodboring gribble used to destroy wooden piers along the coasts of the North Atlantic by rasping with its file-like mandibles, a major pest until concrete replaced wood. Isopods are also to be found in freshwater, and sea slaters live at the top of the intertidal zone, indicating the evolutionary route of the better known woodlice to life on land.

The direct development characteristic of reproduction in the superorder Peracarida (see above) avoids the release of planktonic larvae, and is a major adaptive preparation for terrestrial life. Woodlice have behavioral responses, for example to changes in humidity, to avoid desiccatory conditions and often therefore select damp microhabitats. Members of the pill bug genus *Armadillidium* are able to roll up, as the generic name suggests.

Woodlice have adapted their respiratory organs to take up oxygen from air. Members of the genus *Oniscus* show only little change from the ancestral aquatic isopod arrangement of pleopods as gills. Each of the five pairs of pleopods is two-branched and overlaps the one behind. The exopods of the first pair are extensive enough to cover all the remaining pleopods. The innermost fifth pleopods therefore lie in a humid microchamber and the endopods are well supplied with blood to act as the respiratory surface. *Porcellio* and *Armadillidium* species tolerate dryer conditions than can *Oniscus* species and use the outlying exopods of the first pair of pleopods for respiration. The danger of desiccation is reduced, for these exopods have intuckings of the cuticle (pseudotracheae) as sites of respiratory exchange.

Most isopods are scavenging omnivores, some tending to a diet of plant matter, especially the woodlice which contain bacteria in the gut to digest cellulose. Of the more carnivorous, *Cirolana* species can be an extensive nuisance to lobster fishermen, devouring bait in lobster pots.

Isopods have also evolved into parasites. Ectoparasitic isopods attach to fish with hooks and pierce the skin with their mandibles to draw the blood on which they feed. Those isopods (e.g., *Bopyrus* species) that live in the gill chambers of decapod crustaceans (crabs, shrimps, lobsters) are more highly modified and may cause galls. Some isopods even hyperparasitize rhizocephalan barnacles, themselves parasitic on decapods.

Beach fleas and Other Amphipods

SUPERORDER PERACARIDA; ORDER AMPHIPODA

Amphipods are laterally compressed peracarids that often lie on their flattened sides. The most familiar are the beach fleas or sand hoppers found on sandy shores, and the misnamed "Freshwater shrimp" common in European streams. Amphipods are mainly marine and bottom dwelling (benthic), but some are free swimming (pelagic).

Like the isopods, amphipods have no carapace, their eyes are stalkless, they have one pair of maxillipeds, seven pairs of single-branched walking legs (pereiopods), and a brooding pouch on the underside. Unlike the isopods, they have gills on the thoracic legs, and on the six abdominal segments there are usually three pairs of pleopods and three pairs of backward-directed uropods.

Most amphipods are scavengers of detritus, able to both creep using the thoracic pereiopods and swim with the abdominal pleopods. Many burrow or build tubes and feed by scraping sand grains or filtering plankton with bristle-covered limbs. Beach fleas, which have a remarkable ability to jump, burrow at the top of sandy shores or live in the strand line. This proximity to dry land has helped terrestrial amphipods volve in moist forest litter, although to a more limited extent than in isopods.

Amphipods of the family Hyperiidae are pelagic carnivores that live in association with gelatinous organisms, such as jellyfish, on which they prey. *Phronima* species are often reported to construct a house from remains of salp tunicates in which the animal rears its young. Skeleton shrimps (family Caprellidae) are predators of hydroid polyps, and the atypical, dorsoventrally flattened whale lice are ectoparasites of whales. PSR

The 6 Classes of Crustaceans

A TOTAL OF 420 CRUSTACEAN SPECIES ARE threatened in some measure, with 1 species – the Socorro isopod (*Thermosphaeroma thermophilum*; order Isopoda) – Extinct in the Wild. 56 species are Critically Endangered, including the Florida fairy shrimp (*Dexteria floridana*; order Anostraca), *Atlantasellus cavernicolus* (order Isopoda), Noel's amphipod (*Gammarus desperatus*; order Amphipoda), Tree hole crab (*Globonautes macropus*), and the Big-cheeked cave crayfish (*Procambarus delicatus*; order Decapoda); 73 are Endangered, including the Peninsula fairy shrimp (*Branchinella alachua*; order Anostraca), *Lepidurus packardi* (order Notostraca), Clifton cave isopod (*Caecidotea barri*; order Isopoda), Hay's spring amphipod (*Stygobromus hayi*; order Amphipoda), and the Red-black crayfish (*Cambarus pyronolus*; order Decapoda); 280 species are Vulnerable, including 5 species of fairy shrimp (genus *Branchinella*; order Anostraca), Wilken's stenasellid (*Mexistenasellus wilkensi*; order Isopoda), and 37 species of the cave amphipod *Stygobromus* (order Amphipoda); and 10 species are classified as Lower Risk.

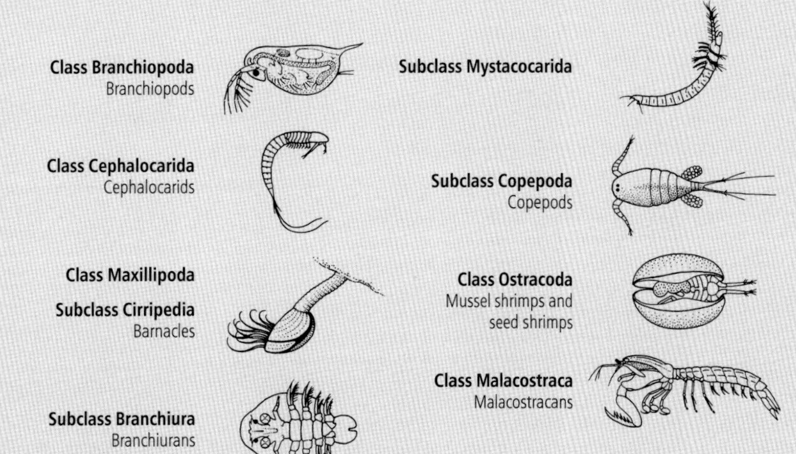

CLASS BRANCHIOPODA

Branchiopods
About 1,001 species in 23 families, 3 orders, and 2 subclasses.

SUBCLASS SARSOSTRACA

FAIRY SHRIMPS ORDER ANOSTRACA
180 species in 7 families. Genera and species include: brine shrimps (*Artemia* species)

SUBCLASS PHYLLOPODA

TADPOLE SHRIMPS ORDER NOTOSTRACA
11 species in 1 family. Genera and species include: *Triops* species.

ORDER DIPLOSTRACA
810 species in 15 families. Suborders: clam shrimps (suborder Spinicaudata), including *Cyzicus* species; water fleas (suborder Cladocera), including *Daphnia* species.

CLASS REMIPEDIA

Remipedes
12 species in 2 families and 1 order (Nectiopoda). Species include: *Speleonectes lucayensis*.

CLASS CEPHALOCARIDA

Cephalocarids
9 species in 4 genera and 1 family (Hutchinsoniellidae) and 1 order (Brachypoda). Species include: *Hutchinsoniella macracantha, Lightiella serendipita*.

CLASS MAXILLOPODA

Maxillopods
Over 9,705 species in 6 subclasses.

SUBCLASS THECOSTRACA

INFRACLASS FACETOTECTA
Arguably the biggest remaining mystery of crustacean classification. Single genus *Hansenocaris* accommodating curious "y-larvae" not yet assigned to a particular maxillopodan taxon.

INFRACLASS ASCOTHORACIDA
45 species in 6 families and 2 orders (Laurida, Dendrogastrida), very small parasites on echinoderms and soft corals. Species include: *Ascothorax* species.

BARNACLES INFRACLASS CIRRIPEDIA
980 species in 3 superorders. Adults parasitic or permanently attached to substratum, most with long, curved filter-feeding "legs."
Superorder Acrothoracica: 50 species in 3 families and 2 orders. Small (0.4mm/0.16in) filter-feeders boring into calcareous substrates (e.g. shells). Species include: *Trypetesa* species.
Superorder Rhizocephala: 230 species in 9 families and 2 orders. Internal parasites of decapods (lobsters, shrimps, crabs). Species include: *Sacculina carcini*.
Superorder Thoracica: 700 species in 29 families and 2 orders. Filter-feeding barnacles on rocks and other substrates. Up to 75cm (30in) long. Species include: goose barnacles (*Lepas* species), acorn or rock barnacles (*Balanus* species) and whale barnacles (*Conchoderma* species).

SUBCLASS TANTULOCARIDA

29 species in 5 families. Microscopic (0.15mm/0.006in) ectoparasites on deep sea bottom-dwelling crustaceans. Species include: *Basipodella harpacticola*.

SUBCLASS BRANCHIURA

150 species in 1 order. Bloodsucking fish lice some 7mm (0.3in) long, ecoparasitic on marine or freshwater fish. Species include: *Argulus foliaceus*.

TONGUE WORMS SUBCLASS PENTASTOMIDA

90 species in 9 families and 2 orders. Wormlike parasites in respiratory tracts of carnivorous vertebrates (especially reptiles). Genera and species include: *Linguatula* species.

SUBCLASS MYSTACOCARIDA

11 species in 1 family. Minute (0.5mm/0.02in) in sediments of seabed feeding on detritus. Species include: *Derocheilocaris typicus*.

COPEPODS SUBCLASS COPEPODA

Over 8,400 species in 220 families and 10 orders. Mostly minute components of plankton and sediment fauna. Orders and species include: order Calanoida with 2,300 species in 43 families, including *Calanus finmarchicus*; order Harpacticoida with 2,800 species in 54 families, including *Harpacticus* species; order Cyclopoida with 450 species in 15 families, including *Cyclops* species; order Poecilostomatoida with 1,320 species in 61 families of fish lice, including *Ergasilus sieboldi*; order Siphonostomatoida with 1,430 species in 40 families of fish lice, including *Caligus* species.

CLASS OSTRACODA

Mussel shrimps and Seed shrimps
5,650 species in 43 families and 4 orders. Species include: *Cypris* species.

CLASS MALACOSTRACA

Malacostracans
23,000 species in 497 families, 16 orders, and 3 subclasses.

SUBCLASS PHYLLOCARIDA

ORDER LEPTOSTRACA
3 families containing 25 species of pelagic or bottom-dwelling feeders on detritus. Species include: *Nebalia bipes*.

SUBCLASS HOPLOCARIDA

ORDER STOMATOPODA
17 families containing 350 species of bottom-dwelling carnivorous mantis shrimps. Species include: *Squilla empusa*.

SUBCLASS EUMALACOSTRACA

SUPERORDER SYNCARIDA
Order Anaspidacea: 15 species in 4 families. S Hemisphere freshwater bottom-dwelling detritus feeders up to 5cm (2in) long. Species include: *Anaspides tasmaniae*.
Order Bathynellacea: 100 species in 2 families, inhabiting freshwater sediments feeding on detritus, blind, 0.5–1.2mm (0.02–0.04in) long. Species include: *Parabathynella neotropica*.

SUPERORDER PERACARIDA
9 orders, including:
Order Thermosbaenacea: 9 species in 4 families of sediment-dwellers in marine and fresh water, and in hot-spring groundwater. About 3mm (0.12 in) long. Species include: *Thermosbaena mirabilis*.
Order Mysida: (Opossum shrimps): 4 families. Free-swimming and bottom-dwelling filter feeders 1–3cm (0.4–1.2in) long. Species include: *Neomysis integer*.
Order Isopoda (Sea slaters, woodlice, pill bugs, and sow bugs): 4,000 species in 120 families. Marine, freshwater, and terrestrial, typically omnivorous crawlers but some parasites (e.g. *Bopyrus* species) and wood-boring gribbles (*Limnoria* species). 0.5–1.5cm (0.2–0.6in) long. Genera include: *Armadillidium, Oniscus, Porcellio*.
Order Amphipoda (Sand hoppers and beach fleas): 6,000 species in 155 families. Mostly marine bottom-dwelling scavengers, 0.1–1.5cm (0.04–0.6in) long. Species include: skeleton shrimps (*Caprella* species), whale lice (*Cyamus* species), and the "freshwater shrimp" (*Gammarus pulex*).

SUPERORDER EUCARIDA
Order Euphausiacea (Krill): 85 species in 2 families of marine pelagic filter feeders, 0.5–5cm (0.2–2in) long. Species include: whale krill (*Euphausia superba*).
Order Amphionidacea: 1 marine free-swimming species, *Amphionides reynaudii*, up to 3cm (1.2in) long.
Order Decapoda (shrimps, lobsters, crabs): 10,000 species in 151 families.*

*See p. 53 for Table of Decapods.

SPECIAL FEATURE

THE ECONOMIC IMPORTANCE OF CRUSTACEANS

Human food and foulers of vessels

THE ECONOMIC IMPORTANCE OF CRUSTACEANS is twofold – their commercial value as food items and the costs caused by their effects as foulers of ships or coastal structures.

Crustaceans, especially the decapods (shrimps, lobsters, crabs), are important as food products, whether cropped from the wild or reared by aquacultural processes. Whale krill of the Antarctic Ocean are now being harvested and processed as food, not only for humans, but more particularly for agricultural livestock. In the Gulf of Mexico and off the southeastern United States, fishing boats trawl for the Brown shrimp, which, with the White shrimp and the Pink shrimp, make up the world's largest shrimp fishery. In Southeast Asia, in Indonesia and the Philippines, and in Taiwan, shrimps of the same genus *Penaeus* have been reared for food in brackish ponds for 500 years. Popular species are the Banana prawn, the Indian prawn, the Giant tiger prawn, the Green tiger prawn, and the Yellow prawn. The shrimps are trapped in pools at high tide and cultured for several months, often with mullet or milkfish. In Singapore ponds are usually constructed in mangrove swamps and in India rice paddies are used. Typical yields vary from 300 to 1,600kg of edible shrimp per hectare (1,650–8,700lb/acre). In Japan there is an intensive culture of the Kuruma shrimp. Eggbearing females are supplied by fishermen and shrimps are reared over two weeks, from eggs through successive larval stages with differing food requirements, to postlarvae before transfer to production ponds. Harvesting takes place 6–9 months later at yields that may attain 6,000kg per ha (32,600lb/acre).

In British waters the Caridean shrimp (also called the Pink shrimp) is fished by beam trawl in spring and summer in the Thames estuary, the Wash, Solway Firth, and Morecambe Bay – the Thames fishery dating back to the 13th century. The Brown shrimp is fished off Britain, Germany, Holland, and Belgium; boats work a pair of beam trawls and total landings reach 30,000 tons a year. *Palaemon adspersus* is taken off Denmark and off south and southwest Britain.

Another caridean, the tropical Indo-Pacific freshwater prawn *Macrobrachium rosenbergi*,

Right *Barnacle fishing on the coast of Galicia, Spain. Risking life and limb, fishermen clamber over tidal rocks in high surf and use poles to lever off their their tightly clinging catch. Their reward is the high price that barnacles command in local restaurants.*

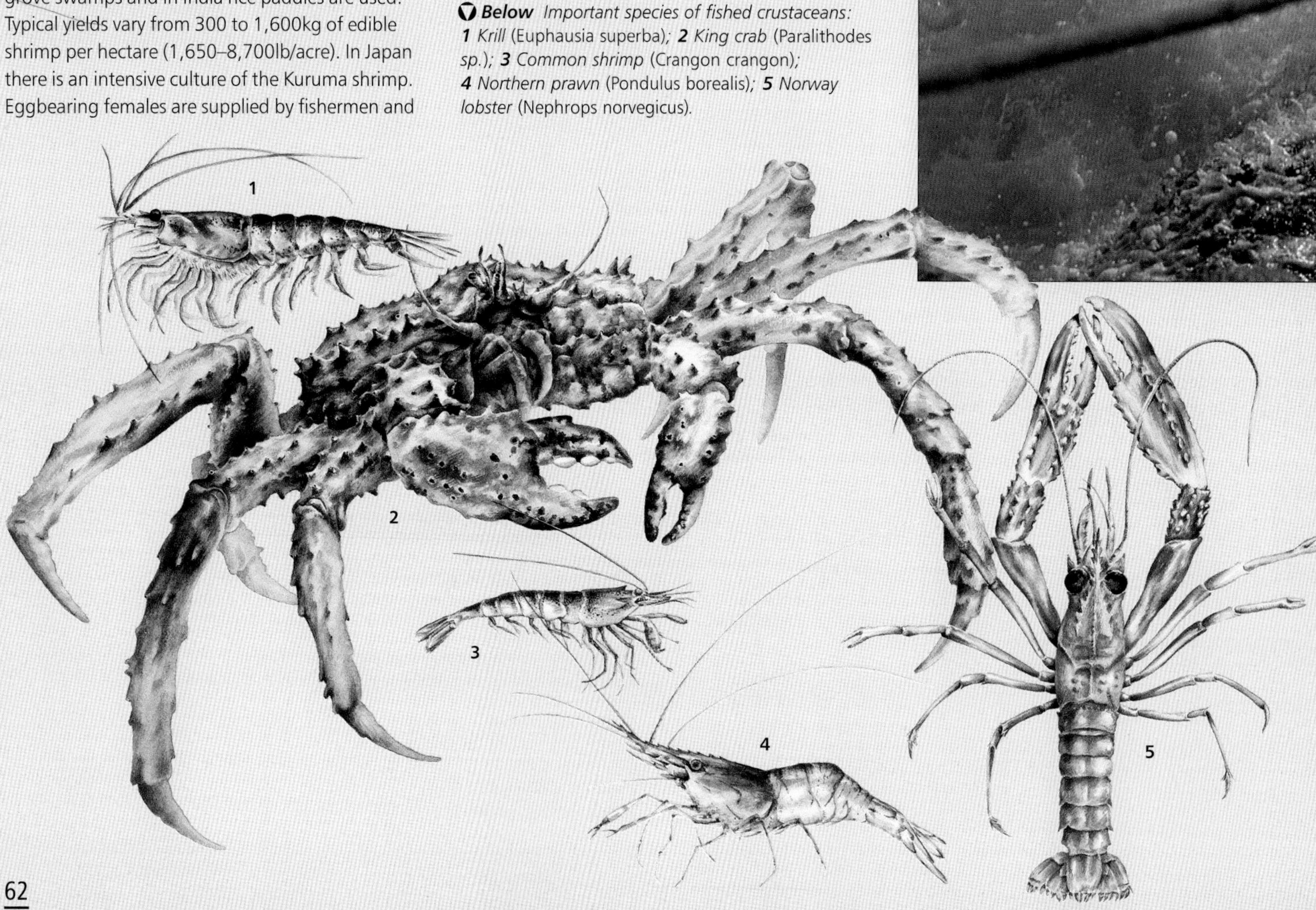

Below *Important species of fished crustaceans:* ***1*** *Krill* (Euphausia superba)*;* ***2*** *King crab* (Paralithodes sp.)*;* ***3*** *Common shrimp* (Crangon crangon)*;* ***4*** *Northern prawn* (Pondulus borealis)*;* ***5*** *Norway lobster* (Nephrops norvegicus).

is a giant reaching 25cm (10in) long; it is therefore attractive to aquaculturists, but its larvae are hard to rear and dense populations do not occur naturally.

Lobsters and crabs are usually caught in traps, enticed by bait to enter via a tunnel of decreasing diameter opening into a wider chamber. Most are predators or scavengers: lobsters prefer rotting bait but crabs are attracted to fresh fish pieces. Crab or lobster meat usually consists of muscle (white meat) from the claws and legs, and lobster abdominal muscle is also used. The hepatopancreas (brown meat) may be taken in some species. The meat is processed as a paste or canned, or the crustacean may be sold fresh or frozen.

The Dublin Bay prawn or Norway lobster supplies scampi – strictly the abdominal muscle. It burrows in bottom mud and is collected by trawling. Most lobsters, however, are caught in lobster pots. American lobsters and European lobsters are of commercial value in temperate seas but are replaced by spiny lobsters in warmer waters: *Panulirus argus* and *P. interruptus* are trapped off Florida and California, respectively, *P. versicolor* in the Indo-Pacific and *Jasus* species off Australia.

Freshwater crayfish are consumed with enthusiasm in France. The habit has transferred to Louisiana, United States, where between 400,000 and 800,000kg (from 880,000 to 1,960,000lb) of wild Red and White crayfish are trapped each year, and a further 1.2 million kg (2.6 million lb) reared in artificial impoundments.

The Edible crab is taken in pots off European coasts. The British annual catch reaches 6,500 tonnes, boats laying out 200–500 pots daily in strings of 20–70 buoyed at each end. The crabs are usually sold to processing factories, to be killed by immersion in freshwater or by spiking, boiled in brackish water, and then cooled to room temperature to set the meat, which is extracted by hand picking or with compressed air. The related Dungeness crab is fished off the west coast of North America, and the Blue crab is taken off the east coast by trap or line or by fishing with a trawl net.

On the negative side, one year's growth of fouling organisms can increase the fuel costs of ships by 40 percent, and barnacles are the most important foulers. Their presence impedes the smooth flow of water over a ship's hull and their shell plates disrupt paint films, so promoting corrosion. Barnacles are cemented to the substratum and necessitate expensive dry docking for removal. Barnacles have swimming larvae that disperse widely, ready to alight on passing ships. Barnacles and tube-building crustaceans may also clog water-cooling intake pipes of industrial installations by the coast, such as power stations.

The commercial importance of woodboring crustaceans, such as gribbles, has decreased since concrete has replaced wood as a major pier construction material. PSR

Horseshoe Crabs

The subphylum Chelicerata, of which horseshoe crabs form a small part, is more commonly represented by its terrestrial members – the arachnids (spiders, scorpions, and their relatives). The horseshoe crabs' primordial appearance hints at the fact that they form one of the most ancient surviving groups of organisms.

All chelicerates are typified by having a pair of pincerlike mouthparts (chelicerae), which are positioned in front of the mouth opening. They have no biting jaws, and two distinct parts of the body are recognizable, the prosoma (front part) and opisthosoma (rear part). Since horseshoe crabs fit this description and in spite of their common names and sea-dwelling habit, they are therefore more closely related to terrestrial arachnids than to crustaceans. They have remained more or less unchanged for some 300 million years, since the Permian period. The fact that extant species occur in two widely separated areas of the world (the eastern seaboard of the USA, and South and Southeast Asia) suggests that horseshoe crabs are relicts of a far more extensive distribution.

Horseshoe crabs have a protective hinged carapace, the domed shield of horseshoe of which covers the prosoma and the first part of the opisthosoma; a long caudal spine protrudes behind. They have compound eyes on the carapace and median simple eyes. Beneath the dark brown carapace lie the chelicerae and five pairs of walking legs, comparable in evolution to the chelicerae, pedipalps, and four pairs of walking legs of spiders.

Horseshoe crabs live on sandy or muddy bottoms in the sea, plowing their way through the upper surface of the sediment. During burrowing, the caudal spine levers the body down, while the fifth pair of walking legs act as shovels, the form of the carapace facilitating passage through the sand. The animal also uses the caudal spine to right itself if accidentally turned over.

Horseshoe crabs are essentially scavenging carnivores. There are jawlike extensions on the bases of the walking legs, used to trap and macerate prey, such as clams and worms, before it is passed forward to the mouth. The stout bases of the sixth pair of legs can also act like nutcrackers to open the shells of bivalve mollusks, which are seized during burrowing.

The appendages on the rear part (opisthosoma) are much modified. The first pair forms a protective cover over the remainder, each of which is expanded into about 150 delicate gill lamellae resembling the leaves of a book, in an adaptation unique to horseshoe crabs. The movement of the appendages maintains a current over the respiratory surfaces of the gill books which are well supplied with blood, which in turn drains back to the heart for pumping around the body. Small horseshoe crabs can swim along upside down, using the gill books as paddles.

At night at particular seasons of the year, male and female horseshoe crabs congregate at the intertidal zone for reproduction. The female lays 200–300 eggs in a depression in the sand, to be fertilized by an attendant clasping male. The eggs hatch after several months as trilobite larvae, so called because of their similarity to the fossil trilobites of the Paleozoic era (545–248 million years ago). Initially 1cm (0.4in) long, the larvae develop to reach maturity in their third year. PSR

Above *During spring tides in Delaware Bay, New Jersey, thousands of male horseshoe crabs gather on the shoreline to await the arrival of females for spawning. The annual migration of shorebirds to this location has evolved to coincide with the resulting superabundance of crab eggs, on which the birds feed.*

Right Limulus polyphemus, *the horseshoe crab species found in US waters, is threatened by human activity. Since the late 1990s, overharvesting of the crabs for use as whelk and eel bait has caused a major decline in their numbers.*

FACTFILE

HORSESHOE OR KING CRABS

Subphylum: Chelicerata (part)

Class: Merostomata

Order: Xiphosura

4 species in 3 genera: *Trachypleus gigas, T. tridentatus, Carcinoscorpius rotundicauda, Limulus polyphemus.*

Distribution Marine, Atlantic coast of N America (genus *Limulus*), coasts of SE Asia (*Tachypleus, Carcinoscorpius*).

Fossil record First appear in the Devonian period, 417–354 million years ago.

Size Larvae about 1cm (0.4in) long, adults up to 60cm (2ft) long.

Features Pair of pincerlike mouthparts (chelicerae) in front of mouth; carapace covers the back of the prosoma (front portion of body) and part of the opisthosoma (rear portion of body); prosoma has 6 pairs of appendages: chelicerae and 5 pairs of walking legs; opisthosoma bears 6 pairs of flattened limbs (a modified genital operculum or lid, and 5 pairs of leaflike or lamellate gillbooks for respiration) and a rear spine; excretion is by coxal glands at limb bases; circulatory system well developed; sexes distinct, with external fertilization; larva has three divisions, like a trilobite.

Sea Spiders

DESPITE THEIR COMMON NAME, THIS CLASS *of marine arthropods is related only distantly to terrestrial spiders. Sea spiders are for the most part extremely small, from 1mm to 1cm (0.04-0.39 in). A few larger species occur mainly in the abyssal zone – for example, some species of the genus* Colossendeis *have a leg span of 70cm (28in).*

Above *A sea spider feeding on a soft coral. It displays the spindly legs, reduced body, and skeletal appearance that is characteristic of the group as a whole. Indeed, the name of the class, Pantopoda, means literally "all legs."*

Exclusively marine animals, sea spiders are to be found from the intertidal zone to the deep sea, down to an extreme depth of 7,000m (23,000ft). They have an exaggerated hanging stance, like larger examples of their terrestrial namesakes, and are typically to be found straddling their sessile prey. Sea spiders grip the substratum with claws as they sway from one individual of their colonial prey to another without shifting leg positions. Although they mostly move slowly over the seabed, many sea spiders are capable swimmers.

Sea spiders typically have four pairs of long legs arising from lateral protuberances along the sides of the small, narrow trunk. There are some ten- and twelve-legged species. The paired chelifores and palps border the proboscis, the size and shape of which varies between species (this feeding structure is not found in terrestrial spiders).

The diet of sea spiders comprises soft-bodied invertebrates, for example cnidarians (jellyfishes, soft corals, and sea anemones), bryozoans, small ragworms and lugworms, hydroid polyps, sponges, and sea mats. The customary feeding technique is either to suck up their prey's body tissues through the proboscis or to cut off pieces of tissue from the prey with the chelifores, then transfer them to the mouth at the tip of the proboscis. A few sea spiders feed on algae.

In addition to the walking legs, there is a further pair of small legs, which is particularly well developed in males. As eggs are laid by the female they are fertilized and collected by the male, which cements them on to the fourth joint of each of its small eggbearing (ovigerous) legs, where they are brooded. The eggs hatch later as protonymphon larvae and develop through a series of molts, adding appendages to the original three larval pairs, the forerunners of the chelifores, palps, and ovigerous legs.

Body colors are variable, being generally white to transparent, but bright red in deep-sea species. The high surface-area-to-volume ratio of the narrow body means that it is only a short distance to the outside from anywhere in the body, and respiratory gases and other dissolved substances can be moved efficiently by diffusion. There is therefore no need for specialized excretory, osmoregulatory, or respiratory organs. The lack of storage space in the body does, however, mean that reproductive organs and gut diverticula have to be partly accommodated in the relatively large legs. PSR

FACTFILE

SEA SPIDERS

Subphylum: Pycnogonida

Class: Pantopoda

c.1,000 species in 70 genera and 8 families.

Distribution Marine, worldwide, from shores to ocean depths.

Fossil record Appear first in the Devonian period, 417–354 million years ago.

Size Narrow, short body 0.1–6cm (0.04–2.5in) long.

Features Cephalon has a tubular proboscis with mouth at tip, paired chelifores (claw-bearing appendages), and palps; usually 4 pairs of long legs on protuberances at the side, adult heart tubular, within blood-containing body cavity (hemocoel); there are no excretory, respiratory, or osmoregulatory organs; sexes distinct, with external fertilization; extra, egg-carrying, legs well developed on males; males brood eggs, which hatch into protonymphon larvae.

FAMILY AMMOTHEIDAE
Species include: *Achelia* species.

FAMILY COLOSSENDEIDAE
Species include: *Colossendeis colossea*.

FAMILY ENDEIDAE
Species include: *Endeis mollis, E. straughani*.

FAMILY NYMPHONIDAE
Deepsea, including 100 species of *Nymphon*.

FAMILY CALLIPALLENIDAE
Species include: *Callipallene tiberi, Pseudopallene brevicollis*.

FAMILY PHOXICHILIDIIDAE
Species include: *Phoxichilidium femoratum, Pallenopsis oscitans*.

FAMILY PYCNOGONIDAE
Species include: *Pycnogonum* species.

FAMILY TANYSTYLIDAE
Species include: the ringed sea spider (genus *Tanystylum*).

Priapulans, Kinorhynchs, Loriciferans, and Horsehair Worms

This disparate grouping contains four different phyla of marine worms. Owing to similarities in their cuticles, kinorhynchs and loriciferans are seen by biologists as closely related. Lacking the well-defined, noninversible mouth cone that characterizes these two phyla, priapulans are seen as their sister phylum in the grouping cephalorhynchs. Horsehair worms are closely related to nematode worms but, unlike nematodes, do not parasitize mammals.

Priapulans

PHYLUM PRIAPULA

The priapulans are ancient marine animals about which very little is known. They are found mostly in colder waters. They are wormlike, but their bodies are not segmented as in annelid worms. In the past, they have sometimes been considered alongside echiurans and sipunculids; however, they are not related to these groups.

The priapulans comprise a small phylum of uncertain affinities, composed exclusively of marine species inhabiting soft sediments. The more familiar forms are the relatively large family Priapulidae, but recently the small sand-living genus *Tubiluchus* and the small tube-dwelling *Maccabeus* have been described.

The trunk of priapulans bears warts and spines, and although it may have rings, no true segmentation is present. A tail consisting of one or two projections may be present, although there is no tail in *Halicryptus* or *Maccabeus*. In *Priapulus* the tail is in the form of a series of vesicles that constantly change shape and volume and may be involved in respiration. In *Acanthopriapulus* the tail is muscular, bears numerous hooks, and may serve as an anchor during burrowing. In *Tubiluchus* the tail is a long, tubular structure.

The pharynx is eversible and muscular, armed with numerous teeth or, in *Maccabeus*, bristles (setae). In the Priapulidae, the pharynx is used to capture living prey, but in *Tubiluchus* a scraping function is more likely. The intestine is in the trunk, the anus opening at its rear, with no part of the gut in the tail.

The sexes are separate, with differences in size between the sexes in *Tubiluchus*, probably associated with copulation.

The free-living larval stage of priapulans may be extremely long, the larva feeding in surface sediment layers for two years or more before changing to the adult condition.

The Priapula are considered to be related to the Kinorhyncha because of the body armature, lack of cilia, and the molting of the cuticle. PRG/AC

Kinorhyncha

PHYLA KINORHYNCHA

Some authorities used to group the kinorhyncha along with the gastrotrichs, roundworms, horsehair worms, rotifers, and priapulids in one phylum, known as the Aschelminthes, thanks to certain common developmental and structural features. Others believe that they are best treated as a phylum in their own right, since the criteria for grouping them together are debatable. This is especially true in the light of DNA-sequencing studies, which indicate that the kinorhynchs now have less in common with the rotifers.

These animals show bilateral symmetry and a body cavity formed from the blastocoel (formerly referred to as a pseudocoel) lying between the body wall and the gut. It is not a true coelom. The animals are not segmented and the body is supported by a layer of skin (cuticle), which is shed periodically when the animals molt. The alimentary tract is usually "straight through," with a mouth, esophagus, intestine, rectum, and anus. Excretion is carried out through structures called protonephridia. Reproductive strategies vary from group to group, although the sexes are usually separate. Development is along protostome lines.

The kinorhynchs are poorly known marine "worms" reaching up to 1mm (0.04in) in length. They probably exist worldwide but most of the 150 or so known species have been recorded from European coastal sands and muds – a reflection probably of where people have searched for them rather than of their actual distribution. Their bodies appear segmented but the divisions are only "skin-deep," so their segmented condition is the subject of some debate, and they are quite unrelated to the annelid worms or arthropods. They creep about in sand and mud using the spines of the head to gain a hold, while the rear of the body is contracted forward. Then the tail spines dig in and the head is advanced and so the process is repeated. The head is not well developed. Male and female kinorhynchs have a similar appearance and sexual reproduction occurs year round. A larva emerges at hatching, which molts several times before becoming an adult.

Loriciferans

PHYLUM LORICIFERA

Despite their small size, the Loriciferan body contains many cells – up to 10,000 – and comprises some complex structures, including muscle bands, protonephridia for excretion (actually located within the reproductive organs), a long gut, a nervous system, and reproductive organs. The head bears a number of spinelike structures, arranged in rings, and other ornaments, as well as an invertible oral cone, which carries the mouth. In some species the mouth is equipped with stylets, which presumably assist in feeding.

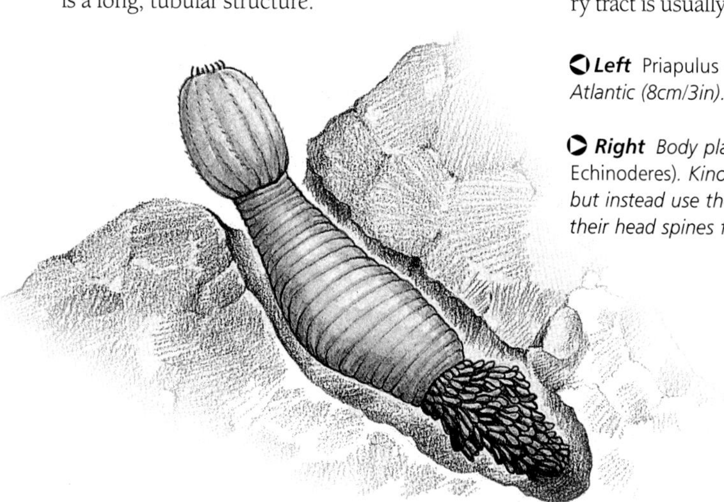

Left Priapulus caudatus, *a priapulan, North Atlantic (8cm/3in).*

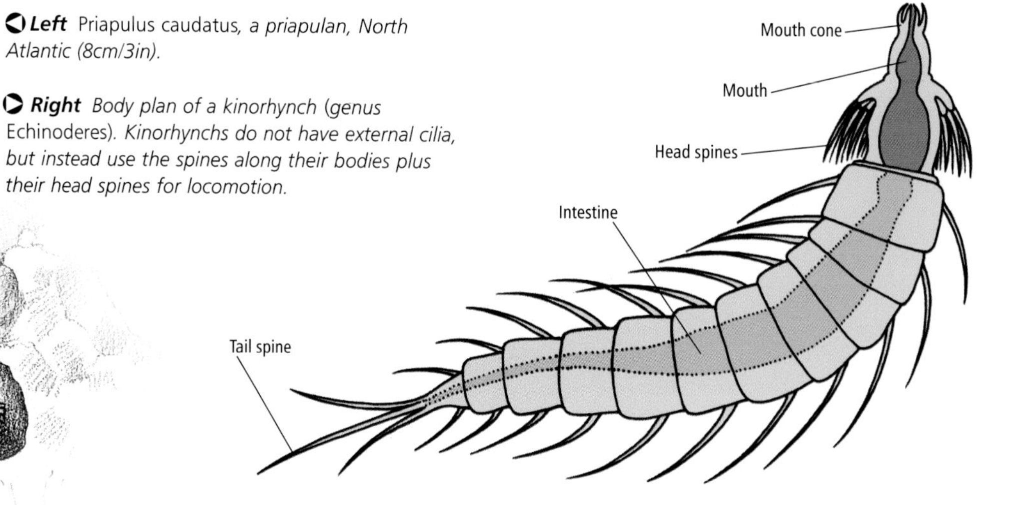

Right *Body plan of a kinorhynch (genus* Echinoderes). *Kinorhynchs do not have external cilia, but instead use the spines along their bodies plus their head spines for locomotion.*

Loriciferans exist as separate males and females, each with a distinct appearance. The males carry a pair of testes and the females a pair of ovaries. Fertilization is believed to take place internally, but little is known of the early development of the young. In most types a special larval type known as a Higgins larva forms. It is able to move around using specially developed "toes," somewhat similar to the toes of the adult body. The larval cuticle is shed periodically and eventually molts into an adult. In some deep-sea species the larva become precociously sexually mature (neoteny) and can produce more larvae, thus speeding up the life cycle. In others the larva may pass through a dormant period before metamorphosing into adults.

Horsehair Worms

PHYLUM NEMATOMORPHA

Adult horsehair worms typically live in the soil around ponds and streams and lay their eggs in the water, attaching them to water plants. The larvae are parasitic and usually attack insects. One mystery about their life cycle is that many of the insects acting as hosts for the horsehair worm larvae are terrestrial, not aquatic, for example, crickets, grasshoppers, and cockroaches. It may be that these animals become infected by drinking water containing the larvae. The development inside the host can take several months, and the larvae usually leave the host insect when it is near water to lead a free existence in the moist soil. The adults probably do not feed. AC

Above *A "Gordian knot" of horsehair worms (*Gordius *sp.). These worms are often found in drinking troughs used by horses, a fact that gave rise to the myth that they were horsehairs come to life.*

Below *Body plan of an adult female loriciferan (*Nanaloricus mysticus*). In this diagram, the abdomen, which is covered by a girdle of spiny plates known as a lorica, has been cut away to show the ovaries. For protection, the animal can retract its head, neck, and thorax into the lorica.*

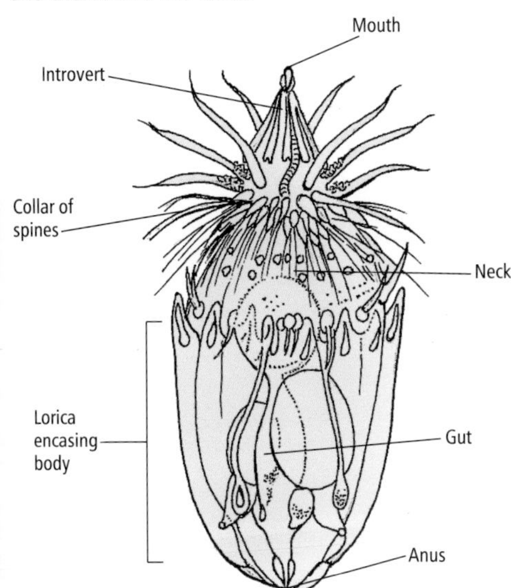

FACTFILE

PRIAPULANS, KINORHYNCHS, ETC.

Phyla: Priapula, Kinorhyncha, Loricifera, Nematomorpha.

About 496 species across the 4 phyla.

PRIAPULANS Phylum Priapula
16 species in 6 genera and 3 families. **Distribution:** exclusively marine, intertidal, and subtidal. **Size:** 0.55mm–20cm (0.03–16in). **Features:** unsegmented blastocoelomate worms; body divided into 2 or 3 regions; introvert at front with terminal mouth and eversible pharynx; trunk may have tail attached; possess chitinous cuticle, which is molted periodically; cilia lacking; sexes separate; produce free-living lorica larva.

FAMILY PRIAPULIDAE
Genera include: *Priapalus, Prialuposis, Acanthopriapulus, Halicryptus.*

FAMILY TUBILUCHIDAE
Genus *Tubiluchus.*

FAMILY MACCABEIDAE (CHAETOSTEPHANIDAE)
Genus *Maccabeus.*

KINORHYNCHS Phylum Kinorhyncha
About 150 species in the class Echinoderida. **Distribution:** worldwide in marine coastal areas, mostly living between sediment particles. **Fossil record:** none. **Size:** less than 1mm (0.04in). **Features:** free-living or epizoic; blastocoelomate, body segmented (13 segments) and covered in spines (no cilia); excretion via a pair of protonephridia, each fed by a single flame cell. Genera include: *Antigomonas, Echinoderes.*

LORICIFERANS Phylum Loricifera
About 10 species in 1 order and 2 families. **Distribution:** exclusively marine, inhabiting sediments like anoxic sands and muds usually in depths of 300–8,000m.
Fossil record: according to some authorities suitable habitats for these animals have existed since the Cambrian era, 500 m.y.a. **Size:** minute 115–383 microns, 0.115–0.383 mm (around 0.015 inches). **Features:** adults: free living, unsegmented, and probably blastocoelomate, bilaterally symmetrical animals, with bodies divisible into head, neck, thorax, and abdominal regions. The abdomen is encased in a strongly developed cuticle known as a lorica, hence the phylum name, meaning "corset bearing." The lorica is made from 6 platelike structures and is also ornamented with minute projections, the number and disposition of which can vary between the sexes. The first three parts can be withdrawn, almost telescopically, into the abdomen.

HORSEHAIR WORMS Phylum Nematomorpha
About 320 species. Also called Gordian or threadworms. **Distribution:** worldwide; adults free living and primarily aquatic in freshwater; juveniles parasitic in arthropods; *Nectonema* is an aberrant pelagic form found in marine coastal environments. **Fossil record:** none. **Size:** 5–100cm (2–39in) long. **Features:** unsegmented worms with thick external cuticle, which is molted; front end slender; body cavity blastocoelomate (a pseudocoel); through gut with mouth and anus present, but may be degenerated at one or both ends; lack circular muscles, cilia, excretory system; sexes separate with genital ducts opening into common duct with gut (cloaca). Genera include: *Gordius, Nectonema, Paragordius.*

Roundworms

ROUNDWORM INFESTATION IN HUMANS affects some one thousand million people worldwide; elephantiasis (lymphatic fliariasis) is a disfiguring disease of the tropics inflicted upon over 250 million individuals. These diseases are just two of the many caused by roundworms (Nematoda).

The nematodes are among the most successful groups of animals. They have exploited all forms of aquatic environment and many live in damp soil. Some are even hardy enough to survive in hostile environments, such as hot springs, deserts, and cider vinegar. Furthermore, by parasitizing both animals and plants, roundworms have widely extended their habitat ranges. Thus they are significant food and crop pests and the agents of disease in plants, animals, and humans.

Free-livers and Parasites

FORM AND FUNCTION

There are many types of nematodes, but these animals are all quite similar. They are typically worm shaped. Their bodies are not divided into segments or regions and generally taper gradually toward both the front and back. The mouth lies at the front and the anus almost at the tail.

Left *The body plan of a nematode, showing in cross-section the internal organs of the worm in the region of the pharynx.*

Right *Magnified image of* Trichinella spiralis *roundworms. Mammals, including humans, are infested by these parasites if raw meat containing the cysts of the larval worm is ingested.*

In cross section, nematodes are perfectly circular and their outer skin is protected by a highly impermeable and resistant cuticle, the secret of the success of this group. The cuticle is complex in structure and is made up of several layers of different chemical composition, each with a different structural layout. The precise form of the cuticle varies from species to species, but in the common gut parasite of humans and domestic animals, *Ascaris*, the cuticle comprises an outermost keratinized layer, a thick middle layer and a basal layer of three strata of collagen fibers that cross each other obliquely. They flex with respect to each other and allow the animal to make its typical wavelike movements, brought about by longitudinal muscle fibers. The high pressure of fluid maintained in the blastocoelomate body cavity (blastocoelom) counteracts the muscles and keeps the worms' cross section round at all times.

Below *A female roundworm* (Mermis *sp.*) *above ground in damp weather. Adults of the family* Mermithidae *are free living, but the larvae are parasitic on various insects, including grasshoppers. This fact has been expolited for the purpose of pest control.*

FACTFILE

ROUNDWORMS

Phyla: Nematoda

About 25,000 species in 2 classes.

ROUNDWORMS, EELWORMS, THREADWORMS, OR NEMATODES Phylum Nematoda

About 25,000 named species in 2 classes, but many may remain to be discovered. **Distribution:** worldwide; mostly free living in damp soil, freshwater, marine: some parasites of plants and animals. **Fossil record:** sparse; around 10 species recognized in insect hosts from Eocene and Oligocene (54–26 million years ago). **Size:** from below 0.05mm to about 1m (0.02in–3.3ft) but mostly between 0.1 and 0.2mm (0.04–0.008in). **Features:** round unsegmented worms sheathed in resistant external cuticle, which is molted from time to time; body tapers to head and tail; show blastocoelomate (pseudocoelomate) cavity between gut and body wall; mouth and anus present; lack circular muscles, cilia, excretory flame cells, and a circulatory system; nervous system quite well developed; sexes separate.

CLASS APHASMIDA

About 14 orders; genera include: *Trichinella*.

CLASS PHASMIDA

About 6 orders; genera include: *Ascaris, Loa, Mermis, Onchocera, Wucheria*.

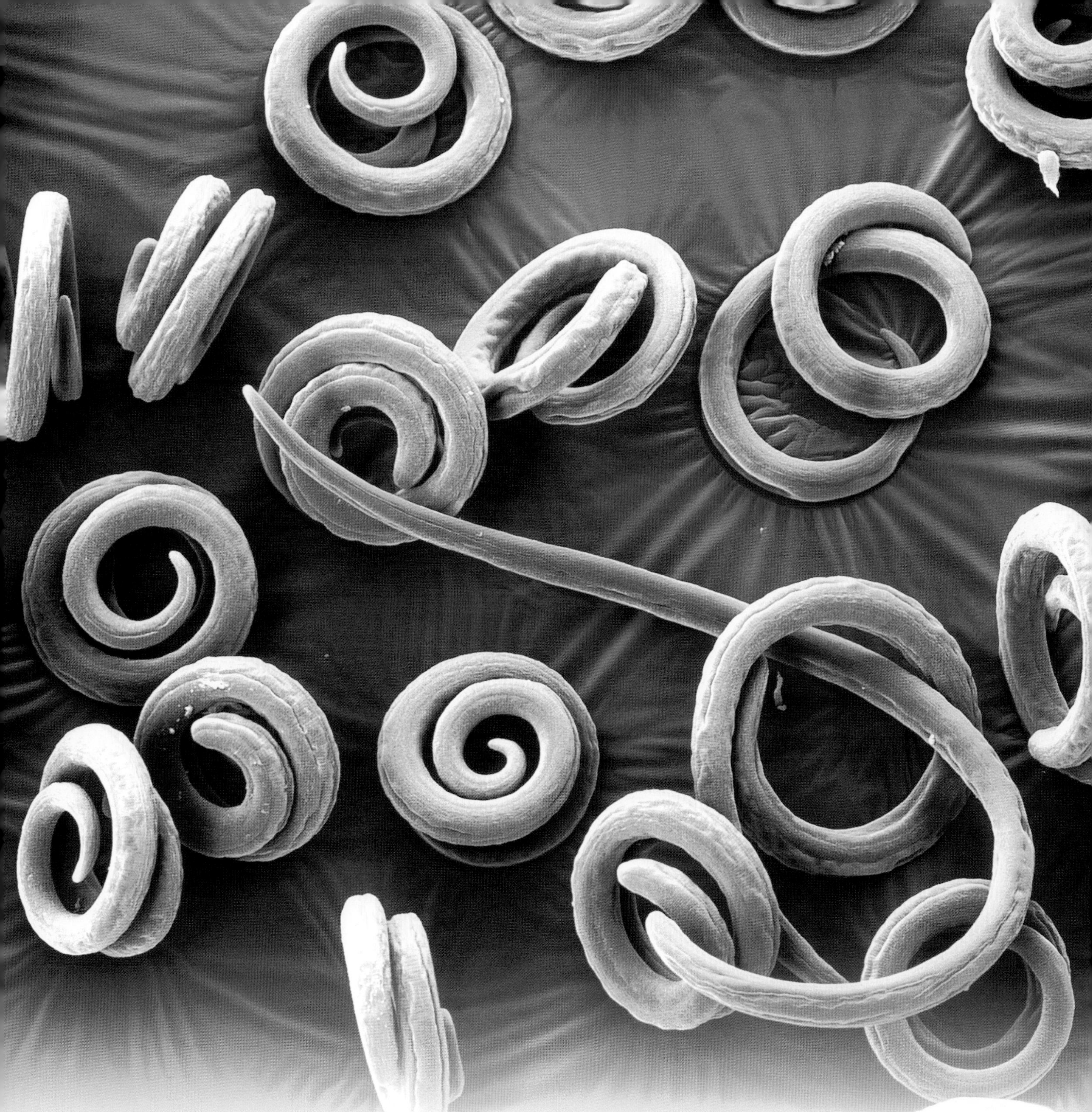

Both free-living and parasitic forms may have elaborate mouth structures associated with their food-procuring activities and ways of life. The gut has to function against the fluid pressure of the blastocoelom and in order to do so may depend on a system of pumping bulbs and valves.

In primitive forms, excretion is probably carried out by gland cells on the lower surface located near the junction of the pharynx and the intestine. In the more advanced types, there is an H-shaped system of tubular canals with the connection to the excretory pore situated in the middle of the transverse canal.

The nervous system of nematodes provides a very simple brain encircling the foregut and supplying nerves forward to the lips around the mouth. Other nerves are supplied down the length of the body. The body bears external sensory bristles and papillae.

Free-living nematodes are mostly carnivores feeding on small invertebrates, including other nematodes. Some aquatic forms feed on algae and some are specialized for piercing the cells of plant tissue and sucking out the contents. Others are specialized to consume decomposing organic matter such as dung and detritus and/or the bacteria that are feeding on these substances. Nematodes parasitic in humans and animals are responsible for many diseases and are therefore of considerable medical and veterinary significance. In humans, among other diseases they cause river blindness (*Onchocera* species), roundworm (*Ascaris lumbricoides*), and elephantiasis (*Wucheria* species). Very spectacular is the eye worm *Loa*, which can sit on the cornea. Nematode diseases are more common in the tropics.

Parasitic nematodes feed on a variety of body tissues and fluids, which they obtain directly from their hosts. Human roundworm is one of the largest and most widely distributed human parasites. Infections are common in many parts of the world and the incidence of parasitism in populations

may exceed 70 percent; children are especially vulnerable to infection. The adult parasites are stout, cream-colored worms that may reach a length of 30cm (1ft) or more. They live in the small intestine, lying freely in the cavity of the gut and maintaining their position against peristalsis (the rhythmic contraction of the intestine) by active muscular movements.

Debilitating Diseases

Human Infection

The adult female has a tremendous reproductive capacity and it has been estimated that one individual can lay 200,000 eggs per day. They have thick, protective shells, do not develop until they have been passed out of the intestine, and for successful development of the infective larvae, a warm, humid environment is necessary. The infective larva can survive in moist soil for a considerable period of time (perhaps years), protected by the shell. Humans become infected by accidentally swallowing such eggs, often with contaminated food or from unclean hands. The eggs hatch in the small intestine and the larvae undergo an involved migration around the body before returning to the intestine to mature. After penetrating the wall of the intestine, the larvae enter a blood vessel and are carried in the bloodstream to the liver and thence to the heart and lungs. In the latter they break out of the blood capillaries, move through the lungs to the bronchi, are carried up the trachea, swallowed, and thus return to the alimentary canal. During this migration, which may take about a week, the larvae molt twice. The final molt is completed in the intestine and the worms become mature in about two months.

An infected person may harbor one or two adults only and, as the worms feed largely on the food present in the intestine, will not be greatly troubled, unless the worms move from the intestine into other parts of the body. Large numbers of adults, however, give rise to a number of symptoms and may physically block the intestine. As in trichinosis, migration of the larvae around the body is a dangerous phase in the life cycle and, where large numbers of eggs are swallowed, severe and possibly fatal damage to the liver and lungs may result. Chronic infection, particularly in children, may retard mental and physical development.

Elephantiasis is a disfiguring disease of humans restricted to warm, humid regions of the world and occurs in coastal Africa and Asia, the Pacific, and in South America.

The blood of humans infected with the parasite contains the microfilaria stage of the worm, that is, the fully developed embryos still within their thin, flexible eggshells, which have been released from the mature female worms. During the day the microfilariae accumulate in the blood vessels of the lungs, but at night, when mosquitoes are feeding, the microfilariae appear in the surface blood vessels of the skin and can be taken up by the insects as they feed. The daily appearance and disappearance of the microfilariae in the peripheral blood is controlled by the activity pattern of the infected person and is reversed when the person is active at night and asleep during the day.

It is an impressive example of the evolution of close interrelationships between parasites and their hosts and ensures maximum opportunity for the parasite to complete its life cycle. Microfilariae

Left Eggs of the parasitic filarial nematode worm Toxascaris leonina, *which commonly affects dogs and cats. This parasite has a very simple life cycle; the second stage larvae are ingested by the dog or cat, and will mature in the animal's intestinal tract.*

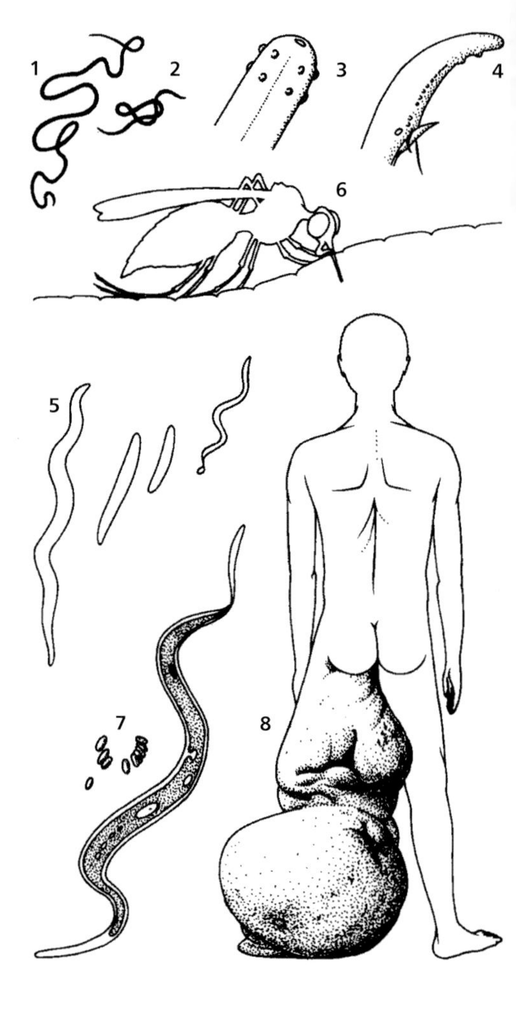

Right The elephant-like skin folds that characterize advanced cases of parasitic infection give the disease elephantiasis (properly known as lymphatic filariasis) its common name. Elephantiasis is caused by roundworms of the genera Wuchereria *or* Brugia. *The adult male* **1** *and female* **2** *worm parasitize the lymph ducts of their victim, to which they attach themselves, the female by papillae* **3** *and the male by a spine* **4**. *The adults produce larvae called microfilariae* **5,** *which are sucked up from the blood of an infected person by mosquitoes* **6** *when they feed. Inside the insect the microfilariae develop into infective larvae* **7,** *which are injected into the blood of another person when the mosquito feeds. There they move to the lymphatic system, completing the cycle. Adults form the tangled mass of worms that cause the build-up of lymphatic fluids resulting in the characteristic swellings* **8**.

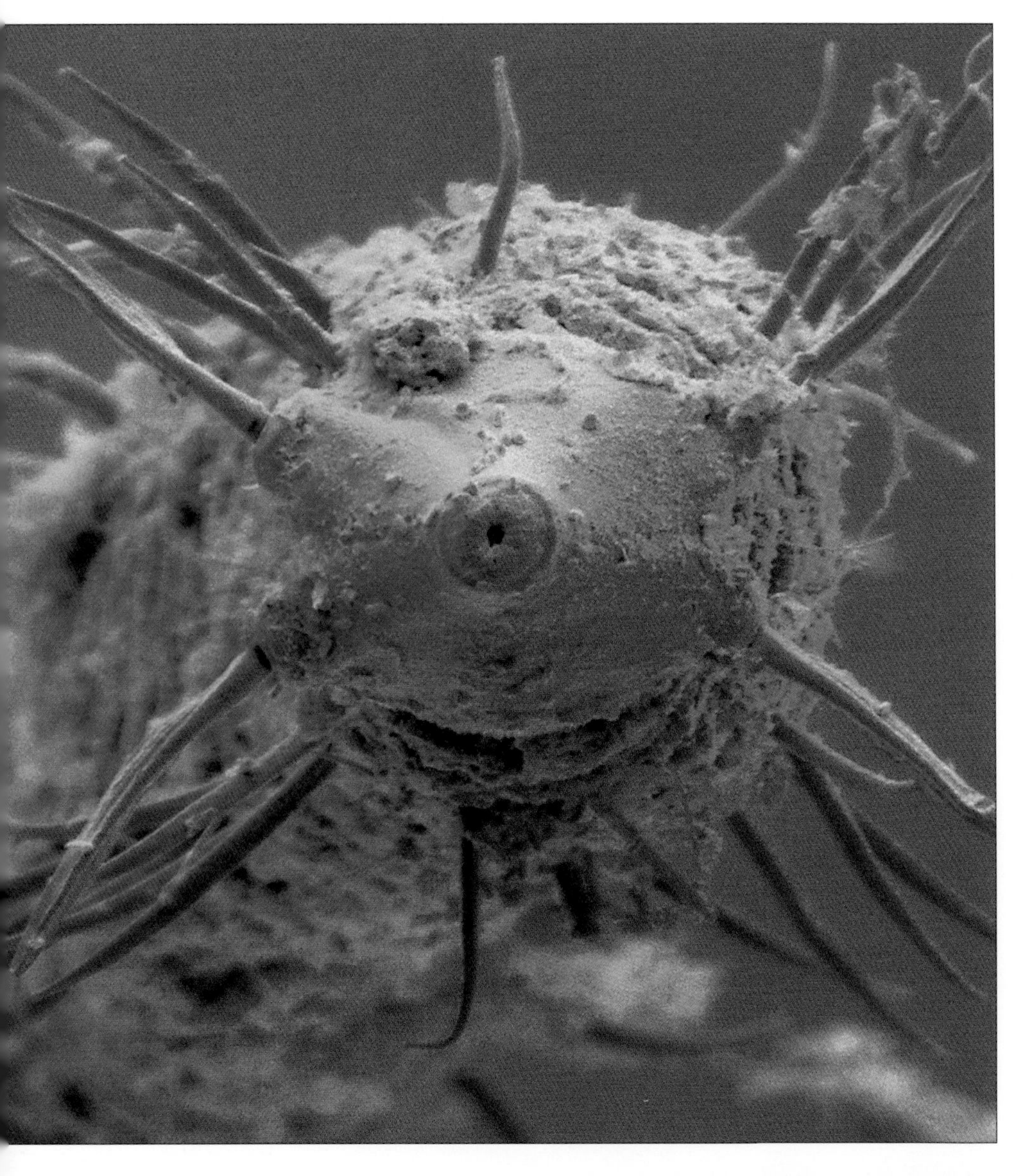

taken up by a mosquito undergo a period of development in the body muscles of the insect before becoming infective to humans. As the mosquito feeds, larvae may once again enter the human host.

The adult worms may reach a length of 10cm (4in), but are very slender. They live in the lymphatic system of the body, often forming tangled masses, and their presence may cause recurrent fevers and pains.

In long-standing infections, however, far more severe effects may be seen, caused by a combination of allergic reactions to the worms and the effects of mechanical blockage, causing accumulation of lymph in the tissues. Certain regions of the body are more commonly affected than others, notably the limbs, breasts, genitals, and certain internal organs, which become swollen and enlarged. The skin in these areas becomes thickened and dry and eventually resembles that of an elephant. In severe cases the affected organ may reach an enormous size and thus engender debilitating or even fatal secondary complications.

Drug treatment for the elimination of the worms is useful in the early stages of infection, but little can be done where chronic disease has produced true elephantiasis. Indeed, the parasites may no longer be present at that time. AC

Left *A marine roundworm (order Desmoscolecida). These worms, which live among sand grains on coastal beaches, are covered with spines that protect them from abrasion and enable them to withstand the pull of incoming and outgoing tides.*

Below *A microscopic nematode worm is lassoed by the predatory fungus* Arthrobotrys abchonia. *The fungus forms a three-celled ring some 20–30 microns in diameter; when the worm enters the ring, the cells inflate instantly and hold it fast.*

EELWORM TRAPS

In a reversal of the roles played by certain gnat larvae and the fungal fruiting bodies on which they feed above ground, soil-living nematodes face the hazard of predatory fungi. Eelworms, active forms that thread their way through the soil particles, fall prey to over 50 species of fungi, whose hyphal threads penetrate the eelworms' bodies.

There are several ways by which the hyphae penetrate the eelworms. Some fungi form sticky cysts, which adhere to the eelworm, then germinate and enter its body. Other fungi form sticky threads and networks, like a spider's web, which trap the eelworms.

Some fungi form lassolike traps. They consist of three cells forming a ring on a side branch of a hypha. An eelworm may merely push its way into the ring and become wedged or the trap may be "sprung," the three cells suddenly expanding inward in a fraction of a second, to secure the eelworm (see right).

It is interesting that these fungi do not need to feed on eelworms: they develop traps only if eelworms are present in the soil. AC

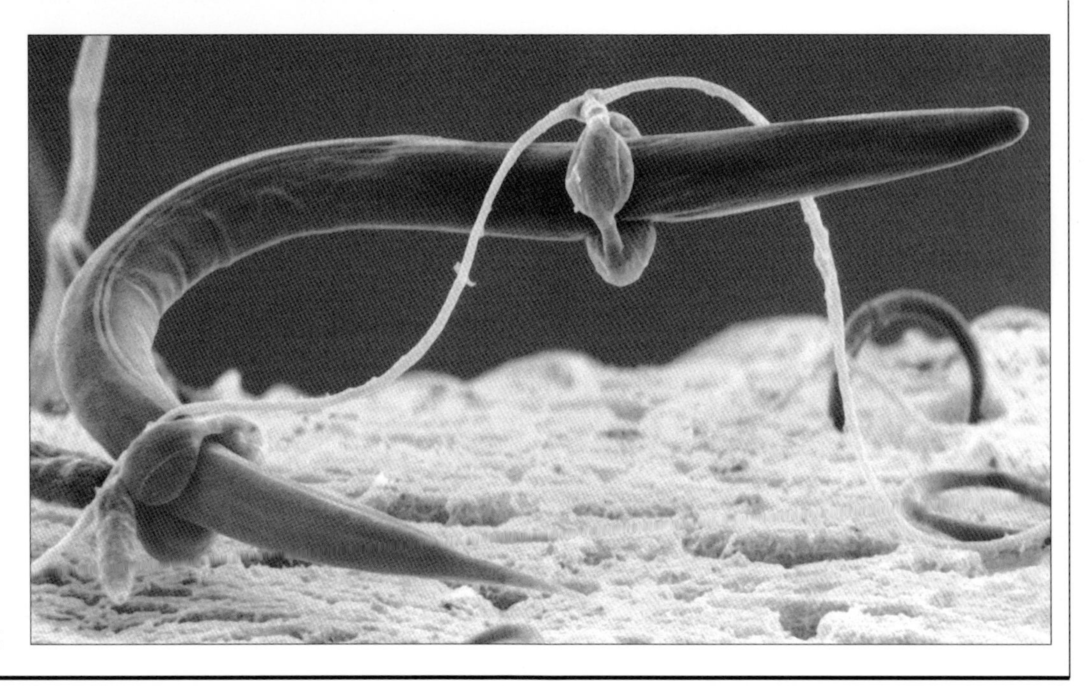

Flatworms and Ribbon Worms

FLATWORMS ARE THOUGHT TO HAVE BEEN THE first animals to evolve distinct front and rear ends. The most remarkable groups are the parasitic flukes and tapeworms, but many species are free-living. Ribbon worms are commonly considered as being most closely related to the turbellarian flatworms, but are more highly organized. As might be expected for these completely soft-bodied groups, the fossil record is virtually nonexistent.

Flatworms are the simplest animals with a three-layered (triploblastic) arrangement of body cells. Their flat shape derives from the fact that, having no body cavity other than the gut, they must respire by diffusion, and so no cell can be too far from their exterior surface. The ribbon worms (phylum Nemertea) are almost entirely free-living, marine worms, often highly colored, with only a few forms living on land or in freshwater, and one group as external parasites on other invertebrates.

Flatworms

PHYLUM PLATYHELMINTHES

With taxonomy constantly under review, the position of some groups of aquatic invertebrates is not always static. This is certainly the case with the order Acoela, which has until recently been considered as a flatworm within the class Turbellaria. Some scientists now believe that these gutless flatworms should be put in a new solitary phylum – Acoelomorpha (see also Gnathostomulids etc.).

The turbellarians are a widely distributed group of mostly free-living flatworms with a ciliated epidermis. They usually have a gut but no anus, and are classified by the shape of the digestive tract. Best known are the planarians, which may be black, gray, or brightly colored, live in fresh or sea water, and feed on protozoans, small crustaceans, snails, and other worms. *Convoluta* has no gut and depends on symbiotic algae for its food. Some members of the order Rhabdocoela (e.g. *Temnocephala* spp.) attach to crustaceans and other invertebrates and feed on free-living organisms, but most have become parasitic, such as *Fecampia* species, which live inside crustaceans.

Some planarians regenerate complete worms

Left *An Indo-Pacific species of planarian,* Pseudoceros ferrugineus, *off Komodo Island, Indonesia. Brightly colored flatworms such as these are often mistaken for nudibranchs (sea slugs).*

from any piece. The more sophisticated parasitic flatworms, however, are unable to replace lost parts. In general the lower the degree of organization of an animal, the greater its ability to replace lost parts. Even a small piece cut from such an animal usually retains its original polarity: a regenerated head grows out of the cut end of the piece that faced the front end in the whole animal. The capacity for regeneration decreases from front to rear: pieces from forward regions regenerate faster and form bigger and more normal heads than pieces from further back. In some planarians only pieces from near the front are able to form a head; those further back effect repair but do not regenerate a head.

The head of a planarian is dominant over the rest of the body. Grafts of head pieces reorganize the adjacent tissues into a whole worm in relation to themselves. Grafts from tail regions, on the other hand, are generally absorbed. However, the dominance of the head over the rest of the body is limited by distance, for example, if the animal grows to a sufficient length. This is what happens when planarians reproduce asexually: the rear part starts to act as if it were "physiologically isolated" and then finally constricts off as a separate animal.

True parasitism is universal in the two other classes, Trematoda and Cestoda. The majority of monogenean trematodes are external parasites living on the outer surface of a larger animal, many on the gills of fishes. A few inhabit various internal organs and are true internal parasites, such as *Polystoma* in the urinary bladder of a frog and *Aspidogaster* in the pericardial cavity of a mussel species. The digenean trematodes, including the flukes, are all internal parasites. The adults inhabit in most cases the gut, liver, or lungs of a vertebrate animal, swallowing and absorbing the digested food, blood, or various secretions of their host. Among these are *Fasciola* species in the liver of sheep and cattle, *Schistosoma* in the blood of humans and cattle, and *Paragonimus* in the lungs of humans. The internal parasites are parasitic throughout the greater part of their life. After an initial short period as a free-living ciliated larva known as a miracidium, the young enters a state of parasitism as a sporocyst or redia in a second host, and after a second free interval as a tadpole-shaped cercaria, may enter the body of a third host to become encysted. The second host is often a mollusk, and the cercaria may complete its life cycle by becoming encysted in the same animal or a fish.

The cestodes are the most modified for a parasitic existence, as the adults remain internal parasites throughout life. They invariably inhabit the gut of a vertebrate. The intermediate host is often also a vertebrate – commonly the prey of the final host. As an adult, *Taenia crassicollis* is parasitic in the intestine of the cat; the cysticercus larval stage occurs in the livers of rats and mice. The adult tapeworm *Echinococeus granulosus* inhabits the gut of dogs and foxes; its hydatid cyst larval stage may be found in the liver or lungs of almost any mammal, but especially sheep and sometimes humans.

Flatworms are bilaterally symmetrical animals without true segmentation. The shape is leaflike or ribboned in the planarians, or cylindrical in some rhabdocoels. While a distinct head is rarely developed, there is often a difference marking the anterior end – the presence of eyes, a pair of short tentacles, or a slight constriction to form an anterior lobe. The mouth is on the underside, mostly toward the front. In some polycladids there is a small ventral sucker on the underside, and in some rhabdocoels there is an adhesive organ at the front and rear. In *Temnocephala* species, a row of tentacles is often present. The trematodes closely related to the Turbellaria in internal organization resemble them externally, with further

FACTFILE

FLATWORMS AND RIBBON WORMS

Phyla: Platyhelminthes, Nemertea

About 19,400 species in total. Platyhelminthes contains 3 classes and Nemertea contains 2 classes.

FLATWORMS

Phylum Platyhelminthes

About 18,500 species in 3 classes (some experts believe that the two subclasses of trematodes warrant elevation to class status). **Distribution:** worldwide; free-living aquatic; parasitic in invertebrates and vertebrates. Fossil record: earliest in Mesozoic era 225–65 million years ago. **Size:** microscopic (1μm) to 4m (13ft), mostly 0.01–1cm (0.04–0.4in). **Features:** bilaterally symmetrical, usually flattened, possibly the first animals to have evolved front and rear ends and upper and lower surfaces; triploblastic, having a third layer of cells (mesoderm), which, by itself or in combination with ectoderm or endoderm, gives rise to organs or organ systems; without skeletal material, true segmentation, apparent body cavity (acoelomate), or blood-vascular system; defined head with a concentration of sense organs and a central nervous system; excretory vessels originate in ciliary flame cells; gut may be absent, rudimentary, or highly developed but there is no anus; sexual and asexual reproduction. male and female organs often in the same animal, often hermaphrodite; development sometimes direct, sometimes accompanied by metamorphosis.

RIBBON WORMS (OR PROBOSCIS WORMS)

Phylum Nemertea

About 900 species in 2 classes. **Distribution:** worldwide; freeliving, mostly marine but some in freshwater and damp soil. **Fossil record:** as for platyhelminthes. **Size:** 0.2cm–26m (0.08in–85ft). **Features:** bilaterally symmetrical, unsegmented worms with flattened elongated body; with a ciliated ectoderm; scientists disagree over status of the body cavity , may be truly coelomate; similar to flatworms but more highly organized; eversible proboscis in a sheath; gut with an anterior mouth, distinct lateral diverticula, and anus at rear; there is a blood-vascular system and excretory vessels with ciliary flame cells; reproduce mostly by sexual means, occasionally asexual (by fragmentation), sexes separate; development in some direct, in others accompanied by metamorphosis.

See The Classes and Orders of Flatworms and Ribbon Worms ▷

MEMORY AND LEARNING IN PLANARIANS

Planarians are capable of learning and have a memory. They acquire a conditioned reflex in response to a bright light followed by an electric shock. If animals trained in this way are cut into two, both halves regenerate into a whole organism, which retains its acquired learning, indeed both "heads" and "tails" show as much retention as uncut animals.

Similar results are found in planarians taught to find their way through a simple maze. Further experiments indicate that sexually mature worms fed on trained animals learn quicker than "control" animals kept on a normal diet. Two-headed planarians, which have been produced by surgery, learn more quickly than others, while animals whose brains have been removed are incapable of learning.

The memory of the animals cannot therefore be located in its brain and nervous system alone; it must be represented by chemical changes in cells throughout the body. It has been suggested that in the head of the planarian, memory is retained by neuron circuitry, whereas in the rest of the body it is retained in the form of a chemical imprint. This hypothesis has been confirmed by an experiment in which trained worms were cut in two and made to regenerate in a liquid containing a chemical "memory eraser" known as ribonucleic acid ASE. The "heads" were not affected by it, but the "tails" forgot eveything they had learned. Planarians show, beyond reasonable doubt, inheritance of acquired learning in animals that reproduce asexually. GD

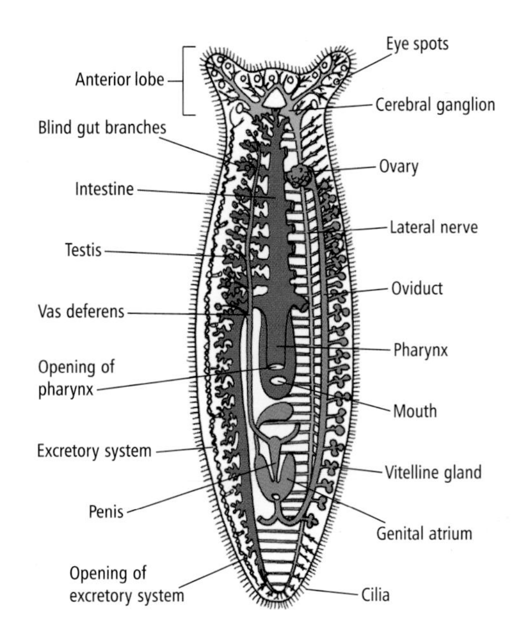

Left Body plan of a planarian flatworm (genus Planaria)*, showing the digestive, excretory, reproductive, and nervous systems. The blind gut is highly branched in order to transport food to all parts of the body.*

modifications to accommodate the parasitic existence. They are generally leaflike with a thicker, more solid body. Suckers on the underside fix the parasite to its host. Usually there is a set at the front, or a single sucker surrounding the mouth and a rear set, or a single large rear sucker. Among trematodes, monogeneans often have more numerous suckers. Cestodes are ribbonlike, the anterior end is, in most cases, attached to the host by means of suckers and hooks placed on a rounded head (scolex). The hooks are borne on a retractile process, the rostellum. In the order Pseudophyllidea a pair of grooves takes the place of suckers and there are no hooks. In many cestodes parasitic in fishes, the head bears four prominent flaps, the bothridia. In *Tetrarhynchus* species there are four long narrow rostella covered with hooklets. The cestodes are mouthless, and nothing distinguishes upper and lower surfaces. The body or strobila, which is narrower at the front, is made up of a series of segmentlike proglottides which become larger toward the rear.

The outer surface of the body wall of parasitic flatworms is differently modified in the three classes and the underlying layers of muscles are also differently arranged. A characteristic of flatworms is the mesenchyme, a form of connective tissue filling the spaces between the organs. There is an alimentary canal in the turbellarians (except the Acoela) and in the trematodes, which also have a muscular pharynx and an intestine. There is no gut in the cestodes, which take in nutrients through the surface of the body wall. Flatworms have a bilateral nervous system involving nerve fibers and nerve cells. The degree of development of the brain varies in the different groups, being greatest in the polycladids and some monogenean trematodes: there is a grouping of nerve cells at the front end into paired cerebral ganglia, especially in the free-living forms. Sense organs in adult free-living turbellarians include light, chemosensory and chemotactic receptors, and they may also occur in the free-living stages of trematodes and cestodes. An excretory system exists in nearly all flatworms except the Acoela. There is usually a main canal running down either side of the body, with openings to the exterior. Opening into the twin canals are small ciliated branches that finally end in an organ known as a flame cell. There is no circulatory-vascular system.

Both male and female reproductive organs occur in the one animal, which is usually hermaphrodite, one of the exceptions being the trematode genus *Schistosoma*. The reproductive organs are most complex in the parasitic forms. The testes are often numerous, their united ducts leading into a muscular-walled penis resting in the genital opening. The female reproductive organs comprise the ovaries, which supply the ova, and the vitellarium, which supplies the ova with yolk and a shell. The ovaries discharge their ova into an oviduct, which later forms a receptacle where fertilization occurs. The oviduct next receives the vitalline ducts, which lead into the genital atrium. The location and arrangement of the genital opening in relation to the exterior are such to prevent self-fertilization and ensure cross-fertilization. Development in rhabdocoels and monogenean trematodes is direct. In digenean trematodes, cestodes and some planarians a metamorphosis occurs. Asexual reproduction occurs commonly

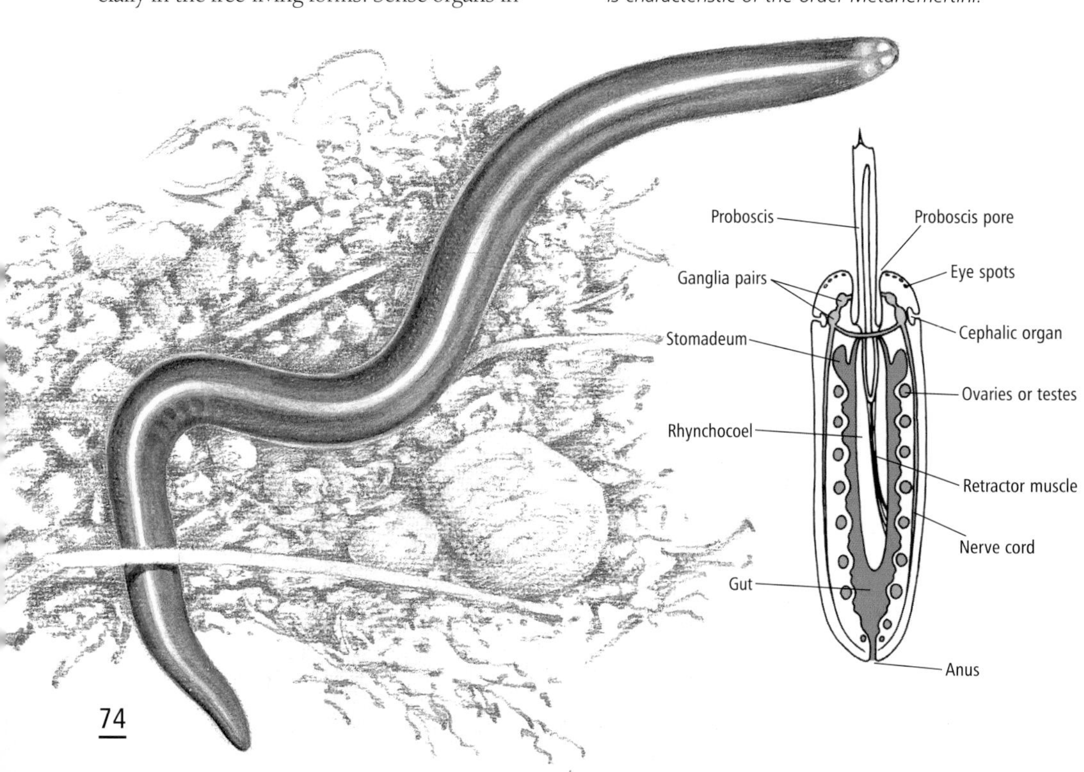

Below left *The Rodrigues nemertine* (Geonemertes rodericana) *is a rare terrestrial ribbon worm. Living on the Indian Ocean islands of Rodrigues and Mauritius, it eats small invertebrates and (unlike its marine relatives) uses its proboscis to aid locomotion.*

Below right *Body plan of a ribbon worm, showing the proboscis in prey-capture position, outside of the rhynchocoel. The piercing barb, or stylet, on its tip is characteristic of the order Metanemertini.*

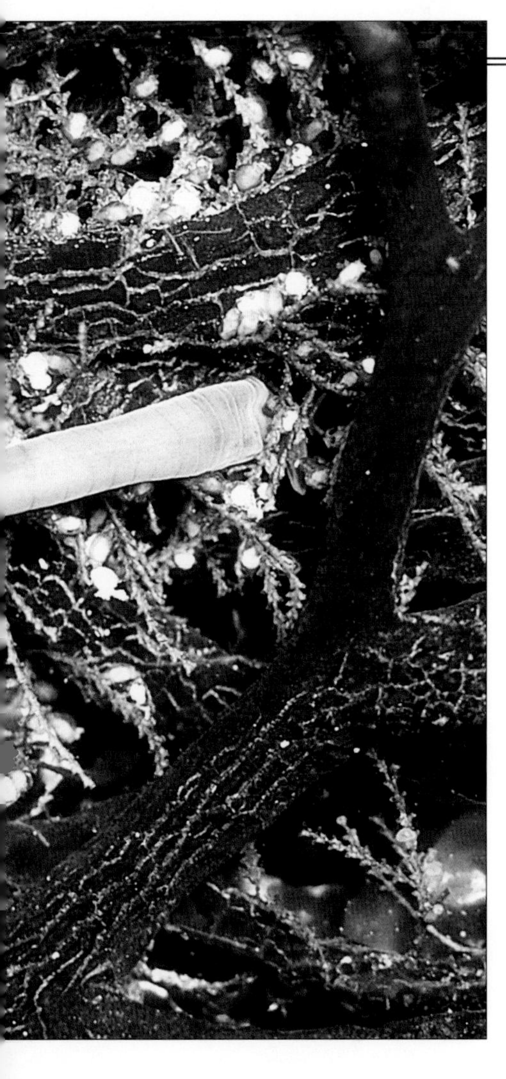

Above *A ribbon worm foraging. Ribbon worms are almost entirely carnivorous, feeding on small crustaceans and annelid worms. Some species have a toxic secretion in the proboscis that helps immobilize prey.*

in the turbellarians by a process of budding. Other planarians may fragment into a number of cysts, each of which develops into a new individual.

Ribbon Worms

PHYLUM NEMERTEA

Ribbon worms are often found burrowing in sand and mud on the shore or in cracks and crevices in the rocks. Some are able to swim by means of undulating movements of the body. Nearly all are carnivorous, either capturing living prey, mostly small invertebrates, or feed on dead fragments. The body is nearly always long, narrow, cylindrical or flattened, unsegmented, and without appendages. The entire surface is covered with cilia and with gland cells secreting mucus, which may form a sheath or tube for the creature. Beneath the ectoderm cell layer are two or three layers of muscles. There is no true coelom or body cavity, the space between organs being filled with mesenchyme. The proboscis, the most characteristic organ, lies in a cavity (rhynchocoel) formed by the muscular wall of the proboscis sheath. The proboscis may be everted for feeding. In members of the order Metanemertini – *Drepanophorus* species – there are stylets on the proboscis, providing formidable weapons. In other nemerteans, such as *Baseodiscus*, there are no stylets, but the prehensile proboscis can be coiled around its prey and conveyed to the mouth underneath the front end. The gut is a straight tube from mouth to anus, but various regions are recognizable in some species. There is a blood-vascular system, the blood being generally colorless, with corpuscles. The excretory system resembles that of ribbon worms, but the nervous system is more highly developed. The brain is composed of two pairs of ganglia, above and below, just behind or in front of the mouth. Certain ganglia are probably related to special sense organs on the anterior end of the worm, and most ribbon worms have eyes. From the brain a pair of longitudinal nerve cords runs back down the body and there is a nerve net, the complexity of which varies in the three orders. In most species the sexes are separate. The ovaries and testes are situated at intervals between the intestinal ceca and each opens by a short duct to the surface. Most ribbon worms develop directly, but some have a pelagic larva stage (pilidium), ending with a remarkable metamorphosis – the young adult develops inside the larva, from which it emerges.

Although very common, ribbon worms generally are of little economic or ecological importance. GD/AC

Right *Head of a* Hymenolepis nana *tapeworm. The smallest of the cestodes, this is the only tapeworm species that parasitizes humans without an intermediate host.*

The Classes and Orders of Flatworms and Ribbon Worms

FLATWORMS PHYLUM PLATYHELMINTHES

Freeliving flatworms
Class Turbellaria

About 4,500 species. Orders include:
Order Acoela*; genera include: *Convoluta*
Order Nemertodermatida*; genera include: *Meara*.
Order Rhabdocoela; genera include: *Dalyellia, Fecampia, Mesostoma, Temnocephala*.
Order Tricladida (planarians); genera include: *Dendrocoelum, Planaria, Polycelis, Procerodes, Rhynchodemus*.
Order Polycladida (planarians); genera include: *Planocera, Thysanozoon*.
The Lake Pedder planarian (*Romankenkius pedderensis*; order Tricladida) was recently declared as extinct by the IUCN.

*Note: There is now consensus among some scientists that this order may constitute a new phylum – the Acoelomorpha (see also Gnathostomulids etc.).

Parasitic flatworms
Class Trematoda

About 9,000 species.
SUBCLASS MONOGENEA*
MONOGENEANS
Order Monopisthocotylea; genera include: *Gyrodactylus*.
Order Polyopisthocotylea; genera include: *Polystoma*.
SUBCLASS ASPIDOGASTREA
Genera include: *Aspidoyaster*
SUBCLASS DIGENEA*
FLUKES, DIGENEANS
Order Strigeatoidea; genera include: *Alaria, Schistosoma*.
Order Echinostomida; genera include: *Echinostoma, Fasciola*.
Order Opisthorchiida; genera include: *Heterophyes, Opisthorchis*.
Order Plagiorchiida; genera include: *Paragonimus, Plagiorchis*.

* Some authorities believe these groups warrant class status.

Tapeworms
Class Cestoda

About 5,000 species.
SUBCLASS CESTODARIA
Order Amphilinidea; genera include: *Amphilina*.
Order Gyrocotylidea; genera include: *Gyrocotyle*.
SUBCLASS EUCESTODA
Order Tetraphyllidea; genera include: *Phyliobothrium*.
Order Proteocephala; genera include: *Proteocephalos*.
Order Trypanorhyncha; genera include: *Tetrarhynchus*.
Order Pseudophyllidea; genera include: *Dibothriocephalus, Ligola*.
Order Cyclophyllidea; genera include: *Echinococcus, Taenia, Hymenolepis*.

RIBBON WORMS PHYLUM NEMERTEA

Unarmed proboscides
Class Anopla

Order Hoplonemertea; genera include: *Amphiphorus, Hubrechtonemertes*.

Armed proboscides
Class Enopla

Order Bdellonemertea; genera include: *Malocobdella*.

PATHOGENIC PARASITES

Life cycles and medical significance of digenean flukes

FEW ANIMALS AFFECT PEOPLE SO ADVERSELY AS flukes of the subclass Digenea. They include creatures that cause one of the most prevalent human diseases – schistosomiasis or bilharzia, and others that cause serious losses of livestock. These digenean flukes can be divided into blood flukes (e.g. *Schistosoma* spp.), lung flukes (e.g. *Paragonimus westermani*), liver flukes (e.g. *Fasciola* and *Opisthorchis* spp.), and intestinal flukes (e.g. *Fasciolopsis buskii*, *Heterophyes heterophyes*). Liver flukes can cause serious losses of sheep and cattle.

Other groups of parasitic flatworms contain representatives that, directly or indirectly, are harmful to people. Some monogenean parasites of fish often cause serious losses in fish-farming stocks kept in overcrowded conditions. Human cestode disease caused by adults of the tapeworm species *Taenia solium*, *T. saginata*, and *Diphyllobothrum latum* is relatively nonpathogenic, but hydatid disease produced by the hydatid cysts of *Echinococcus granulosus*, cysticercosis caused by the cysticerus larval stage of *T. solium*, and sparganosis caused by the plerocercoid larvae of the genus *Spirometra* can be pathogenic.

The Common liver fluke (*Fasciola hepatica*) of sheep and cattle, almost worldwide in its distribution, is replaced by *F. gigantica* in parts of Africa and the Far East. Young flukes live in the liver tissue, feeding mainly on blood and cells, while the adults live in the bile ducts. The flukes are hermaphrodite, with male and female organs in the same worm. Large numbers of eggs are produced and pass out in the feces onto the pasture. After a variable period, depending on the temperature, a miracidium hatches out. Moving on its cilia, this first-stage larva seeks out and penetrates the appropriate snail host, which in Europe and the United States is the Dwarf pond snail (*Lymnaea truncatula*), an inhabitant of temporary pools, ditches, and wet meadows. The miracidium develops in its secondary host into a sporocyst, which produces two or three generations of multiple rediae. Each redia produces free-living cercariae, the number being determined by temperature. From one miracidium therefore many thousands of infective cercariae are produced. The tadpole-shaped cercariae leave the snail and may encyst on grass as metacercariae until eaten by a suitable primary host, which may be a sheep, cow, donkey, horse, camel, or even a human – eating, for example, infected watercress.

Spectacular losses due to acute fluke infestation can occur in sheep but are rarer in cattle. In some years the annual economic losses from fascioliasis in cattle and sheep in the United Kingdom, for example, have been estimated to be in the region of £10m and £50m, respectively. Annual total losses in The Netherlands, the United States, and Hungary can reach proportionally equivalent figures. It is possible in some countries to predict or forecast outbreaks of fascioliasis by using a meteorological system relying on monthly rainfall, evaporation, and temperature data. In some countries chemical control of the snail host is possible, but in others reliance has been placed on periodic and strategic dosing of infected animals with drugs such as rafoxanide, oxyclozanide, and nitroxynil. In the early 1980s a new breakthrough in treatment was made with the development of triclabendazole, which is very effective against early immature and mature *F. hepatica* in sheep and cattle.

Schistosomiasis or bilharzia affects the health of over 250 million people in over 70 developing countries, and it is estimated that another 600 million people are at risk. There are three principal schistosome species affecting humans. *Schistosoma haematobium* causes disease of the bladder and reproductive organs in various parts of Africa, especially Egypt, and the Middle East; *S. mansoni*, affecting mainly the liver and intestines, occurs in many parts of Africa, the Middle East, Brazil, Venezuela, Suriname, and some Caribbean islands; *S. japonicum*, also affecting the liver and intestines and sometimes the brain, occurs in China, Japan, the Philippines, and parts of Southeast Asia.

Schistosomes live in the blood vessels feeding on the cells and plasma. They do little damage themselves but are prolific egglayers. The eggs pass through the veins into the surrounding tissues of the bladder, intestines or liver, causing severe damage. The long cylindrical adult females live permanently held in a ventral groove of the more muscular males, laying eggs that eventually pass out in the host's urine or feces. Adult schistosomes possess adaptations that enable them to evade the immunological defenses of the host and

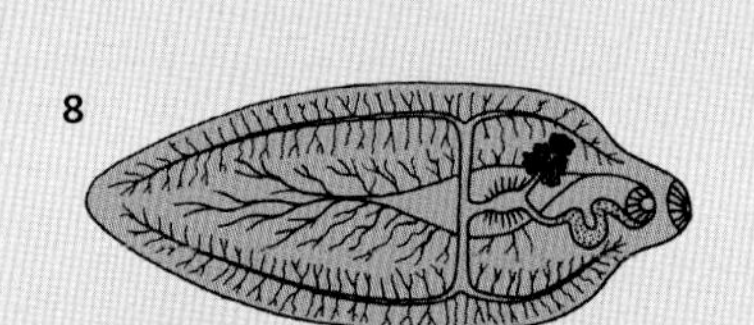

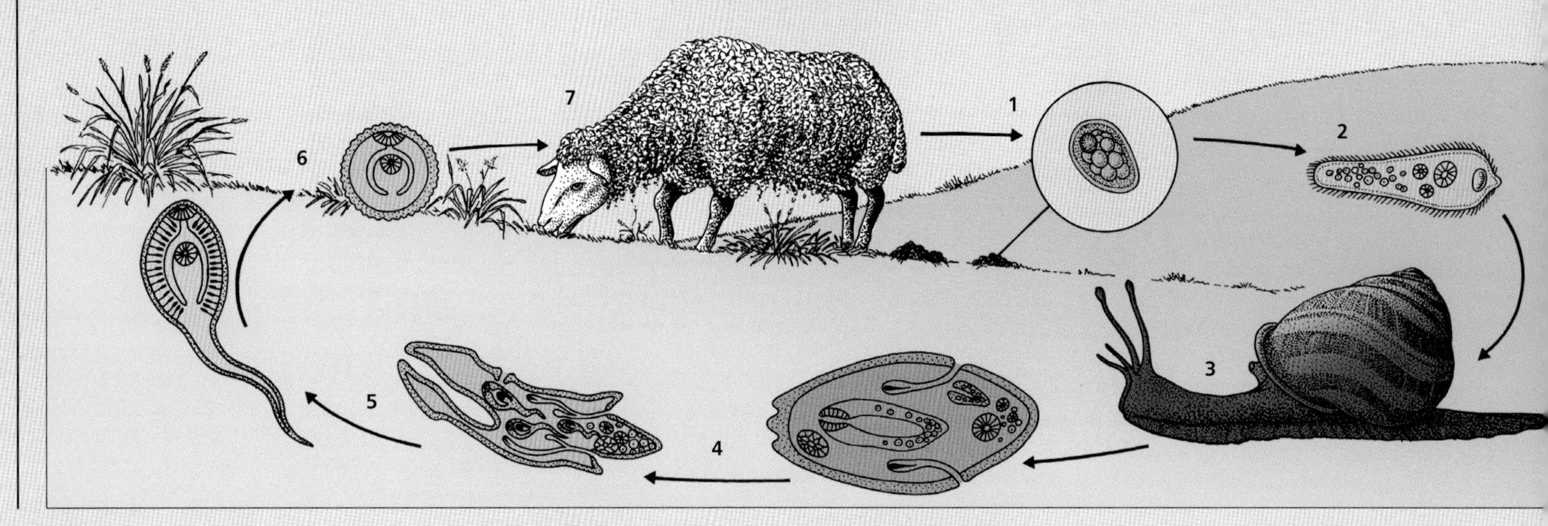

Below *Life cycle of the Liver fluke* (Fasciola hepatica) *in sheep:* ***1*** *Eggs of flukes living in the sheep are passed from the body in droppings;* ***2*** *In wet or damp pastures, the eggs hatch into miracidia larvae;* ***3*** *Free-swimming miracidium penetrates the snail* Lymnaea truncata, *where it develops into a sporocyst;* ***4*** *The sporocyst develops into redia larvae in the digestive gland of the snail;* ***5*** *The redia develop into cercaria larva, which leave the snail via the pulmonary aperture;* ***6*** *The cercariae encyst on the grass;* ***7*** *Encysted cercariae are eaten by sheep where they migrate to the liver, where the fluke* ***8*** *matures within 10–14 weeks.*

they may live and reproduce in one person for up to 30 years. Should the eggs reach freshwater they hatch, releasing a free-swimming miracidium which actively seeks out and penetrates the appropriate intermediate freshwater snail host – a pulmonate (e.g. *Bulinus*, *Physopsis*) or, for *S. japonicum*, a prosobranch snail. The parasite multiplies asexually inside the snail and after a period of between 24 and 40 days has produced a large number of infective cercariae. They escape from the snail and need to burrow through human skin or into another suitable host for the parasite to develop into either a male or female worm.

The geographical range of each species of schistosome is confined to the distribution of suitable snail hosts. The most important factor affecting the spread of schistosomiasis is the implementation of water-resource development projects in developing countries – primarily the construction of hydroelectric dams and also irrigation systems. The Aswan High Dam in Egypt and Lake Volta in Ghana have already aggravated and increased the spread of a disease already endemic in these countries. The debilitating diseases schistosomiasis and malaria are the two most prevalent of the world and have received much attention from the World Health Organization (WHO), which stimulates research into the basic problems related to the spread and control of these important diseases. The WHO claims that schistosomiasis could be eliminated by a combination of clean-water programs and the intensive use of drugs such as metrifonate and praziquantel against the appropriate schistosomes. Some experts believe that a vaccine is needed too, and some promising developments are being made in that direction as a result of applying new technology using monoclonal antibodies. GD/AC

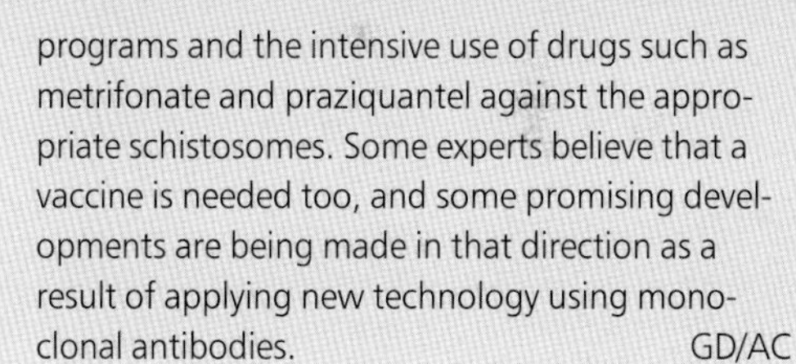

Above The parasitic blood fluke Schistosoma japonicum. *Symptoms of bilharzia include anaemia, diarrhoea, dysentery, cirrhosis of the liver, and enlargement of the liver and spleen.*

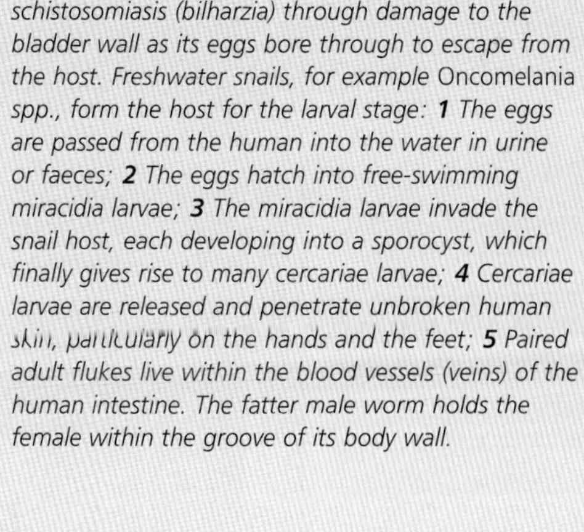

Left Life cycle of the human blood fluke (Schistosoma *spp.*)*, which cause urinary or vesical schistosomiasis (bilharzia) through damage to the bladder wall as its eggs bore through to escape from the host. Freshwater snails, for example* Oncomelania *spp., form the host for the larval stage:* ***1*** *The eggs are passed from the human into the water in urine or faeces;* ***2*** *The eggs hatch into free-swimming miracidia larvae;* ***3*** *The miracidia larvae invade the snail host, each developing into a sporocyst, which finally gives rise to many cercariae larvae;* ***4*** *Cercariae larvae are released and penetrate unbroken human skin, particularly on the hands and the feet;* ***5*** *Paired adult flukes live within the blood vessels (veins) of the human intestine. The fatter male worm holds the female within the groove of its body wall.*

Mollusks

THE DIVERSITY OF MOLLUSKS IS EXPRESSED IN their exploitation for food, pearls, and dyes, in their role as hosts to pathogens and parasites, and as garden pests. This great variety is reflected in the range of body forms and ways of life. Mollusks include coat-of-mail shells or chitons, marine, land, and freshwater snails, shell-less sea slugs and terrestrial slugs, tusk shells, clams or mussels, octopuses, squids, cuttlefishes, and nautiluses.

In addition to the 90,000–100,000 living species, the many extinct species include the ammonites and belemites. Mollusks now live throughout the world, living in the sea, in freshwater, in brackish water, and on land. Apart from those that float, swim weakly (e.g. sea butterflies) or powerfully (e.g. squids), or burrow (e.g. clams), mollusks live either attached to or creeping over the substrate, whether seabed, ground, or vegetation.

Mantled Mollusca

FORM AND FUNCTION

The molluskan body is soft and is typically divided into head (lost in the bivalves), muscular foot, and visceral hump containing the body organs. There are no paired, jointed appendages or legs, a feature distinguishing mollusks from the arthropods. Most species have their soft parts protected by a hard calcareous shell.

Two key features of the molluskan body are the mantle, an intucking of skin tissue that forms a protective pocket, and the toothed tongue or radula.

Right *Flaps from the mantle extending to envelop the glossy shell of a Mole cowrie* (Cypraea talpa). *These flaps protect the shell from abrasion, and continually deposit enamel onto it.*

Below *Alarmed by a diver, a Giant Pacific octopus* (Octopus dofleini) *emits a cloud of ink. Cephalopods such as this have highly developed brains and eyes.*

Mollusks have a gut with both mouth and anus, associated feeding apparatus, a blood system (usually with a heart), nervous system with ganglia, reproductive system (which in some is very complex), and excretory system with kidneys. The epidermal (skin) tissues of mollusks are generally moist and thin, and prone to drying out. Gills are present in most aquatic species, which are used to extract oxygen from water. In most bivalves and some gastropods, however, the gills are also used in feeding, when they strain out organisms and detritus from the water or silt with minute flickering cilia on the gills. These particles are then conveyed by tracts of cilia to the mouth.

In land and some freshwater snails, the walls of the mantle cavity act as lungs, exchanging respiratory gases between the air and body. Many mollusks have a free-floating (pelagic) larval stage, but this is absent in land and some freshwater species.

Lack of an internal skeleton has kept most mollusks to a relatively small size. Cephalopods have achieved the greatest size in the Giant squid, which can be 20m (60ft) long, including tentacles. Giant ammonites with shells up to 2m (6.6ft) across existed in the Jurassic period 195–135 million years ago. The largest living bivalves are the tropical giant clams, which can reach 1.5m (4.5ft) in shell length. A substantial number of species measure less than 1cm (0.4in). Some of the smallest, like the tiny gastropod *Ammonicera rota*, are only 1mm (0.04in) long when fully grown.

Inbuilt Protection

THE MANTLE

The back of a mollusk is covered by a fold of skin, the mantle, which forms a pocket housing the gills, osphradium (a chemical sensory organ), hypobranchial gland (secreting mucus), anus, excretory pore, and sometimes the reproductive opening. This special feature of mollusks has been adapted in many different ways and is present in all molluskan classes.

The cells of the mantle, particularly at the thickened edge of the mantle skirt, secrete the shell and may also produce slime, acids, and ink for defense. Mucus for protection and for cohesion of food particles is secreted by the gill and the hypobranchial glands. Products of the mantle can be defensive, acting to deter predators. The purple gland in the mantle of the sea hare expels a purple secretion when the animal is disturbed. Several species of dorid sea slugs or sea lemons can expel acid from glands in the mantle, while on land the Garlic snail gives off a strong aroma of garlic from cells near the breathing pore.

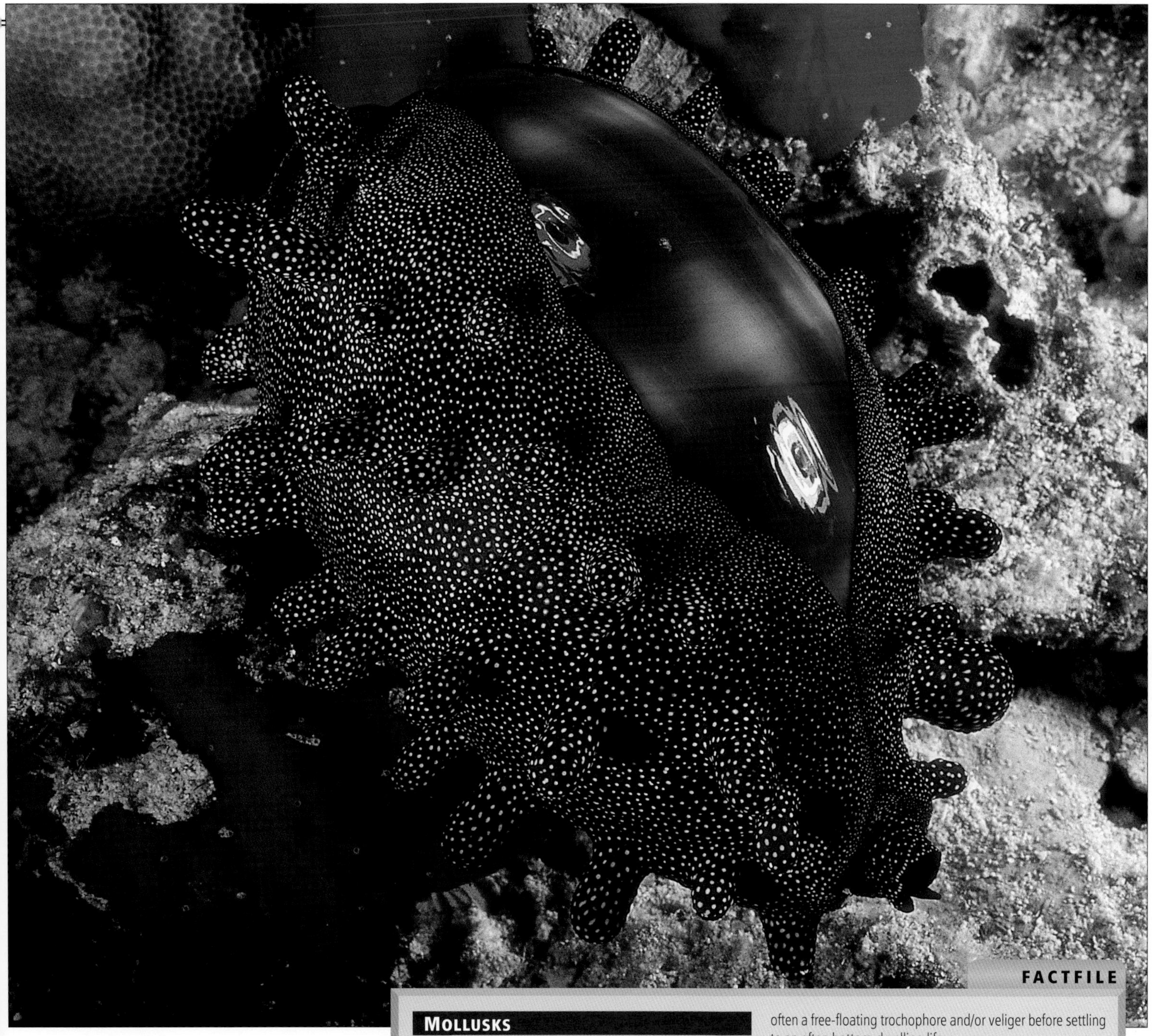

The mantle wall may be visible and in some sea slugs it is brightly colored and patterned – acting as either warning coloration or camouflage. The glossy and colored shells of cowries are usually hidden by a pair of flaps from the mantle.

Protection of the delicate internal organs was probably an early function of the mantle, which also provides a space into which the head and foot can be retracted when the animal withdraws into its shell. Within the mantle cavity, the gills are protected from mechanical damage (from rocks, coral etc.) as well as from silting. At the same time the gills must have ready access to sea water from which to extract oxygen. In some mollusks, special strips of mantle tissue are developed as tubular siphons that help separate two currents of water that pass over the gills – the inhalant and exhalant water currents. Fleshy lobes to the mantle, as in freshwater bladder snails of the genus

FACTFILE

Mollusks

Phylum: Mollusca

About 90,000–100,000 species in 7 classes.

Distribution Worldwide, primarily aquatic, in seas mainly, some in freshwater and brackish water, some on land.

Fossil record Appear in Cambrian rocks about 530 million years ago, modern classes distinct by about 500 million years ago; abundant extinct nautiloids, ammonites, belemites.

Size Often smaller than crustaceans, but ranging from tiny 1mm (0.04in) gastropods to Giant squid up to 20m (60ft) long.

Features Body soft, typically divided into head (lost in bivalves), muscular foot, and visceral hump containing body organs; protected by hard calcareous shell in most species; no paired jointed appendages; fold of skin (mantle) protects soft parts; mouthparts include usually a toothed tongue (radula); breathing mostly by gills, which may serve also in filter feeding; heart usually present; cephalopods have veins; nervous system of paired ganglia; sexes typically separate (not land species), fertilization external or by copulation; young develop via larval stages (not land or freshwater species) including often a free-floating trochophore and/or veliger before settling to an often bottom-dwelling life.

Monoplacophorans Class Monoplacophora
25 species of deepsea segmented limpets.

Solenogasters, chaetoderms Class Aplacophora
c. 370 wormlike marine species in 2 subclasses.

Chitons or coat-of-mail shells
Class Polyplacophora
c. 1,000 species.

Slugs and snails (gastropods) Class Gastropoda
c. 60,000–75,000 species in 3 subclasses.

Tusk or tooth shells Class Scaphopoda
c. 900 species of marine sand burrowers.

Clams and mussels (bivalves)
Class Bivalvia (or Pelecypoda)
c. 20,000 species in 7 orders in sea, brackish, and freshwater

Cephalopods Class Cephalopoda
c. 900 marine, mostly pelagic, species in 3 subclasses.

See The 7 Classes of Mollusks ▷

Physa, may function as extra respiratory surfaces.

Fertilization of eggs may take place within the mantle cavity of bivalves, and eggs are brooded there in, for example, the small pea shells of freshwater habitats.

The versatile molluskan mantle can become muscular and serve in locomotion. Some sea slugs employ their leaflike mantle lobes in swimming. Some scallops, such as the Queen scallop, swim by expelling water from the mantle cavity.

Shedding Teeth

THE RADULA

The radula, a toothed tongue, is typically present in all classes of mollusk except the bivalves. It is secreted continuously in a radula sac and is composed of chitin, the polysaccharide also found in arthropod exoskeletons. The oldest teeth are toward the tip: when a row of teeth becomes worn, they detach and are often passed out with the feces. A new row of teeth then moves into position. Inside the mouth is an organ, the buccal mass, which contains and operates the radula during feeding. The radula is carried on a rod of muscle and cartilage (odontophore) that projects into the mouth cavity, while further complexes of muscles and cartilage in the walls of the buccal mass operate the radula, usually in a circular motion.

The form of the radula depends on feeding habits and is used in identifying and classifying individual species. Herbivores, such as land snails and slugs, have a broad radula with many small teeth, while carnivores, including whelks, have a narrow radula with a few teeth bearing long pointed cusps. Limpets, which browse algae off rocks, have an especially hard rasping radula and a few very strong teeth in each row; they leave scratch marks on the surface of rock.

In the carnivorous cone shells, each tooth is separated from the membrane, and is a harpoonlike structure delivered into the body of the prey (often a fish or a worm) to facilitate penetration of an accompanying nerve poison. The tiny sacoglossan sea slugs feed on threadlike algae – their radula teeth are adapted to pierce individual algal cells.

A Wealth of Variety

THE SHELL

Mollusks usually hatch from the egg complete with a tiny shell (protoconch) that is often retained at the apex of the adult shell. This calcareous shell, into which the animal can withdraw, is often regarded as a hallmark of a mollusk. For the living mollusk, the shell provides protection from predators and mechanical damage, while on land and on the shore it helps prevent loss of body water. Empty shells have long been a source of fascination, and many people collect them.

New growth occurs at the shell lip in gastropods and along the lower or ventral margin in bivalves. Shell is secreted by glandular cells, particularly along the edge of the mantle. This fact is easy to demonstrate in young land snails, by painting waterproof ink along the outside of the lip. After a few days new shell will be seen in front of the ink mark. Newly secreted shell is thin, but gradually attains the same thickness as the rest. Although, in the event of damage, a repair can be made further back from the mantle edge, it will be a rough patch and not contain all the layers of normal shell. When growth stops for a while during cold weather, in drought, or a time of starvation, a line forms on the shell, which continues to be visible after resumption of growth. A number of mollusks (e.g. cockles) normally have regular marks recording interruptions of their growth.

The cross-lamellar component of the shell, revealed by high magnification, consists of different layers of oblique crystals, each layer with a different orientation, rather like the structure of radial car tyres. This is thought to give greater strength without extra weight or bulk.

Shell is mostly composed of calcium carbonate, in calcite form (in the prismatic layer) and in argonite form (in the cross-lamellar layers), together with some sodium phosphate and magnesium carbonate. The mineral component of the shell is laid down in organic crystalline bodies in a matrix of fibrous protein and polysaccharides (conchiolin) secreted by the mantle. Snails can store calcium salts in cells of the digestive gland (hepatopancreas).

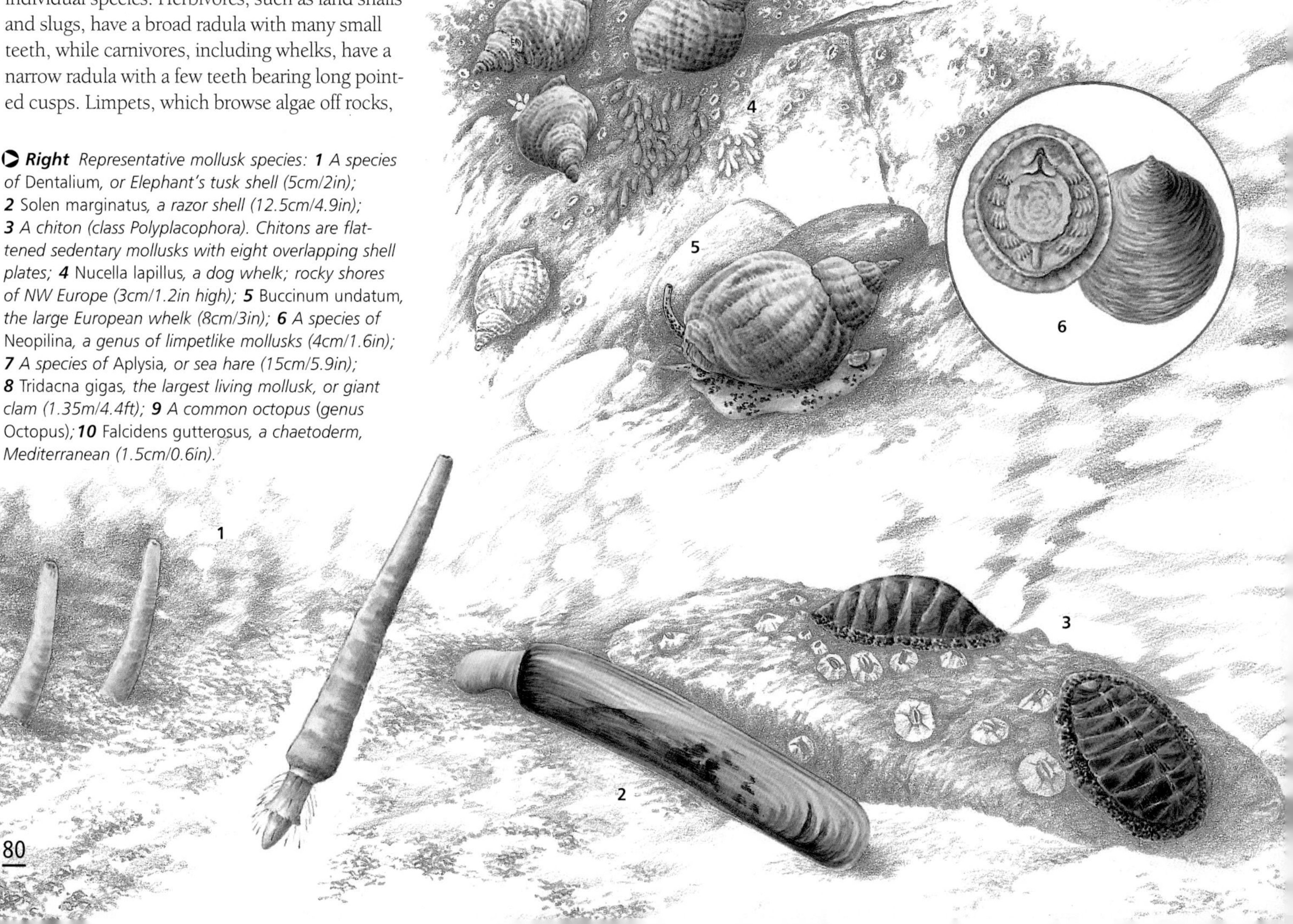

Right *Representative mollusk species:* ***1*** *A species of* Dentalium, *or Elephant's tusk shell (5cm/2in);* ***2*** Solen marginatus, *a razor shell (12.5cm/4.9in);* ***3*** *A chiton (class Polyplacophora). Chitons are flattened sedentary mollusks with eight overlapping shell plates;* ***4*** Nucella lapillus, *a dog whelk; rocky shores of NW Europe (3cm/1.2in high);* ***5*** Buccinum undatum, *the large European whelk (8cm/3in);* ***6*** *A species of* Neopilina, *a genus of limpetlike mollusks (4cm/1.6in);* ***7*** *A species of* Aplysia, *or sea hare (15cm/5.9in);* ***8*** Tridacna gigas, *the largest living mollusk, or giant clam (1.35m/4.4ft);* ***9*** *A common octopus (genus* Octopus*);* ***10*** Falcidens gutterosus, *a chaetoderm, Mediterranean (1.5cm/0.6in).*

When needed for growth or shell repair, the salts are transported to the mantle by migratory cells.

Molluskan shells show great variation in shape, size, thickness, sculpture, surface texture, and shine. Marine examples are often thick and heavy, while land snails, lacking the support of water, tend to have thinner shells.

The spirally coiled shells of gastropods range from tall and spindle shaped to flat and disklike; the body whorl containing the animal itself may be small or enlarged to occupy most of the shell; and likewise the aperture or mouth of the shell can be open or constricted and armed with a range of teeth or ribs – in whelks and other carnivorous sea snails there is a groove (siphonal canal) to house the siphon. With the shell apex uppermost, the mouth of the shell in most mollusks is on the right-hand (dextral) side, but some species have the mouth on the left (sinistral). Some genera, such as Hawaiian tree snails and *Amphidromus* (both tropical land snails), and the temperate-zone whorl snails, have both dextral and sinistral examples. In some normally dextral species, a very occasional sinistral species may be found.

The nautilus, an exception among living cephalopods, has a light, brittle, spiral shell. In section this is seen to be divided behind the outermost body chamber by thin walls into progressively smaller earlier body chambers. Each wall (septum) has a central perforation, which in the living animal is traversed by a threadlike extension of the body – the siphuncle – which extends to the shell apex. The pressure of gas in the chambers affects buoyancy.

Many shells are strongly sculpted into ribs, lines, beading, knobs, or spectacular spines. Such detail, prized by shell collectors, is also used in the identification of species. The surface of the shells is sometimes rough, but in certain examples, such as cowries and olive shells, it may be smooth and glossy. Many tropical shells are very colorful and may also have attractive patterns and markings.

A number of mollusks in different groups have a reduced shell or none at all. Some sea slugs retain external shells, others have thin internal shells, and some (the nudibranchs) like land slugs are without any shell. Shell-less sea slugs have evolved to swim as well as crawl, to squeeze into small crannies, and to develop secretions and body color as means of defense. An external shell is also lacking in some parasitic gastropods, in the wormlike aplacophorans, and in most cephalopods.

Convergent Development

The Evolution of Mollusks

The success of the mollusks has been due to their adaptability of structure, function, and behavior. Mollusks are thought to be derived from either an ancestor of the Platyhelminthes (planarians, flukes, tapeworms) or from the arthropod annelid line, the latter having trochophore larvae, as do most mollusks.

There is fossil evidence of mollusks in some of the oldest fossil-bearing rocks, dating back over 530 million years to the Cambrian period. Mollusks soon evolved in different directions and the modern classes had largely separated by the end of the Cambrian, 490 m.y.a. The earliest fossils are all of marine species. Land snails appeared in the coal-measure forests of 300 m.y.a., but land snail fossils are rare until deposits of the Tertiary (65–1.8 m.y.a.).

The original molluskan shell was probably caplike. Many families, such as limpets, have reverted to that form from the spiral coiling that was widespread in gastropods (and still is) and in the few remaining shelled cephalopods. Nautiloids (now represented by only six living species) gave rise to some 3,000 known fossil species, many of which flourished in the Paleozoic seas, but they dwindled in the Mesozoic (248–65 m.y.a.) when ammonites expanded. After a successful period, ammonites suddenly vanished in their turn at the end of the Cretaceous, along with the dinosaurs.

There is no central theme to molluskan evolution. Different groups of mollusks have adapted to similar habitats, often adopting similar characteristics (convergent evolution). Among bivalves, for example, both true piddocks and the False piddock bore into mudstone, but, although the outsides of their shells are similar, the latter is more closely related to venus shells and its shell-hinge teeth are quite different. The bivalve shell of members of the class Bivalvia even has its counterpart in a different class, the bivalve gastropods.

Isolated islands, as in the Pacific, show a high level of speciation (evolution of new species) and forms that are endemic – for example the Partula snails (limited to that island) – because of the separation of the snail population from other larger populations. In Hawaii there are even local color forms of tree snails in isolated valleys.

Above *Mollusks exhibit a wide diversity of shell shapes. The Wavy-top shell* (Astraea undulosa; *main picture*)*, derives its name from the pronounced undulations at the edge of its shell. The cross section of a Pearly nautilus* (Nautilus pompilus; *inset*) *shows the chambers into which this animal's distinctively beautiful shell is divided.*

Gills and Lungs

RESPIRATION AND CIRCULATION

Mollusks originated in the sea – and their basic method of breathing is by gills that extract dissolved oxygen from the surrounding water. The typical molluskan gill (ctenidium) consists of a central axis from which rows of gill filaments project on either side (bipectinate). Blood vessels enter the filaments and the surface of each filament is covered with cells bearing cilia; some of these cells create a current in the water with their long cilia and others pick up food particles. An inhalant current of water enters the gill on one side, passes between the filaments, and goes out as an exhalant current on the other side.

In some mollusks the gill serves only for respiration, while in others (e.g. most bivalves) the two enlarged gills have a dual role of feeding as well as respiration. Gills are delicate structures, which need to be kept clear of clogging particles: the ciliary devices evolved for cleaning the gills later became adapted for feeding. The mantle is often developed at the rear end into a tube (siphon), which projects and takes in water, testing it with sensory cells and tentacles on the way; in some bivalves (e.g. the tellinids) there is a separate siphon for the exhalant current.

Blood is pumped around the molluskan body by a heart, and is distributed to the tissues by arteries but, in all except cephalopods, it has to make a slow passage back through blood spaces (hemocoel).

Monoplacophorans and chitons have several pairs of gills in the mantle cavity. In prosobranch or operculate gastropods there is typically one pair of gills, but in the more highly evolved groups the gill is reduced and lost on one side, so winkles and whelks have only one gill in the mantle cavity. The more primitive prosobranchs (e.g. slit shells, ormers, and slit and keyhole limpets) still have two gills. These limpets have lost the typical molluskan ctenidium, replacing it by numerous secondary gills of different structure, which hang down around the mantle cavity. Most prosobranchs are aquatic, breathing by gills, although some have adapted to life on land and breathe by a vascularized mantle cavity or lung.

During periods of drought, land prosobranchs can spend long spells of time inactive inside the shell, sealed off by an operculum. Some tropical land species of the family Annulariidae have developed a small tube of shell material behind the operculum that enables the snail to obtain air when the operculum is enclosed. The tropical apple snails live in stagnant water, and they have long siphons that reach above the surface of water to breathe air.

Among the sea slugs and bubble shells, the primitive shelled species such as *Acteon* and the sea hares have a gill within the mantle cavity, but in the shell-less forms there are either secondary gills, as in the sea lemons, or respiration takes place directly through the skin, which may have its surface area increased by numerous papillae.

The pulmonates (land snails and slugs, pond snails), as their name implies, are essentially lung-breathers, having lost the gill, and breathe air from a highly vascularized mantle wall. The finely divided blood vessels of the respiratory surface can often be seen as silvery lines through the thin shell. The entrance to the mantle cavity is sealed off and opens by a breathing pore (pneumostome) whenever an exchange of air is needed. The pulmonate pond snails usually breathe air and come up to the surface to open the breathing pore above water.

Some deepwater pond snails of lakes have reverted to filling the mantle cavity with water and no longer use air. While freshwater prosobranchs breathe by gills and are more often found in the oxygenated waters of rivers and streams, the pulmonates, breathing air, are better adapted for living in the stagnant, deoxygenated water of ponds and ditches.

Bivalves are excusively aquatic and breathe by gills. In primitive families, including the nut shells, the gills are relatively small, and are used only for breathing, not feeding.

Cephalopods breathe by gills in the mantle cavity. The system is highly efficient since there is a greater flow of water through the mantle cavity due to its use in jet propulsion, and faster blood circulation through a closed blood system with veins as well as arteries. The branchial hearts of most cephalopods are not present in nautiluses, which rely instead on duplication of gills to meet their respiratory needs.

Above and below *Consisting of four folds of tissue, the gills of oysters (above, a Rock oyster,* Chama *sp.) serve both for respiration, pumping water through the mantle cavity, and for feeding, gathering food particles. The octopus' siphon has a triple function, not only drawing water across the gills, but also propelling it through the water and dispensing ink for defense.*

Left *The Blue sea slug* (Glaucus atlanticus) *preys voraciously on the Portuguese man-of-war (as here) and other cnidarians. When devouring these jellyfishes, it transfers their stinging cells into its cerata (hornlike projections from its body surface) and redeploys them for its own defense.*

Herbivores, Carnivores, Parasites

FEEDING

Primitive mollusks probably fed on small particles, the macrophagous habit (eating large particles) developing only later. Most mollusks, except bivalves, feed using the radula (see above). The bivalves and some gastropods (e.g.slipper limpets of the USA, ostrich foot of New Zealand, and river snails of Europe) are ciliary feeders, either straining food from seawater (filter feeding) or sucking in sludge off the bottom (deposit feeding).

Another typical molluskan feature, which is found in some prosobranchs and bivalves, is the stomach, its wall protected by chitinous plates from a pointed, forward-projecting style that winds round and brings the string of food from the esophagus into the stomach. The style also secretes digestive enzymes.

The gastropods include browsing herbivores, ciliary feeders, detritus feeders, carnivores, and parasites. The muscular mouthparts (buccal mass) and gut show adaptions reflecting this variety. Limpets, top shells, and winkles browse on algae and other encrustations on rocks, while the Flat periwinkle eats brown seaweed and the Bluerayed limpet rasps at the fronds and stems of oarweed (*Laminaria* species). Some will eat carrion, while others attack live animals. The whelks, murexes or rock shells, volutes, olives, cones, auger, and turret shells (neogastropods) are specialized for a carnivorous diet – the shell has a siphonal canal housing a siphon that directs water over taste cells in the osphradium (a chemoreceptor in the mantle cavity), which helps in the detection of food. Certain carnivorous gastropods, like the Dog whelk and the necklace shells, drill holes in the shells of other mollusks, which are then consumed. The murex or rock shells bore mechanically, but necklace shells use acid to soften the shell before excavating with the radula.

The sea slugs and bubble shells include herbivores, suspension feeders, carnivores, and parasites, but the majority are carnivores feeding on encrusting marine animals. The sea slug *Melibe leonina*, from the west coast of the USA, is an active swimmer and adapted to feeding on crustaceans, which it catches with the aid of a large cephalic hood: the radula is absent. Shelled opisthobranchs, like species of canoe shells, lobe shells, and cylindrical bubble shells, feed on animals in the sand, including mollusks. The inside of the gizzard is lined with special plates, which they use to crush the prey. The curious pyramidellids with small coiled shells are parasitic in a range of marine animals.

Pulmonates are chiefly herbivorous, although many of them feed on dead rather than living plant material, and pond snails often consume detritus and mud on the bottom. In these animals the radula is broad, with large numbers of small teeth. There is no style nor chitinous gizzard plates, as found in bubble shells. The smaller land snails retain a microphagous diet, while a few species from different families have become carnivorous. They include the shelled slugs that eat earthworms, and a number of tropical land snails, such as the family Streptaxidae and genus *Euglandina*, which eat other snails. Some glass snails (family Zonitidae) have carnivorous tendencies and the large glass snail *Aegopis verticillus* of eastern Europe readily eats land snails.

The more primitive bivalves feed on detritus, which is pushed into the mouth by labial palps, but most modern bivalves are ciliary feeders, making use of phytoplankton, while others take in detritus from the surface of the substrate with siphons. In the more adanced bivalves, the gills are used for filtering food and conveying it to the mouth by wrapping it in mucus and passing it along food grooves to the mouth. The crystalline style and stomach plates are well developed in bivalves. The woodboring shipworms have cellulase enzymes used to digest wood shavings.

Above *Young of the Australian octopus* (Octopus australis) *hatching. After attaching her eggs to the roof of her lair, the female octopus aerates them with gentle water jets from her siphon.*

Below *An egg ribbon of the Spanish dancer nudibranch* (Hexabranchus sanguineus). *These striking agglomerations, which may contain thousands of eggs, are laid on reefs by this species of sea slug.*

The curious group of bivalve septibranchs – clams of the family Cuspidariidae – have lost the gill and have reduced labial palps and style. They are scavengers, sucking juices of dead animals. Species of the bivalve genus *Entovalva* are parasites inside sea cucumbers, and some of the freshwater mussels are parasitic in fish in the early stages of their life history.

The carnivorous cephalopods catch mostly fish, although the slower-moving octopus takes crustaceans. The radula is relatively small, but the prey is seized by jaws with a hard beak. In cephalopods, enzymes are secreted by gland cells into the tubules of the digestive gland where extracellular digestion takes place: this contrasts with the ingestion of food particles by cells of the digestive gland (intracellular digestion) in other mollusks. Extracellular digestion is a feature in which cephalopod body organization is in advance of the rest of the mollusks and parallels the situation in vertebrates.

Regulating Waste

EXCRETION

In mollusks there is a kidney next to the heart, which extracts nitrogenous waste from the blood. The excretory duct runs alongside the rectum to the pore at the mantle edge in pulmonates, but in prosobranchs there is a simple opening on the side of the kidney directly into the mantle cavity. In some mollusks certain minerals are selectively resorbed. There is little water regulation in marine mollusks but considerable activity in those of freshwater. Land mollusks conserve their water and little goes out with the excreta, nitrogenous waste being in an insoluble crystalline form and often stored in the kidney. Bivalves give off their nitrogenous waste as ammonia or its derivatives. In some opisthobranchs excretory waste is discharged into the gut.

Egglayers

BREEDING

Eggs of mollusks vary considerably. Some are shed into water before fertilization as in bivalves. Many mollusk eggs are very small but those of cephalopods are large and yolky. When fertilization is internal, elaborate egg cases may be secreted and very often gastropods lay eggs in large clutches. Some winkles and water snails deposit eggs in a jellylike matrix often attached to vegetation. Necklace shells form stiff collars of egg cases, which are large for the size of the mollusk, while whelks and murexes deposit eggs in leathery capsules attached to rocks and weed. Land snail eggs tend to be buried in soil: some are contained in a transparent envelope but others have a limy eggshell, and the eggs of one of the large African land snails, *Archachatina marginata*, are of the size and appearance of a small bird's egg. Egg masses of sea slugs can be quite spectacular when found in rock pools.

In aquatic mollusks there is usually a planktonic larva, the primitive trochophore, of short duration, and/or the characteristic veliger larva that develops from it (often within the egg capsule), with shell and ciliary lobes. Some mollusks retain the egg inside the body of the female, or in the capsule, from which the young emerge as miniature adults. The veliger lives in the plankton,

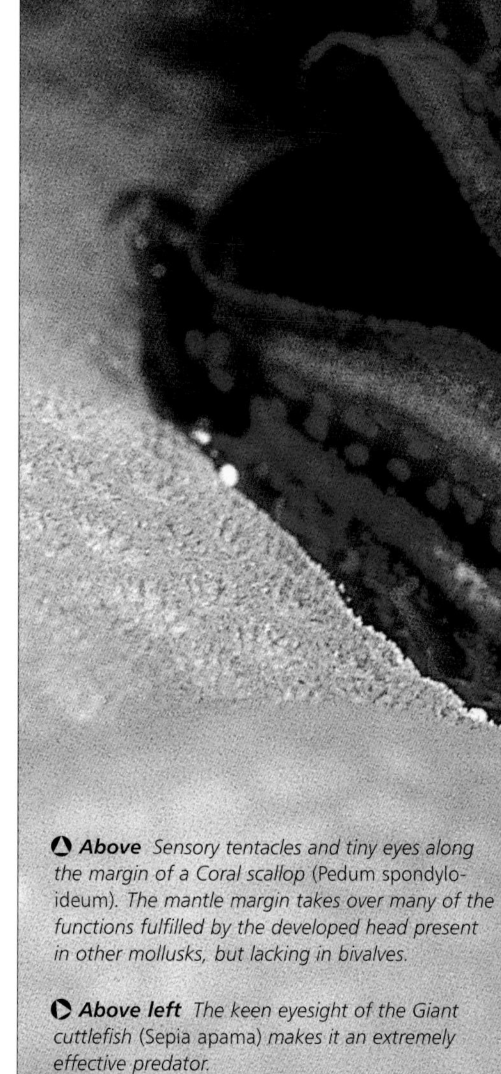

Above *Sensory tentacles and tiny eyes along the margin of a Coral scallop* (Pedum spondyloideum). *The mantle margin takes over many of the functions fulfilled by the developed head present in other mollusks, but lacking in bivalves.*

Above left *The keen eyesight of the Giant cuttlefish* (Sepia apama) *makes it an extremely effective predator.*

feeding on algae for a day to several months before settling. Vast numbers of molluskan larvae are present in the oceanic plankton and many of them are eaten by other members of the zooplankton or by filterfeeders, or perish when they fail to find a suitable habitat in which to settle.

In land and freshwater prosobranchs, the veliger stage is suppressed, and a snail hatches directly from the egg. Pulmonates also lack the veliger stage, and floating larvae occur in only some freshwater bivalves. The pelagic larva was important in establishing the freshwater Zebra mussel that first came to Britain in the first half of the 19th century. Freshwater or river mussels brood the eggs in the gills and release them as glochidia larvae parasitic on fish.

In sea snails and limpets (prosobranchs) and in sea slugs and bubble shells (opisthobranchs) the veliger is most varied. Chitons and more primitive prosobranchs such as slit shells and ormers have a trochophore larva, with a horizontal band of ciliated cells, which lasts for only a few days. Mollusks with veligers in the plankton for several weeks have a better opportunity for dispersal. At metamorphosis the ciliated lobes (velum) by which the veliger swims and feeds are engulfed, and the mollusk ends its planktonic life and sinks to the bottom. Bivalve veliger larvae also occur in the plankton, but in some, such as the small midshore *Lasaea rubra*, the young hatch as bivalves and establish themselves near the parent colony. Following the bivalve veliger stage is an intermediate pediveliger when the larva searches for a suitable place to settle; if none is found, the velum can be reinflated and the larva is carried to other sites.

Advanced and Sensitive

NERVOUS SYSTEM AND SENSE ORGANS

The brain and eye of the cephalopod are the most highly developed of any invertebrate. The molluskan nervous system essentially consists of pairs of ganglia (masses of nerve tissue), each ganglion linked by nerve fibers. In the more primitive groups, the individual ganglia are well separated, but in more highly evolved mollusks, such as land snails and whelks, there is both a shortening of the connectives, bringing the ganglia into closer association, and a concentration of most of the ganglia in the head. The ring of ganglia around the front part of the gut (esophagus) in mollusks, compares with the nervous system of annelids and arthropods. The main pairs of ganglia in mollusks are the cerebral, pleural, pedal, parietal, visceral, and buccal (receiving impulses from the head, mantle, foot, body wall, and internal organs), and there is a pedal "ladder" arrangement in monoplacophorans, chitons and the more primitive prosobranchs such as slit shells. In prosobranch gastropods there is the further complication of torsion.

The majority of mollusks are sensitive to light, which can be detected by sensors in the shell plates of chitons, the black eyespots associated with the tentacles of most gastropods, and the extremely elaborate cephalopod eye. The balance of the animal is maintained by special fluid-filled sacs, known as statocysts, containing mineral grains. Prosobranchs have a special chemosensory organ, the osphradium, but there are less specialized patches of chemosensory cells in other mollusks. The terrestrial slugs have a sense of smell, which they use to locate food.

Mollusks are also sensitive to touch: the suckers of octopuses can discriminate texture and pattern (see Learning in the Octopus), while the lower pair of tentacles of land pulmonates are largely tactile and function in feeling the way ahead.

The Foot

MOVEMENT

Some mollusks (e.g. mussels and oysters) anchor themselves to one place, but most move around in pursuit of food, for mating, and to escape enemies. The octopus crawls using suckers on its arms, modifications of the foot. Despite the mollusks' slow image, some, like cephalopods, can swim surprisingly fast.

Usually the foot is the organ involved in locomotion, which involves gliding over the surface of seabed, rock, or plant. Land snails and slugs, particularly, lay down a lubricating and protective

film of slime or mucus – the silvery trails seen on garden paths and walls. Some species "leap" with long stretches of the foot; the head lunges forward and attaches to ground ahead. Lobes of the foot (parapodia), are often developed for swimming in sea slugs, and the pelagic thin-shelled *Limacina* and sluglike *Clione* species also have swimming "wings." Sea slugs, freed from the restrictions of a shell, can swim by lateral movements using muscles in the body wall.

Planktonic larvae and some small gastropods move primarily by the beating action of cilia. Veliger larvae of gastropods and bivalves have minute, hairlike cilia on the lobes of the enlarged

Right *Gastropods, such as the Roman snail* (Helix pomatia) *move forward by creating undulating contractions of the foot. This species can cover around 7cm/min.; aquatic species are generally faster.*

velum, which are used for feeding, respiration, and locomotion. They are also able to adjust their depth by retreating into the shell to sink, then re-expanding the velum to halt the descent. Many pond snails move by cilia on the sole of the foot, and can glide along the surface film of water by this method. On land, where the body weight is not supported by water, snails moving by ciliary means are more likely to be the smaller species. Bivalves may swim by shell-flapping (scallops), they may "leap" across the surface (cockles), or burrow. Cockles use the pointed foot for moving the shell across the surface of the sand. Most bivalves, however, use the foot (which can be of considerable size) for burrowing; it is pushed forward and expanded by blood entering the pedal hemocoel (blood space); the muscles of the foot also contract and it then changes shape, often forming an anchor. Further contraction brings the shell down into the sand; as the shell closes, water jetted out of the mantle cavity can help to loosen the sand ahead. The digging cycle then starts again.

The Ubiquitous Mollusk

ECOLOGY

Mollusks have colonized the sea, freshwater, and land. Tropical regions tend to have a more diverse fauna than temperate belts, although temperate New Zealand has one of the richest molluskan land faunas.

Marine habitats can include rocky, coral, sandy, muddy, boulder and shingle shores and also the transitions between freshwater, sea, and land found in salt marshes, brackish lagoons, mangrove swamps, and estuaries. Beyond the molluskan fauna of the shores is that of the ocean, with communities below the lowtide mark in comparatively shallow water of continental shelves. There is a mosaic of different types of communities on the seafloor, relating mostly to differences in bottom materials. Certain mollusks, like squids, can form part of the free-swimming animal population (nekton) in open water. The veliger larvae of most marine mollusks float passively in the upper waters of the sea, part of the plankton. A few prosobranchs (violet sea snails and heteropods) and opisthobranchs (pteropods or sea butterflies) spend their entire adult lives on or just below the surface as part of the pelagic community. Deep waters of the abyss were once thought to be devoid of life but investigations have revealed a limited but characteristic fauna. With some exceptions (e.g. squids) abyssal mollusks are small.

Freshwater habitats colonized by mollusks include running waters of streams and rivers, still waters of lakes, ponds, canals, and temporary waters of swamps, all with their own range of species. There are both bivalves and gastropods living in freshwater, the latter including both prosobranchs and aquatic pulmonates. Foreign species can spread dramatically like the freshwater fingernail clam *Corbicula manilensis*, introduced from Asia to the USA, where it now clogs canals, pipes, and pumps. Only the gastropods have successfully colonized land; they include both prosobranchs and pulmonates, although the latter, as both slugs and snails, are the most common in temperate climates.

In the food chains of the sea, mollusks are eaten by other mollusks, as well as by starfish and by bottom-living fish such as rays. Some starfish are notorious predators of commercial mollusks such as oysters and mussels. Some whales eat large quantities of squid. On the shore seabirds probe in mud for mollusks, which can form a substantial part of their diet. Such predation by animals that are part of the natural ecosystem can usually be tolerated by mollusks, as they can be prolific breeders.

A few mollusk species have adopted the parasitic way of life. They are nearly all gastropods, with a few bivalves. Among the prosobranchs are parasites on the exterior of the host (ectoparasites), such as needle whelks (eulimids), which are parasites of echinoderms but look like normal gastropods. Internal parasites (endoparasites) are less active and have reduced body organs and less shell. The caplike genus *Thyca* lives attached to the underside of starfish, in the radial groove under the arms. *Stilifer*, which penetrates the skin of echinoderms, has a shell but it is enclosed in fleshy flaps of proboscis (pseudopallium) outside the skin of the host. The further inside the host a parasite is, the more the typical molluskan structure is lost.

Empty gastropod shells are often used by hermit crabs. Commensalism, in which one partner feeds on the food scraps of the other, is shown by the small bivalves *Devonia perrieri*, *Mysella bidentata*, and *Montacuta* species ,which live with sea cucumbers, brittle stars, and heart urchins, respectively. Symbiosis is shown by the presence of algae (zooxanthellae) in the tissues of sea slugs such as *Elysia* and *Tridachia* species and also in the mantle edge of the Giant clam.

Mollusks are also hosts to their own parasites, many of which may have become established via commensalism. Most parasites of mollusks are larvae of two-suckered flukes. Two commercially and medically important parasites are the liver fluke of sheep and cattle (*Fasciola hepatica*) and blood flukes or bilharzia of humans (see Pathogenic Parasites). Mollusks can also be parasitized by arthropods. Familiar examples are the small white mites *Riccardoella limacum* found crawling on the skin of slugs and snails and the small pea crab, *Pinnotheres pisum*, living in the mantle cavity of the Edible mussel.

Mollusks have long been used in human culture for food, fishing bait and hooks, currency, dyes, pearl, lime, tools, jewelry and ornament. Mother-of-pearl buttons were once manufactured from the shells of freshwater mussels, particularly

1

2

in the USA, where these mussels were originally common in the rivers. Most pearls come from marine pearl oysters, but at one time fine pearls were obtained from the Freshwater pearl mussel, which occurred notably in upland rivers of Wales, Scotland, and Ireland. Pearls are now cultured commercially by inserting a "seed pearl" inside the mantle skirt of the oyster.

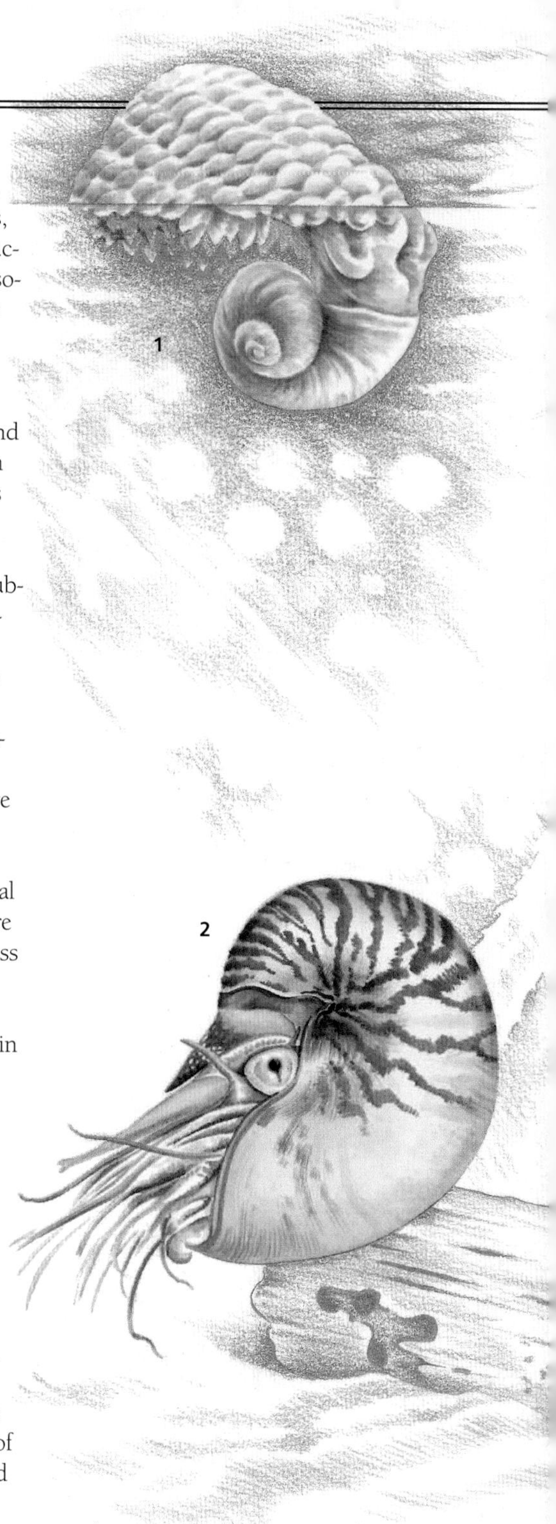

Mediterranean cooks are famous for their seafood dishes, which utilize gastropods such as necklace shells, top shells, ormers, murexes, and occasionally limpets; bivalves such as mussels, scallops, date mussels, venus shells, carpet shells, wedge shells, and razor shells; and also cephalopods, including cuttlefishes, squids, and octopuses. The traditional "escargot" of French cuisine is the pulmonate Roman snail. On the east coast of the USA the hardshell clams used by

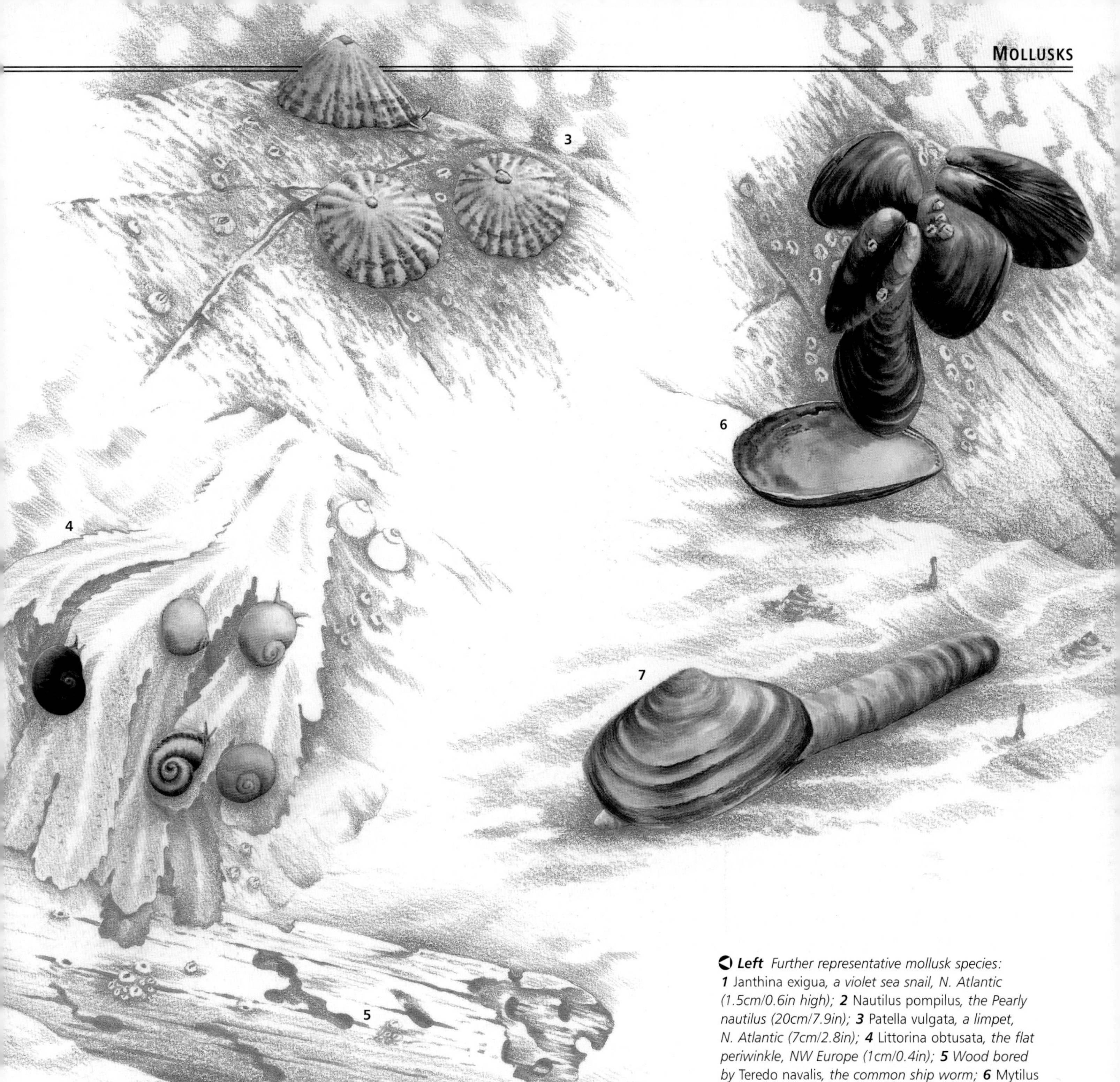

Left *Further representative mollusk species:* **1** Janthina exigua, *a violet sea snail, N. Atlantic (1.5cm/0.6in high);* **2** Nautilus pompilus, *the Pearly nautilus (20cm/7.9in);* **3** Patella vulgata, *a limpet, N. Atlantic (7cm/2.8in);* **4** Littorina obtusata, *the flat periwinkle, NW Europe (1cm/0.4in);* **5** *Wood bored by* Teredo navalis, *the common ship worm;* **6** Mytilus edulis, *the Edible mussel. It lives attached to rocks on the lower shore (10cm/3.9in);* **7** Mya arenia, *a clam or gaper (15cm/5.9in).*

cooks are quahogs (introduced from waste from the galleys of transatlantic liners to the Solent in southern England), while softshell clams are from a range of genera including gapers, carpet shells and trough shells.

Both slugs and snails can be pests of agriculture and horticulture. The mollusks are controlled by biological, cultural and chemical methods, the latter being the ones most usually employed.

A few marine mollusks also affect human activity. Bivalves like the shipworm bore into marine timbers, and gastropods – including the slipper limpet and oyster drill – are pests of oyster beds.

The major threats to mollusks are habitat destruction and pollution – the latter being more important for aquatic species, which are subject to crude oil spills, heavy metals, detergents, fertilizer in agricultural runoff, and acid rain from distant industry. Native land snails of deciduous wood land in the American Midwest, for example, often cannot cope with the more rigorous conditions of cleared land. In consequence, much North American farmland has been colonized by European mollusks, especially slugs, introduced with plants.

In addition to the harvesting of natural populations and measures limiting trade in shells, others aim to prevent introductions of non-native species. The Red List compiled by the World Conservation Union (IUCN) collates data on endangered species that can be used in their conservation.

Little is known of the molluskan fauna of some of the potentially richest and most threatened habitats, many of which are delicate and intolerant of disturbance. Hundreds of snail species are likely to be exterminated before they are even described and studied.

The 7 Classes of Mollusks

THIS TAXONOMIC ACCOUNT INCLUDES ALL species, genera, and families mentioned in the text. For reasons of space, divisions such as suborder sand superfamilies, important in some groups, have been omitted. The sequence of families reflects their relationships. The Gastropoda and Bivalvia are the 2 classes that between them contain around 1,048 threatened species.

CLASS MONOPLACOPHORA

Monoplacophorans

25 species, including *Neopilina galathea*.

CLASS APLACOPHORA

Solenogasters, Chaetoderms

About 370 species in 2 subclasses: Solenogastres, including *Epimenia verrucosa*; and Caudofoveata.

CLASS POLYPLACOPHORA

Chitons or Coat-of-mail shells

About 1,000 species, including the Giant Pacific chiton (*Amicula stelleri*) and *Mopalia* species.

CLASS GASTROPODA

Slugs and Snails (gastropods)

About 60,000–75,000 species in 3 subclasses. In this class alone there are over 1,000 threatened species, mostly within the Monotocardia, Basommatophora, and Stylommatophora.

PROSOBRANCHS OR OPERCULATES
SUBCLASS PROSOBRANCHIA

ORDER DIOTOCARDIA

Families, genera and species include: Slit shells (family Pleurotomariidae), including *Pleurotomaria* species; ormers and abalones (family Haliotidae), including *Haliotis* species; slit and keyhole limpets (family Fissurellidae), including *Diodora*, *Fissurella* species, Great keyhole limpet (*Megathura crenulata*); true limpets (family Patellidae), including Common limpet (*Patella vulgata*), Bluerayed limpet (*Patina pellucida*); top shells (family Trochidae), including Thick top shell (*Monodonta lineata*); turban shells (family Turbinidae), including Tapestry turban (*Turbo petholatus*); pheasant shells (family Phasianellidae), including Pheasant shell (*Tricolia pullus*), Australian pheasant shell (*Phasianella australis*).

ORDER MONOTOCARDIA

Mesogastropods: Apple snails (family Ampullarlidae), including *Pila*, *Pomacea* species; river snails (family Viviparidae), including *Viviparus viviparus*; winkles or

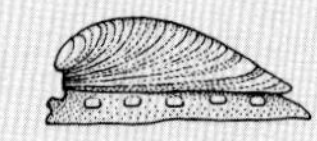
Class Monoplacophora
Monoplacophorans

Class Aplacophora
Solenogasters, chaetoderms

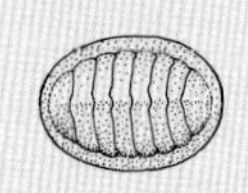
Class Polyplacophora
Chitons or coat-of-mail shells

Class Gastropoda
Slugs and snails (gastropods)

periwinkles (family Littorinidae), including Dwarf winkle (*Littorina neritoides*), Flat winkle (*L. littoralis*), Edible winkle (*L. littorea*); roundmouthed snails (family Pomatiidae), including *Pomatias elegans*; spire snails (family Hydrobiidae), including Jenkins' spire shell (*Potamopyrgus jenkinsi*); family Omalogyridae, including *Ammonicera rota*; vermetids (family Vermetidae), including *Vermetus* species; sea snails (family Janthinidae), including Violet sea snail (*Janthina janthina*); family Styliferidae (parasites), including *Stylifer*, *Gasterosiphon* species; family Eulimidae (parasites), including *Eulima*, *Balcis* species; family Entoconchidae (parasites), including *Entoconcha*, *Entocolax*, *Enteroxenos* species; slipper limpets, cup-and-saucer and hat shells (family Calyptraeidae), including Atlantic slipper limpet (*Crepidula fornicata*); family Capulidae (parasites), including *Thyca* species; ostrich foot shells (family Struthiolariidae), including *Struthiolaria* species; conch shells (family Strombidae), including Pink conch shell (*Strombus gigas*); cowries (family Cypraeidae), including Money cowrie (*Cypraea moneta*), Gold ringer (*C. annulus*); necklace and moon shells (family Naticidae), including *Natica* species. About 369 species are threatened: 3 *Aylacostoma* species are Extinct in the Wild, including *A. stigmaticum*; 40 species are Critically Endangered, including 7 species of the pebblesnail *Somatogyrus*; and 100 species are Endangered, including *Beddomeia fallax*, *Graziana klazenfurtensis*, *Jardinella pallida*, *Lanistes solidus*.

Neogastropods: Whelks (Buccinidae), including Edible whelk (*Buccinum undatum*); dog whelks (family Nassariidae), including *Bullia tahitensis*; spindle shells (family Fasciolariidae), including *Fasciolaria* species; rock shells or murexes (family Muricidae), including *Murex* species, Common dog whelk (*Nuceila lapillus*), Oyster drill (*Urosalpinx cinerea*); volutes (Volutidae), including *Voluta* species; olives (family Olividae), including *Oliva* species; turret shells (family Turridae); cones (family Conidae), including Courtly cone (*Conus aulicus*), Geographer cone (*C. geographicus*), Marbled cone (*C. marmoreus*), Textile cone (*C. textile*), Tulip cone (*C. tulipa*); auger shells (family Terebridae), including *Terebra* species.

7 species are listed as threatened by the IUCN: 4 *Conus* species are Vulnerable, including *C. africanus*, and 3 are Lower Risk/Near Threatened, including *Latiaxis babelis* and 2 *Ranella* species.

Right *The Panama Horse conch (Fasciolaria princeps), a gastropod, is an aggressive predator that feeds on other mollusks.*

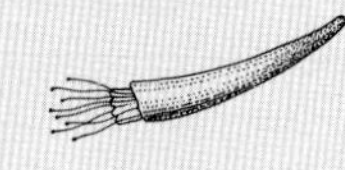
Class Scaphopoda
Tusk or tooth shells

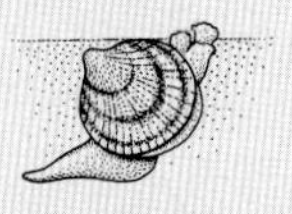
Class Bivalvia (Pelecypoda)
Clams and mussels (bivalves)

Class Cephalopoda
Cephalopods

SEA SLUGS AND BUBBLE SHELLS
SUBCLASS OPISTHOBRANCHIA

BUBBLE SHELLS ORDER BULLOMORPHA

Acteon shells (family Acteonidae), *Acteon* species; bubble shells (family Hydatinidae), including *Hydatina* species; cylindrical bubble shells (family Retusidae), including *Retusa* species; bubble shells (family Bullidae), including *Bullaria* species; lobe shells (family Philinidae), including *Philine* species; canoe shells (family Scaphandridae), including *Scaphander* species.

ORDER PYRAMIDELLOMORPHA

Pyramid shells (family Pyramidellidae).

ORDER THECOSOMATA

Sea butterflies or pteroptods (family Spiratellidae), including *Limacina* species.

ORDER GYMNOSOMATA

Sea butterflies or pteropods (family Clionidae), including *Clione* species.

ORDER APLYSIOMORPHA

Sea hares (family Aplysiidae), including *Aplysia* species.

ORDER PLEUROBRANCHOMORPHA

ORDER ACOCHLIDIACEA

Family Hedylopsidae, including *Hedylopsis* species; family Microhedylidae, including *Microhedyle* species.

ORDER SACOGLOSSA

Bivalve gastropoids (family Julidae), including *Berthelinia limax*; family Elysiidae, including *Elysia*, *Tridachia* species; family Stiligeridae, including *Hermaea* species; family Limapontiidae, including *Limapontia* species.

SHELL-LESS SEA SLUGS
ORDER NUDIBRANCHIA

Sea slugs (family Dendronotidae), including *Dendronotus* species; sea lemons (suborder Doridacea); family Tethyidae, including *Melibe leonina*; family Aeolidiidae, including Common gray sea slug (*Aeolidia papillosa*); floating sea slugs (family Glaucidae), including *Glaucus atlanticus*, *G. marginata*.

LUNGBREATHERS OR PULMONATES
SUBCLASS PULMONATA

TROPICAL SLUGS
ORDER SYSTELLOMMATOPHORA

Genera include: *Veronicella*.

POND AND MARSH SNAILS
ORDER BASOMMATOPHORA

Operculate pulmonates (family Amphibolidae), including *Salinator* species; dwarf pond snails (family Lymnaeidae), includin *Lymnaea truncata*; bladder snails (family Physidae), including *Physa* species, Moss bladder snail (*Aplexa hypnorum*); ramshorr snails (family Planorbiidae), including *Buli nus*; freshwater limpets (family Ancylidae).

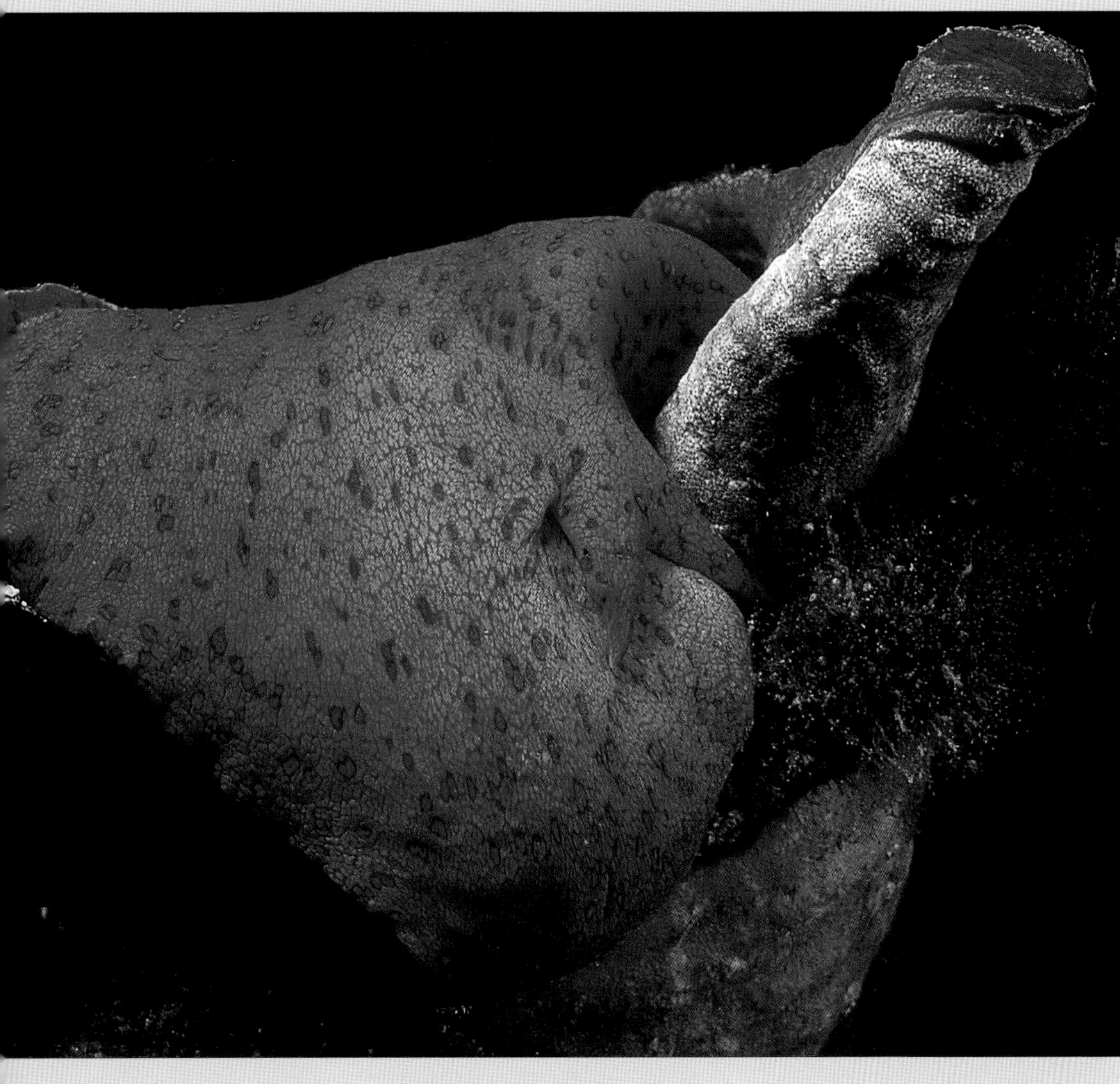

About 39 species are endangered: 8 are Critically Endangered, including *Gyraulus cockburni, Lantzia carinata*; 5 are Endangered, including *Afrogyrus rodriguezensis, Glyptophysa petiti*; 15 are Vulnerable, including *Bulinus nyassanus*.

LAND SNAILS AND SLUGS
ORDER STYLOMMATOPHORA

Hawaiian tree snails (family Achatinellidae), including *Achatinella* species; whorl snails (family Vertiginidae), including *Vertigo* species; African land snails (family Achatinidae), including *Archachatina maryinata*; family Oleacinidae, including *Euglundina* species; shelled slugs (family Testacellidae), including *Testacella* species; glass snails (family Zonitidae), including Garlic snail (*Oxychilus alliarius*), *Aegopis verticillus*; family Limacidae, including Gray field slug (*Deroceras reticulatum*), Great gray slug (*Limax maximus*); family Chamaemidae, including *Amphidromus* species; family Helicidae, including Brownlipped snail (*Cepaea nemoralis*), Common garden snail (*Helix lucorum*), Roman snail (*H. pomatia*), Desert snail (*Eremina desertorum*), *Eobania vermiculata, Otaila lactea*; carnivorous snails (family Streptaxidae).
About 620 species are threatened: 9 species (all genus *Partula*) are Extinct in the Wild, including *P. hebe*; 116 species are Critically Endangered, including 24 species of *Achatinella*, 9 species of *Opanara*, and 7 species of *Belgrandiella*; 101 species are Endangered including *Ampelita julii, Bellamia robertsoni, Hirasea insignis, Thapsia snelli*, and *Victaphanta compacta*.

CLASS SCAPHOPODA

Tusk or Tooth shells

About 900 species. Family Dentaiiidae, including Elephant tusk shell (*Dentalium elephantinum*); family Cadulidae, including *Cadulus* species.

CLASS BIVALVIA (PELECYPODA)

Clams and Mussels (bivalves)

About 20,000 species in 7 orders. 157 species are listed as threatened by the IUCN.

ORDER PROTOBRANCHIA

Includes nut shells (family Nuculidae), including *Nucula* species.

ARK SHELLS, DOG COCKLES
ORDER TAXODONTA

Includes dog cockles (family Glycimeridae), including *Glycimeris* species.

ORDER ANISOMYARIA

Mussels (family Mytilidae), including Date mussels (*Lithophaga* species), Edible mussels (*Mytilus edulis*), *Botulus, Fungiacava* species; pearl oysters (family Ptetiidae), including *Pinctada martensil*; scallops (family Pectinidae), including *Pecten* species, Queen scallop (*Aequipecten opercularis*); file shells (family Limidae), including *Lima* species; saddle oysters (family Anomiidae), including Window oyster (*Placuna placenta*), oysters (family Ostreidae), including *Ostrea, Crassostrea* species.

ORDER SCHIZODONTA

Freshwater pearl mussel (family Margaritiferidae), *Margaritifera margaritifera*; freshwater or river mussels (family Unionidae), including *Unio, Anodonta, Lampsilis* species; tropical freshwater mussels (family Aetheriidae), including *Aetheria* species.
Around 52 species are Critically Endangered, including the Oyster mussel (*Epioblasma capsaeformis*), Marshall's mussel (*Pleurobema marshalli*); 28 are Endangered, including the Georgia spiny mussel (*Elliptio spinosa*), Dwarf wedge mussel (*Alasmidonta heterodon*); 8 species are Vulnerable.

ORDER HETERODONTA

Cockles (family Cardiidae); giant clams (family Tridacnidae), including *Tridacna gigas, T. crocea*; pea shells (family Sphaeriidae), *Pisidium* species; Fingernail clam (family Corbiculidae), *Corbicula manilensis*; Adantic hardshell clam (family Arctidae), *Arctica islandica*; Zebra mussel (family Dreissenidae), *Dreissena polymorpha*; Ruddy lasaea (family Erycinidae), *Lasaea rubra*; family Galleomatidae, including *Devonia perrieri*; montagu shells (family Montacutidae), including *Montacuta* species, *Mysella bidentata*; venus and carpet shells (family Veneriidae), including Venus shells (*Venus* species) – quahog or Hard-shell clam (*V. mercenaria*), smooth Venus (*Callista* species) – carpet shells (*Venerupis* species); False piddock (family Petricolidae), *Petricola pholadiformis*, oval piddocks (*Zirfaea* species); wedge shells or bean clams (family Donacidae), including *Donax* species; Tellins (family Tellinidae), including *Tellina folinacea*.
4 species of *Tridacna* are classed as Vulnerable by the IUCN, including *T. gigas* and *T. rosewateri*.

ORDER ADEPEDONTA

Trough shells (family Mactridae), including *Spisula* species; razor shells (family Solenidae), including *Ensis* species; gapers (family Myidae), including *Mya, Platyodon* species; Rock borer or Red nose (family Hiatellidae), *Hiatella arctica*; flask shells (family Gastrochaenidae), including *Gastrochaena* species; piddocks (family Pholadidae), including *Pholas* species, wood piddocks (*Xylophaga* species); shipworms (family Teredinidae), including *Teredo* species.

SEPTIBRANCHS
ORDER ANOMALODESMATA

Dipper clams (family Cuspidariidae), including *Cuspidaria* species.

CLASS CEPHALOPODA

Cephalopods

About 900 species in 3 subclasses.

NAUTILOIDS
SUBCLASS NAUTILOIDEA

Includes Pearly nautilus, *Nautlius* species.

AMMONITES†
SUBCLASS AMMONOIDEA

Including Giant ammonite (*Titanites titan*).

SUBCLASS COLEOIDEA

CUTTLEFISHES ORDER DECAPODA

Cuttlefishes, including *Sepia, Sepiola* species; squids, including *Loligo* species, flying squid (*Onycoteuthis* species), Giant squid (*Architeuthis harveyi*); also extinct belemites.

ORDER VAMPYROMORPHA

Including *Vampyroteuthis infernalis*.

OCTOPUSES ORDER OCTOPODA

Including *Octopus* species, Paper nautilus (*Argonauta* species).

Monoplacophorans

CLASS MONOPLACOPHORA

First of the seven classes in the phylum, the Monoplacophora is a small group of primitive mollusks that were originally thought to have gone extinct about 400 m.y.a. The flattish caplike shell of monoplacophorans resembles that of limpets, and the groups used to be classified with the gastropods.

In 1952 living monoplacophorans were collected 3,750m (11,700ft) down in a Pacific Ocean deepsea trench off South America, and a new species of monoplacophoran, *Neopilina galathea*, was described. The shell is pale, fairly thin, caplike in shape and about 2.5cm (1in) long. On its inner surface, instead of the single horseshoe-shaped muscle scar of limpets, there are several pairs of muscle scars in a row on either side.

The particularly interesting feature of *Neopilina galathea* is the repetition of pairs of body organs: there are eight pairs of retractor muscles attaching the animal to the shell, 5–6 pairs of gills, 6–7 pairs of excretory organs, a primitive ladderlike pedal-nervous system with 10 connectives across, and two pairs of gonads. Although there is a parallel in chitons (see below), in monoplacophorans this repetition is taken much further. In consequence, it has been suggested that mollusks evolved from an annelid/arthropod ancestor, rather than from an unsegmented flatworm, and that the original segmentation of the body was lost during early molluskan evolution. Some other researchers disagree, considering the repetition of body organs to be a more recent, secondary character, rather than a primitive one.

In *Neopilina galathea* there is a molluskan radula and posterior mantle cavity and anus, showing that torsion (the twisting of body organs found in gastropods) did not occur in the monoplacophorans. Since the original discovery, further living species of monoplacophorans have been found in deepwater trenches in other parts of the world such as Aden, Yemen, and now five living species are described.

Solenogasters and Chaetoderms

CLASS APLACOPHORA

The shell-less aplacophorans are wormlike creatures that live in the mud of marine deposits, usually offshore. This small and little-understood group was once classified with the chitons, but is now placed in a group of its own. Indeed, recent research suggests that the class Aplacophora should be divided into two classes, the solenogasters (class Solenogastres) and the chaetoderms (class Caudofoveata).

Aplacophorans do not have an external shell, although there may be tiny, pointed calcareous spicules in the skin, sometimes of a silvery, "fur-like" appearance. In common with other mollusks, most possess a radula, a mantle, mantle cavity, a foot, and a molluskan-type pelagic larva. There are dorsoventral muscles crossing the body, reminiscent of similar structures in flatworms, flukes, and tapeworms.

Most species are small, but the solenogaster *Epimenia verrucosa* can reach 30cm (12in) in length. Solenogasters are fairly mobile and can twist their bodies around other objects; the foot is reduced to a ventral groove. Chaetoderms (named for their spiny skin), live in mud, moving like earthworms but spending much of their time in a burrow. The radula is often reduced in this group.

Above *Hatching eggs of the Garden snail* (Helix aspersa). *Newly hatched snails have a tiny, fragile shell, and take about two years to reach maturity.*

Left *Most chitons are herbivorous, but this species,* Craspedochiton laqueatus, *is a carnivore that feeds primarily on small crustaceans.*

Chitons or Coat-of-mail Shells

CLASS POLYPLACOPHORA

Chitons have a dinstinctive oval shell consisting of eight plates bounded by a girdle. They are exclusively marine and, with the exception of a few deepwater species, mostly limited to shores and continental shelves. Beneath the shell with its low profile and stable shape, the animal attaches to the rock by a suckerlike foot. The plates of the shell are well articulated – chitons can roll up into a ball when disturbed. The articulations are also an advantage when moving over the uneven surface of rocks.

Although there are no obvious eyes, chitons are sensitive to light through receptors in the shell. They are found mostly on rocky shores. When a boulder is turned over, chitons on the underside quickly move down again out of the light.

Chitons have remained substantially unchanged since the Cambrian period 543–490 m.y.a. The different species are identified by the relative width of the girdle protecting the mantle, by the sculpturing of the shell valves and the bristles they bear, and also the teeth and the surfaces of the joins between the valves. Most chitons are 1–3cm (0.4–1.2in) long, but some, like the Giant Pacific chiton can reach 20–30cm (8–12in). The larger and more spectacular species are found on the Pacific coast of the USA and off Australia.

Chitons are browsers, rasping algae and other encrusting organisms off the rock with the hard teeth of the radula. One family, the Mopaliidae, is carnivorous, feeding on crustaceans and worms.

The anatomy of the chiton is closer to that of the primitive ancestor than the more highly evolved gastropods. The mantle cavity and anus are situated toward the rear. The mantle extends forward and houses several pairs of gills, rather more than in most other classes of mollusks. Chitons resemble monoplacophorans and the more primitive groups of prosobranchs, such as the ormers and the top shells, in the ladderlike nervous system with paired ventral nerves and cross-connections. Sexes are separate in the chitons, and the eggs are fertilized externally.

Slugs, Snails, and Whelks (Gastropods)

CLASS GASTROPODA

This largest class of mollusks contains three-quarters of the living species and shows the greatest variation in body and shell form, function, and way of life. Unlike the other classes, which are all aquatic, gastropods have also adapted very successfully to life on land and they have achieved greater diversity in freshwater than the bivalves. Gastropods occur in all climatic zones of the world, colonizing the sea, brackish and freshwater habitats, and land. They are among the earliest molluskan fossils.

In the more primitive prosobranchs, gastropods with a caplike shell, such as slit shells and ormers, which have a trochophore larva with only a short pelagic phase, torsion occurs after the larva has settled. In the more advanced prosobranchs, such as winkles and whelks, the newly hatched veliger larva already has the mantle cavity to the front. Sea slugs and bubble shells do not retain torsion in adult life; loss of shell was influential in the development of this trend. Land and pond snails and slugs have retained torsion, but their nervous system is not twisted.

For the larva, torsion may provide a space into which the animal can quickly contract, enabling it to sink out of dange, or to reach the bottom for settling when it metamorphoses. Advantages of torsion for the adult may include the use of mantle cavity sensors for testing the water ahead or possibly the intake of cleaner water not stirred up by the foot, and providing a space into which the head can be withdrawn. Besides the looping of the

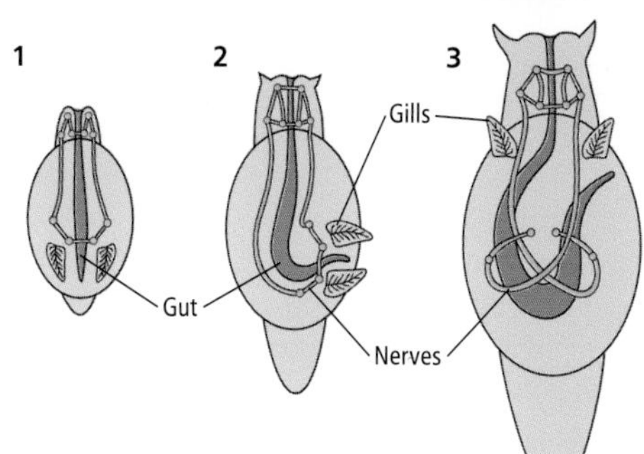

Below Body plan of a lung-breathing freshwater snail. Characteristic of gastropods are the flat, creeping foot sole, the distinct head with tentacles, mantle, and coiled shell made of one piece (univalve). In most gastropods, rotation (torsion) of the body in the developing embryo (**1–3**) has brought the opening of the mouth cavity, anus, and other organs to the front, and the nervous system is twisted. Sea slugs and Bubble shells lose torsion in adult life, and the nervous system of land gastropods is not twisted.

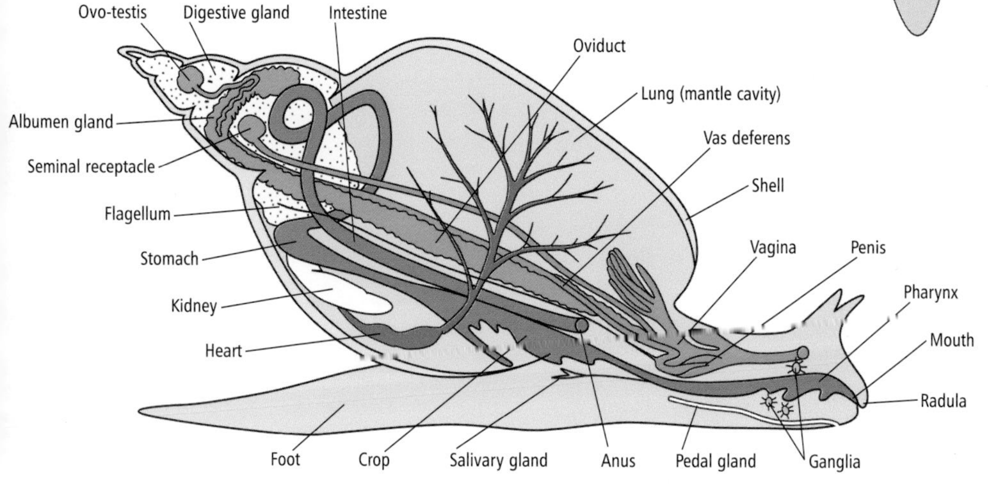

alimentary canal and reorientation of the reproductive organs, torsion causes the twisting of the prosobranch nervous system (streptoneury).

The spiral coiling of the gastropod shell is a separate phenomenon from torsion. Coiling occurs also in some of the cephalopods (e.g. nautiluses), which do not exhibit torsion, and is a way of making a shell compact. If the primitive caplike shell had become tall it would have been unstable as the animal moved along. Spiral coiling is brought about by different rates of growth on the two sides of the body. During their evolutionary history, various spirally coiled mollusks in the gastropods, extinct ammonites, and nautiloids have uncoiled, producing loosely coiled shells, tubular forms, or, in for example limpets, a return to the caplike shape.

Prosobranchs

SUBCLASS PROSOBRANCHIA

Prosobranchs include most of the gastropod seashells – limpets, top shells, winkles, cowries, cones, and whelks – as well as a number of land and freshwater species. The prosobranchs, or operculates, have separate sexes, unlike the two other subclasses of gastropods, which are hermaphrodite, and the mouth of the shell is usually, except in limpet forms, protected by a lid or operculum (see below). Aquatic prosobranchs have a gill in the mantle cavity, together sometimes, especially in the carnivorous groups, with a chemical sense organ (osphradium) and a slime-secreting hypobranchial gland. The mantle and associated structures exhibit torsion and as a result the principal nerves are twisted into a figure-8 in the more highly evolved orders (Mesogastropoda and Neogastropoda).

In marine species there is usually a planktonic trochophore or veliger larval stage that helps to distribute the species.

Prosobranchs live in seawater, brackish water, freshwater, and on land and have an ancient fossil history going back to the Cambrian. Terrestrial prosobranchs are more abundant and varied in the tropical regions than in temperate zones. Shells vary in shape from the typical coiled snail shell to the caplike shells of limpets and the tubular form of the warm-water vermetids (family Vermetidae) that look from the shell more like marine worm-tubes than mollusks. This diverse group exploits most opportunities in feeding from a diet of algal slime, seaweed, detritus, suspended matter and plankton (ciliary feeders), to terrestrial plants, dead animal matter, and other living animals.

The lid, or operculum, is secreted by glands on the upperside of the back of the foot. It is the last part of the animal to be withdrawn and therefore acts as a protective trapdoor. It keeps out predators and also prevents water loss in land prosobranchs and intertidal species. The operculum is present in the veliger larvae, even in limpets and slipper limpets, which later lose the operculum.

Above *The Warted egg cowry* (Calpurnus verrucosus) *lives only on Leather coral; the spots on its mantle blend perfectly with the coral for camouflage.*

Right *An interesting defensive strategy is adopted by the Rough keyhole limpet* (Diodora aspera). *When threatened by a starfish, it raises itself and extends its mantle over its foot and the edge of its shell, which makes it hard for the starfish to gain any purchase.*

The opercula of most prosobranchs are horny, but hard calcareous ones are found in some of the turban and pheasant shells. The thick operculum of the Tapestry turban shell is green and may be up to 2.5cm (1in) across: this is the "cat's eye" used for jewelry. The much smaller pheasant shell *Tricolia pullus* from northern Europe and the Australian pheasant shell and others have conspicuous white calcareous opercula.

The different shapes of opercula usually fit the form of the mouth of the shell. Shells with a narrow aperture, such as cones, for example, have a tall, narrow operculum. In some species, such as the whelk *Bullia tahitensis*, there are teeth on the operculum. In the Pink conch shell these teeth are thought to be defenses against attack by predators that include tulip shells. In some species, particularly the land prosobranchs, the operculum may indeed be small, enabling the animal to retreat further inside its shell.

Sea slugs and Bubble shells (Opisthobranchs)

SUBCLASS OPISTHOBRANCHIA

Sea slugs and bubble shells are hermaphrodite – both male and female reproductive systems function in the same individual – and usually have a reduced shell or none at all. Bubble shells do have

CONES – THE VENOMOUS SNAILS

The 400–500 species of cone shells are found mostly in tropical and subtropical waters. These great favorites among collectors are unusual exceptions among mollusks in that they are directly harmful to humans.

Cones are carnivores, taking a range of prey, from marine worms to sizable fish, and those feeding on fish are most dangerous. In common with other carnivorous gastropods, such as whelks, cones have a proboscis. This muscular retractable extension of the gut carries the mouth, radula, and salivary gland forward to reach food in confined spaces or for other reasons at a point distinct from the animal. When the probing, extended proboscis of a cone touches a fish, it embeds one of the harpoonlike teeth on the tip of the radula into the prey, accompanied by a nerve poison that paralyzes the fish: the cone swallows the fish whole.

The Geographer cone, Textile cone (RIGHT), and Tulip cone have been known to kill humans, while others like the Courtly cone, Marble cone, and *Conus striatus* can cause an unpleasant, although not fatal, sting. JEC

a normal external shell, which in the Acteon shell looks very like that of a prosobranch, as it is fairly solid with a distinct spire. Most of the bubble shells, such as *Hydatina*, *Bullaria,* and canoe shell species, have an inflated shell that consists mostly of body whorl with little spire, is rather brittle, and houses a large animal. Other opisthobranchs have a reduced shell that is internal, for example the thin bubble shells of *Philine* and *Retusa* species. Sea hares have a simple internal shell plate in the mantle that is largely horny. A few species have a bivalve shell (see below). The rest of the sea slugs have lost their shell altogether. They include *Hedylopsis* species with hard spicules in the skin, *Hermaea* species and other sacoglossans such as *Limapontia*, and the large group of the nudibranchs (meaning "exposed-gills") or sea slugs, including the sea lemons, the family Aeolidiidae and Dendronotidae, and many others.

The bodies of opisthobranchs, particularly nudibranchs, can be very colorful. Although they may function as warning coloration or camouflage, little is known of the function of such bright colors, which are less vivid at depth under water. Some sea lemons emit acid from glands in the mantle as a defense against predators.

Opisthobranchs reproduce by laying eggs, often in conspicuous egg masses. The eggs hatch to veliger larvae.

Some species of bubble shells, such as *Retusa* and Acteon shells, have a blunt foot, which they use to plow through surface layers of mud or sand. The round shell-less sea lemons creep slowly on the bottom with the flat foot, but many of the sea slugs are agile and accomplished swimmers, capable of speed. Sea butterflies or pteropods swim in surface waters of the oceans.

In 1959 a malacological surprise came to light – an animal with a typical bivalve shell but a gastropod body, complete with flat creeping sole and tentacles. This was *Berthelinia limax*, which was found living on the seaweed *Caulerpa* in Japan. Other bivalve gastropods have since been discovered, also on seaweed, in the Indo-Pacific and Caribbean as well as Japanese waters. They are classed with the sea slugs of the order Ascoglossa.

Bivalve gastropods have a single-coiled shell in the veliger larva. In mature shells this is sometimes retained at the prominent point (umbone) of the lefthand valve. This development of a bivalve shell in gastropods is an example of convergent evolution rather than evidence of an ancestor shared with the class Bivalvia.

Lungbreathers or Pulmonates

SUBCLASS PULMONATA

In pond snails, land snails, and slugs, the mantle wall is well supplied with blood vessels and acts as a lung. In parallel with this specialization, the lungbreathing snails have specialized in colonizing land and freshwater, although a few continue to live in marine habitats. Like the sea slugs and bubble shells, pulmonates are hermaphrodite, with a complex reproductive system: the free-swimming larval stage is lost in land and freshwater species and, except in the marine genus *Salinator*, there is no operculum.

The shell is usually coiled, although the varied shapes include the limpet form, and in several unrelated families the shell is reduced or lost altogether, leading to the highly successful body form of slugs. The thin shell of land snails still offers some protection against drying out, but is more portable and requires less calcium than do the shells of marine gastropods. Both snails and slugs further conserve body water by being active chiefly at night and by their tendency to seek out crevices. The shell-less slugs are freed from restriction to calcareous soils and can also retreat into deeper crevices.

The body is differentiated into a head with tentacles (one or two pairs), foot, and visceral mass. The mantle and mantle cavity are at the front (still showing signs of torsion) but the entrance is sealed off except at the breathing pore (pneumostome), which can open and close. The mouthparts and their muscles (buccal mass) may incorporate both a radula and a jaw. Pulmonates are predominantly plant feeders although there are a few carnivores.

The subclass may be divided into three superorders. In the mostly tropical Systellommatophora, the mantle envelops the body. The pond snails (superorder Basommatophora) have eyes at the

Right *The Blue dragon sea slug* (Pteraeolina ianthina) *is common in Southeast Asian and Australian waters. If disturbed, it sheds several of its cerata in the defensive strategy known as autotomy.*

base of their two tentacles, while the land snails and slugs have two pairs of tentacles and eyes at the tips of the hind pair.

Pulmonates succeed in less stable environments than the sea by their opportunistic behavior and the fact that they can enter a dormant state during adverse periods of cold (hibernation) or drought (estivation). A solidified plug of hardened mucus (epiphragm) can seal off the mouth of the shell and in some species, such as the Roman snail, becomes hardened. Unlike the operculum of a prosobranch, the epiphragm is neither permanent nor attached by muscle tissue to the animal.

Tusk shells or Tooth shells

CLASS SCAPHOPODA

Tusk shells are a small group of around 900 species that are entirely marine and live buried in sand or mud of fairly deep waters. Only their empty shells are to be found on the beach. Tusk shells or scaphopods occur in temperate as well as tropical waters: the large Elephant tusk shell can be up to 10–13cm (4–5in) long. There are two families, the Dentaliidae, which include the large examples more commonly found, and the Siphonodentaliidae (e.g. *Cadulus* species), which are shorter, smaller, and less tubular. The oldest fossil tusk shells known date from the Ordovician period 490–443 m.y.a. Like the chitons, they have changed little and show very little diversity of body form and way of life.

The shell is tubular, tapering, curved, and open at either end. Scaphopods position themselves in the sand with the narrow end protruding above the surface, and through this pass the inhalant and exhalant currents of seawater, usually in bursts rather than as a continuous flow. The broader end of the shell is buried in the sand. From it the head and foot emerge: the foot creates a space in front into which the animal extends the tentacles of the head that pick up detritus, foraminiferans, and other microorganisms from the sand. The tentacles are sensory as well as collecting food and conveying it to the mouth. Food can be broken up by the radula, and the shells of forminiferans are further crushed by plates in the gizzard.

The anatomy of scaphopods is rather simple. The tube is lined by the mantle. There are no gills, oxygen being taken up by the mantle itself, which may have a few ridges with cells bearing tiny hair-like cilia that help to create a current. Oxygen may also be taken in through the skin of other parts of the body. There are separate sexes, fertilization takes place externally in the sea, and the egg hatches into a pelagic trochophore larva.

BIVALVE BORERS

An important number of bivalves, from seven different superfamilies, have adapted from burrowing into soft sand and mud to boring into hard surfaces like mudstone, limestone, sandstone, and wood – the woodboring habit being the most recent to evolve.

Boring developed from bivalves settling in crevices, which they subsequently enlarged – one of the giant clams, *Tridacna crocea*, does this. Rock borers of the genus *Hiatella*, while able to use crevices, also erode tunnels of circular section. They push the shell hard against the wall of the burrow by pressure of water in the mantle cavity. At low tide, the red siphons can be seen protruding from holes in the rock low on the shore – they are sometimes known as red noses.

Most rock borers make their tunnels mechanically by rotation of the shell, often aided by spikes on the shell surface, which erodes the rock. The foot may attach the animal to the end of the burrow, as in piddocks. Closure of the siphons helps keep up fluid pressure. Rock raspings are passed out from the mantle as pseudofeces. While some borers form a tunnel of even width, flask shells (*Gastrochaena* spp.) are surrounded by a jacket of cemented shell fragments and live in a rock tunnel that is narrower at the entrance. The contracted siphons dilate outside the shell to form an anchorage during boring.

The date mussels of warm seas and other bivalves have an elongated smooth shell. They make round burrows in limestone by an acid secretion from mantle tissue, which is applied to the end of the burrow. The thick, shiny brown periostracum protects the shell from the mollusk's own acid.

Among rock borers, *Botula*, *Platyodon* (gapers), and false piddocks drill in clays and mudstone, *Fungiacava* species in coral, and rock borers, flask shells, piddocks, and oval piddocks in rock. The piddocks are recognized by a projecting tooth (apophysis) inside, to which the foot muscles are attached.

Wood piddocks and shipworms burrow into wood. Wood in seawater is only a transitory habitat, and boring bivalves have thus adapted in many ways to make the most of what may be a short stay – high population densities, early maturation, prolific breeding, and dispersal by pelagic veliger larvae.

Wood piddocks use the wood only for protection, feeding on plankton by normal filtration of seawater, while shipworms exploit the wood further by ingesting the shavings, from which with the aid of cellulase enzymes, they obtain sugar. Shipworms also feed by filtration, but in some species where most food comes from the wood, gills are reduced in size. JEC

Clams, Mussels, Scallops (Bivalves)

CLASS BIVALVIA (OR PELECYPODA)

Members of this, the second largest class of mollusks with around 20,000 living species, are recognized by their shell of two valves that articulate through a hinge plate of teeth and a horny ligament, which may be inside or outside the shell.

The bivalve shell can vary considerably in shape, from circular, as in dog cockles, to elongate, as in razor shells. It can be swollen (e.g.

Right *Body plan of a bivalve. The bivalve body consists of the mantle, often extended into one or two siphons, the visceral mass, and relatively small foot. It lacks a developed head. Usually the siphons and foot can be seen protruding from the shell, but in mussels* (Mytilus, *illustrated here*), *they remain largely inside.*

Left *Donax gouldii bean clams mass on a California beach. Their superabundance varies – in some years, there are thousands, in others hardly any.*

Above *Jet swimming by a scallop,* Pecten maximus, *to escape from a starfish. It flaps its two shell valves, expelling jets of water from its mantle cavity.*

cockles) or flat (e.g. tellins) and can have radial or concentric shell sculpture (ridges, knobs, and spines), bright colors, and patterns.

The shell is closed by adductor muscles passing from one valve to the other. Where these attach to the shell, distinctive muscle scars are formed on the inside of the shell. The muscle scars are very important in identifying and classifying bivalves. In fossils they can provide clues to the way of life of long-extinct species (fossil bivalves are known from the late Cambrian period). Some bivalves, such as oysters and scallops, have a single centrally placed muscle scar (monomyarian), but most have two adductor scars, one at each end of the valves, which may be of similar size, as in cockles, or of different sizes, as in mussels. Also on the inner surface of the shell is the pallial line, the scar of attachment of the lobes of mantle lining the shell, which runs from one adductor muscle scar to the other. In those bivalves with small projecting mantle tubes (siphons), living on the surface or in shallow burrows, the pallial line is unbroken and parallel to the ventral margin of the shell, opposite the hinge. In the shells of burrowers (for example, venus shells and tellins), which have a long siphon, the pallial line is indented to provide an extra area for attachment of the muscles involved in contracting the siphon when the animal withdraws.

In most bivalves, the pair of gills is large and fills the mantle cavity, performing a dual role of respiration and feeding. The primitive nut shells, however, have small gills that are respiratory only; nut shells shovel detritus into the mouth with a pair of labial palps. The carnivorous and more highly evolved dipper clams (septibranchs, e.g. *Cuspidaria* species) have replaced gills with a wall that controls water flow into the mantle cavity. They feed on very small crustaceans and worms drawn in with water.

The bivalve reproductive system is very simple. The sexes are usually separate, although some, like oysters, do alter sex during their lives. The eggs are fertilized externally, in the sea or in the mantle cavity, by sperm taken in with surrounding water. There is a pelagic bivalve veliger larva in most species.

Most species live in the sea, but some have colonized brackish water and freshwater. The adults lead a relatively inactive life buried in the substrate, or firmly attached to rock by cement or byssus threads, or boring into stone and wood. A few, like scallops and file shells, flap the valves and swim by jet propulsion.

Heart
Kidney
Digestive gland
Ganglion
Stomach
Muscle
Mouth
Exhalant (dorsal) siphon
Muscle
Inhalant (ventral) siphon
Palps
Gills
Shell
Mantle
Foot
Ganglia
Intestine
Gonad

Octopuses, Squids, Cuttlefishes, and Nautiluses

CLASS CEPHALOPODA

Cephalopods are quite different from the rest of the mollusks in their appearance and their specializations for life as active carnivores. The estimated 900 living species are all marine and include pelagic forms, swimmers of the open sea, and bottom-dwelling octopuses and cuttlefish. While octopuses can be found in rock crevices on the lower shore, most cephalopods usually occur further out and some penetrate deep abyssal waters,

like *Vampyroteuthis infernalis*, which lives 0.5–5km (0.3–3 miles) below the surface.

Cephalopods that flourished in the seas of the Mesozoic period over 65 m.y.a. included nautiloids, ammonites, and belemnites. With the exception of nautiluses, these groups, most of which possessed shells, are now extinct.

Most modern cephalopods are descended from the extinct belemnites, which had internal shells.

Cephalopods are typically good swimmers, catching moving fish, and have evolved various buoyancy mechanisms. They are very responsive to stimuli, owing to special giant nerve fibers (axons) with few nerve-cell junctions (synapses). Thus messages can pass quickly to and from the brain. (Giant axons are also found in annelid worms and some other invertebrate groups.) The well-developed cephelopod eye focuses by moving its position rather than changing the shape of the lens. The high metabolic rate of cephalopods is also aided by a particularly efficient blood system with arteries and veins (other mollusks have arteries only) and extra branchial hearts.

All cephalopods except nautiluses have an ink sac opening off the rectum, which contains ink, the original artists' sepia. Discharged as a cloud of dark pigment, it confuses an enemy. A cephalopod can also change color while escaping. Body color and tone are changed by means of pigment cells (chromatophores) in the skin. They are operated under control of the nervous system by muscles radiating from the edge of the chromatophore which, can contract it, concentrating the pigment. Stripes and other patterns appear in the skin of cephalopods under certain conditions.

The possibility that these are a means of communication, for example in recognizing sex, is being investigated. In bottom-living species like cuttlefish (*Sepia* species), the chromatophores function as camouflage.

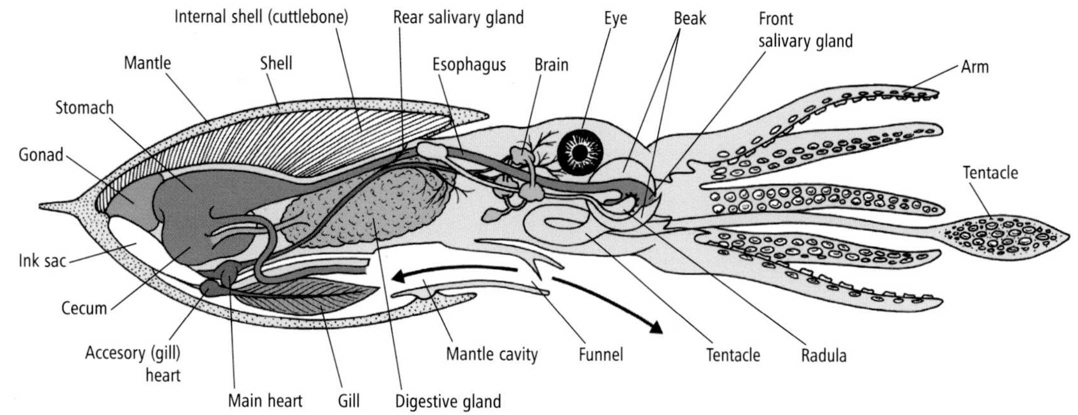

Above *California market squid* (Loligo opalescens). *mating. Grasping the females from below, the males use their third arm (hectocotylus) to insert the spermatophore into the females' mantle cavity.*

Left *Cephalopod body plan. A prominent head region is marked by mouth, eyes, arms, and a cartilage-protected brain.*

Right *The Paper nautilus* (Argonauta nodosa) *is not in fact a nautilus, but an octopus. The 'shell' is a thin egg-case produced by the female.*

Below *With a mantle length of 30cm (12in), the Maori octopus* (Octopus maorum) *is the largest octopus found in Antipodean waters.*

The sexes are separate and the male fertilizes the female by placing a sperm package (spermatophore) inside her mantle cavity, where the sperm are released. The yolky eggs are large and there is often a pelagic stage like a miniature adult. Cuttlefish come to inshore waters to breed and after egg laying the spent bodies may be washed up on beaches.

Nautiluses have a brown and white coiled shell. When the nautilus is active, some 34 tentacles protrude from the brown and white coiled shell, and to one side of these are the funnel and hood. The hood forms a protective flap when the nautilus retreats into its shell.

Cuttlefishes are flattened and usually have a spongy internal shell. They often rest on the sea bottom but also may come up and swim. They have 10 arms, eight short and two longer ones, as in squids, for catching prey. The internal shell or "bone" is often found washed up on the seashore.

Squids are torpedo-shaped, active, and adapted for fast swimming. Unlike cuttlefishes, which are solitary, squids move around in shoals in pursuit of fish. The suckers on the 10 arms may be accompanied by hooks. The internal shell or pen is reduced to a thin membranous structure. Oceanic or flying squids can propel themselves through the surface of the water.

Octopuses have adapted to a more sedentary lifestyle, emerging from rock crevices in pursuit of prey. They can both swim and crawl, using the eight arms. Female octopuses often brood their eggs. The female Paper nautilus is rather unusual in secreting, from two modified arms, a large, thin shell-like eggcase in which she sits protecting the eggs. The male of the species is very small, only one-tenth of the size of the female, and does not produce a shell. JEC/AC

LEARNING IN THE OCTOPUS

Octopuses have memory and are capable of learning. A food reward coupled with a punishment of a mild electric shock has been used successfully to train octopuses to respond to sight and touch. Sight is important to cephalopods in the recognition of prey. The well-developed eye of cephalopods approaches the acuity of the vertebrate eye more closely than that of any other invertebrate. Presented with two distinct shapes, one leading to a food reward and the other to an electric shock, the octopuses in the above test learned the "right" one after 20–30 trials, although they found some shapes easier to recognize than others, and mirror images of the same shape difficult to separate.

Similar experiments using only the tactile stimulus of cylinders with different patterns of grooves, have shown that octopuses distinguish between and respond differently to rough and smooth objects, different degrees of roughness, and objects with differing proportions of rough and smooth surfaces. Touch is perceived by tactile receptors in the octopus's suckers. Octopuses do not distinguish between objects of different weight, but they can recognize sharp edges and distinguish between inanimate objects and food.

Other researches into the brain and nervous systems of octopuses have led to important advances in knowledge of how nervous systems work in general. JEC

Sipunculans

SIPUNCULANS OCCUPY A VARIETY OF MARINE habitats. Many live in temporary burrows in soft sediment, while others can bore into chalky (calcareous) rocks, as in the genera Phascolosoma *and* Aspidosiphon, *or corals, as in* Cloeosiphon *and* Lithacrosiphon. *Some, such as* Phascolion, *inhabit empty gastropod shells. Most that have been studied so far are found in shallow waters.*

All sipunculans have an extensible introvert up to ten times the length of the trunk, with the mouth at its tip, which can be withdrawn completely into the trunk. It allows feeding at the substrate surface while the body remains protected, and can be regenerated readily if lost. The introvert is extended by fluid pressure brought about by contraction of the musculature of the trunk wall. In some cases, for example *Sipunculus* species, the circular and longitudinal muscle layers may be arranged in separate bundles. Up to four special muscles retract the introvert. The majority of sipunculans consume sediment, although some are thought to filter material from the water using cilia on their tentacles.

Excretion occurs through the one or two nephridia, which connect the large fluid-filled coelomic space to the exterior, opening on the front of the trunk. They also serve as storage organs for gametes immediately before spawning. Sipunculans have no blood vascular system; rather, the coelomic fluid performs the circulatory function.

FACTFILE

SIPUNCULANS

Phylum: Sipuncula

About 250 species in 17 genera.

Distribution Worldwide. exclusively marine; intertidal down to 6,500m (21,000ft).

Size Trunk 1–50cm (0.4–20in); introvert half to 10 times as long as trunk.

Features Unsegmented worms with coelom; body divided into introvert and trunk; mouth at tip of retractable introvert; gut U-shaped with anus usually at front of trunk; possess 1–2 excretory nephridia; ventral nerve cord unsegmented; trunk wall of outer circular and inner longitudinal muscles; asexual reproduction rare; usually unisexual; larvae free swimming. Families: Sipunculidae, including genus *Sipunculus*; Golfingiidae, including genera *Golfingia, Phascolion, Themiste*; Phascolosomatidae, genera *Phascolosoma, Fisherana*; Aspidosiphonidae, including genus *Aspinosiphon*.

Externally, sipunculans may have hooks or spines on the introvert and small protrusions (papillae) and glandular openings on the introvert and trunk. In many boring forms, a horny or calcareous shield is present at the front end of the trunk, protecting the animal as it lies retracted in its burrow.

Asexual reproduction is known in only two species, sexual reproduction with separate sexes being the rule. Some populations of *Golfingia minuta* are hermaphrodite. In all sipunculans studied, much of the germ cell development occurs free in the coelomic fluid. At spawning, eggs and sperm are released via the nephridia, and fertilization occurs in the sea. The mode of larval development, however, varies between species and may include a free-swimming trochophore-type larva, or may be direct, with juvenile worms emerging from egg masses. In several species, the floating stage occupies a period of months, and long-distance transportation of such larvae across the Atlantic has been suggested.

The genus *Golfingia* received its remarkable name from two famous 19th-century zoologists, Ray Lankester and W. C. McIntosh, who discovered a specimen of an animal new to science while playing golf on the Old Course at St Andrews, Scotland. This sipunculan was about the size of a human finger. PRG/AC

Above *Phascolosoma annulatum, with its introvert extended. This proboscis-like structure enables sipunculans to feed while keeping their bodies buried.*

Below Phascolion strombi, *a sipunculan that lines empty shells (e.g. that of* Turitella) *with mud and forms a burrow inside.*

Echiurans

IN COMMON WITH THE SIPUNCULANS AND priapulans, echiurans are only really known to scientists – hence the lack of common names for these animals. They are wormlike but their bodies are not segmented as in annelid worms. The small hooks at their posterior end gives the phylum its scientific name, which literally means "spine-tails."

Although they live mainly in marine environments, a few echiuran species penetrate into brackish waters, and many members of the family Bonelliidae are found at very great depth.

Echiurans are delicate, soft-bodied animals that live either in soft sediments or under stones in semipermanent tubes, or inhabit crevices in rock or coral. Food, usually in the form of surface detritus, is caught up in mucus and transported by cilia down the muscular proboscis to the mouth. In *Urechis* species, a net of mucus is used to filter bacteria out of water pumped through the burrow, and the net is then consumed along with the food.

In *Urechis* the rear end of the gut is modified for respiration, rhythmic contractions of the hind gut drawing in and expelling water. This modification may be linked to the absence of a blood circulatory system in this genus, the coelomic fluid containing blood cells having taken over this function. In all other echiurans, the blood circulatory system is separate from the coelom.

Right Thalassema neptuni, *with its body buried in the substrate.*

Below Metabonellia haswelli *extends its forked proboscis and generates mucus. In some species, the proboscis may measure as much as 1.5m (almost 5ft).*

FACTFILE

ECHIURANS OR SPOON WORMS

Phylum: Echiura

About 140 species in 34 genera.

Distribution Worldwide, largely marine but with a few brackish water species; intertidal down to 10,000m (33,000ft).

Size Length usually between 3 and 40cm (1.2–16in; trunk length excluding proboscis), occasionally up to 75cm (30in).

Features Unsegmented worms with coelom; trunk with muscular proboscis at front; mouth at base of proboscis; gut coiled; anus terminal; blood vascular system usually present; unsegmented ventral nerve cord; usually a pair of bristles (setae) just behind mouth; one to several excretory nephridia and a pair of anal excretory organs; sexes separate with extreme sexual dimorphism in some; produce free-swimming trochophore larva in some others. Families: Bonellidae, including genus *Bonellia;* Echiuridae, including genus *Echiurus, Thalassema;* Ikedaidae, including genus *Ikeda;* Urechidae, including genus *Urechis.*

The musculature of the trunk wall includes an outer circular layer and an inner longitudinal layer. An additional oblique layer may be present, its position being a key feature in defining families.

Reproduction in echiurans is always sexual, the sexes being separate. Mature gametes are collected into the excretory nephridia just prior to spawning, in certain cases by complex coiled collecting organs. Where fertilization occurs in the water (families Echiuridae, Urechidae), a free-swimming trochophore larva results, gradually changing to the adult morphology. In the Bonelliidae, where the male lives in or on the female, fertilization is presumed to be internal and larval development takes place essentially on the sea floor. In *Bonellia*, if the larva comes into contact with an adult female, it tends to become male, and if not, female. PRG/AC

Segmented Worms

WORMS THAT CAN SUCK BLOOD, WORMS that can be 3m (10ft) long, and worms that are vaguely reminiscent of a strip of rag – leeches, earthworms, and ragworms. These organisms are representatives of the three groups of worms known collectively as annelids. Annelids are characterized by a long, soft body and a cylindrical or somewhat flattened cross section.

An annelid worm is basically a fluid-filled cylinder, with the body wall comprising two sets of muscles, one circular, the other longitudinal. The fluid-filled cavity (the coelom) is effectively a hydrostatic skeleton (in which fluid cannot be compressed), on which the muscles work antagonistically to produce changes in width and length of the animal.

Annelids are divided into a number of segments, which in some forms give rise to a ringed appearance of the body, hence the name "ringed worms" sometimes applied to this group. Although some annelids can be broken up into individual segments and each of these can regenerate from the front or back into a completely new worm, the segments of which the normal adult is composed are not really independent. The gut, vascular system, and nervous system run from one end of the segmental chain to the other and coordinate the whole body. The series of segments is bounded by a nonsegmental structure at each end: the prostomium and a pygidium. The mouth opens into the gut immediately behind the prostomium, while the gut terminates at the anus on the pygidium. The prostomium and the pygidium are variously adapted: the prostomium, lying at the front, naturally develops organs of special sense, such as the eyes and tentacles, and hence contains a ganglion or "primitive brain," which connects with the nerve cord that runs from one end of the body to the other. The segments, although primitively similar, are also variously adapted, and the annelids demonstrate the variety of changes that can be rung on this apparently simple and monotonous body plan.

Earthworms

CLASS CLITELLATA

Earthworms and other oligochaete worms have a worldwide distribution. Earthworms may be found in almost any soil, often in very large numbers, and sometimes reaching a great size (*Megascolecides* of Australia, reaches 3m (10ft) or more in length). The commonest European earthworms are *Lumbricus* and *Allolobophora* species. The aquatic oligochaetes, sometimes known as blood-

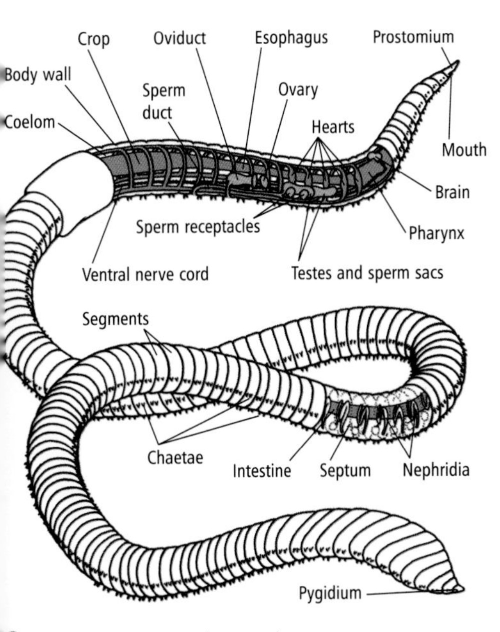

Above *Structure of an earthworm, cut away to show the internal structure, in particular the digestive system, reproductive organs, and the nerve and circulatory systems.*

Left *An earthworm pulling a leaf down into its burrow. Leaves are used to block the burrow entrance and also as food.*

FACTFILE

SEGMENTED WORMS

Phylum: Annelida

Classes: Clitellata, Polychaeta, Clitella

About 16,500 species in 3 classes (some authorities put earthworms and leeches in one class).

Distribution Worldwide in land, water, and moist habitats.

Fossil record Polychaete tubes (burrows) known from Precambrian (over 600m years ago).

Size Length 1mm–3m (0.04in–10ft).

Features Body typically elongate, divided into segments and bilaterally symmetrical; body cavity (coelom) present; gut runs full length of body from mouth to anus; nerve cord present in lower part of body cavity; excretion via paired nephridia; appendages, when present, never jointed.

EARTHWORMS Class Clitellata; order Oligochaeta
About 6,000 species in 284 genera. **Distribution:** worldwide in terrestrial, freshwater, estuarine and marine habitats. **Size:** 1mm–3m (0.04in–10ft) long. **Features:** no head appendages; few bristlelike chaetae usually in 4 bundles per segment; coelom spacious and compartmented; has blood vascular system, well developed body-wall musculature and ventral nerve cord with giant fibers in some; hermaphrodite with reproductive organs confined to a few segments; glandular saddle (clitellum) present, which secretes cocoon in which eggs develop; no larval stage.

RAGWORMS, LUGWORMS, AND BEARD WORMS Class Polychaeta
About 10,000 species in c. 80 families. **Distribution:** worldwide, essentially marine. **Size:** 1mm–2m (0.04–6.6ft) long. **Features:** morphology very variable; various feeding and sensory structures may be present at front end; body segments with lateral appendages (parapodia) bearing bristle-like chaetae; coelom spacious; has blood vascular system, nephridia, and ventral nerve cord with giant nerve fibers in some; usually sexual reproduction, with or without free-swimming larval phase.

Includes the specialized family of beard worms (Pogonophoridae). 150 species in all oceans, usually at considerable depths. **Size:** length 5cm to 1.5m (2in–5ft), usually 8–15cm (3.2–6in); diameter up to 3mm (0.1in). **Features:** bilaterally symmetrical, solitary, tube-dwelling worms; body divided into 3 zones – head (cephalic) lobe bearing tentacles, trunk, and posterior opisthosoma; gill slits, digestive tract, and anus all lacking; protostome development.

LEECHES Class Clitella; order Hirudinea
About 500 species in 140 genera. **Distribution:** worldwide, mainly freshwater, some terrestrial or marine. **Size:** 5mm–12cm (0.2–4.7in) long. **Features:** 33 segments; suckers at front and rear; coelom much reduced; blood vascular system often restricted to remaining coelomic spaces (sinuses); hermaphrodite with one of mating pair transferring sperm; glandular saddle (clitellum) produces cocoon; no larval stage.

See The 3 Classes of Segmented Worms ▷

DARWIN AND EARTHWORMS

In 1881 Charles Darwin published *The Formation of Vegetable Mould through the Action of Worms*, summarizing 40 years of observation and experiment. After discussing the senses, habits, diet, and digestion of earthworms, he concluded that they live in burrow systems and feed on decaying animal and plant material, as well as quantities of soil. They produce casts, forming a layer of vegetable mold, thereby enriching the surface soil.

He noted that earthworms often plug their burrow entrances with small stones or leaves, the latter also serving as a food source. Darwin saw signs of intelligence in the way in which leaves are grasped. When the leaf tip is more acutely pointed than the base, most are dragged by the tip, but where the reverse is true, the base is more likely to be grasped. Pine needles are almost invariably pulled by the base, where the two elements join.

The habit of producing casts at the burrow entrance results in the gradual accumulation of surface mold, and Darwin estimated that annually some 20–45 tonnes of soil per hectare (8–18 tonnes per acre) may be brought to the surface in pasture. This soil soon buries objects left on the soil surface, and may aid the preservation of archeological sites, but under certain conditions cast formation may also hasten soil erosion. Burrowing affects the aeration and drainage of the soil, which is beneficial to crop production.

Darwin failed to distinguish between earthworm species (only relatively few British earthworms make permanent burrows or produce surface casts), but he was one of the first to recognize their important ecological role. He firmly established their beneficial effects on soil, which have subsequently been largely confirmed. PRG

worms because of their deep red color, are smaller than earthworms and generally simpler in structure. Some are found in the intertidal zone, under stones, or among seaweeds. But many of the aquatic genera are found in freshwater habitats, living, for example, in mud at the bottom of lakes (*Tubifex*). Such worms commonly have both anatomical and physiological adaptations that equip them to withstand the relatively deoxygenated conditions commonly found in polluted habitats.

Oligochaetes are remarkably uniform. They entirely lack appendages, except for gills in a few species. Some families are restricted to one type of environment, for example where there are fungi and bacteria associated with the breakdown of organic material; some may eat just microflora.

Oligochaetes are exclusively hermaphrodite, with reproductive organs limited to a few front segments. Male and female segments are separate, with segments containing testes always in front of those with ovaries (the reverse is true in leeches). The clitellum, a glandular region of the epidermis (outer skin), is always present in mature animals. It secretes mucus to bind together copulating earthworms and produces both the egg cocoon and the nutritive fluid it contains. The exact position of the clitellum and the details of the reproductive organs and their exterior openings is of fundamental importance in the classification and identification of oligochaetes.

During copulation, two earthworms come together, head to tail, and each exchanges sperm with the other. The sperm are stored by each recipient in pouches called spermathecae until after the worms separate. A slimy tube formed by the clitellum later slips off each worm, collecting eggs and the deposited sperm as it goes, and is left in the soil as a sealed cocoon. The eggs are well supplied with yolk and are also sustained by the albumenous fluid surrounding them. In earthworms there is a tendency for the eggs to be provided with less yolk and rely more on the albumen. Development of the earthworms is direct (unlike most polychaetes), that is, there are no larval stages.

Parthenogenesis (development of unfertilized eggs) and self-fertilization are known in some species, and the aquatic members of the Aeolosomatidae and Naididae almost exclusively reproduce asexually.

Some earthworms are potentially longlived, but the majority of oligochaetes probably have one- to

Right *An earthworm pair* (Lumbricus terrestris) *mating. Earthworms are hermaphrodite and during mating mutually exchange sperm.*

two-year life cycles. Cocoon production occupies much of the year, interrupted by unfavorable environmental conditions, such as dryness, or by a fixed dormancy in some earthworms, which is apparently unrelated to outside conditions.

An earthworm moves by waves of muscular contraction and relaxation, which pass along the length of the body, so that a particular region is alternately thin and extended or shortened and thickened. A good grip on the walls of the burrow is aided by the spiny outgrowths of the body wall called chaetae, which project from the body wall. Chaetae are present in both polychaetes and oligochaetes; but, as their names imply, polychaetes have many chaetae in each segment, whereas oligochaetes have relatively few. Furthermore, the number of chaetae in each segment of an oligochaete is not only small but constant. The earthworm *Lumbricus*, for example, has eight chaetae in each segment. In the aquatic oligochaetes the chaetae may be longer and more slender than those of an earthworm.

The earthworm feeds either on leaves pulled into the burrow with the aid of its suctorial pharynx, or by digesting the organic matter present among the particles of soil it swallows when burrowing in earth otherwise too firm to penetrate. A muscular gizzard near the front end of the gut serves to break up compacted soil particles into smaller ones, with the result that digestion and absorption of the organic material within the intestine is more efficient. Undigested matter is extruded from the anus on to the surface of the soil as the familiar worm casts. These casts give some indication of the valuable effects earthworms have upon the soil (see Darwin and Earthworms).

Ragworms and Lugworms

CLASS POLYCHAETA

In contrast to the other annelids, polychaetes, such as ragworms, have extremely diverse forms and biology, although a common pattern is usually recognizable for each family. Polychaetes are almost all marine and they are often a dominant group, from the intertidal zone to the depths of the oceans. Planktonic, commensal, and parasitic forms are also found.

Basically, a polychaete is composed of a series of body segments, each separated from its neighbor by partitions (septa). Externally, each segment bears a pair of bibbed muscular extensions of the body wall known as parapodia, which contain internal supports (acicula), two bundles of chaetae, and a pair of sensory tentacles (cirri). The parapodia are most highly developed in active crawling forms, for example, ragworms.

Many polychaete families have an eversible pharynx, which is most conspicuous in those species that are carnivorous, feed on large pieces of plant material, or suck body fluids of other organisms. The pharynx in such families may be armed with jaws. Families with feeding tentacles or crowns use cilia to collect food from the sediment surface or water column and convey it to the

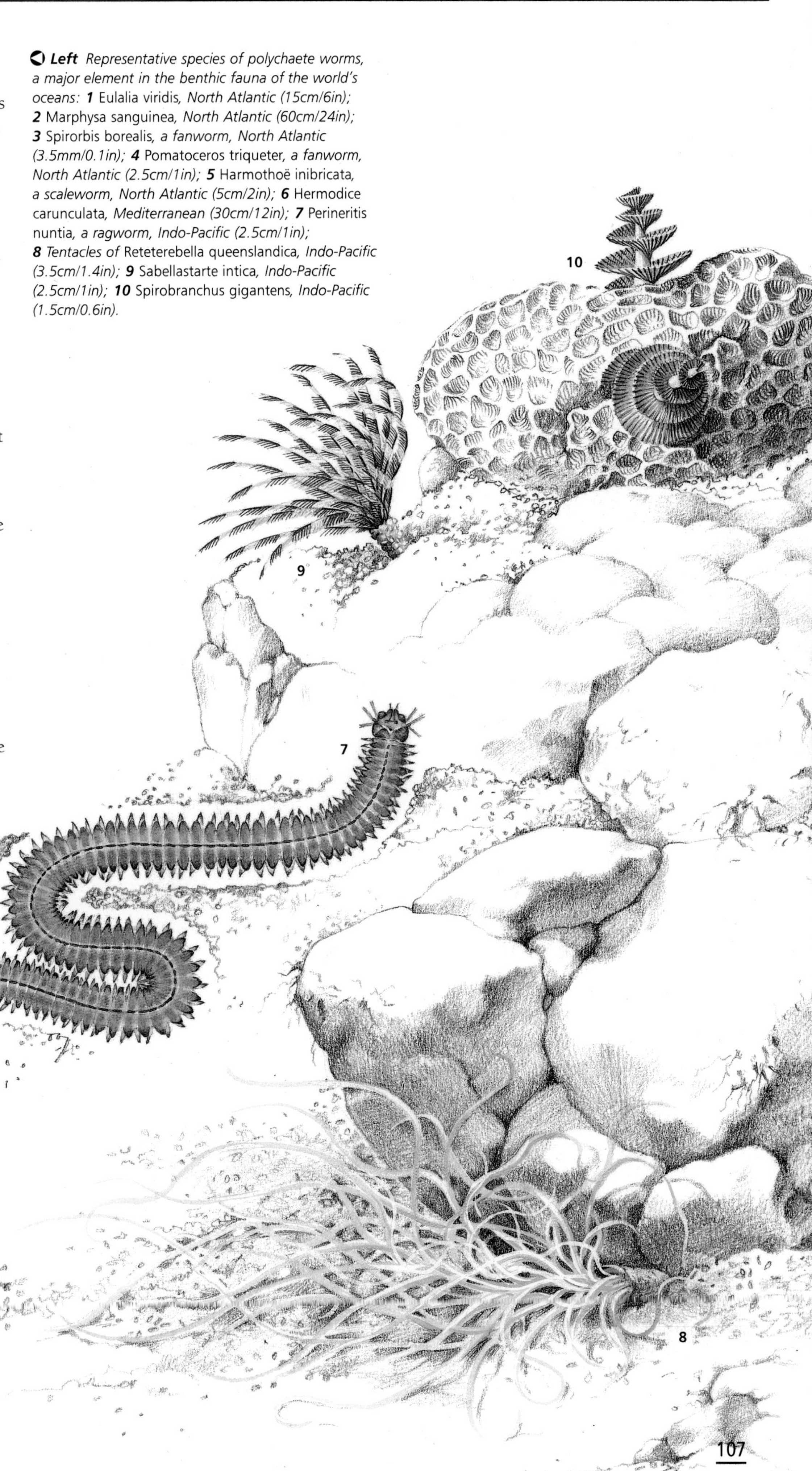

Left *Representative species of polychaete worms, a major element in the benthic fauna of the world's oceans:* ***1*** Eulalia viridis, *North Atlantic (15cm/6in);* ***2*** Marphysa sanguinea, *North Atlantic (60cm/24in);* ***3*** Spirorbis borealis, *a fanworm, North Atlantic (3.5mm/0.1in);* ***4*** Pomatoceros triqueter, *a fanworm, North Atlantic (2.5cm/1in);* ***5*** Harmothoë inibricata, *a scaleworm, North Atlantic (5cm/2in);* ***6*** Hermodice carunculata, *Mediterranean (30cm/12in);* ***7*** Perineritis nuntia, *a ragworm, Indo-Pacific (2.5cm/1in);* ***8*** *Tentacles of* Reteterebella queenslandica, *Indo-Pacific (3.5cm/1.4in);* ***9*** Sabellastarte intica, *Indo-Pacific (2.5cm/1in);* ***10*** Spirobranchus gigantens, *Indo-Pacific (1.5cm/0.6in).*

mouth. Many other forms ingest sediment, selecting food particles to a greater or lesser degree.

Small polychaetes respire through the body surface, but larger forms often have gills, which tend to be concentrated near the front in tube-dwelling forms; feeding appendages may also serve a respiratory function. In many tube and burrow dwellers, water is drawn through the tube past the gills by special respiratory movements. Gills are usually associated with the parapodia, and may be simple or threadlike or branched in structure.

The blood system consists of dorsal and ventral longitudinal vessels, connected by segmental vessels and capillaries or by blood sinuses around the gut. It is reduced or absent in many of the smaller forms, and even in some large ones, for example members of the Glyceridae, its function being taken over by the coelomic fluid. Loss of the blood system is often associated with reduction or loss of walls (septa), allowing free passage of coelomic fluid. A variety of blood pigments may be present in polychaetes, either in cells or free in the blood.

Active crawling forms often show a reduction in the circular muscle layer of the body wall, with corresponding development of the parapodia and their musculature and reduction or loss of the septa. This change is correlated with a switch from peristaltic locomotion (as in earthworms) to the use of parapodia as the main propulsive organs.

Connecting the coelomic fluid to the exterior are a pair of excretory nephridia and a pair of genital ducts, the coelomoducts, in each segment. The nephridia pass through a septum, the external pore occurring in the segment behind that in which the nephridium originates. In most polychaetes, the coelomoducts and nephridia are fused in fertile segments to produce urinogenital ducts, although often only at sexual maturity. Fertilization is generally external, and may lead to a free-swimming larva, which may or may not feed, or development may be entirely on the sea floor (benthic). Adults protect the brood in a number of species; this activity is not restricted to those with benthic development. Polychaetes may be hermaphrodite, eggs may develop into larva within the body (viviparous), or internal fertilization may take place. Asexual reproduction, by fragmentation or fission followed by regeneration, occurs in some 30 species.

Ragworms are familiar through their use by sea anglers. Some species are carnivorous, feeding on dead or dying animals; some are omnivorous; others feed only on weed. Polychaete worms predominate among bait species. In Europe, ragworm and lugworm are used most commonly for bait, and catworms are occasionally collected. In North America the sandworm and the bloodworm are widely used, as are the long beachworms (species of the family Onuphidae) in Australia. All these species live in intertidal soft sediment, although ragworms are often found in muddy gravel.

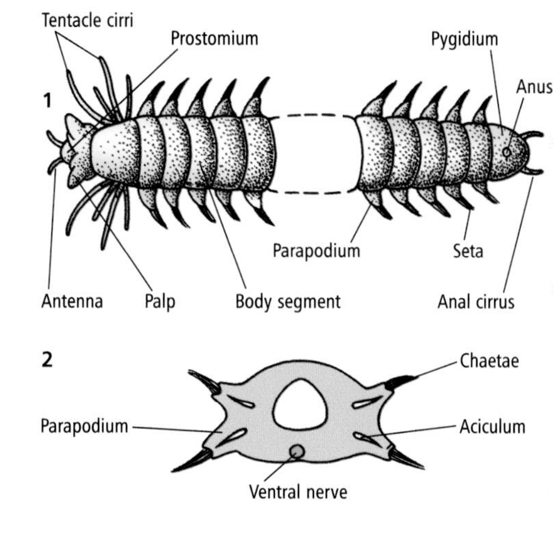

◀ ***Left*** *Commonly known as the Social feather duster or Cluster duster, the polychaete tubeworm* Bispira brunnea *(family Sabellidae) is found in abundance in the Caribbean and the Gulf of Mexico.*

▶ ***Right*** *The Bearded fireworm* Hermodice carunculata *is a voracious predator that attacks living soft corals, engulfing their tips with its eversible pharynx. It is seen here on a Purple sea fan* (Gorgonia ventalina).

Ragworms live for protection in the mud or muddy sand, from which they emerge or partially emerge to feed on the plant and animal debris on the surface. The burrows may be located by the holes on the surface of the mud. The burrows are U-shaped or at least have two openings, for the worms must irrigate them to respire. They do this by undulating their bodies, an action that serves not only to renew the water within the burrow, but enables them to detect in the incoming water signs of food in the vicinity. Vast populations may occur in suitable habitats. Some species can tolerate the stringent conditions in estuaries and some are found in very low salinities.

Scaleworms also occur on the seashore and below the low-water mark. Their backs are covered or partly covered by a series of more or less disk-shaped scales. The scales are usually dark and overlap slightly. They are protective not only in providing concealment but also in their ability to luminesce. The luminescent organs are under nervous control, so that the animal can "flash" if alarmed. Not all scaleworms do this, however.

Another polychaete worm is the sea mouse, which lives below low tide level on sandy bottoms. It is often brought up by fishermen in the dredge. The body is covered with a fine "felt" of silky chaetae, and it is to this apparently furry appearance that its common name is due.

Sea mice live for the most part just beneath the surface of the sand or mud. They are rather lethargic although they can scuttle quite rapidly for a short distance when disturbed. If the matted "felt" over the back is removed, a series of large disk-shaped scales will be seen overlapping the back: the sea mouse is nothing more than a large scaleworm in which the characteristic scales are concealed by the hairlike chaetae.

Fanworms are the most elegant of all the marine polychaete worms. Pinnate or branched filaments radiate from the head to form an almost complete crown of orange, purple, green, or a combination of these colors. The crown is developed from the prostomium. It forms a feeding organ and, incidentally, a gill. The remainder of the body is more or less cylindrical.

All fanworms secrete close-fitting tubes, which provide protection. Although their often gaily colored crowns must tempt predatory fish, they can all contract with startling rapidity. These startle responses are made possible by relatively enormous giant nerve fibers, which run from one end of the body to the other within the main nerve cord. In *Myxicola* this giant fiber occupies almost the whole of the nerve cord. Almost 1mm across, it is one of the largest nerve fibers known. Giant nerve fibers are associated with particularly well-developed longitudinal muscles, which enable the worm to retract promptly when danger threatens. Other movements are relatively slow.

Beard Worms

CLASS POLYCHAETA; FAMILY POGONOPHORIDAE

The evolutionary position of the beard worms has been extensively researched. Most biologists now concur that the beard worms' possession of a coelom and the segmentation of the opisthosoma make them close relatives of the annelid worms. Consequently, they are placed in the specialized family Pogonophoridae, in the class Polychaeta.

Beard worms are the most unlikely of animals. They are a zoological curiosity of comparatively recent discovery. Their anatomy was not fully known until 1963, when the first whole specimens were obtained. The name Pogonophoridae comes from the Greek *Pogon* ("beard") – an allusion to the shaggy tentacles carried on the front of the body.

These long, thin animals – often 500 times as long as broad – live in tubes that are generally completely buried in the mud or ooze of the ocean floor. There are two exceptions to this:

◀ ***Left*** *Diagram of a polychaete worm, showing:* ***1*** *the whole body and* ***2*** *a cross section. The basic structure is a series of body segments, each separated from its neighbor by the septa.*

▶ ***Right*** *Heads of polychaete worms:* ***1*** Nothria elegans, *in which the prostomium is well developed and bears a small number of sensory antennae;* ***2*** *member of the family Terebellidae, with feeding appendages in the form of extensible tentacles;* ***3*** *member of the family Sabellidae, with a stiff crown;* ***4*** *member of the family Sabellariidae, in which specialized chaetae form part of a lidlike opercular structure.*

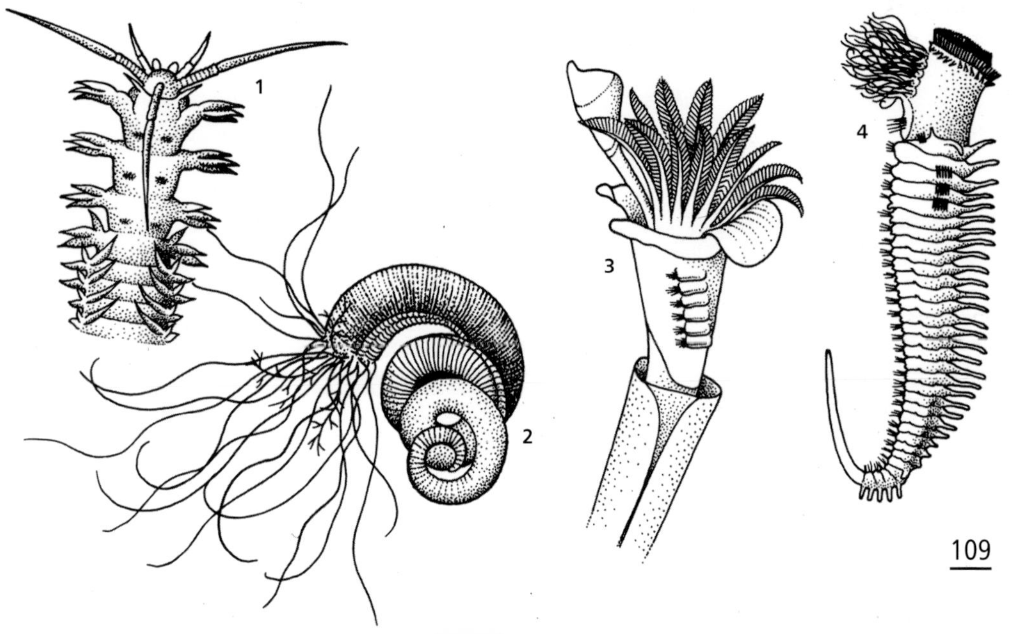

The 3 Classes of Segmented Worms

Earthworms

Class: Clitellata; order Oligochaeta

About 6,000 species in 284 genera. Families: Aeolosomatidae, 4 genera; Alluroididae, 4 genera; Dorydrilidae, 1 genus; Enchytraeidae, 23 genera, including *Enchytraeus, Marionina, Achaetus*; Eudrilidae, 40 genera; Glossoscolecidae, 34 genera; Haplotaxidae, 1 genus; Lumbricidae, 10 genera, including earthworms (*Lumbricus, Allolobophora, Eisenia*); Lumbriculidae, 12 genera; Megascolecidae, 101 genera, including *Megascolex, Pheretima, Dichogaster*; Moniligastridae, 5 genera; Naididae, 20 genera, including *Chaetogaster, Dero, Nais*; Opistocystidae, 1 genus; Phreodrilidae, 1 genus; Tubificidae, 27 genera, including some bloodworms (*Tubifex* species), *Peloscolex*. 6 species are threatened: *Phallodrilus macmasterae* (family Tubificidae) is Critically Endangered; 4 species are classed as Vulnerable, including the Washington giant earthworm (*Driloleirus americanus*; family Megascolecidae) and the Giant Gippsland earthworm (*Megascolides australis*; family Megascolecidae); and *Lutodrilus multivesiculatus* (family Lutodrilidae) is Lower Risk Near Threatened.

Ragworms, Lugworms and Beard Worms

Class: Polychaeta (includes aberrant groups myzostomes and beard worms)

In the region of 10,000 species in around or so 80 families. Families include: sea mice (Aphroditidae); Arenicolidae, including lugworm (*Arenicola marina*); Cirratulidae; Eunicidae; Glyceridae, including some bloodworms (*Glygcera* species); Nephthyidae, including catworms (*Nephthys* species); Nereididae, including ragworm or sandworm (*Nereis virens*); beachworms (Onuphidae); Opheliidae; beard worms (Pogonophoridae), including *Lamellibrachia, Lamellisabella, Sclerolinum, Spirobrachia*; Sabellariidae; scaleworms (Polynoidae); fanworms (Sabellidae); Serpulidae; Spionidae; Syllidae; Terebellidae (including some bloodworms). *Mesonerilla prospera* (family Nerillidae) is Critically Endangered.

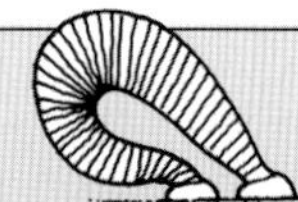

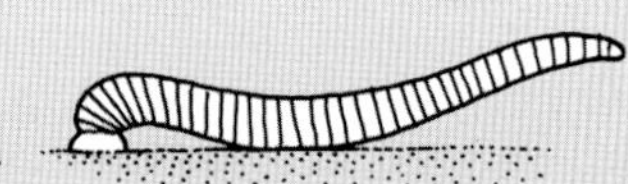

Above Leech suckers attach alternately to the substrate, are lifted, and move forward. This action removes the need for chaetae and a fluid-filled coelom, features of other annelid worms.

Leeches

Class: Clitella, order Hirudinea

About 500 species in 140 genera. Families: Acanthobdellidae, genus *Acanthobdeila*; Americobdellidae, genus *Americobdeila*; Erpobdellidae, including *Erpobdeila*; Glossiphoniidae, including *Glossiphonia, Theroemozon, Placobdeila*; Haemadipsidae, including *Haemadipsa*; Hirudidae, including Medicinal leech (*Hirudo medicinalis*), *Haemopis*; Pisciolidae, including *Pisciola, Brancheilion, Ozobranchus*; Semiscolecidae, including *Semiscolex*; Trematobdellidae, including *Trematobdella*; Xerobdellidae, including *Xerobdella*. The Medicinal leech is classed as Lower Risk Near Threatened.

Scierolinum brattstromi lives inside rotten organic matter, such as paper, wood, and leather in Norwegian fjords; and *Lamellibrachia barhami* builds tubes that project from the sediment.

Beard worms have a front head (cephalic) lobe bearing between one and 100 or more tentacles. Immediately behind the head is the forepart, or bridle, which appears to be important in tube building. Behind the bridle lies the trunk, which comprises the main part of the body. It is covered with minute tentacles (papillae), which are arranged in quite regular pairs toward the front. They become more irregular further back along the trunk and some may be enlarged. These papillae are thought to enable the worm to move inside its tube, and they are associated with plaques, hardened plates that probably assist in this respect. Further along the trunk there is a girdle where the skin is ridged; rows of bristles, similar to those of annelid worms, occur on the ridges. They probably help the worm maintain its position inside the tube.

The distinct rear region of the body, the opisthosoma, is easily broken from the rest of the body, and thus its presence was not appreciated for many years. It comprises between five and 23 segments, each of which carries bristles larger than the ones on the girdle. The opisthosoma probably protrudes from the bottom of the tube in life and acts as a burrowing organ.

The tube is made from chitin; depending on the species, lengths range from several centimeters to about 1.5m (5ft). Beard worm tubes are very narrow, never exceeding 3mm (0.1in) in diameter.

One of the most peculiar features of the beard worms is the lack of a gut within the body. It is probable that nutrients are absorbed directly from the environment as organic molecules.

The sexes are separate and the sperms are released in packets (spermatophores) that are trapped by the tentacles of the females. The packets gradually disintegrate, and the freed sperms fertilize the eggs, which develop into larvae while still inside the female tube. When large enough, they move out and settle nearby.

Leeches

CLASS CLITELLA

Leeches are easily recognized. Their soft bodies are ringed, usually without external projections and with a prominent, often circular, sucker at the rear. At the front there is another sucker around the mouth, and although this may be quite prominent, as in fish leeches, often it is not. The skin is covered with a thin cuticle onto which mucus is secreted liberally. In front there is usually a series

Left Large beard worms thrive near hydrothermal vents on the ocean floor, in temperatures up to 40°C. Their large surface-area-to-volume ratio favors the uptake of nutrients, which they obtain from commensal bacteria around the vent.

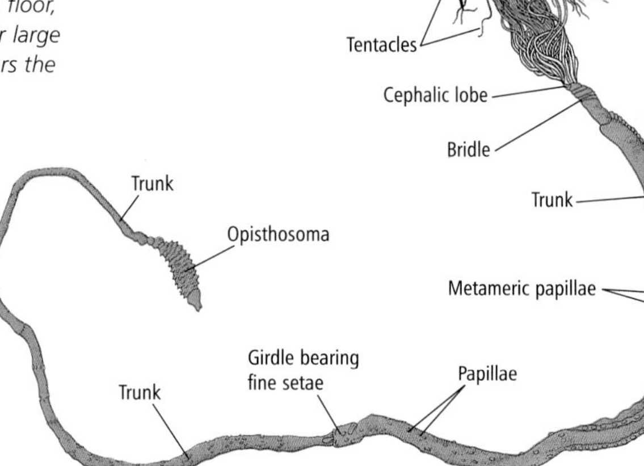

Right Beard worm body plan. A typical pogonophoran, showing the whole worm, most of which is normally buried in its narrow tube in the sea bed.

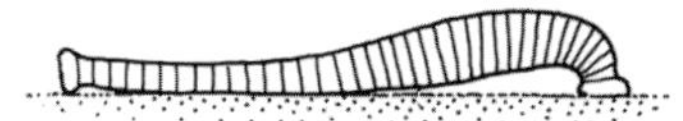
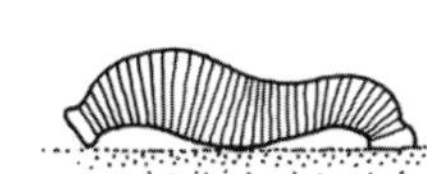
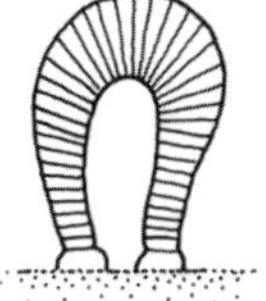
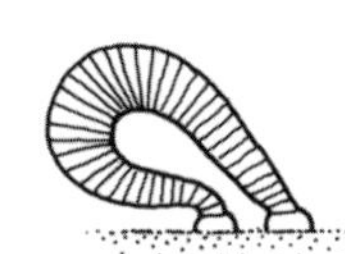

of paired eyes.

Leeches lack chaetae (except in the Acanhobdellidae) and have a fixed number (33) of segments. Like oligochaetes, they possess a clitellum, involved in egg cocoon production.

Leeches are divisible into two types: those that have an eversible proboscis (Glossiphoniidae and Fisciolidae) and those that have a muscular sucking pharynx, which may be unarmed (Erpobdellidae, Trematobdellidae, Americobdellidae, Xerobdellidae) or armed with jaws (Hirudidae, Haemadipsidae).

Not all species feed in the same way as the Medicinal leech (see Bloodsucker – The Medicinal Leech). Some species feed on other invertebrates, either sucking their body fluids (e.g. Glossiphoniidae) or swallowing them whole (e.g. Erpobdellidae). The Hirudidae, including the Medicinal leech, and the terrestrial Haemadipsidae feed on vertebrate blood. The Pisciolidae are mostly marine, living on fishes' body fluids.

As well as circular and longitudinal muscles in the body wall, diagonal muscle bundles form two systems spiralling in opposite directions between these layers and opposing both. Muscles running from top to underside are well developed and are used to flatten the body, especially during swimming and movements associated with respiration.

The coelomic space is invaded largely by tissue, leaving only a system of smaller spaces (sinuses). The internal funnels of the excretory nephridia open into this system and may be closely associated with the blood system, playing a part in blood fluid production. The blood vascular system is intimately connected with the coelomic sinuses and may be completely replaced by them.

At copulation, one animal normally acts as sperm donor, and in the jawed leeches, a muscular pouch (atrium) opening to the outside acts as a penis for sperm transfer to the female gonadal pore. In other leeches, sperm packets formed in the atrium attach to the body wall of the recipient, and the spermatozoa make their way through the body tissues by a poorly understood mechanism. Fertilization is internal or occurs as gametes are released into the cocoon, which is secreted by the clitellum and slipped over the leech's head. A small number of zygotes are put in each cocoon, which is full of nutrient fluid secreted by the clitellum. In the Glossiphoniidae, cocoons are protected by the parent, and the juveniles may spend some time attached to the parent. Most leeches only go through one or two breeding periods, with one- to two-year life cycles.

Leeches are generally accepted as having been derived from oligochaetes, specializing as predators or external parasites. The fish parasite *Acanthobdella* shows many features intermediate

BLOODSUCKER – THE MEDICINAL LEECH

The most familiar leech is the Medicinal leech, native to Europe and parts of Asia. Its use in bloodletting in the 18th and 19th centuries led to its introduction to North America. However, it is now thought to be extinct there, as well as in Ireland. The practice of using leeches for bloodletting to cure almost all conceivable ills ("vapors and humors") has long been discredited, but leeches' ability to drain blood makes them highly valuable allies in plastic surgery, helping restore circulation to grafted tissues.

The Medicinal leech eats blood, usually mammalian but also that of amphibians and even fishes. Once attached to a potential blood source by means of its suckers, its three jaws come into contact with the skin. Each is shaped like a semicircular saw, with numerous small teeth. Sawing action of the teeth results in a Y-shaped incision, through which blood is drawn by the sucking action of the muscular pharynx. Glandular secretions, released through each tooth, dilate the host's blood vessels, prevent coagulation of the blood, and act as an anesthetic.

A feeding Medicinal leech may take up to five times its body weight in blood, which passes into its spacious crop. Water and inorganic substances are rapidly extracted and excreted through the excretory cells (nephridia), but digestion of the organic portion may take 30 weeks. It was once thought that the Medicinal leech entirely lacked digestive enzymes, relying on bacteria in its own gut for digestion. However, although the involvement of bacteria is still recognized, the leech does produce its own enzymes. The time taken to digest this highly nutritious food means that leeches need to feed only infrequently – not much more than one full meal per year may be sufficient to permit growth.

Although little is known of its life history, the Medicinal leech is thought to take at least three years to mature. Cocoons, containing 5–15 eggs, are laid in damp places 1–9 months after copulation: they hatch 4–10 weeks later. PRG

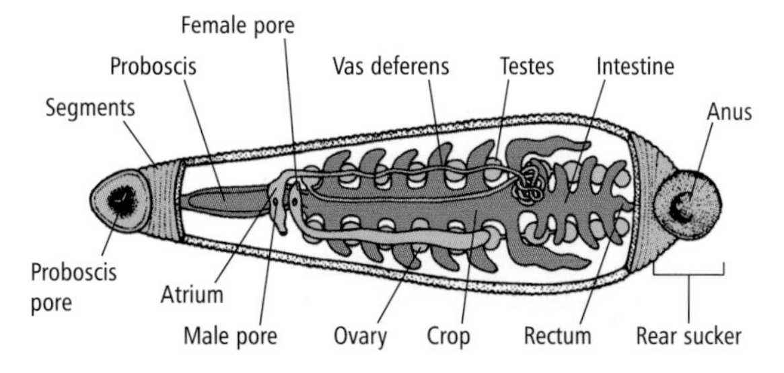

Left *Structure of a leech. Leeches are protandrous hermaphrodites, with one pair of ovaries and 4–10 pairs of testes, each gonad lying in a sac in the body cavity (coelom).*

Below *A swimming Horse leech (Haemopis sanguisuga). This species does not suck blood but feeds on smaller worms, such as earthworms.*

Rotifers, or Wheel Animalcules

Borne on a retractable disk, the crown of cilia known as the corona is the rotifers' most conspicuous structure. When the cilia beat they resemble a wheel spinning in opposite directions when viewed under the microscope. Thus the early microscopists termed rotifers "wheel animalcules."

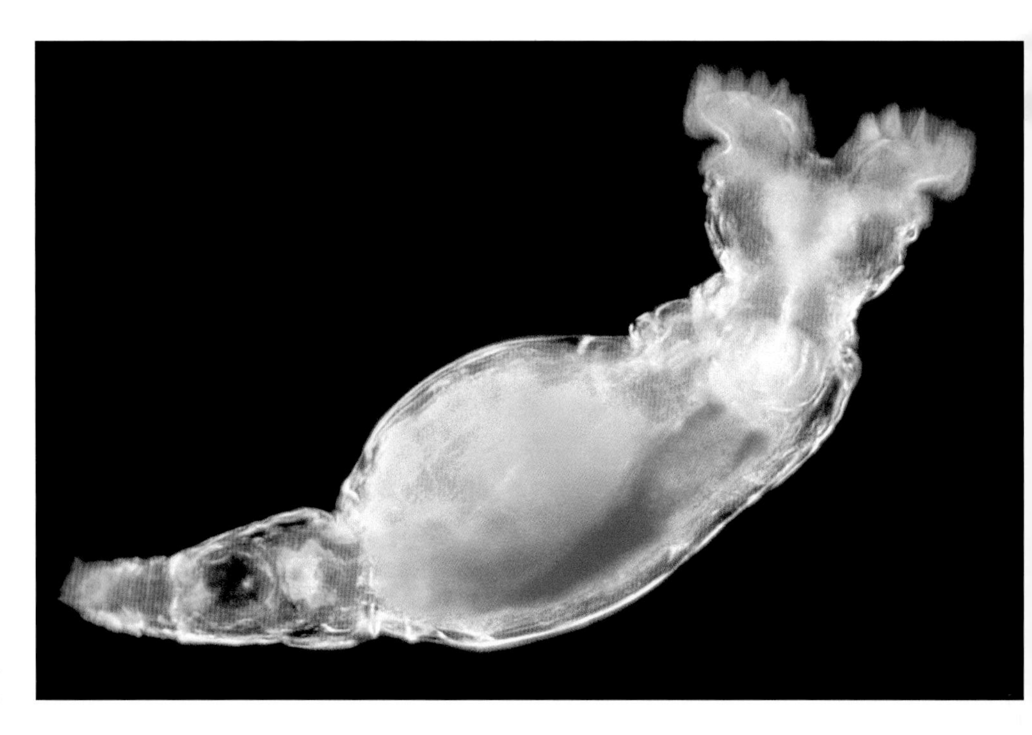

Above *Little animals of great endurance, Antarctic rotifers* (Philodina gregaria) *have survived being frozen for over a hundred years. They can withstand immersion in liquid nitrogen.*

Rotifers constitute quite a large phylum with many species living in freshwater and soils all over the world. They inhabit lakes, ponds, rivers, and ditches as well as gutters, puddles, the leaf axils of mosses and higher plants, and damp soil. Most are free living but a few form colonies and some live as parasites. A smaller number of species are marine.

A Hidden Life

BIOLOGY

Rotifer bodies are generally divided into a head, a long middle section or trunk housing most of the viscera, and a tailpiece or foot ending in a gripping toe. The head is not well developed in comparison with those of the higher animals but it does bear an eyespot, sometimes colored red. The beating cilia on the corona draws in a stream of water to the head, bringing in other microorganisms as food, which are then passed into the muscular esophagus. The chewing structure is termed the mastax, and the teeth and jaws are made of strong cuticle for chewing and grinding the food. In some predatory species long teeth may protrude out of the mouth and help seize protozoans and other species of rotifer.

The trunk houses the internal organs, including the mastax. Its contents appear complex: they include elements of the gut, excretory system, and reproductive system, as well as the musculature for retracting the corona. The tailpiece is important for posture and stance – many rotifers use it to cling to fronds of aquatic plants or other substrates while they feed or rest. The tail may also serve as a sort of rudder.

The breeding strategies of rotifers typically include a phase of reproduction where females lay unfertilized eggs that develop and grow into other females (parthenogenetic). When the sexual season dictates or when adverse environmental conditions prevail, some of the females lay eggs that need to be fertilized by a male. At the same time others lay smaller eggs that develop into males for this purpose. The males then fertilize the new females by injecting them with sperm through the body wall. The resulting fertilized eggs are tolerant of harsh conditions and can withstand drought, an important attribute for species dwelling in temporary pools of water. They can also serve as a dispersal phase when blown about by the wind. Once hatched, only females develop from the eggs. Survival over difficult times is also helped by the ability of many rotifers to tolerate water loss and to shrivel up into small balls and remain in a state of suspended animation referred to as cryptobiosis, or "hidden life."

Despite their obscurity, rotifers are of ecological and direct economic importance. Millions of them may exist in a small body of water and their populations may rise and fall rapidly. They are crucial members of numerous aquatic food webs. Soil-dwelling rotifers aid soil breakdown. AC

FACTFILE

ROTIFERS, OR WHEEL ANIMALCULES

Phylum: Rotifera

Classes: Monogononta, Digononta, Bdelloidea

About 1,800 species in 100 genera and 3 classes.

Distribution Worldwide, mainly freshwater, some marine, and damp soils.

Fossil record None.

Size Microscopic, 0.04–2mm (0.002–0.08in).

Features Solitary or colonial; body cavity a pseudocoel (now described as blastocoelomate); head bears a crown of cilia; trunk houses internal jaws (mastax); tailpiece in some forms bears gripping toes; excretion by means of flame cell protonephridia; body sometimes encased in a lorica, males often reduced or absent, females sometimes capable of reproduction by themselves.

CLASS MONOGONONTA
Genera include: *Keratella, Hexarthra, Collotheca.*

CLASS DIGONONTA (OR SEISONIDA)
Genus: *Seison.*

CLASS BDELLOIDEA
Genera include: *Philodina, Rotaria.*

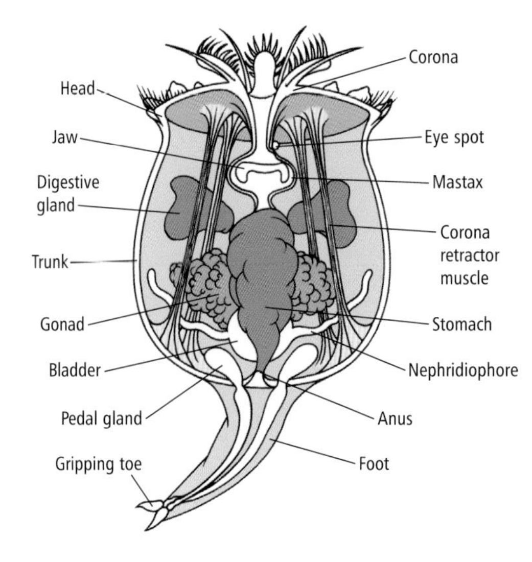

Below *Body structure of the most common rotifer genus* Brachionus. *It has a distinct cuticle and bilateral symmetry. Its two toes can clearly be seen on the foot.*

Spiny-headed Worms

WORMS IN THE ACANTHOCEPHALA ARE ALL parasites living in the intestines of various groups of vertebrates. Spiny-headed worms are particularly successful at parasitizing bony fishes and birds but they have completely failed to conquer cartilaginous fishes (skates and rays).

Unlike the parasitic flatworms (flukes and tapeworms) and the parasitic roundworms (nematodes), spiny-headed worms are of little medical or economic significance. However, they are known to cause ill-health to domestic livestock. Acanthocephalans' relative insignificance is probably because the insects and crustaceans that serve as secondary hosts to the juveniles are not eaten by humans. In a few species, another vertebrate may serve as a secondary host.

Two-host Parasites

BIOLOGY

Spiny-headed worms are elongate, dorsoventrally flattened forms that are typically dull in color. They completely lack a gut. Food digested by the host's gastric system is absorbed across the body wall and nourishes them. The body wall is composed of a fibrous epidermis that contains channels, sometimes referred to as lacunae, which are not connected to the interior or the exterior of the animal. They probably function to circulate absorbed food materials. The front is equipped with a reversible proboscis clad with hooks. The proboscis is extended to attach the worm to the lining of the host's gut. It is retracted by muscles into a special proboscis sac and everted by fluid pressure by a reduction in the proboscis sac volume. Beside the sac are two bodies called lemnisci, filled with small spaces and thought to be food storage areas. There are excretory nephridia in some spiny-headed worms. The nervous system is simple; there is a ventral mass of nerve cells at the front of the body from which arise longitudinal nerve cords.

Spiny-headed worms need two parasites where one is used as an intermediate invertebrate host for their larval stages. The sexes are separate, and males are often larger than females. Internal fertilization takes place with the male using a penis to transfer sperm to the female. The fertilized eggs develop within the female blastocoelom until they reach a larval stage encased in a shell. These larvae are liberated and pass out with the feces of the host. If eaten by the appropriate secondary host, they emerge from the egg cases, penetrate the secondary host's gut wall, and come to lie in its blood space. They remain here until the secondary host is eaten by the primary host in a prey-predator relationship or by casual accident, as when water containing a small infected crustacean is drunk. At this point the primary host is reinfected. AC

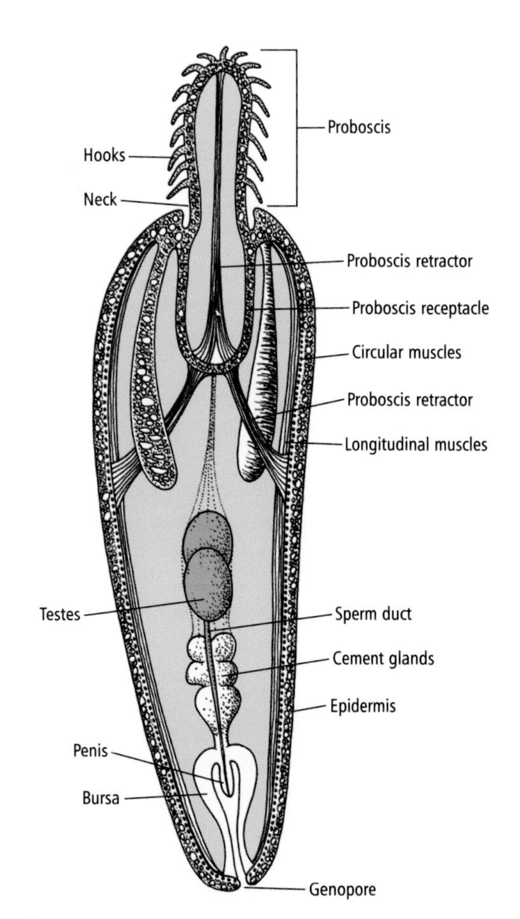

Above *Body structure of a spiny-headed worm showing the eversible pharynx. Note the numerous hooks on the proboscis, which are used to help the worm embed itself deeply in its host's tissues.*

FACTFILE

SPINY-HEADED WORMS

Phylum : Acanthocephala

Classes: Arahiacanthocephala, Palaeacanthocephala, Eoacanthocephala

About 1,100 species in 3 classes.

Distribution Widespread as gut parasites of terrestrial, freshwater, and marine vertebrates; arthropods are intermediate hosts.

Fossil record Only 1 species from Cambrian (600–500 million years ago).

Size About 1mm–1m (0.04in–3.3ft) but mainly under 20mm (0.8in).

Features Worms lacking a mouth, gut and anus; cavity a blastocoelom; peculiar protrusible proboscis ("spiny head") armed with curved hooks; sexes separate.

CLASS ARAHIACANTHOCEPHALA
2 orders; genera include: *Gigantorhynchus*.

CLASS PALAEACANTHOCEPHALA
2 orders; genera include: *Polymorphus*.

CLASS EOACANTHOCEPHALA
2 orders; genera include: *Pallisentis*.

ZOOLOGICAL HIGHLIGHT OF THE 1990S

In 1995 a new species, *Symbion pandora,* was discovered living symbiotically on the mouthparts of Norway lobsters (*Nephrops norvegicus*) in northern Europe. The organism caused excitement as it was so unusual that it could not be classified into any of the existing metazoan phyla. A new phylum was suggested and named Cycliophora, a Greek derivative referring to the circular mouth ring. Although originally believed to be affiliated with the Ectoprocta and Endoprocta, a recent DNA review of the phylum has revealed that its closest relatives may well be members of the phyla Rotifera and Acanthocephala.

S. pandora is marine, inhabiting waters in northwest Europe and the coastal areas of Norway, Sweden, and Denmark. Females grow to 0.35mm (0.14in) long, while the dwarf males reach only 0.0085mm (0.0003in) in length.

Adults are attached to their host's setae (bristles) by an adhesive pad. The body is acoelomate and divided into an anterior, ciliated, moveable, buccal funnel, an oval trunk, and a posterior adhesive disc. The trunk and disk are covered by cuticle. They suspension feed by generating currents in the buccal funnel using the cilia carried there. The gut is U-shaped and ciliated along its whole length. There is a distinct stomach. The regions between the gut, other organs, and the body wall are filled with packing tissue. The dwarf males live attached to the female and lack a gut; presumably they derive sustenance from the female. Diffusion is probably responsible for the exchange of respiratory gasses and the circulation of dissolved food substances around the body. Females have a nervous system.

The life cycle is little understood and may involve alternation between sexual and asexual phases. The females produce a Pandora larva, possibly by asexual means, which leave the mother's body and then eventually settles on the same individual host and subsequently develops into a new female. Sexual reproduction appears complex and synchronized with the molting cycle of the crustacean host. A crucial aim for the species is to infect the same or new hosts after they have molted and cleared their previous cycliphoran symbionts, which remain attached to the old exoskeletons. Scientists still have a lot to learn about *S. pandora*. AC

Endoprocts

Endoprocts were once classified with the ectoprocts (moss animals) in one phylum, the Polyzoa. Yet several significant differences (for instance, in the type of body cavity, lophophore structure, and reproductive stages) have persuaded most authorities to regard them as separate phyla.

However, above all it is the excretory system that forms the single most significant contrasting character; in the ectoprocts (phylum Bryozoa) the anus opens outside the ring of tentacles, not within the calyx, while in the endoprocts the anus (proctodeum in anatomical terms) lies within the ring of tentacles. This explains the meaning of the two names: Endoprocta – anus inside; Ectoprocta – anus outside.

Tentacular Feeders

FORM AND FUNCTION

Endoprocts are small inconspicuous animals that are usually less than 1mm (0.04in) in length. Most inhabit the sea but one genus lives in freshwater. They are sedentary and live attached to hard surfaces or other organisms by a stalk. In the latter case they live as commensals, neither receiving benefit from nor giving benefit to the host. Some of the commensal species, such as *Loxosomella phascolosomata*, are catholic and dwell on various marine invertebrates, such as the sipunculids *Golfingia* and *Phascolion* as well as on some bivalve shells. Others, like *Loxomespilon perenzi*, are host-specific, living on the polychaete worm *Sthenelais*. Some are even specific to certain parts of the host, such as the exhalant pores or oscula of sponges or the segmental appendages or parapodia of polychaete worms. Preferences for these specific habitats among endoprocts may reduce competition for survival between endoproct species and place individuals where they can benefit most from water currents.

Individual endoprocts are called zooids and live, according to species, either as solitary animals or in a colony where many zooids are linked together by creeping rootlike stolons. The zooids have a fairly simple structure, each one consisting of a cuplike body called the calyx supported on a stalklike peduncle. The calyx and peduncle are comprised of body wall tissue, which is soft, flexible, and cloaked with a thin protective layer of cuticle under which lies a thin layer of epidermis and a thin layer of muscle. The muscle brings about the nodding movements of the calyx on the peduncle. Towards its top the calyx is slightly constricted by a rim, above which radiate about 40 hollow tentacles. They can be folded over the top of the calyx and partially retracted so that the muscular web of tissue that connects their bases affords them some protection in unfavorable circumstances. The tentacles are covered with cilia, minute beating threads projecting from the skin.

On top, inside the circlet of tentacles, the calyx is penetrated by the mouth and anus, between which runs a U-shaped gut. The space inside the calyx is filled with cells, and thus the animals are functionally acoelomate, although this space may originally have been formed from the blastocoel in development. The reproductive organs lie close to the gut and their short ducts discharge into a fold, the atrium, in the top of the calyx; embryos may be brooded in the atrium. Endoprocts lack an internal circulatory system and have no special respiratory structures; oxygen dissolved in the surrounding water simply diffuses into the zooid. A pair of nephridia are responsible for excretion of nitrogenous waste and they discharge through a single nephridiophore on the top of the calyx just behind the mouth.

Endoprocts are suspension feeders utilizing small organic particles and microorganisms borne in the water currents to supply them with food. The cilia on the tentacles drive water in between

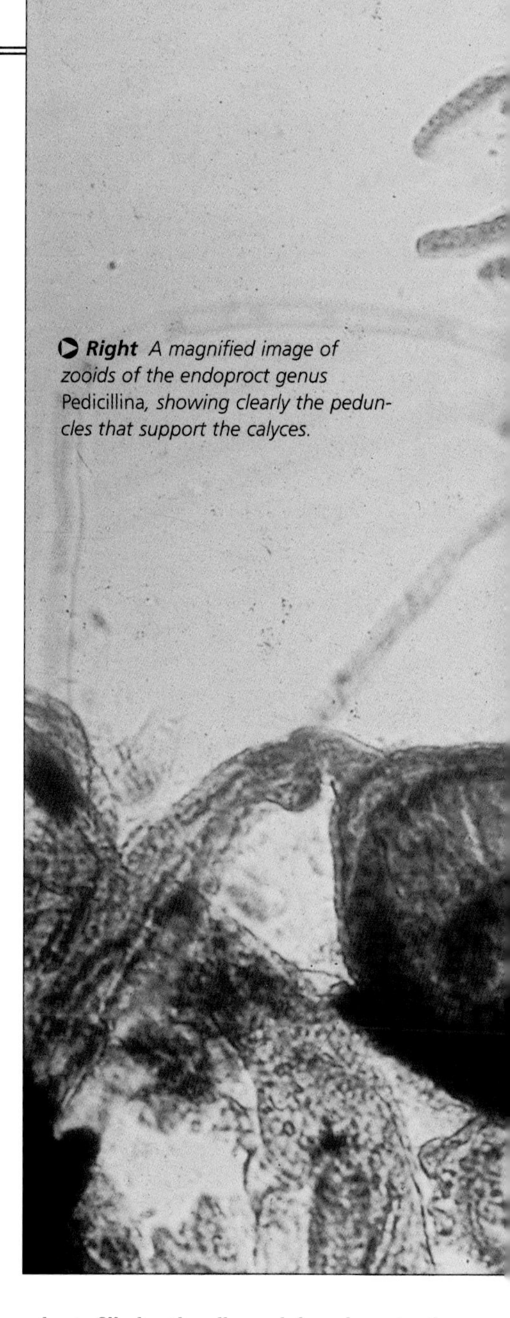

Right *A magnified image of zooids of the endoproct genus* Pedicillina, *showing clearly the peduncles that support the calyces.*

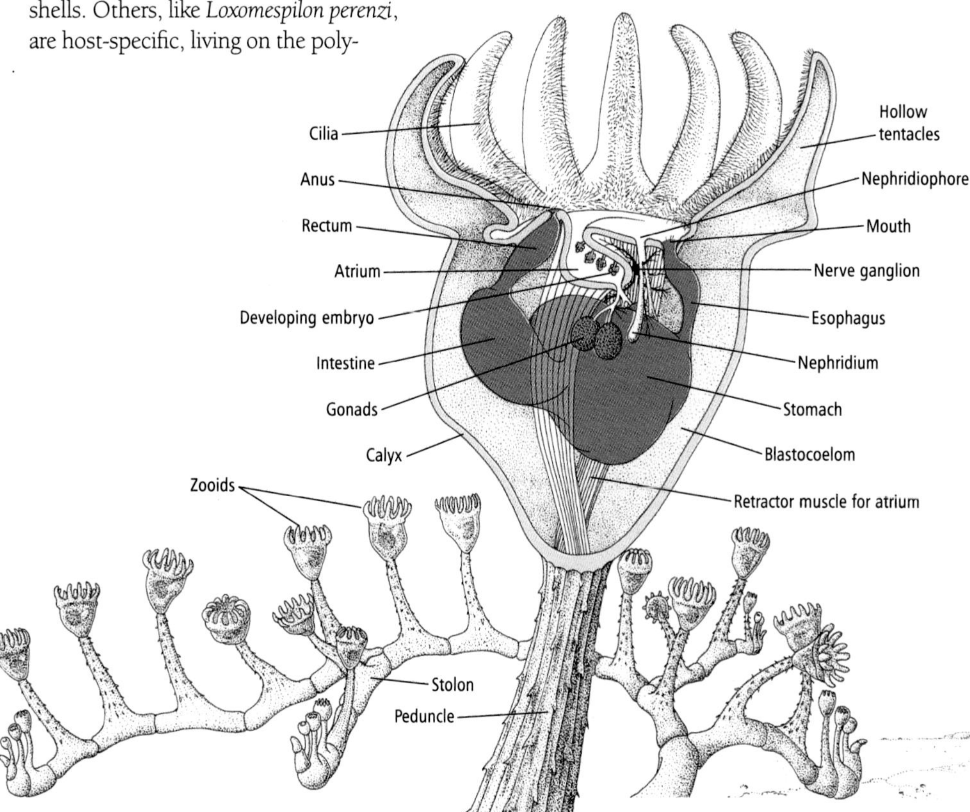

Left *A section showing the internal structure of a zooid (genus* Pedicillina*). In the background is part of a colony of the same organism, showing each zooid linked by a creeping stolon.*

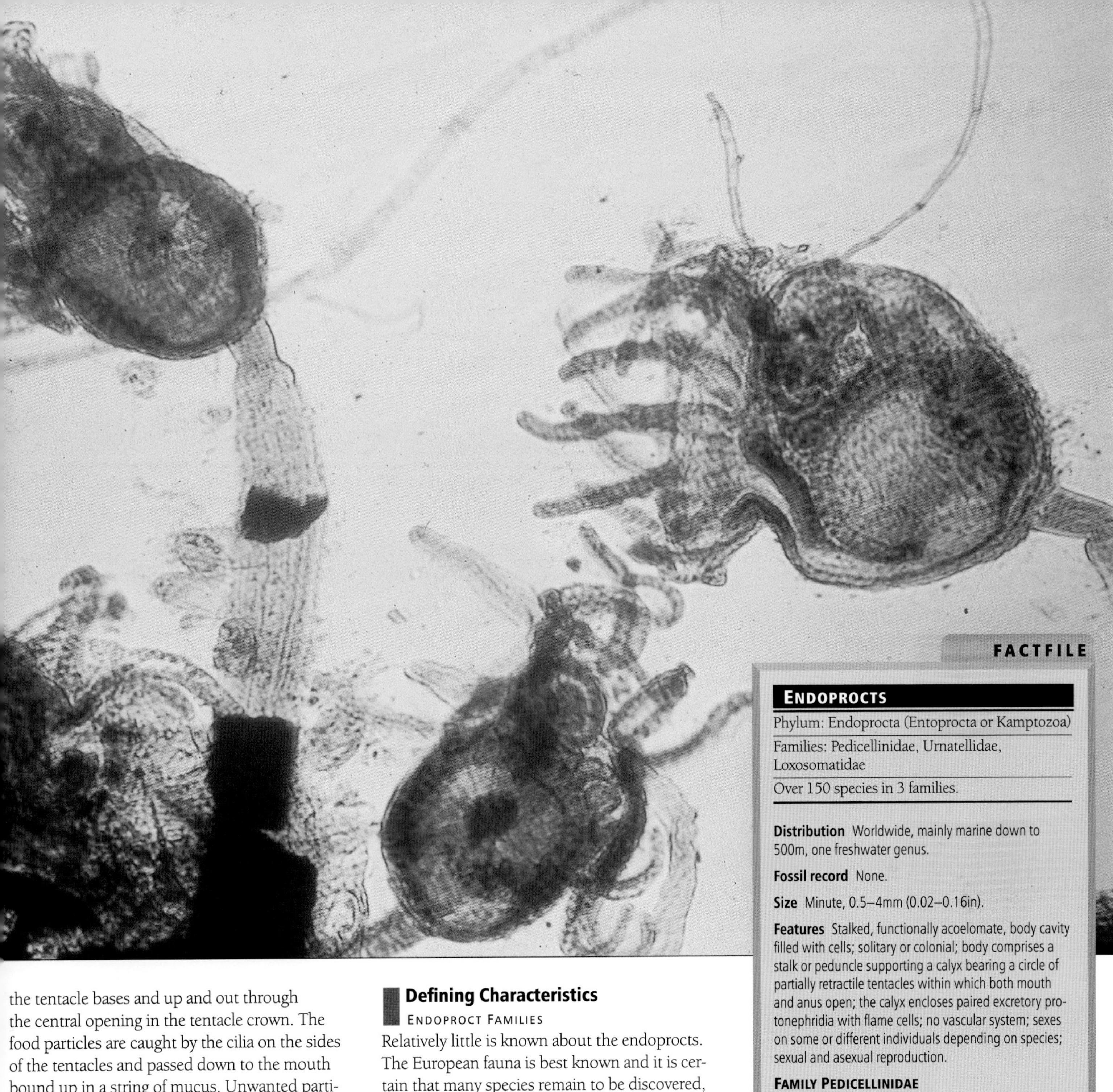

FACTFILE

ENDOPROCTS

Phylum: Endoprocta (Entoprocta or Kamptozoa)

Families: Pedicellinidae, Urnatellidae, Loxosomatidae

Over 150 species in 3 families.

Distribution Worldwide, mainly marine down to 500m, one freshwater genus.

Fossil record None.

Size Minute, 0.5–4mm (0.02–0.16in).

Features Stalked, functionally acoelomate, body cavity filled with cells; solitary or colonial; body comprises a stalk or peduncle supporting a calyx bearing a circle of partially retractile tentacles within which both mouth and anus open; the calyx encloses paired excretory protonephridia with flame cells; no vascular system; sexes on some or different individuals depending on species; sexual and asexual reproduction.

FAMILY PEDICELLINIDAE
Includes: *Pedicellina*, *Myosoma*, *Barentsia*, and 3 other genera.

FAMILY URNATELLIDAE
Sole genus *Urtanella*.

FAMILY LOXOSOMATIDAE
Includes: *Loxosoma*, *Loxocalyx*, *Loxosomella*, *Loxomespilon*, *Loxostemma*.

the tentacle bases and up and out through the central opening in the tentacle crown. The food particles are caught by the cilia on the sides of the tentacles and passed down to the mouth bound up in a string of mucus. Unwanted particles are flicked into the water current leaving the tentacle crown.

Many species undergo asexual reproduction by budding. The style of budding, for example directly from the calyx or from the peduncle (as well as the pattern of growth form in colonial types), is characteristic of particular genera. Budding from the calyx is customary in solitary forms, such as *Loxosomella*. When the buds reach an advanced stage they drop off to occupy a new site and live independently.

Some endoproct species are hermaphrodite, while others have separate sexes. Fertilization is believed to be within each zooid. The resulting embryos are then brooded in the atrium until they are released as free-swimming larvae. Following the planktonic phase, the larvae settle and grow into new zooids.

Defining Characteristics

ENDOPROCT FAMILIES

Relatively little is known about the endoprocts. The European fauna is best known and it is certain that many species remain to be discovered, particularly in the tropics.

The three families of endoprocts can be distinguished by the form of the zooids and the growth habit. In the Pedicellinidae there are no solitary species. Here the calyx is separated from the peduncle by a diaphragm so that when conditions are unfavorable or when damaged by predators, such as small grazing arthropods, the calyx can be shed. It can be regrown: examples of this family are frequently seen showing a range of regenerating calyces.

The Urtanellidae contains only one genus, *Urtanella*, which is the only freshwater one. Here the stolon is small and disklike and the calyces may be shed and show regeneration frequently. One species occurs in both western Europe and the eastern USA; a second is known from India.

The Loxosomatidae, which are the most abundant endoproct species, are all solitary, usually with a short peduncle attached by a broad base with a cement gland or a muscular attachment disk. The former type cannot move, being cemented down, but the latter can detach and reattach themselves as conditions require. The peduncle and calyx are continuous, and there is no diaphragm. The calyx cannot be shed in the Loxosomatidae. AC

Horseshoe Worms

UNLIKE THE MOSS ANIMALS, TO WHICH THEY are probably related, the horseshoe worms form a small and relatively insignificant phylum. These worms are the least diverse and least familiar of the larger-bodied marine phyla. Only some 11 species are known as adults and they are not of any ecological or commercial importance.

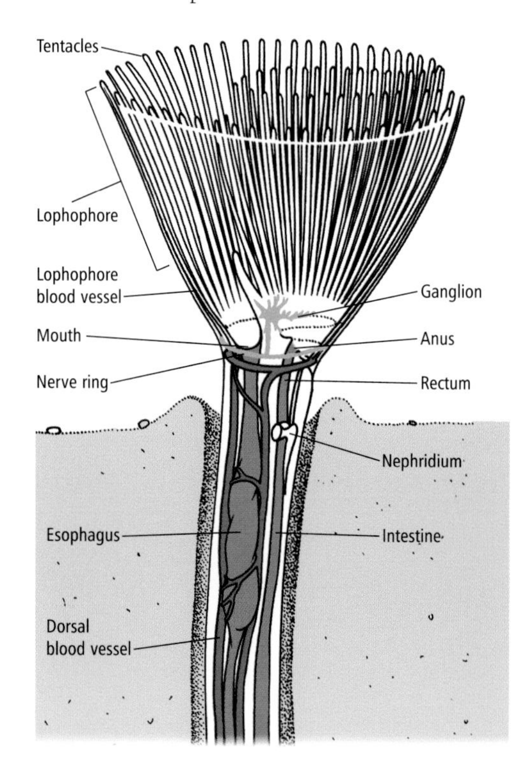

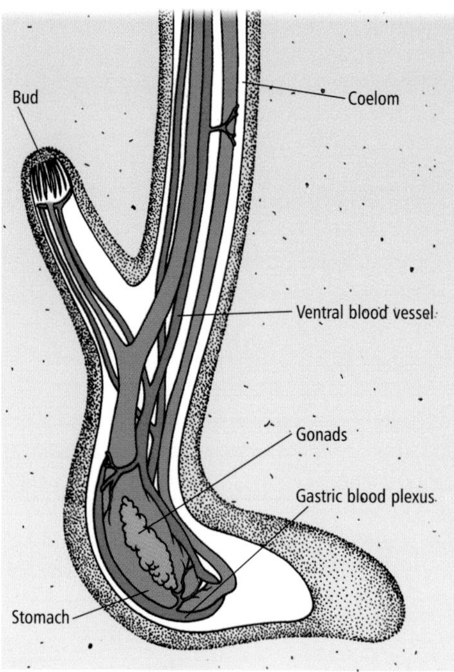

Horseshoe worms live attached to the substrate, in secreted tubes of chitinlike substance that quickly become decorated with fragments of shells and grains of sand. These animals are small and obscure and are normally discovered only by accident when, for example, items of substrate are being examined carefully. They live in shallow temperate and tropical habitats. None dwell in freshwater. Their known distribution in America, Australia, Europe, and Japan is more likely a reflection of where marine biologists are active in discovering them rather than an accurate delineation of where horseshoe worms dwell. A number of larvae, technically known as actinotrochs, have been collected, the adults of which are unknown. It is certain that the development of marine biology will see an increase in the number of known species.

Life in the Tube

FORM AND FUNCTION

The horseshoe worm body is essentially wormlike. At the front is a well developed horseshoe-shaped lophophore, which in plan resembles two crescents, a smaller one set inside a larger one with their ends touching. The mouth lies between the two crescents. From these crescentlike foundations, a number of ciliated tentacles arise to form the lophophore crown; in some species the crescent tips may be rolled up as spirals, thus increasing the extent of the lophophore. The beating of the cilia drives currents of water down the tentacles and out between their bases. Any small food particles are rolled up with secreted mucus and passed to the mouth.

The gut itself is U-shaped, a long esophagus leads back from the mouth to the stomach, which is situated near the rear of the animal. From this, the intestine leads back to the front, and the rectum opens by the anus, which is situated on a small protrusion (papilla) just outside the lophophore. The body cavity, a true coelom, houses paired excretory nephridia, the openings of which lie near the anus, and the gonads. Most species are hermaphrodites, and the sex cells are liberated into the coelom and find their way to the exterior via the excretory ducts.

The body wall consists of an outer tissue (epithelium) covering a thin sheath of circular muscles, which make the animal long and thin when they contract. Inside the circular muscles is a thick layer of longitudinal muscle, which is responsible for contracting and shortening the body. The muscles enable the animal to move inside the tube and are controlled by a simple nervous system. There is a nerve ring at the front set in the epidermis at the base of the lophophore. From this nerves arise and innervate the lophophore tentacles and the muscles of the body wall.

One interesting feature of the horseshoe worms is their clearly defined blood system with red pigment (hemoglobin) borne in corpuscles in the otherwise colorless blood. Blood is carried forward by a dorsal vessel that lies between the limbs of the gut and back to the posterior by a ventral vessel. Small branches penetrate each tentacle of the lophophore. The blood is moved around the system by contractions that sweep along the dorsal and ventral vessels.

One species, *Phoronis ovalis*, is known to reproduce asexually and to establish large aggregations by budding. In sexual reproduction, the eggs are generally fertilized in the sea. Typically, an actinotroch larva develops, which lives in the plankton for several weeks before undergoing rapid change to settle on the sea bed, form a tube and take up the adult mode of life.

Left *Internal structure of a* Phoronis *species. As tube-dwellers, horseshoe worms have a U-shaped gut and there is a tiny gap betweeen the mouth and anus.*

Right *A colony of horseshoe worms feeding. Hemoglobin is visible in the blood vessels of some of the individuals.*

Oligomerous Worms

EVOLUTION AND TAXONOMY

One aspect of interest that zoologists attach to the horseshoe worms is their evolutionary position in general and their relationship in particular with the phyla of moss animals and lampshells. In the course of development, embryos of species in all these phyla show a form of development where the body becomes divided into three sections, each with its own region of body cavity (coelom). As the animals develop the front section becomes progressively reduced and all but disappears. The middle region, with its own coelom, forms the lophophore and the rear section forms the bulk of the body of the adult. Zoologists have described these animals as oligomerous, that is they have few sections or segments. Thus they are quite different from groups like the annelid worms and the arthropod phyla where there are many segments to the body (metamerous). The general pattern of development in these groups is inclined toward that of the protostome groups, for example, the annelid worms, but occasional features, such as the way in which the coelom is formed, are inconsistent with this, suggesting that in evolutionary terms they may occupy a position among the coelomate groups intermediate between the annelids and the arthropod phyla on the one hand, and the echinoderms on the other. AC

Above *Actinotroch larvae like this* Phoronis *species swim with their hood upright and ciliated tentacles spread. The ciliary band at the rear acts as a propeller.*

FACTFILE

HORSESHOE WORMS

Phylum: Phoronida

About 11 species (more known as larvae only) in 2 genera.

Distribution Marine, bottom-dwelling, mainly from shallow temperate and tropical seas.

Fossil record none

Size 0.6–25cm (0.2–10.73 in).

Features Bilaterally symmetrical wormlike animals living permanently in a secreted tube; body cavity a coelom; mouth surrounded by a ciliated tentacle crown (lophophore); gut U-shaped; anus not enclosed by lophophore; closed circulatory system containing red blood corpuscles; paired nephridia also serve as ducts for release of reproductive cells.

GENUS *PHORONIS*
10 species including *P. ovalis, P. architecta, P. gracilis, P. pallida*.

GENUS *PHORONOPSIS*
One species – *P. harmeri*.

Moss Animals

Moss animals are an important group of sedentary aquatic invertebrates. They live attached to the substrate, either on rocks or empty shells, or on tree roots, weeds, or other animals where they can find a hold. They are of interest because of their number and diversity. Some types, for example Flustra, may dominate conspicuous regions of the seabed and have particular groups of organisms living alongside them. Others, such as Zoobotryon, are important because they can foul structures such as piers, pilings, buoys, and ships' hulls.

Most moss animals are marine, inhabiting all depths of the sea from the shore downward, in all parts of the world. A smaller group inhabit freshwater, where they are relatively inconspicuous. Moss animals are colonial; many individuals termed zooids live together in a common mass. In a colony, the first individual (formed from the larva, itself the result of sexual reproduction) is the founder. From the founder, all other individuals are produced by asexual reproduction and thus share a common genetic constitution.

Colonial Cooperation

FORM AND FUNCTION

In many colonies of moss animals, the individuals are all similar and all participate in feeding and sexual reproduction. Each individual is housed in a protective cup known as a zoecium, whose walls are reinforced with gelatinous horny or chalky secretions. The particular nature of these individual cup walls is conferred on the texture of the colony as a whole, so that some, such as *Alcyonidium,* may feel soft and pliant ,while others (e.g. *Pentapora*) feel hard, sharp, and stiff. Equally, the form of the colonies varies. Some genera, such as *Electra*, adopt a flat encrusting habit, while others such as *Myriapora* grow up from the substrate in order to exploit stronger water currents flowing above the substrate.

Each feeding individual in a colony is constructed on a similar plan. There is no head as such. A crown of hollow tentacles, filled by the coelom, forms a food-collecting and respiratory surface. (This parallels the tentacles of lampshells and horseshoe worms.) This crown is termed the lophophore. At the center of the lophophore, the mouth opens and leads in to a simple U-shaped gut with an esophagus and stomach. The anus lies close to, but outside the lophophore – the ectoproct condition (compare the Endoprocts, where the anus also lies inside the ring of tentacles).

The lophophore can be withdrawn for protection by a set of muscles and it can be extended by other muscles, which deform one wall of the cup. This wall is often the upper or frontal wall, which includes a flexible, frontal membrane. When special muscles pull the frontal membrane in, fluid pressure in the coelom increases and the lophophore is protracted. Many interesting evolutionary developments have been explored by moss animals to ensure that a flexible frontal membrane to the cup, or some other means of adjusting the cup volume, is retained, while making sure that

Below *An encrusting colony of bryozoans (genus* Electra*). Such colonies depend on currents flowing close to the rocks to bring them their food.*

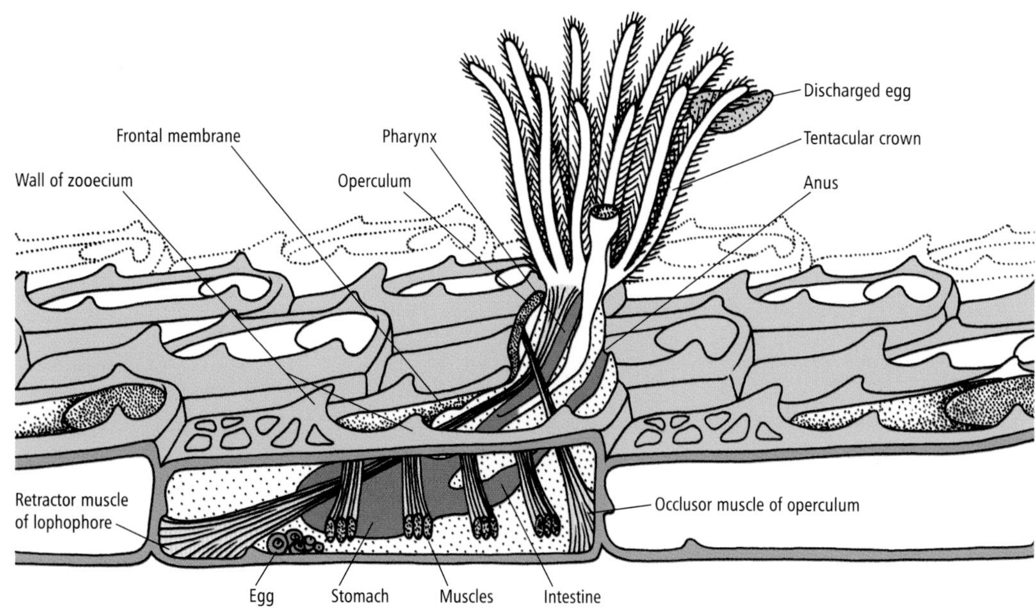

FACTFILE

Moss Animals or Sea Mats

Phylum: Bryozoa (or Ectoprocta)

Classes: Gymnolaemata, Phylactolaemata, Stenolaemata

About 4,000 species in about 1,200 genera and 3 classes.

Distribution Worldwide, some freshwater but mainly marine; sedentary bottom-dwelling animals found in all oceans and seas at all depths but mainly between 50–200m (165–655ft).

Fossil record From Ordovician period (500–440 m.y.a.) to recent times; about 15,000 species.

Size Individuals 0.25–1.5mm (0.01–0.06in), colonies 0.5–20cm (0.2–8in) wide.

Features Microscopic colonial animals attached to substrate, living within a secreted tube or case (zoecium); body cavity a coelom; mouth encircled by hairy (ciliated) tentacle crown (lophophore); gut U-shaped; anus outside lophophore; no nephridia or circulatory system.

Class Gymnolaemata
2 orders – Ctenostomata and Cheilostomata. Marine. Genera include: *Alcyonidium*, *Memranipora*.

Class Phylactolaemata
1 order Plumatellida. Fresh water. Genera include: *Cristatella*, *Plumatella*.

Class Stenolaemata
4 living orders – Trepostomata, Cryptostomata, Fenestrata, Cyclostomata – 3 extinct. Marine. Genera include: *Crisia*.

Above *Close-up of the individual zooids of* Bugula flabellata. *This erect, foliose bryozoan is a highly evolved moss animal, equipped with aviculariae.*

Left *Colonies of* Canda folifera *species of bryozoans off the island of Bali, Indonesia. The bodies of moss animals contain no circulatory or excretory mechanisms; since they are of such small volume, these roles are performed by diffusion.*

Below *The forms of some genera of colonial moss animals:* **1** Flustrellidra; **2** *and* **3** Bugula; **4** Alcyonidium; **5** Pentapora; **6** Cellaria; **7** Flustra; **8** Myriapora; **9** Sertella; **10** Cupuladria.

predators are deterred from gaining entry through what is potentially a weak spot. Clearly the requirements of rigid protection and flexibility are not easily reconciled. The animal's epidermis secretes the horny or gelatinous cuticle, and in the chalky ectoprocts this cuticle, in turn, is reinforced with calcium salts.

The lophophore tentacles are covered with beating cilia to make efficient filter-feeding organs. When the lophophore is withdrawn, the hole it goes in by may be covered over with a flaplike operculum or lid, which is present in some groups but not in others.

Specialized Zooids

Breeding and Feeding

Most animals are hermaphrodite, with male and female organs developing from special regions of the body cavity's inner lining. Ripe sperm and eggs are liberated into the coelom and find their way out of the body by separate pores. In many species the fertilized egg is brooded in a large chamber or ovicell, but in some a free-swimming larva (often called a cyphanautes) spends a period in the plankton feeding on minute algae before settling, metamorphosing, and founding a new colony.

In some more highly evolved moss animals (class Gymnolaemata, order Cheilstomata), colonies contain several forms. Here, some zooids have given up feeding and are supplied with food by others. Rather, their role is to defend the colony against predators like amphipod crustacea and against clogging by silt. There are two such forms: aviculariae and vibraculae. The first are named for their resemblance to tiny birds' heads with snapping beaks. Modified into tiny jaws with delicate sense organs, the cup and the operculum can seize small animals. The others have a large, elongated, bristle-like operculum that generates a current, vibrating to and fro to sweep away silt. AC

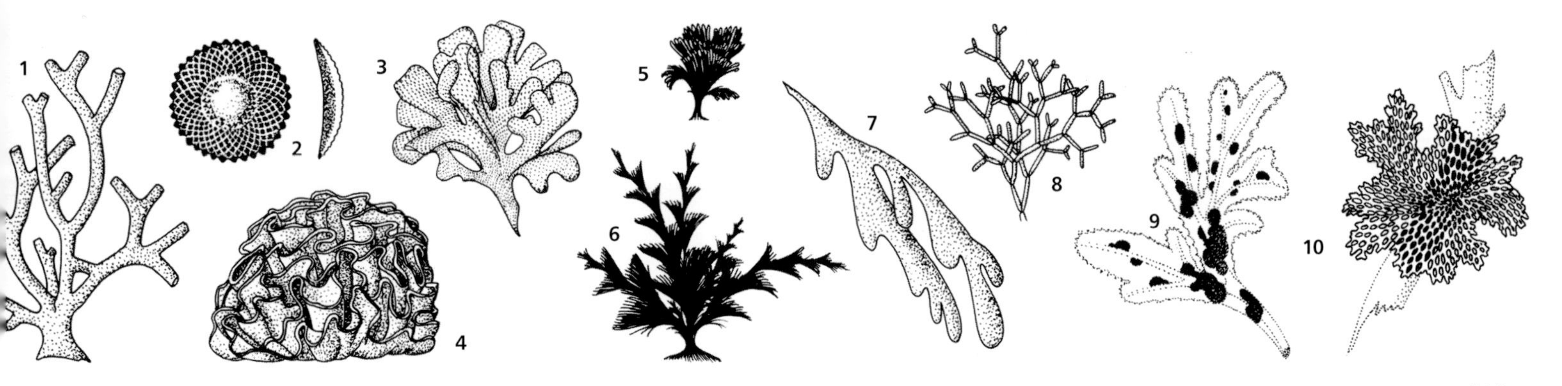

Lampshells

THE COMMON NAME "LAMPSHELL" EXPRESSES the superficial resemblance of the shells of some of these animals to Roman oil lamps. Lampshells are really animals of the past, for although they are present in modest numbers in various marine habitats the world over, nowhere do they now dominate the seas of the world as they did in the late Paleozoic. Over 30,000 fossil species of lampshells have been identified between the Cambrian and recent times.

Shelf Dwellers

DISTRIBUTION

Lampshells resemble bivalve mollusks, and indeed were classified with them until the middle of the 19th century. Their bodies are ensheathed in a mantle and enclosed by two hinged valves (shells) that protect the soft animal within. The two lobes of the mantle secrete the shells and also enclose and protect the crown (lophophore). However, there are some startling differences between lampshells and mollusks which begin with the issues of symmetry and orientation.

Lampshells now live mainly on the continental shelves either attached to rocks or other shells (for example *Crania*), or some (for example *Lingula*) live in burrows in mud. Attachment may be direct or by a cordlike stalk. Some lampshells live in very shallow water, even on shores. Few prosper at any depth. The distribution of these animals is sporadic but where they do occur they may exist in great numbers.

Complex Creatures

FORM AND FUNCTION

The anatomy of lampshells is variable and quite complex. A salient feature is that the upper (dorsal) valve is smaller than the lower (ventral) one. In most forms the shells are both convex, and the apex of the lower one may be extended to give the effect of the spout of the Roman lamp that the common name alludes to. In some of the burrowing forms the shells are more flattened. The shells themselves may be decorated variously with spines, and concentric growth lines that are fluted or ridged; their colors range from orange and red to yellow or gray.

The two valves are hinged along the rear line and the manner of their contact forms the basis for a division of the phylum into two classes: Inarticulata and Articulata. In inarticulate lampshells, the valves are linked by muscles in such a way that they can open widely, but in articulate lampshells the valves carry interlocking processes that limit the extent of the gape. In inarticulate lampshells as many as five pairs of muscles control the movements of the valves, while in articulates two or three sets of muscles are involved.

There are other differences between the two classes. In inarticulates the shells are formed from calcium phosphate and the gut terminates in a blind pouch, there being no anus. In articulates the shell is made of calcium carbonate and the gut is a "through" system with an anus. The stalk, or pedicle, by which most species are attached to the substrate is itself attached to the lower valve. The animals often position themselves so that the lower valve is uppermost, thus adding confusion to the ideas of orientation and symmetry. In a few species the pedicle has been completely lost and here the animals are attached directly to the substrate by the lower valve, with the upper valve uppermost.

The body of the animal lies within the valves sheathed in mantle tissue, which are extensions of the body wall. A true body cavity (coelom) lies

FACTFILE

LAMPSHELLS

Phylum: Brachiopoda

Classes: Articulata, Inarticulata

About 335 species in 69 genera and 2 classes.

Distribution Worldwide, marine.

Fossil record Very extensive, early Cambrian (about 600 million years ago) to recent, with greatest profusion in the late Paleozoic (up to 225 m.y.a.).

Size Shell length usually from less than 1mm to 4cm (0.3–1.5in) plus stalk, 9 cm in some extreme cases. Some extinct species reached 30cm (12in).

Features Stalked animals with bilaterally symmetrical shells in two halves (bivalve), comprising dorsal and central valves; body cavity a coelom; circulatory system open with contractile dorsal vessel; excretion via one or two pairs of nephridia, which also serve to release the reproductive cells; conspicuous loop-shaped ciliated respiratory and filter-feeding tentacle crown (lophophore).

CLASS ARTICULATA

3 living orders, 4 extinct. Valves locked at rear by a tooth and socket; lophophore has internal support; anus absent. Genera include: *Terebratella*.

CLASS INARTICULATA

2 living orders, 3 extinct. Valves held together by muscles only; lophophore lacks an external skeleton; anus present. Genera include: *Lingula*.

Right *A lampshell* (Terebratulina retusa). *Inside the shells, it is possible to make out the lophophore, with its numerous fine tentacles. This species occurs at depths ranging from 15m to 1500m, and particularly favors habitats that are sheltered from wave action and strong currents.*

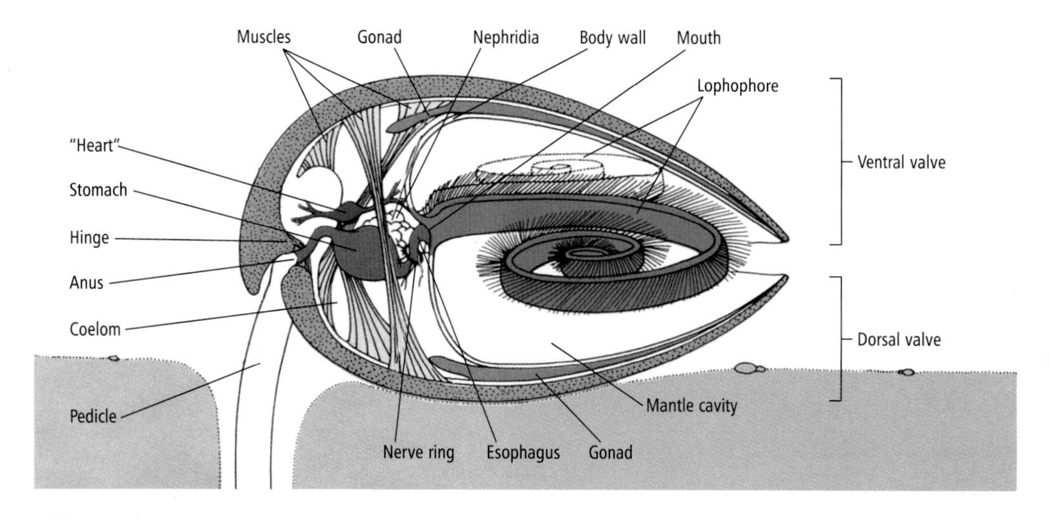

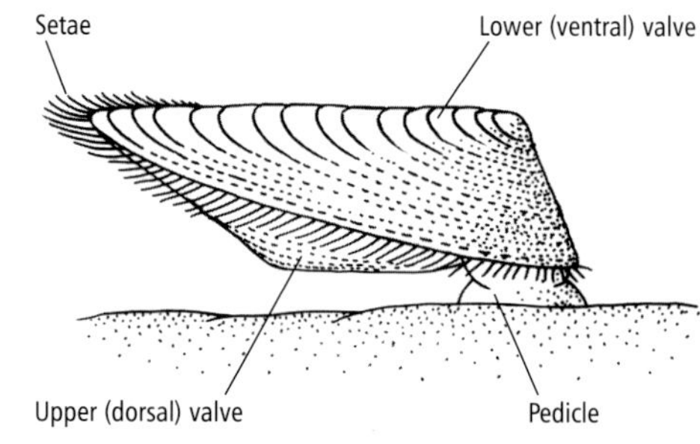

Left and above *Body plan of a lampshell, showing the internal organs. The external view shows its attachment to the substrate via the pedicle.*

between the body wall and the gut. The epidermis of the exposed inner mantle surface is hairy (ciliated). The lophophore is suspended in the mantle cavity and consists of a folded crown of hollow tentacles surrounding the mouth. It is supported by the upper valve of the shell and in some species there is a special skeletal structure to carry it. For feeding, the valves gape to the front, thus allowing water to flow over the lophophore.

The individual tentacles, which are hollow and ciliated on the outside, can reach to the front margin of the valves. They tend to be held close to the upper valve. Suspended food material is trapped in mucus and swept to the mouth of the lampshell via a special groove. Special currents carry away rejected particles. The mouth leads to an esophagus, which in turn leads on to the stomach.

There is an open circulatory system with a primitive pumping heart sited above the stomach. Blood channels supply the digestive tract, tentacles of the lophophore, the gonads, and the nephridia. There is no pigment and few blood cells. Circulation of absorbed food is thought to be the main function of the blood system, while oxygen transport would appear to be undertaken by the coelomic fluid.

For excretion, lampshells have one or two pairs of nephridia, which discharge through pores situated near the mouth. The nervous system is rather simple, with a nerve ring around the esophagus and a smaller dorsal and larger ventral ganglion. From these ganglia, nerves supply the lophophore, mantle lobes, and the various muscles that control the valves.

There are very few hermaphrodite lampshells. There are generally up to four gonads per individual and the ripe gametes are discharged into the coelom and leave the body via the nephridiopores. The eggs are generally fertilized in the sea. In most cases the embryo develops into a free-swimming larva, but in a few species it may be brooded. The larval development and planktonic period vary considerably between species. AC

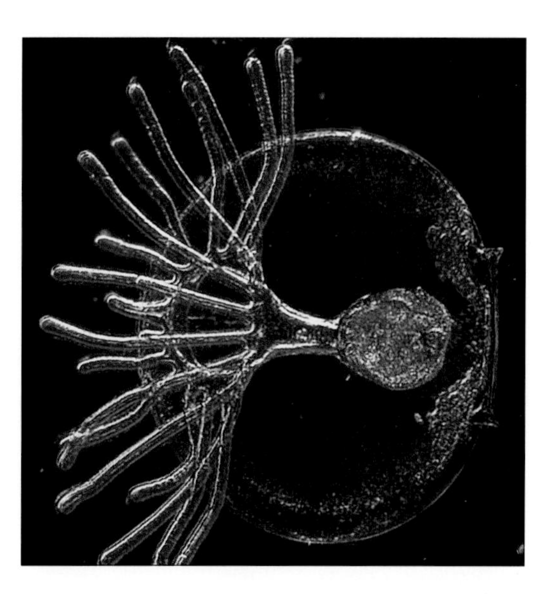

Above *Larva of the genus* Lingula. *The large, disk-like structure marks the development of the mantle and shells, while the developing lophophore crown shows as the group of tentacles.*

Spiny-skinned Invertebrates

Alternatively known as echinoderms, spiny-skinned invertebrates are distinct from all other animal types and easily recognizable. The name echinoderm means "spiny-skinned," as most members of the group have defensive spines on the outside of their bodies.

Echinoderms are found only in the sea, never having evolved to cope with the problem of salt balance that life in freshwater would impose on them. As adults they virtually all dwell on the seabed, either, like sea lilies, being attached to it, or, like the starfishes, brittle stars, sea urchins, and sea cucumbers, creeping slowly over it. These five groups, or classes, represent the types of echinoderms now found living in the seas and oceans.

FACTFILE

SPINY-SKINNED INVERTEBRATES

Phylum: Echinodermata

About 6,000 species in 3 subphyla and 5 classes (at least 17 in total when including the 12 extinct classes). There is also a fourth subphylum – Homalozoa – but this is extinct.

Distribution Worldwide, exclusively marine.

Fossil record Some 13,000 species known, extensively from pre-Cambrian to recent times.

Size 5mm–1m (0.2in–3.3ft).

Features Adults coelomate (body originally made up of 3 layers of cells), mostly radially symmetrical and 5-sided; no head, body star-shaped or more or less globular, or cucumber-shaped; calcareous endoskeleton gives flexible or rigid support, often extends into spines externally; tentacled tube feet, associated with water vascular system unique to phylum, used in respiration, food gathering, and usually locomotion (not crinoids or ophiuroids); no complex sense organs; nervous tissue dispersed through body; no nephridia; nearly all species sedentary bottom-dwellers; generally separate sexes, fertilization external; fertilized eggs and larvae generally planktonic; larval stages bilaterally symmetrical.

SEA LILIES AND FEATHER STARS

Subphylum Crinozoa

About 650 species; one living class among the 6 classes and most species known from fossils. Sedentary, mostly stalked, at least in young even if adults freeliving; branching main nervous system; anal opening on same surface as mouth.

STARFISHES, BRITTLE STARS, AND BASKET STARS

Subphylum Asterozoa

About 4,000 species (including the sea daisies) in 2 classes. Stemless, mobile, freeliving; mouth surface faces down, nervous system on mouth surface; usually have arms (rays).

SEA URCHINS AND SEA CUCUMBERS

Subphylum Echinozoa

About 1,250 species in 2 living classes (making up the 6 fossil classes). Stemless, mobile, freeliving; mouth surface downward or to front; main nervous system on oral surface; without arms or rays, external madreporite away from undersurface.

See The 3 Subphyla and 5 Classes of Living Echinoderms ▷

Crystalline Skeletons

FORM AND FUNCTION

For animals relatively high on the evolutionary scale, it is remarkable that a head has never been developed. Echinoderms show a peculiar body symmetry known as pentamerism. This is effectively a form of radial symmetry with the body arranged around the axis of the mouth. Superimposed on this radial pattern is a five-sided arrangement of the body – epitomized by the starfish. The result is that the echinoderm body usually has five points of symmetry arranged around the axis of the mouth. These points are very often associated with the locomotory organs, or tube feet.

While five-pointed symmetry, or pentamerism, is displayed largely by most present-day adult echinoderms, it is interesting to note that their larvae are bilaterally symmetrical (symmetrical on either side of a line along the length of the animal), and that their primitive ancestors, which appeared in the Precambrian seas, were also bilaterally symmetrical. The causes of pentamerism are unclear, but some authorities have suggested that it leads to a stronger skeletal framework.

The body of echinoderms shows a deuterostome coelomate level of organization. Thus, they are relatively highly evolved invertebrates with a body constructed originally from three layers of cells.

The echinoderm skeleton is made of many crystals of calcite (calcium carbonate). It is unusual because these crystals are perforated by many spaces in life (reticulate) so that the tissue that forms the crystals invades them. Such a reticulate arrangement leads to a lightening of the crystal

Below *The Western spiny brittle star* (Ophiothrix spiculata) *lives in large groups on the Pacific coast of N and S America. Anchoring itself with one or more arms, it extends the others for filter feeding.*

Right Oxycomanthus bennetti, *a species of feather star or crinoid, traps food in its sticky, brittle, fernlike arms. Crinoids are the only surviving echinoderm group that is primarily sessile.*

structure and hence a reduction in weight of the animal without any loss of strength. One side effect of this crystal structure is that it is easily invaded by minerals after the death of the animal and thus it fossilizes beautifully.

The skeleton supports the body wall, or test. This reinforced structure may be soft (as in sea cucumbers) or hard (sea urchins) but it should never be thought of as a shell because it is covered by living tissue.

The exterior of each class of echinoderm appears different, and so too is the way in which the skeleton has been deployed. In sea cucumbers, the calcite crystals are embedded in the body wall and linked by flexible connective tissue in a way that does not occur in the other classes. In starfish there is sometimes a flexible body wall, but more often the crystals are grouped close together, sometimes being "stitched" together by fibers of connective tissue running through the crystal perforations. Individual crystals may be developed extensively to form spines or marginal plates. Sea urchins have carried the skeletal process further, for in almost all the skeleton is rigid, being composed of many interlocking crystals. At the same time there is a reduction in the soft tissue of the body wall. Sea urchins have some of the most complex arrangements of muscle and skeleton in the phylum, for example, the chewing teeth or "lantern teeth," as Aristotle called them, and the pedicellariae.

In sea lilies and feather stars, and brittle stars, the skeleton is massive and arranged as a series of plates, ossicles, and spines with a minimum of soft tissue. In both these classes the major internal organs, or viscera, are contained in a reduced area, the cuplike body (theca) of the crinoids and the disklike central body of the brittle stars. Here the skeleton reinforces the body wall, which remains flexible; but in the arms, the ossicles become massive, operating with muscles and connective tissue in a way rather reminiscent of the vertebrae of the human backbone. In the arms of both types there is relatively little soft tissue. In sea lilies, the arms branch near their bases into two or more main axes, each bearing lateral branches called pinnacles. The arms of brittle stars branch only in the basket stars.

The drifting echinoderm larvae also have a skeleton, which serves to support their delicate swimming processes.

Sophisticated Hydraulics

WATER VASCULAR SYSTEM AND TUBE FEET

Another unique feature of echinoderms is their water vascular system. It probably arose in the primitive echinoderms as a respiratory system pointing away from the substratum, which could be withdrawn inside the heavily armored test. As the echinoderms became more advanced, the system was arranged around the mouth, but still held away from the substratum. Branching processes developed, forming a system of tentacles that became useful for suspension feeding as well as respiration. In this state the water vascular system is seen in present-day sea lilies and feather stars. Their branched tentacles, also called tube feet (although in the crinoids they have no locomotory role), are arranged in a double row along the upper side of each arm, bounding a food groove, and along the branches of the arms (pinnules). The tube feet can be extended by hydraulic pressure from within the animal, and much of the water vascular system is internal. The tube feet are supplied with fluid from a radial water canal, which runs down the center of each arm, just below the food groove, and which sends a branch into each pinnule. The radial water canal of each arm connects with that of its fellows via a circular canal running around the gullet of the animal. Pressure is generated inside the system by the contraction of some of the tube feet, and also by special muscles in the canal itself that generate local pressure increases to distend the neighboring tube feet. The water vascular system in crinoids is associated with several other tubular networks, notably the hemal and perihemal systems (whose role is less easy to define), and the radial water canal also runs close to the radial nerve cord, which controls the tube feet.

The activities of the tube feet relate to gas exchange (respiration) and food gathering. The tube feet are equipped with mucous glands in crinoids; when a small fragment of drifting food collides with one, the fragment sticks to the tube feet, is bound in mucus and flicked into the food groove, by which it passes down to the central mouth. The tube feet are arranged in double rows alternating with small nondistendable lappets. This arrangement assures their efficient use in feeding.

Crinoids exploit currents of water in the sea. They do not pump water to get their food, but gather it passively. They "fish" for food particles using the tube feet and select mainly those in the 0.3–0.5mm (0.01–0.02in) size range.

In all the remaining groups of echinoderms, the orientation of the body is reversed with respect to the substrate. The tube feet make contact with the ground over which the animals are moving and thus take up an additional role in locomotion. This activity happens in starfishes, sea urchins, and sea cucumbers but not in the basket and brittle stars, where movement is

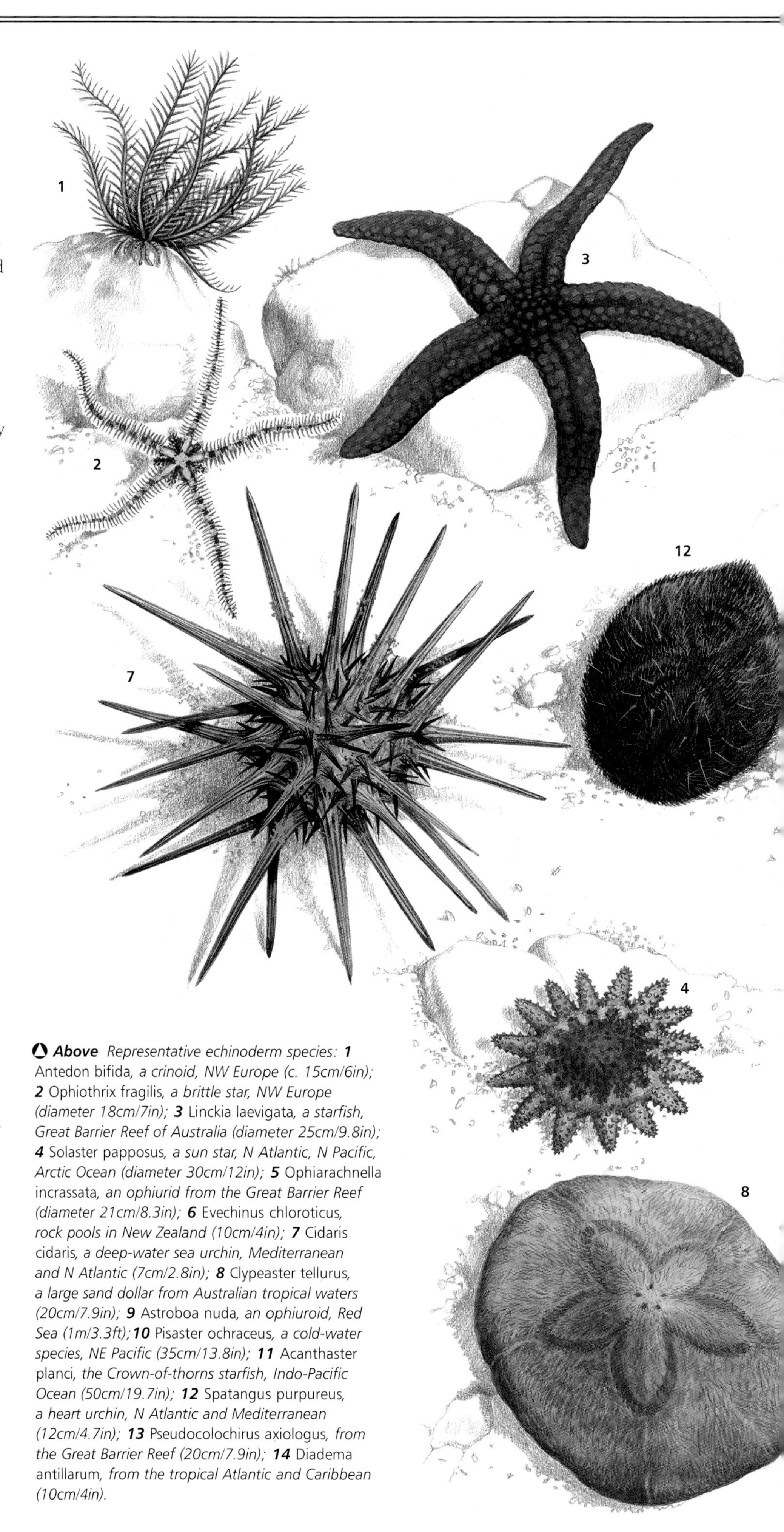

Above *Representative echinoderm species:* **1** Antedon bifida, *a crinoid, NW Europe (c. 15cm/6in);* **2** Ophiothrix fragilis, *a brittle star, NW Europe (diameter 18cm/7in);* **3** Linckia laevigata, *a starfish, Great Barrier Reef of Australia (diameter 25cm/9.8in);* **4** Solaster papposus, *a sun star, N Atlantic, N Pacific, Arctic Ocean (diameter 30cm/12in);* **5** Ophiarachnella incrassata, *an ophiurid from the Great Barrier Reef (diameter 21cm/8.3in);* **6** Evechinus chloroticus, *rock pools in New Zealand (10cm/4in);* **7** Cidaris cidaris, *a deep-water sea urchin, Mediterranean and N Atlantic (7cm/2.8in);* **8** Clypeaster tellurus, *a large sand dollar from Australian tropical waters (20cm/7.9in);* **9** Astroboa nuda, *an ophiuroid, Red Sea (1m/3.3ft);* **10** Pisaster ochraceus, *a cold-water species, NE Pacific (35cm/13.8in);* **11** Acanthaster planci, *the Crown-of-thorns starfish, Indo-Pacific Ocean (50cm/19.7in);* **12** Spatangus purpureus, *a heart urchin, N Atlantic and Mediterranean (12cm/4.7in);* **13** Pseudocolochirus axiologus, *from the Great Barrier Reef (20cm/7.9in);* **14** Diadema antillarum, *from the tropical Atlantic and Caribbean (10cm/4in).*

6
5
10
9
11
14
13

achieved by bending the arms, while the tube feet are still important in respiration and food gathering. In basket stars the tube feet are well developed for suspension feeding in a way that has interesting parallels with crinoids.

Basket stars also exploit currents of water, and arrange their complex branching arms with tube feet as a parabolic net sieving the water currents for particles in the 10–30μm size range. Thus they do not exactly compete with the crinoids in the same habitat. They are able to withstand stronger currents than the crinoids. In the remaining types of brittle stars there is a range of feeding habits. Some, like *Ophiothrix fragilis*, are suspension feeders, often living in huge beds. Others are detritus or carrion feeders. In many species the tube feet, which are suckerless, are very important in transferring food to the mouth and have a sticky mucous coating that helps this activity. Water pressure for their extension is derived partly from the head bulbs associated with each tube foot and partly from the effects of other tube feet retracting.

In starfishes, sea urchins, and sea cucumbers each tube foot is associated with its own reservoir or ampulla. The ampulla is thought to play a role in filling the tube foot with water vascular fluid. It has its own muscle system and connects to the foot by valves to control the flow. However it seems certain that fluid pressure within the water vascular system is also important. The shafts of the tube feet are equipped with muscles for retraction and for stepping movements.

Suckered tube feet occur in all sea urchins and many sea cucumbers. Some asteroids, for example *Luidia* and *Astropecten* species, lack suckers on the tube feet, and most of them burrow in sand. Other starfishes, inhabiting hard substrates, have

Below *An eleven-armed seastar* (Coscinasterias muricata) *regenerating from a severed arm; starfishes can regrow a whole new body if part of the animal's central disk is attached to the dismembered arm.*

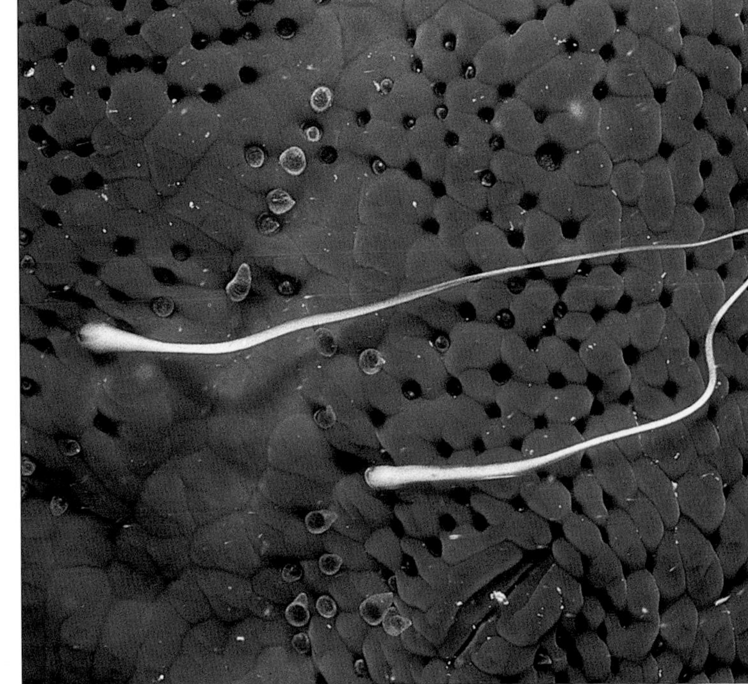

Below *Sperm is released from a spawning male Velvet seastar* (Petricia vernicina). *Seastars can reproduce either asexually (through regeneration) or sexually, with females producing 10–25 million eggs a year.*

Left *The aptly named Biscuit seastar* (Tosia australis) *varies widely in color – ranging from orange and purple to a vivid yellow – and in pattern.*

suckered tube feet and use them for locomotion and for seizing prey. In burrowing sea urchins, such as *Echinocardium* species, some of the tube feet are highly specialized for tunnel building and for ventilating the burrow. In sea cucumbers the ambulacral tube feet may be used for locomotion and respiration, while those surrounding the mouth have become well developed for suspension or deposit feeding and form the characteristic oral tentacles. There are many closely related sea cucumbers, which feed in slightly different ways, each having slightly modified oral tentacles so they can exploit food deposits of detritus particles of different sizes.

The fluid within the water vascular system is essentially seawater with added cellular and organic material. Water vascular fluid is responsible for other tasks apart from driving the tube feet. It transports food and waste material and conveys oxygen and carbon dioxide to and from the tissues of the body. It contains many cells, mainly amoeboid coelomocytes that have a role to play in excretion, wound healing, repair, and regeneration. No excretory organs have been identified in echinoderms.

The water vascular system of starfishes, basket and brittle stars, and sea urchins appears to communicate to the exterior of the animal via a special sievelike plate, the madreporite. In sea cucumbers the madreporite is internal, while in the crinoids it is lacking altogether. It used to be thought that seawater entered and left the water vascular system via the madreporite, but more recent research in sea urchins shows that very little water moves across this special structure. In starfish and sea urchins the madreporite may be associated with orientation during locomotion (see below).

Sensitive and Self-righting

NERVOUS SYSTEM

The nervous system of echinoderms is peculiar to the group. Because of the absence of a head, there is no brain and no aggregation of nerve organs in one part of the body. With the exception of the rudimentary eyes (optic cushions) of starfishes, the balance organs (statocysts) of some sea cucumbers, and the chemosensory receptors of pedicellariae of sea urchins (see Grooming Tools of Echinoderms), there are no complex sense organs in echinoderms. Instead, there are simple receptor cells responding to touch and chemicals in solution. They appear to be widely spread over the surface of the animals. Some authorities suggest that all the external epithelial cells of starfishes and sea urchins may have a sensory function.

In all living echinoderms the main part of the nervous system comprises the nerve cords that run along the axis of each arm close to the radial

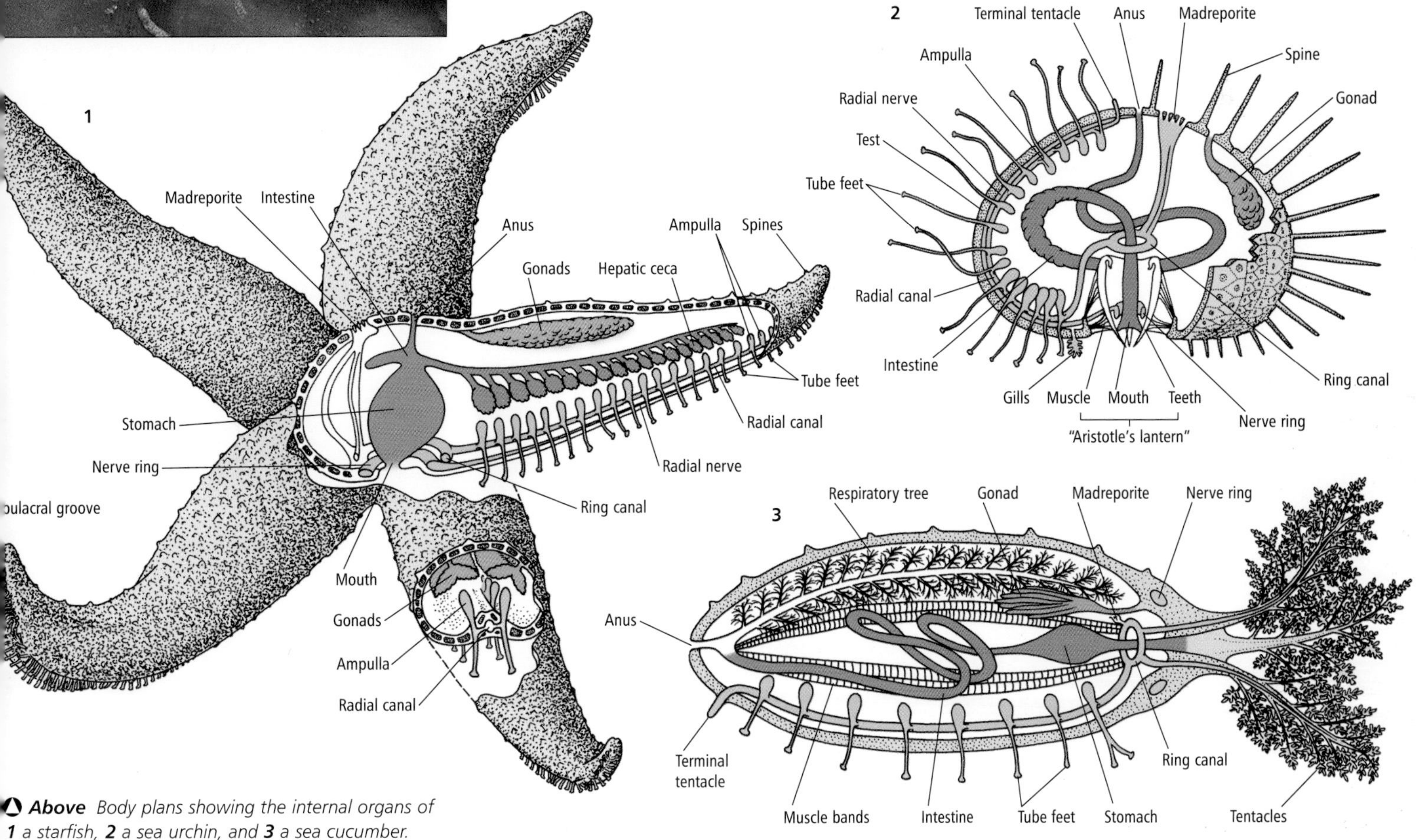

Above *Body plans showing the internal organs of* **1** *a starfish,* **2** *a sea urchin, and* **3** *a sea cucumber.*

Left *A starfish moves in for the kill on a Queen scallop* (Aequipecten opercularis). *Starfishes use their tube feet to pry open the shells of bivalves.*

Below left *The arms of brittle stars are made up of many ossicles that fit together rather like the vertebrae of the chordate spine. Here, a snake star* (Ophiothrix sp.) *is wrapped around a gorgonian sea fan.*

Below *A Southern basket star* (Conocladus australis). *Unlike their relatives in the class Ophiuroidea, the carnivorous brittle stars, basket stars are highly specialized echinoderms that are adapted for suspension feeding on plankton.*

water canal. These radial nerve cords are linked together around the esophagus by a circum-esophageal nerve cord so that the activities of one arm or ray may be integrated with the activities of the others.

The control of the tube feet and body-wall muscles is under the command of each radial nerve cord. The responsibilities for coordinated locomotion and direction of movement lie here, too. In directional terms, echinoderms may move with one arm or ray taking the lead, or even with the space between two acting as a leading edge. Where there is a need to back away, the animal may either go into reverse or turn around. In starfishes and sea urchins there is some evidence to suggest that the space between two rays containing the madreporite may often act as the leading edge, possibly because the madreporite has some as yet unknown sensory function.

Echinoderms are all extremely sensitive to gravity and generally show a well-defined righting response if they are turned upside down. It has been suggested that in sea urchins, small clublike organs known as sphaeridia act as organs of balance. All other echinoderms, apart from a few sea cucumbers, lack the balance organs (statocysts) that are frequently encountered in mollusks and in crustaceans.

In starfishes and sea urchins the outer surface of the body is covered by a well-developed epithelium at the base of which lies a network of nerves. This nerve plexus controls the external appendages of these two groups, which are richly developed in many species. The appendages include various effector organs, movable spines, pedicellariae (minute tonglike grasping organs), and sphaeridia in sea urchins, and paxillae and papulae in starfishes. These organs are concerned with defending the animal against intruders and keeping the delicate skin of the test clean from deposits of silt and detritus.

The basi-epithelial nerve plexus connects with the radial nerve cords of each arm or ray, thereby forming a system of fine nerves that link the receptor site of the epithelium with the various effector organs.

Efficient Predators

FEEDING AND BEHAVIOR

The various groups of existing echinoderms show characteristic patterns of behavior. All echinoderms are sensitive to touch and to waterborne chemicals that signal the presence of desirable prey or of potential predators that must be avoided. Most starfish species are efficient predators, feeding on other invertebrates, such as worms, mollusks, and other echinoderms. Recent research has shown that not only are starfishes and sea urchins able to "smell" the presence of suitable food in the water and move efficiently towards it, but they can discriminate between the scent of undamaged members of their own species and ones that have been injured, and thus escape the attacks of other predators upon them.

For the common European starfish, mussels and oysters are significant prey items. The sun stars *Solaster endeca* and *Crossaster papposus* also feed on bivalves but may attack *Asterias,* too. In tropical waters starfishes exhibit a variety of tastes, but the Crown-of-thorns (*Acanthaster planci*) is well known for its selection of certain species of reef-building coral as prey. All these echinoderm predators move efficiently toward the source of chemical scent in the water. When they have arrived at it they commence attack. Some excitement attended the discovery of a new group of echinoderms in the early 1980s. These deepsea forms, the sea daisies (genus *Xyloplax*) were initially grouped in a new class of their own but are now thought to be aberrant starfishes belonging to the order Spinulosida.

Some of the burrowing starfishes (e.g. *Astropecten* species) ingest their prey of gastropods whole. *Crossaster* species may attack *Asterias* by hanging on to one ray with the mouth and eating it while the prey drags the predator about.

Above *Purple sea urchins* (Stronglyocentrotus pupuratus) *are a key element in the ecosystem of the North American west coast; they keep kelp forests in check, but themselves are preyed upon by sea otters.*

Right *The Fire urchin* (Asthenosoma varium) *is well equipped to defend itself against even large predators such as starfishes. Each of its spines is tipped with a white venom sac.*

Acanthaster feeds on objects too large to be ingested whole, so it everts the stomach membranes through its mouth and smothers the prey with them, digesting the victim outside its body. When the process of digestion is complete, the stomach membranes are withdrawn.

Members of the genus *Asterias*, like other starfishes that prey on bivalves, are able to use their tube feet with their suckers to prize open the valves of a mussel or oyster.

They do this by climbing on to the prey and attaching some tube feet to each valve. The two valves are then pulled apart by the persistent actions of the tube feet. The muscles that keep the shells closed eventually tire, so that they gape ever so slightly. A gap of one or two millimetres is all that is needed for the starfish to insert some of its stomach folds passed out through its mouth. Once this has been done, digestion of the victim begins and in the end only the cleaned empty shell remains.

It is interesting to note that in both tropical starfishes (for example *Acanthaster* species) and temperate species (for example *Asterias*), solitary individuals display a different type of feeding behavior from that of individuals feeding together in groups. In these starfishes, regular feeding tends to be solitary and at night, the individuals being well spaced one from the next. In some populations, individuals gather periodically in large numbers at a superabundant food source and feed by day as well as by night; as a result of such social feeding, the growth rate of individuals considerably surpasses the norm. To what extent these differences are acquired or inherited is not yet clear.

In the sea urchins there is a range of feeding behavior. Many of the round (or regular) echinoids (e.g. the genera *Strongylocentrotus*, *Arbacia*, and *Echinus*) are omnivores. They browse on algae and encrusting animals, such as hydrozoans and barnacles, using their Aristotle's lantern teeth. In many places echinoids are important at limiting the growth of marine plants and compete very successfully with other types of algal grazers, including gastropod mollusks and fish, as recent research in the Caribbean has shown.

The irregular sea urchins, including the sand dollars and heart urchins, are more specialized feeders. Sand dollars live partly buried in the sand and use their modified spines and tube feet to collect particles of detritus for food. They are then passed to the mouth along ciliated tracts. Heart urchins burrow quite deeply in sand and gravels and ingest the substrate entire. They have lost the Aristotle's lantern. As the substrate particles pass

Above *The Pencil sea urchin* (Goniocidaris tubaria) *is a generalist feeder. While algae and seabed detritus make up most of its diet, it will also take carrion. These two urchins are devouring a dead seahorse.*

Overleaf *Close-up of the respiratory gills and tube feet of a Green sea urchin* (Psammechinus miliaris). *This regular echinoid species typically grazes on beds of brown kelp* (Laminaria saccharina).

GROOMING TOOLS OF ECHINODERMS

Between their spines, most starfishes and all sea urchins carry unique small grooming organs like microscopic tongs or forceps. These organs were once thought to be parasites on the tests of the animals that bear them, but are an integral part of the animal. Each grooming organ consists of two or more jaws supported by skeletal ossicles called valves. Some starfish types are directly attached to the test. Others, like those of sea urchins, are carried on stalks, so that the jaws are able to reach down to the surface of the test.

The jaws are caused to open or close by muscles attached to nerves from both the base of the epithelium of the test and from special receptors responsive to touch and certain chemicals. Most of these receptors are situated on the inside of the jaw blades, but some lie on the outside.

In an undisturbed animal, the epithelial nerves may close down most of the pedicellariae (except some on sea urchins; see below), so that they are inactive with the jaws closed. If an intruding organism strays onto the test, such as a small crustacean or a barnacle larva seeking a place to settle, the resultant tactile stimuli cause the pedicellariae to gape open and thus expose the special touch receptors. If these receptors are stimulated, the jaws rapidly snap shut, trapping the intruder. The pedicellariae of starfish (skeletal parts only) **1**, the tridentate (three-jawed) **2**,

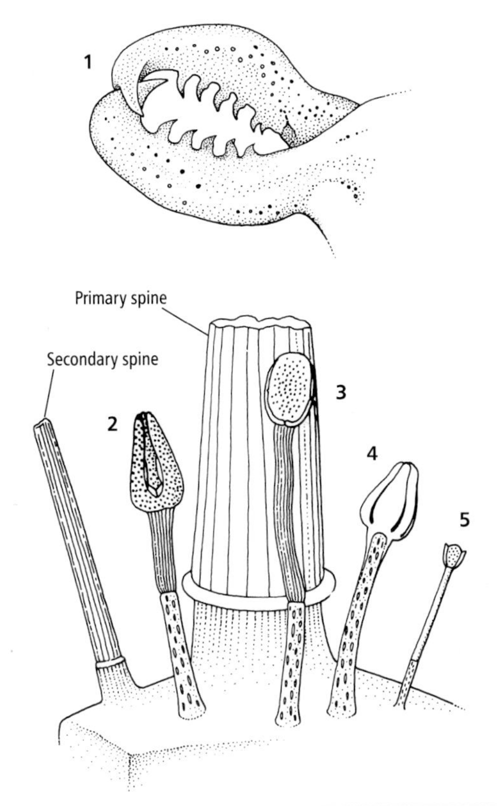

and ophiocephalous (snakehead) **3** ones of sea urchins are all specialized for such activities and often have fearsome teeth to grip their victims.

In the globiferous (roundheaded) pedicellariae **4** of the sea urchin class Echinoidea, there are venom sacs. Here the jaws close only on objects that carry certain chemicals. The venom is injected into the victim via a hollow tooth in many echinoids, and in species such as *Toxopneustes pileosus* it has a powerful effect. Globiferous pedicellariae detach after the venom is injected and remain embedded in the tissue of the intruder. They seem to be deployed mainly by sea urchins as defenses against larger predators such as starfishes.

In sea urchins, sand dollars, and heart urchins the smallest class of pedicellariae are known as trifoliate (three-leaved) **5**. They differ from all the others in having spontaneous jaw movements, "mouthing" over the surface of the test in grooming and cleaning activities. This movement persists even when these pedicellariae are removed from the test.

Far from being the parasites they were once believed to be, pedicellariae therefore perform a highly beneficial function, by keeping the surface of the echinoderm free from other animal or plant organisms. Their complex structure is a good example of the intricacies of echinoderm biology but their various roles are not yet all understood. AC

Left Sea potatoes (*Echinocardium* spp.) are a form of heart urchin. Mass mortality of this species sometimes occurs, either through storms disturbing the seabed or decaying plankton robbing the water of oxygen.

along their guts, any organic material is digested and "clean" substrate is passed out from the anus.

Sea cucumbers have diversified to exploit a number of food sources. In virtually every case the form of their oral tentacles, the specialized tube feet arranged round the mouth, is adapted to gathering food.

Some groups sweep the surface of sand and mud for particles of detritus and thus live as deposit feeders. Others are suspension feeders relying on currents of seawater to sweep suspended particles of food into their oral tentacles. In both cases the size of the sweeping or filtering fronds and the gap between dictates the sizes of the particles collected.

Synchronized Spawning

BREEDING

The majority of echinoderm species are dioecious – the sexes are separate. A few are hermaphrodite, passing through a male phase before becoming functional females. In one genus, *Archaster* from the western Pacific, pseudocopulatory activity occurs, with the one partner climbing on top of the other, but even here fertilization is external. Sperm and eggs are released into the seawater via short gonoducts. In many species, this is almost a casual affair, and the partners do not come close together. However, synchrony of spawning is essential; it is usually governed by water temperature and by chemical stimuli operating between participants.

Antarctic and abyssal echinoderms often brood their eggs, and a direct development of juveniles occurs in brood pouches or between spines. In the remaining species – the vast majority – the fertilized eggs drift in the oceanic plankton and develop through characteristic larval stages, usually feeding on minute planktonic plants such as diatoms and dinoflagellates. After a period of larval life that may range from a few days to several weeks, depending on different species, metamorphosis occurs, and the juvenile echinoderms settle on the seabed.

Unusual Delicacies

CONSERVATION AND ENVIRONMENT

The fascinating and often beautiful echinoderms inhabit all the world's seas, from the intertidal zone down to the ocean abyss. They are also present in all latitudes, from the tropics to the poles. In temperate intertidal zones such as the North Atlantic and North Pacific, starfishes and sea urchins are familiar organisms. In some places, sea urchins may be harvested for use as food, for example *Paracentrotus lividus* in Ireland and the Mediterranean and *Tripneustes gratilla* in Mauritius. Brittle stars and sea cucumbers, although present, are less conspicuous intertidally. The shallow seas overlying the continental shelves are particularly good habitats for echinoderms; coastal currents, rich in nutrients, sweep detritus and plankton for suspension feeders, such as crinoids and sea cucumbers, and nourish prey suitable for starfishes. In tropical areas the development of reefs allows a great diversity of echinoderm species to develop because of the variety of niches. Certain tropical species of sea cucumber are fished commercially and sold to far-eastern markets as *Trepang* or *Beche-de-mer*. Although all groups of echinoderms inhabit the ocean abyss, it is here that the sea cucumbers flourish, often in great densities, moving over the benthic ooze in search of detrital food. Here, too, some highly unusual epibenthic sea cucumbers have taken to a swimming life, moving along in deep currents and collecting food as they go. AC

Above and below Sea cucumbers play a vital role in the ecology of the seabed. Their feeding churns up sediments, so oxygenating the water, and also recycles nutrients. Above, the colorful *Bohadschia graeffei* and below, a *Stichopus* sp. holothurian.

The 3 Subphyla and 5 Living Classes of Echinoderms

SUBPHYLUM CRINOZOA

Crinoids
Class Crinoidea

650 species. Comprises the order Articulata, lacking madreporite, spines and pedicellariae appendages. Genera include: Sea lily (*Metacrinus*), feather star (*Antedon*).

SUBPHYLUM ASTEROZOA

Starfishes
Class Asteroidea

About 2,000 species in 2 subclasses; flattened, star-shaped with 5 unbranched arms which blend into the central body (a few with more); endoskeleton flexible; arms contain digestive ceca; branching madreporite.

SUBCLASS SOMASTEROIDEA

Almost exclusively fossil but including one living genus *Platasterias*.

SUBCLASS EUASTEROIDEA

5 orders include: Platyasterida, including genus *Luidia;* Spinulosida, including the Crown-of-thorns (*Acanthaster planci*) and sea daisies (genus *Xyloplax*); Forcipulatida, including the European starfish (*Asterias rubens*).

Brittle stars, Basket stars
Class Ophiuroidea

About 2,000 species in 3 orders; flattened, 5 sided with long flexible arms rarely branched, clearly demarked from central "control" disk; madrepore on underside; no anus or intestine; tube feet lack suckers.

Order Oegophiurida
Includes the genus *Ophiocanops*.

Basket stars
Order Phrynophiurida
Includes the genus *Gorgonocephalus*.

Brittle stars Order Ophiurida
Includes the genus *Ophiura*.

SUBPHYLUM ECHINOZOA

Sea urchins, Sand dollars, Heart urchins
Class Echinoidea

About 750 species in 2 subclasses; mainly globular or disk-shaped, without arms; covered with numerous spines and pedicecellariae; tube feet usually ending in suckers; endoskeleton comprises close-fitting plates.

SUBCLASS PERISCHOECHINOIDEA

1 living order Cidaroida, including the genus *Cidaris*.

Subphylum Crinozoa
Sea lilies, feather stars

Subphylum Asteroidea

Class Asterozoa
Starfishes

Class Ophiuroidea
Brittle stars, basket stars

Subphylum Echinozoa

Class Echinoidea
Sea urchins, sand dollars, and heart urchins

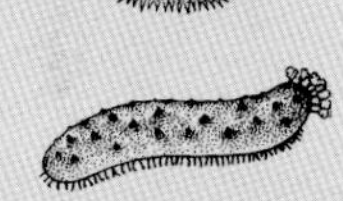

Class Holothuroidea
Sea cucumbers

SUBCLASS EUECHINOIDA

Superorder Diadematacea
Includes genera *Asthenosoma*, *Diadema*.

Superorder Echinacea
Includes genera *Echinus*, *Psammechinus*, *Paracentrotus*, *Toxopneustes*. *Echinus esculentus* is the sole endangered echinoderm species (Lower Risk/Near Threatened).

Upperorder Gnathostomata
Orders: Holectypoida, including *Echinoneus*, *Micropetalon* and Clypeasteroida, including *Rotula*, *Clypeaster*, *Echinocyamus*; Atelostomata with suborders Holasteroidea, including *Pourtalesia*, *Echinosigria*, and Spatangoida, including *Spatangus*, *Echinocardium*, *Brissopis*.

Sea cucumbers
Class Holothuroidea

About 500 species in 3 subclasses; long, saclike, without arms; bilaterally symmetrical; mouth surrounded by tentacles; no spines or pedicellariae; endoskeleton reduced to microscopic spicules or plates, or absent, internal madreporite.

SUBCLASS DENDROCHIROTACEA

Order Dendrochirotida
Genera include: *Cucumaria*, *Thyone*, *Psolus*.

Order Dactylochirotida
Genera include: *Rhopalodina*.

SUBCLASS ASPIDOCHIROTACEA

Order Aspidochirotida
Genera include: *Holothuria*.

Order Elasipodia
Genera include: *Pelagothuria*, *Psychropotes*.

SUBCLASS APODACEA

Order Molpadiida
Genera include: *Moipadia*, *Caudina*.

Order Apodida
Genera include: *Synapta*, *Leptosynapta*.

Acorn Worms and Allies

Hemichordates are a minor phylum of marine invertebrates; they are not abundant and relatively few species are known. Yet they fascinate zoologists, displaying as they do several characters that indicate similarities to the chordates (lancelets, fish, mammals, etc.), a group with which they were once classified. These characters include gill slits and a nerve cord on the upper (dorsal) side of the body. In some species the nerve cord is hollow and similar to those of vertebrates. At no stage of their development, however, does a rudimentary backbone (notochord) appear, which excludes them from the phylum Chordata.

FACTFILE

ACORN WORMS AND ALLIES

Phylum: Hemichordata

Classes: Enteropneusta, Pterobranchia, Planctosphaeroida

About 90 species in 3 classes.

Distribution Marine; acorn worms worldwide, pterobranchs mainly European and N American waters; Planctosphaeroida planktonic, oceanic, possibly a larval stage of an unknown adult.

Fossil record Acorn worms and Planctosphaeroidea, none; pterobranchs possibly rare from ordovician (500–440 million years ago) to present.

Size 2cm to 2.5m (0.8in–8.22ft); pterobranchs as colonies up to 10m (33ft) in diameter with individuals up to 1.4cm (0.6in) although often less.

Features Essentially wormlike with a coelom developed along deuterostome lines; may be solitary (acorn worms) or colonial (pterobranchs) with a body divided into three zones; proboscis, collar, and trunk; nervous system quite well developed; blood circulatory system present; with or without gill slits and tentacles; nephridia lacking; sexes separate. *Planctosphaera pelagica* aberrant and like a larva.

ACORN WORMS Class Enteropneusta
One order. Genera include: *Balanoglossus*, *Glossabalanus*, *Saccoglossus*.

PTEROBRANCHS Class Pterobranchia
Two orders. Genera include: *Cephalodiscus*, *Rhabdopleura*.

PLANCTOSPHAEROIDS Class Planctosphaeroida
One order and one genus containing the sole species *Planctosphaera pelagica*.

Hemichordates are divided into three classes: Enteropneusta (acorn worms); Pterobranchia; and Planctosphaeroida , which have no common names. The pterobranchs are regarded as being more primitive than the acorn worms and some of them lack gill slits and have solid nerve cords. Hemichordates also show similarities to echinoderms; the larva is a tornaria, which has features in common with some echinoderm larvae, for example the asteroid bipinnaria. Intriguingly, the pterobranchs, with their appendages clothed with tentacles, display similarities to the sea mats, horseshoe worms, and lampshells.

Acorn Worms

CLASS ENTEROPNEUSTA

The wormlike acorn worms are the only hemichordates likely to be encountered other than by a scientist, and then only rarely. They grow to 2.5m (5.2ft) in length but are often smaller. Their bodies are made up of three sections, the proboscis at the front, the collar, and the trunk at the rear (which forms the bulk of the body). The group's common name derives from the way in which the proboscis joins the collar, resembling an acorn sitting in its cup. The proboscis is a small conical structure connected to the collar by a short stalk. The collar itself is cylindrical and runs forward to ensheath the proboscis stalk. The collar bears the mouth on its underside. The trunk makes up most of the body length and at the front end this contains

Right The front region of the acorn worm Balanoglossus australiensis, *a species found in waters around Australia, living in sand under stones.*

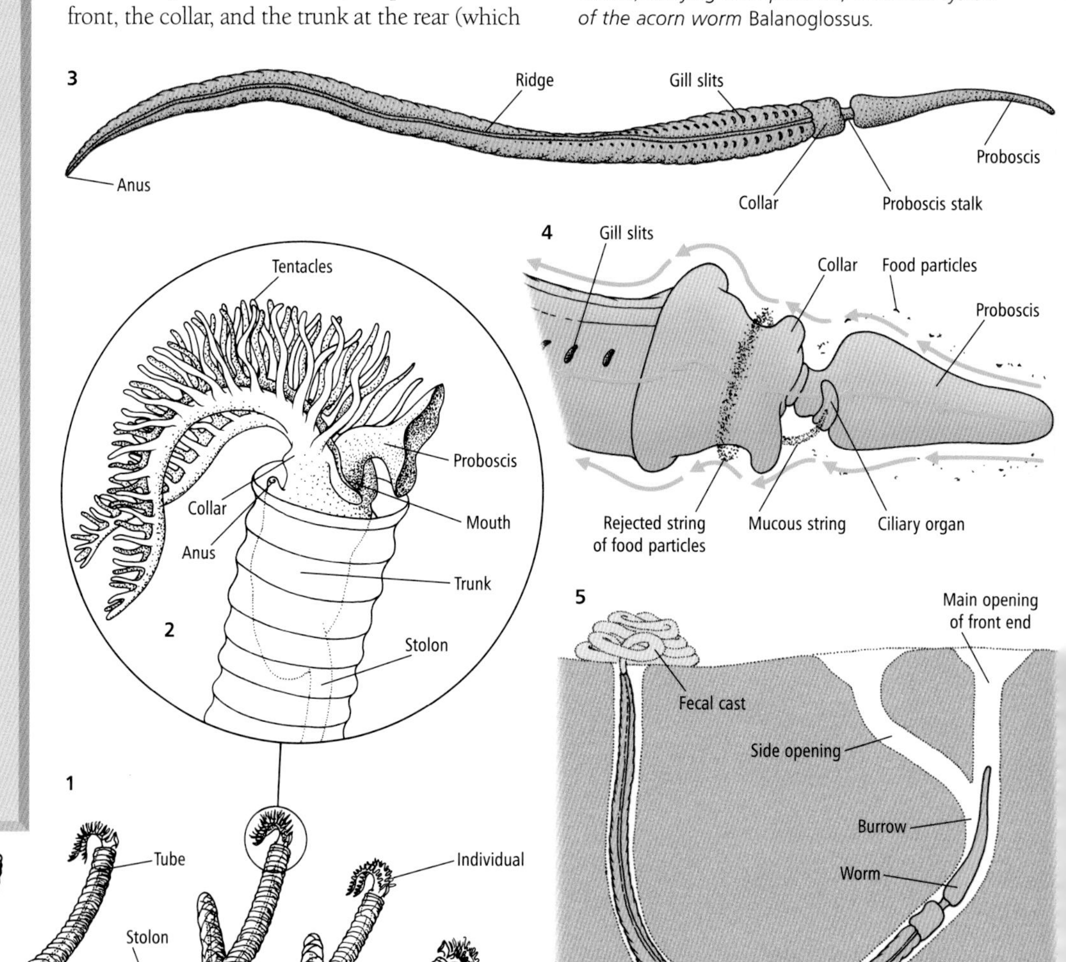

Below Body plans of hemichordates. ***1*** *A colony of individuals of the pterobranch genus* Rhabdopleura, *showing individuals within tubes that are connected by a stolon;* ***2*** *Close-up of the head of* Rhabdopleura *protruding from its tube;* ***3*** *General body plan of the acorn worm* Saccoglossus*;* ***4*** *Front end of the acorn worm* Protoglossus, *showing water currents (as arrows) carrying food particles;* ***5*** *Burrow system of the acorn worm* Balanoglossus.

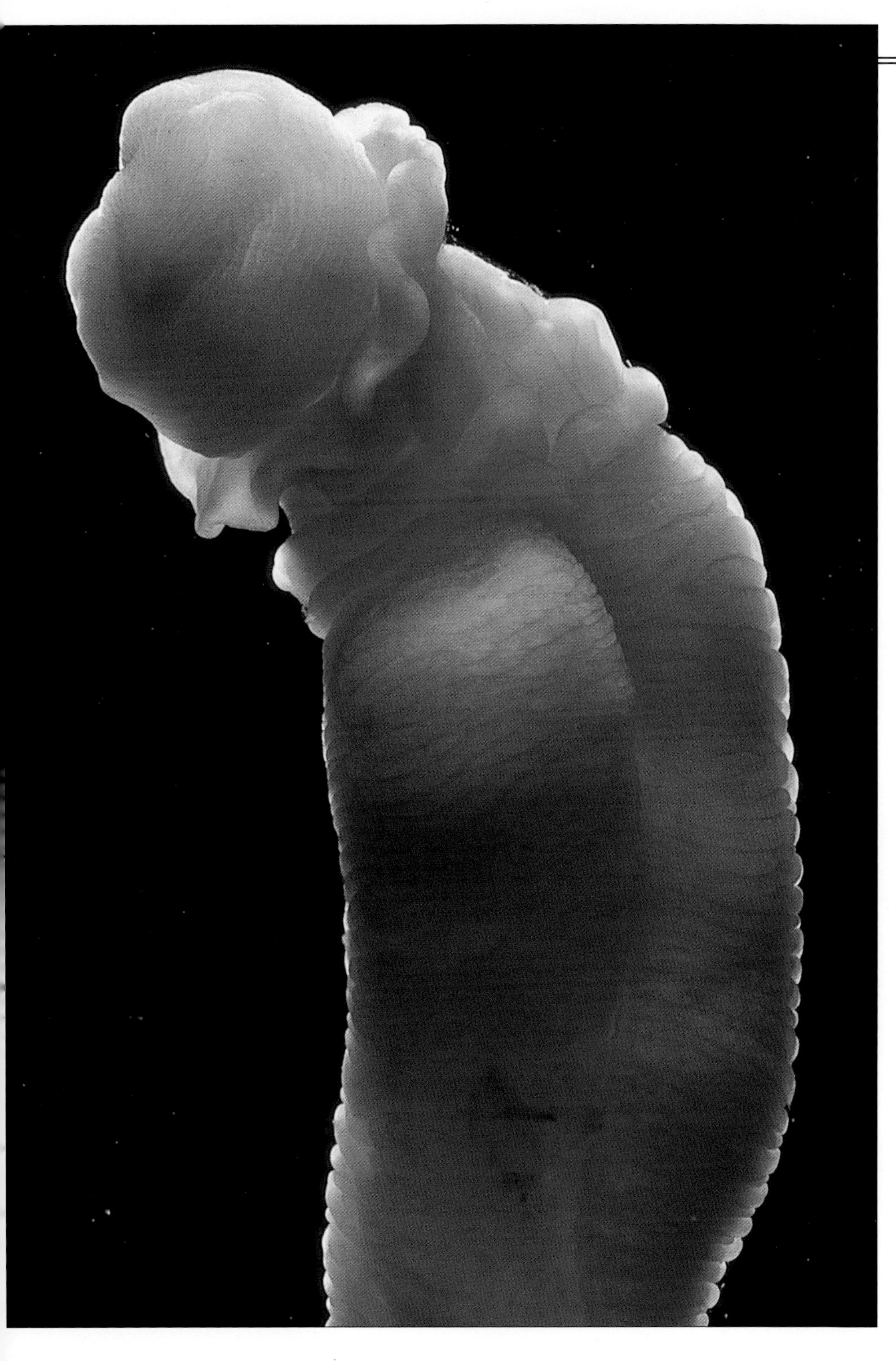

a row of gill slits on each side. A ridge runs down the middle of the back of the trunk. The reproductive organs are borne outside the gill slits; in some forms the trunk is extended as winglike genital flaps to contain them. The body cavity (coelom) is present in all three parts of the body. There is a single cavity in the proboscis, a pair of cavities in the collar and a pair in the trunk.

Acorn worms live in shallow water and some species may be found burrowing in sandy and muddy shores, identifiable by their characteristic coiled fecal casts. The burrows may be more or less permanent. Other species can live under stones and pebbles. Most burrowing acorn worms feed on the organic material in the sand or mud in which they live, simply by ingesting the sediment as an earthworm ingests soil. Some feed by trapping suspended plankton and particles of detritus in the mucus covering on the proboscis. Cilia then pass these particles back to the mouth and the collar plays a role in rejecting unsuitable particles.

The gut runs from the mouth on the collar, via the pharynx of gill slits in the front trunk, to the rear of the trunk where the anus is situated. There is a long, thin, blind-ended branch (diverticulum) running from the gut near the mouth into the proboscis, which once was mistaken for a notochord. The gill slits were originally probably feeding mechanisms, which have since become involved in gas exchange. The cilia they bear pump in water through the mouth and out through the gills.

The acorn worms' nervous system is primitive in comparison to the chordates, and the animals are not highly active. The sexes are separate and the eggs are fertilized in the sea. Early development is like that of the echinoderms. In some species, it proceeds to a tornaria larva, which lives in the plankton before it metamorphoses. Other types develop directly, with a juvenile worm appearing.

Pterobranchs and Planctosphaeroids

CLASSES PTEROBRANCHA AND PLANCTOSPHAEROIDA

The pterobranchs are very different, although their bodies still show the three zones. The proboscis is smaller and shield- or platelike, while the collar has well-developed outgrowths of tentacles. According to the group, there may be two backward-curving arms bearing tentacles (e.g. *Rhabdopleura*) or between five and nine (*Cephalodiscus*). The tentacles may play a food-gathering role.

The gill slits are few and inconspicuous and none is present in *Rhabdopleura*. In this group the gut is U-shaped, with an anus opening on the top side of the collar. Again, the sexes are separate and many individuals may live grouped together.

The solitary species *Planctosphaera pelagica* of the Planctosphaeroida drift in the plankton of the Atlantic and Pacific oceans. Their body form is quite unlike any other hemichordate; this organism might well be the larva of an as yet unknown adult hemichordate living on the ocean floor. AC

THE ENIGMATIC XENOTURBELLA

The primitive marine worm *Xenoturbella*, which DNA sequencing has shown to be related to the hemichordates, may be a common ancestor to all bilaterally symmetrical animals. First described in 1949, *Xenoturbella bocki* was found in marine muds off the Swedish coast. The external surface of the animal is covered with cilia and it is small enough to be confused with some large ciliate protozoans. The taxonomic affinities of *Xenoturbella* were long disputed – it was speculated that it was a flatworm, a cnidarian, or a mollusk.

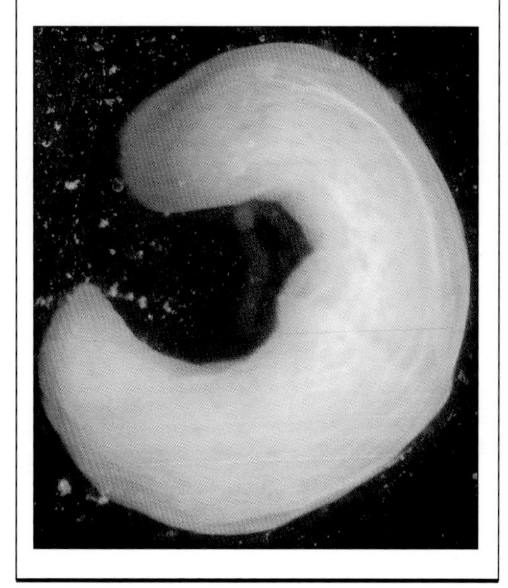

Sea Squirts and Lancelets

VERTEBRATES, FISHES, AMPHIBIA, REPTILES, and mammals (subphylum Craniata or Vertebrata) are the most familiar members of the phylum Chordata. These animals possess a definite backbone and bony braincase. Yet there are a number of lowly chordates that display the phylum's characteristics at a simple level. All aquatic, they are included in the two other subphyla of the chordates, Urochordata and Cephalochordata.

Chordates are "higher animals" possessing a single, hollow, dorsal nerve cord and a body cavity that is a true coelom. However, in Urochordata and Cephalochordata the dorsal nerve cord is present only during the embryonic development of the animal.

Sea squirts and Allies

SUBPHYLUM UROCHORDATA

Adult sea squirts look nothing like vertebrates. They are bottom dwellers that grow attached to rocks or other organisms. Their bodies are encased in a thick tunic made of material that resembles cellulose, the main constituent of plant cell walls. At the top of the body lies the inhalant siphon, and on the side is the exhalant siphon. There is no head. Water is pumped in via the inhalant siphon and passes through the pharynx and out between the gills into the sleevelike atrium. From the atrium it is discharged via the exhalant siphon. The gills serve as a respiratory surface and also act as a filter, extracting suspended particles of food. They are ciliated, the microscopic cilia providing the pumping force to maintain the respiratory and filter current. Acceptable food particles are collected in sticky mucus secreted by the endostyle, a sort of glandular gutter running down one side of the pharynx, and they are then passed into the gut for digestion. Waste products are liberated from the anus, which opens inside the atrium near the exhalant siphon.

Sea squirts may be solitary or colonial. In a colony the exhalant siphons of individuals open into a common cloaca, and each individual retains its own inhalant siphon. Colonial squirts may be arranged in masses, as in *Aplidium* species, where the individuals are hard to recognize, or in encrusting platelike growth, for example *Botryllus* species, where the individuals can be made out easily.

Right *Sea squirt species of different genera (*Polycarpa*, the large central organism,* Didemnum*, and the solitary, translucent-blue* Rhopalaea*) on a reef. Some colonial tunicates can be highly invasive.*

FACTFILE

SEA SQUIRTS AND LANCELETS

Phylum: Chordata

Around 3,020 species in 2 subphyla.

SEA SQUIRTS AND ALLIES
Subphylum Urochordata (or Tunicata)
About 3,000 species in 4 classes. **Distribution:** worldwide; bottom-dwelling and pelagic at all depths and in all oceans. **Fossil record:** very few, from the Cambrian to the Quaternary (600–500 million years ago to recent). **Size:** individuals from less than 1cm (0.4in) to about 20cm (8in) long. **Features:** chordate charactenstics – hollow, dorsal nerve cord, enterocoelic body cavity, gill slits, tail behind anus, and notochord – are all present in the larvae; adults show no segmentation, lack hollow dorsal nerve cord and notochord, and most have the gills surrounded by a large cavity (atrium) and are ensheathed in a test or tunic of tunicin, a substance related to cellulose; hermaphrodite; reproduction often involves budding and an asexual phase; life cycles complex in Thaliacea and Larvacea.

SEA SQUIRTS Class Ascidiacea
Genera include: *Aplidium*, *Botryllus*, *Ciona*, *Clavelina*.

SALPS OR PELAGIC TUNICATES Class Thaliacea
Genera include: *Pyrosoma*, *Salpa*.

CLASS APPENDICULARIA (= LARVACEA)
Sexually mature adults resemble larvae. Genera include: *Fritillaria*, *Oikopleura*, *Stegasoma*.

SORBERACEANS Class Sorberacea
Genera include: *Octanemus*.

LANCELETS Subphylum Cephalochordata (or Acrania)
Fewer than 20 species in 2 families. **Distribution:** temperate to tropical shallow seas , adults bottom-dwellers, burrowing in sand and fine gravels. **Fossil record:** none. **Size:** up to 5cm (2in) long. **Features:** simple fishlike chordates lacking a recognizable head, with hollow dorsal nerve cord similar to vertebrates, but no vertebrae, instead a notochord, muscle blocks segmented; enterocoelic coelom and well-developed gills; can swim; sexes separate; planktonic tornaria larva produced.

FAMILY BRANCHIOSTOMIDAE
Contains one genus, *Branchiostoma* (or *Amphioxus*).

FAMILY ASYMMETRONIDAE
Contains one genus, *Asymmetron*.

The heart is situated in a loop of the gut and services the very simple blood system. Sea squirts are hermaphrodite, each having male and female organs. In some species the eggs are retained within the atrium, where the sex ducts open, and where they are fertilized by sperm drawn in with the feeding and respiratory currents. The embryos are able to develop here in a protected environment. In other species the eggs and sperm are both liberated into the sea where fertilization occurs. The embryo quickly develops into a tadpolelike larva. This process may take from a few hours to a few days, a timescale that makes these animals ideal specimens for observation and experimental embryology.

The "tadpoles" of sea squirts are small, independent animals like miniature frog's tadpoles. They are sensitive to light and gravity, enabling the animal to select an appropriate substrate for settlement, attachment, and metamorphosis. These tadpoles also serve to distribute the species. The ascidian tadpole with its chordate characters of dorsal hollow nerve cord, stiffening notochord supporting the muscular tail, and features of the head imparts much more information about the likely evolutionary position of the sea squirts than does the adult.

Some sea squirts are of economic importance because they act as fouling organisms encrusting the hulls of ships and other marine structures. They are also of considerable evolutionary and zoological interest.

The other urochordates, thaliaceans (salps and others), and the larvalike appendicularians, have characters fundamental to the group but have evolved along pelagic and not bottom-dwelling lines. Members of one thaliacean order, the Pyrosomidae, are commonly found in warmer waters and emit phosphorescent light in response to tactile stimulation. These pelagic animals (for example *Pyrosoma* species) live in colonies, but some other types exist as solitary individuals. They have complex life histories. *Oikopleura* is an appendicularian genus of small animals like the tadpoles of sea squirts. Rhythmic movements of the tail draw water through the openings of the gelatinous "house" in which they live and which is secreted by the body. Members of the class Sorberacea resemble sea squirts, but they are predators living in the deep ocean and they possess dorsal nerve cords as adults.

Lancelets

SUBPHYLUM CEPHALOCHORDATA

Lancelets (named for their elongate, bladelike form) are a minor group of chordates with only two genera and under 20 species. These small, apparently fishlike animals show primitive chordate conditions. The notochord, which in sea squirt larvae merely supports the tail, extends into the head (hence the term "cephalochordata"). There

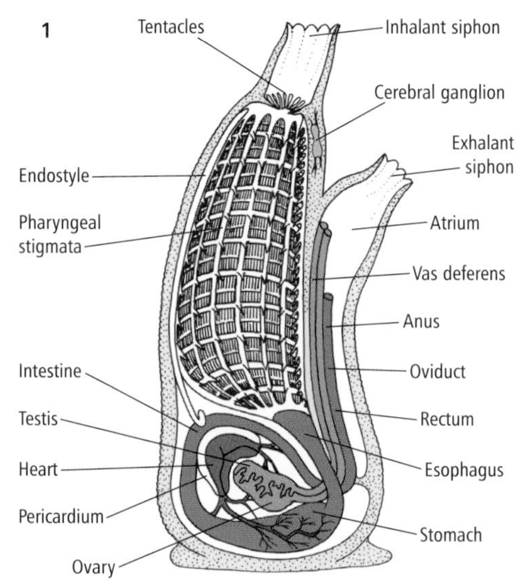

Left *Colonial sea squirts vary widely in form. The Pinecone tunicate* (Neptheis fascicularis) *has clearly defined individuals on peduncles; this is unusual for tunicates, which are normally attached directly to the substrate. Other sea-squirt colonies form a single, gelatine-like mass.*

Right *A colony of fluorescent salps* (Salpa maxima). *Chains of these pelagic tunicates can reach lengths of over 25m (82ft), but are easily broken. Most species live in tropical waters.*

Below *A lancelet of the species* Branchiostoma lanceolatum. *Lancelets usually lie half-buried in the substrate with their ventral side turned upward. However, they emerge from their burrows and swim actively for short periods.*

Bottom *Body plans showing the internal organs of* **1** *a sea squirt,* **2** *its "tadpole" larva, and* **3** *a lancelet.*

is no cranium as in the craniates or vertebrates, and the front end, although quite distinct, lacks the well-developed brain, eyes, and other sense organs as well as the jaws associated with vertebrates.

In species of *Branchiostoma*, a hood extends over the mouth, equipped with slender, tentacle-like cirri instead of jaws. These structures form a sieve that assists in rejecting particles of food too large for the lancelet's suspension-feeding habit.

The oral hood leads to the extensive pharynx via a thin flap of tissue, the velum. The pharyngeal wall is composed of many gill bars, and the action of the cilia situated on those bars draws the water and suspended food particles into the body by way of the oral hood. The gill bars act as a filter for food as well as providing a surface for the absorption of oxygen. When water passes through the gill bars, food particles are trapped and passed down to the floor of the pharynx, where they are swept into the endostyle. In this ciliated gutter, trapped food is collected, bound into mucus strands, and passed into the midgut. The filtered water passes out into the atrium and leaves the body via the atrial pore. The midgut has a blind diverticulum leading forward. Backward, the midgut leads to the intestine and eventually to the anus.

The anatomy of these coelomates is complex compared to most invertebrates and has some unusual features. The excretory system consists of saclike nephridia lying above the gill bars. Each lancelet nephridium has a number of flame cells reminiscent of those present in flatworms, annelids, and mollusks. In evolutionary terms the nephridia (which do not occur in other primitive chordates) are a far cry from the vertebrate kidneys, which must have evolved via a different route.

The muscle blocks are arranged in segments (myotomes) along either side of the body. A hollow dorsal nerve cord runs the length of the animal and shows very little anterior specialization or brain. As it passes back along the body, it branches to supply muscles and other organs.

There is a simple circulatory system with blood vessels passing through the gills to collect oxygen and distribute it to the body. The sexes are separate. In *Branchiostoma* the gonads are arranged on both sides of the body at the bases of the muscle blocks, but in *Asymmetron* they lie only on the right side. Sperms and eggs are released into the atrium by rupture of the gonad wall. They pass out to the sea through the atriopore. Fertilization occurs in the open water, and a swimming larva or tornaria type develops. This larva lives a dual life for several months as it feeds and matures. In the daytime it lies on the seabed but when darkness comes it swims and joins the plankton. When it has attained a length of about 5mm (0.2in), it metamorphoses and becomes bottom-dwelling and more sedentary. The tornaria is the chief distributive phase.

The characters of *Branchiostoma* have helped further an understanding of vertebrate organization, and its simple structure makes it quite easy to identify the basic chordate features, many of which are seen in humans themselves. AC

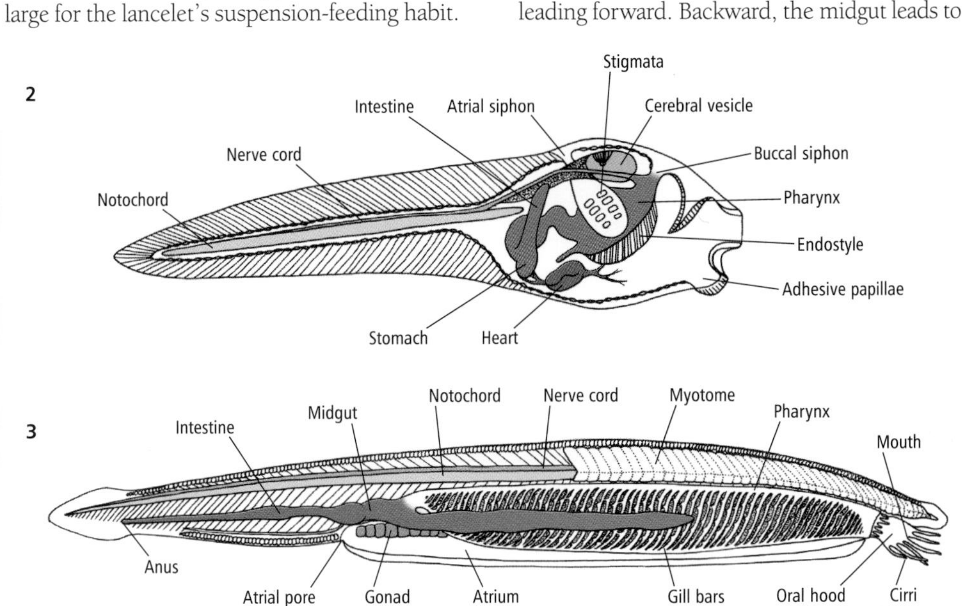

What is a Fish?

IN ALL PROBABILITY LIFE STARTED IN EARTH'S waters around 3,000 million years ago. For a very long time not much seems to have happened. The first known multicellular, or many-celled, invertebrates (animals without backbones) evolved about 600 million years ago. After a much shorter interval (in geological terms), about 120 million years, the first aquatic vertebrates (back-boned animals) – the fishes – appeared. From these early fish groups arose the most familiar animals: birds, reptiles, and mammals.

Over half of the vertebrates now alive are fishes. They do just about everything that the other vertebrates do and also have many unique attributes. Only fishes, for example, make their own light (bioluminescence), produce electricity, and have complete parasitism, as well as having the largest increase in volume from hatching to adulthood.

Different people have different impressions of fishes. To some, the image of the perfect fish is the sharp-toothed shark, elegantly and effortlessly hunting its prey in the sea. To others, fishes are small, entrancing animals kept in home aquaria. For anglers, fish are a cunning quarry to be outwitted and caught. For commercial fishermen, fishes are a mass of writhing bodies being hauled on board the fishing vessel. Biologists regard fishes as representing a mass of challenges concerning evolution, behavior, and form, the study of which results in more questions than answers.

It is the very diversity of fishes, the large number of species ,and the huge numbers of individuals of some species that make them so interesting, instructive, and useful to humans. It is easy to see that the huge shoals of many species (provided they are maintained by careful management) form a valuable food resource for humans and other animals, but it is not so generally realized that fishes are also valuable research animals that may help humans solve many complex surgical and medical problems. For example, in making transplants of hearts and lungs in humans, there are problems of tissue rejection. How much might people learn from the fusion of the male angler fish onto the female's body, seemingly without any rejection problems? Studies on a few small closely related fishes that exist as eyed, colored,

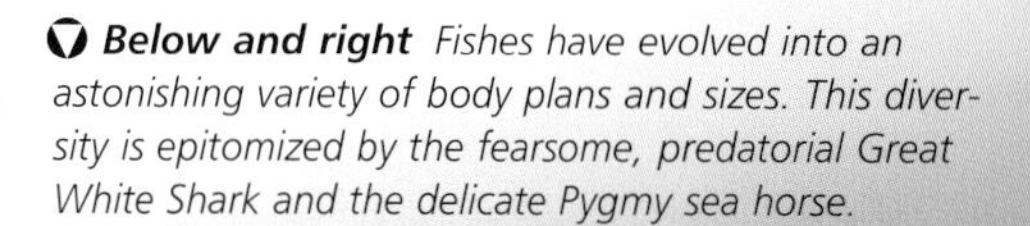

Below and right *Fishes have evolved into an astonishing variety of body plans and sizes. This diversity is epitomized by the fearsome, predatorial Great White Shark and the delicate Pygmy sea horse.*

SUPERCLASSES: AGNATHA AND GNATHOSTOMATA
5 classes; 58 orders; c.469 families

SUPERCLASS AGNATHA (Jawless fishes)
2 families, 12 genera, c.90 species

CLASS CEPHALASPIDOMORPHI

LAMPREYS Order Petromyzontiformes p156
c.40 species in 6 genera and 1 family

CLASS MYXINI

HAGFISHES Order Myxiniformes p160
c.43–50 species in 6 genera and 1 family

SUPERCLASS GNATHOSTOMATA (Jawed fishes)
c.467 families, c 4,180 genera, c. 24,510 species

CLASS CHONDRICHTHYES (Cartilaginous fishes)
34 families, 138 genera, c. 915 species in 14 orders

FRILLED SHARK Order Chlamydoselachiformes p276
1 species *Chlamydoselachus anguil*

SIX- AND SEVEN-GILLED SHARKS
Order Hexanchiformes p276
5 species in 3 genera and 1 family

CATSHARKS Order Scyliorhyniformes p276
87 species in c.15 genera and 3 families

SMOOTH DOGFISH SHARKS Order Triakiformes p276
30 species in 5 genera and 1 family

PORT JACKSON SHARKS
Order Heterodontiformes p278
8 species in 1 genus and 1 family

ORECTOLOBOIDS Order Orectolobiformes p278
31 species in 13 genera and 7 families

REQUIEM SHARKS Order Carcharhiniformes p279
c.100 species in 10 genera and 1 family

GOBLIN SHARKS & ALLIES
Order Odontaspidiformes p279
7 species in 4 genera and 3 families

THRESHER SHARKS & ALLIES Order Isuriformes p281
10 species in 6 genera and 3 families

SPINY DOGFISH & ALLIES Order Squaliformes p281
c.70 species in c.12 genera and 1 family

ANGEL SHARKS Order Squatiniformes p283
10 species in 1 genus and 1 family

SAWSHARKS Order Pristiophoriformes p283
5 species in 2 genera and 1 family

SKATES, RAYS, & SAWFISHES Order Batiformes p284
c.465 species in c.62 genera and 12 families

CHIMAERAS Order Chimaeriformes p289
31 species in 6 genera and 3 families

CLASS ACTINOPTERYGII (Ray-finned fishes)
c.428 families, c.4,037 genera, c.23,600 species in 39 orders

BICHIRS Order Polypteriformes p268
10 species in 2 genera and 1 family

CLASS SARCOPTERYGII (Lobe-finned fishes)
5 families, 5 genera, 8 species, and 3 orders

COELACANTHS Order Coelacanthiformes p269
2 species in 1 genus and 1 family

AUSTRALIAN LUNGFISHES
Order Ceratodontiformes p270
1 species *Neoceratodus forsteri*

SOUTH AMERICAN AND AFRICAN LUNGFISHES
Order Lepidosirenifomes p270
5 species in 3 genera and 3 families

STURGEONS AND PADDLEFISHES
Order Acipenseriformes p162
27 species in 6 genera and 2 families

BOWFIN Order Amiiformes p165
1 species *Amia calva*

GARFISHES Order Semionotiformes p164
7 species in 2 genera and 1 family

TARPONS & ALLIES Order Elopiformes p166
8 species in 2 genera and 2 families

BONEFISHES, SPINY EELS, & ALLIES
Order Albuliformes p167
29 species in 8 genera and 3 families

EELS Order Anguilliformes p168
738 species in 141 genera and 15 families

GULPER EELS & ALLIES
Order Saccopharyngiformes p178
c.26 species in 5 genera and 4 families

HERRINGS AND ANCHOVIES
Order Clupeiformes p182
c357 species in 83 genera and 4 families

BONYTONGUES & ALLIES
Order Osteoglossiformes p184
c.217 species in 29 genera and 6 families

PIKES AND MUDMINNOWS Order Esociformes p186
10 species in 4 genera and 2 families

SMELTS & ALLIES Order Osmeriformes p191
c.241 species in 74 genera and 13 families

SALMON, TROUT, & ALLIES
Order Salmoniformes p196
66 species in 11 genera and 1 family

BRISTLEMOUTHS & ALLIES Order Stomiiformes p204
c.320 species in 50 genera and 4 families

LIZARDFISHES Order Aulopiformes p210
c.225 species in c.40 genera and 15 families

LANTERNFISHES Order Myctophiformes p213
c.25 species in c.35 genera and 2 families

CHARACINS & ALLIES Order Characiformes p215
c.1,340 species in c.250 genera and c.15 families

CATFISHES Order Siluriformes p218
c.2,400 species in c.410 genera and 34 families

CARPS & ALLIES Order Cypriniformes p222
c.2,660 species in c.279 genera and 5 families

NEW WORLD KNIFEFISHES
Order Gymnotiformes p225
c.62 species in 23 genera and 6 families

MILKFISHES & ALLIES
Order Gonorhynchiformes p225
c.35 species in 7 genera and 4 families

TROUT-PERCHES & ALLIES
Order Percopsiformes p229
9 species in 6 genera and 3 families

CUSKEELS & ALLIES Order Ophidiiformes p230
c.355 species in 92 genera and 5 families

CODFISHES & ALLIES Order Gadiformes p231
c.482 species in 85 genera and 12 families

TOADFISHES Order Batrachoidiformes p234
69 species in 19 genera and 1 family

ANGLERFISHES Order Lophiiformes p234
c.310 species in 65 genera and 18 families

SILVERSIDES Order Atheriniformes p240
290 species in 49 genera and 6 families

KILLIFISHES Order Cyprinodontiformes p240
800 species in 84 genera and 9 families

RICEFISHES & ALLIES Order Beloniformes p243
200 species in 37 genera and 5 families

PERCHLIKE FISHES Order Perciformes p244
c.9,300 species in c.1,500 genera and c.150 families

FLATFISHES Order Pleuronectiformes p256
c.570 species in c.123 genera and 11 families

TRIGGERFISHES & ALLIES
Order Tetraodontiformes p258
c.340 species in c.100 genera and 9 families

SEAHORSES & ALLIES Order Syngnathiformes p260
c.241 species in 60 genera and 6 families

SQUIRRELFISHES & ALLIES Order Beryciformes p262
c.130 species in 18 genera and 5 families

DORIES & ALLIES Order Zeiformes p263
c.39 species in c.20 genera and 6 families

PRICKLEFISHES & ALLIES
Order Stephanoberyciformes p264
c.86 species in 28 genera and 9 families

SWAMP EELS & ALLIES
Order Synbranchiformes p264
c.87 species in 12 genera and 3 families

STICKLEBACKS Order Gasterosteiformes p265
c.216 species in 11 genera and 5 families

MAILCHEEKED FISHES Order Scorpaeniformes p266
c.1,300 species in c.266 genera and 25 families

OARFISHES & ALLIES
Order Lampriformes p267
c.19 species in 12 genera and 7 families

surface forms and eyeless, pink-bodied cave types may help scientists understand the relationship between the genetic code and the environment. There are many similar examples.

Fishes, then, are many things to many people and for millennia they have fascinated humanity in many ways, not least because they have conquered an environment that is alien to humans. Yet all these associations between humans and fishes beg the most vital question of all: What is a fish?

What is a Fish?

BASIC PARAMETERS

Incredible as it may sound, there is no such thing as a "fish." The concept is merely a convenient umbrella term to describe an aquatic vertebrate that is not a mammal, a turtle, or anything else. There are five quite separate groups (classes) of fishes now alive – plus three extinct ones – not at all closely related to one another. Lumping these together under the term "fishes" is like lumping all flying vertebrates – namely, bats (mammals), birds, and even the flying lizard – under the single heading "birds," just because they all fly. The relationship between a lamprey and a shark is no closer than that between a salamander and a camel.

However, the fact that "fish" has become hallowed by usage over the centuries as a descriptive term dictates that, for convenience's sake, it will be used here. It is worth remembering, however, that employing this term to describe the five different living groups is equivalent to referring to all other vertebrates as tetrapods (four-legged animals), even if some have subsequently lost or modified their legs.

The five living groups consist of two groups of jawless fishes – hagfishes and lampreys – and three groups with jaws – cartilaginous fishes (sharks and rays), lobe-finned fishes (the coelacanths and lungfishes), and bony fishes (all the rest). The last two groups possess bony, rather than cartilaginous, skeletons.

The five living groups differ widely in their numbers of species. There are about 43 species of hagfishes and about 40 of lamprey. Jawed fishes now predominate: the sharks, rays, and chimaeras comprise about 700 species, whereas the greatest flowering is in the bony fishes, with over 26,000 species.

A Brief History of the Major Groups

EVOLUTION

The first identifiable remains of fishes are small, broken, and crushed plates in rocks of the middle Ordovician era 490–443 million years ago. (Possible traces from the upper Cambrian era, more than 500 m.y.a., have not been confirmed as belonging to fishes.) These plates represent parts of the bony external armor of jawless fishes. Although none of the living jawless fishes has any external protection, large defensive head-shields were not uncommon in the early forms. However, it is not known what the overall body shape of the first known fishes was like.

About 150 million years after they first appeared, the jawless fishes radiated into many widely varying forms, quite unlike the eel-like forms now alive. In some species of the Devonian era (417–354 million years ago), the armor was reduced to a series of thin rods allowing greater flexibility of the body and, in one poorly known group, represented now by mere shadowy outlines on rocks, the armor consisted solely of tiny isolated tubercles (nodules).

Although most of these Devonian jawless fishes were small, the pteraspids, fishes with the front half of the body covered with a massive plate, often with a backward pointing spine, were exceptional and could reach 1.5m (5ft) in length. The cephalaspids, with their shield-shaped head plates, are the best-known fossil jawless fishes (agnathans), a group well known from fossils in the Old Red Sandstone formations of the Devonian. A fortunate discovery of some well-preserved

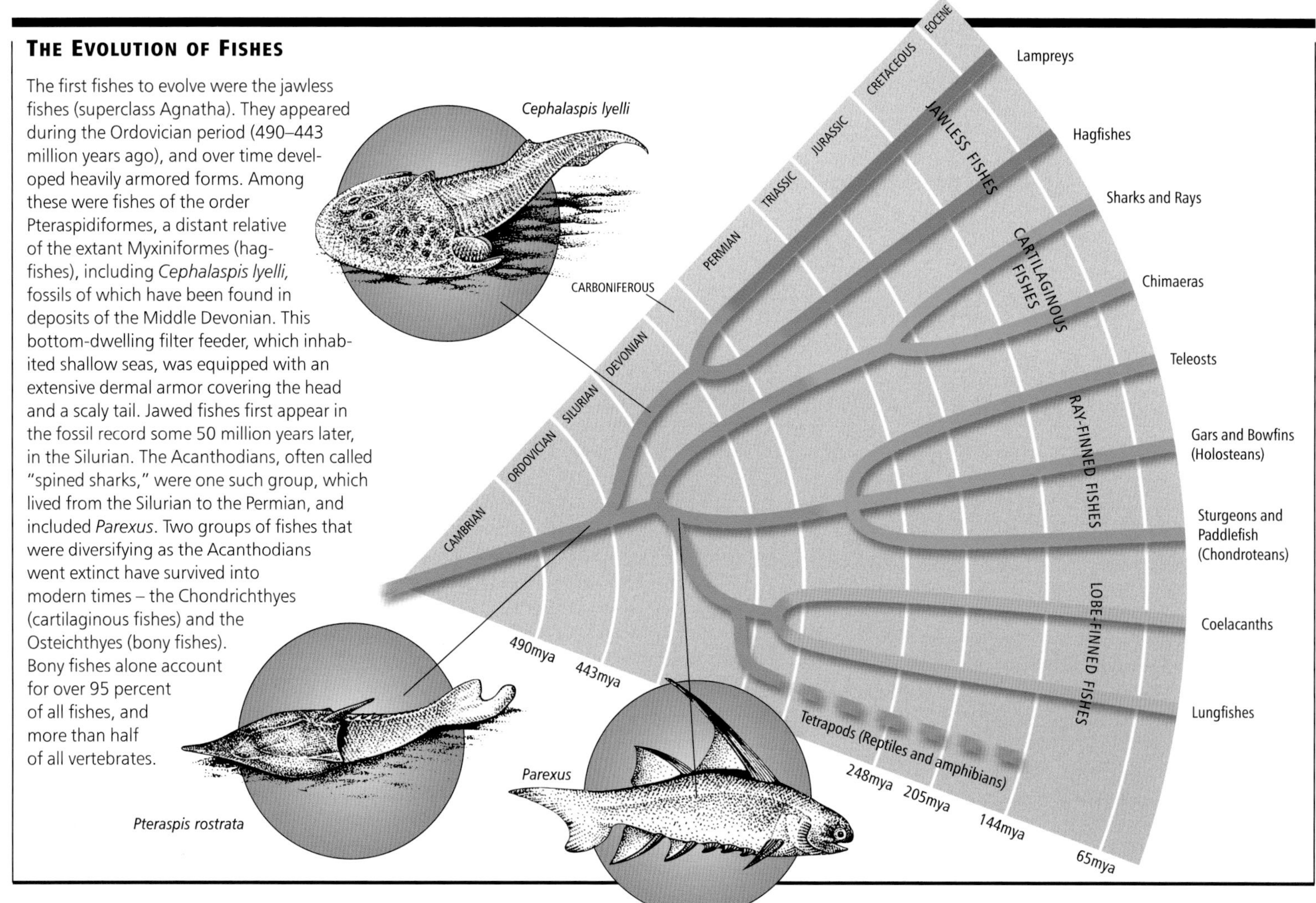

THE EVOLUTION OF FISHES

The first fishes to evolve were the jawless fishes (superclass Agnatha). They appeared during the Ordovician period (490–443 million years ago), and over time developed heavily armored forms. Among these were fishes of the order Pteraspidiformes, a distant relative of the extant Myxiniformes (hagfishes), including *Cephalaspis lyelli,* fossils of which have been found in deposits of the Middle Devonian. This bottom-dwelling filter feeder, which inhabited shallow seas, was equipped with an extensive dermal armor covering the head and a scaly tail. Jawed fishes first appear in the fossil record some 50 million years later, in the Silurian. The Acanthodians, often called "spined sharks," were one such group, which lived from the Silurian to the Permian, and included *Parexus*. Two groups of fishes that were diversifying as the Acanthodians went extinct have survived into modern times – the Chondrichthyes (cartilaginous fishes) and the Osteichthyes (bony fishes). Bony fishes alone account for over 95 percent of all fishes, and more than half of all vertebrates.

Above *The limestone quarries at Solnhofen, in Bavaria, Germany, have yielded a huge variety of fossils. Among them is this Jurassic fish,* Gyrodas circularis, *which strongly resembles the body shape of modern bony fishes.*

cephalaspids buried in fine mud enabled, with careful preparation, the course of their nerves and blood vessels to be discovered.

The first recognizable fossil lampreys have been found in Carboniferous (Pennsylvanian) rocks of Illinois, USA (dating from 325–290 m.y.a.). To date, however. no indisputable fossil hagfishes have been found.

The earliest of the true jawed fishes were the acanthodians. Spines belonging to this group of large-eyed, scaled fishes have been reported from rocks 440 million years old, from the Silurian period. They are thought to have hunted by sight in the upper layers of the water. Some of the largest species, which grew to over 2m (6.5ft) long, have jaws, which suggest that they were active predators, much like present-day sharks. The majority of the acanthodians, though, were small fishes. The earliest acanthodians were marine; the later species lived in freshwater.

Acanthodians had bony skeletons, ganoid scales (typical of primitve fishes and containing several top layers of enamel, followed by dentine – a hard elastic substance also known as ivory – and, finally, by one or more layers of bone), and a stout spine in front of each fin except the caudal. Most species had a row of spines between the pectoral and pelvic fins. The tail was sharklike (heterocercal), that is, the upper lobe was longer than the lower. The tail shape and the presence of spines have led to them being called "spiny sharks," despite the presence of bone and scales. Recent research has suggested that they may, however, be more closely related to the bony fishes. Acanthodians never evolved flattened or bottom-living forms. One hundred and fifty million years after they appeared, Acanthodians became extinct.

The other group of extinct fishes is the placoderms, a bizarre class that may be related to sharks, or to bony fishes, or to both, or to all other jawed fishes; no one knows for certain. The front half of their body was enclosed in bony plates, formed into a head shield that articulated (formed a joint) with the trunk shield. Most had depressed bodies (i.e. flattened) and lived on the bottom. However, one order – the arthrodires – were probably fast-swimming active predators growing to 6m (20ft) long. They probably paralleled the living sharks in the same way that the very depressed rhenanids paralleled rays and skates. Another group of placoderms, the antiarchs, are among the strangest fishes ever to have evolved. About 30cm (1ft) long, they had an armor-plated trunk that was triangular in cross section. The eyes were very small and placed close together on the top of the head. The "pectoral fins" are most extraordinary and unique among vertebrates. Whereas all other vertebrates had developed an internal skeleton, the "pectoral fin" of these antiarchs had changed to a crustacean-like condition, resembling the legs of a lobster – a tube of jointed bony plates worked internally by muscles. The function of these appendages is unknown, but they may have been used to drag the fish slowly over mud or rocks. The antiarchs also had a pair of internal sacs, which have been interpreted as lungs. In another

Fish Body Plan

The oldest vertebrates on Earth, fishes display a number of intriguing physiological adaptations to life in water. The fins give fishes control over their movement by directing forward thrust and providing lift. The swimbladder provides buoyancy. The complex respiratory system – the gills and gill arches – enables fishes to absorb and concentrate the scant oxygen available in the water, while the ultra-sensitive lateral line allows them to detect predators or prey items.

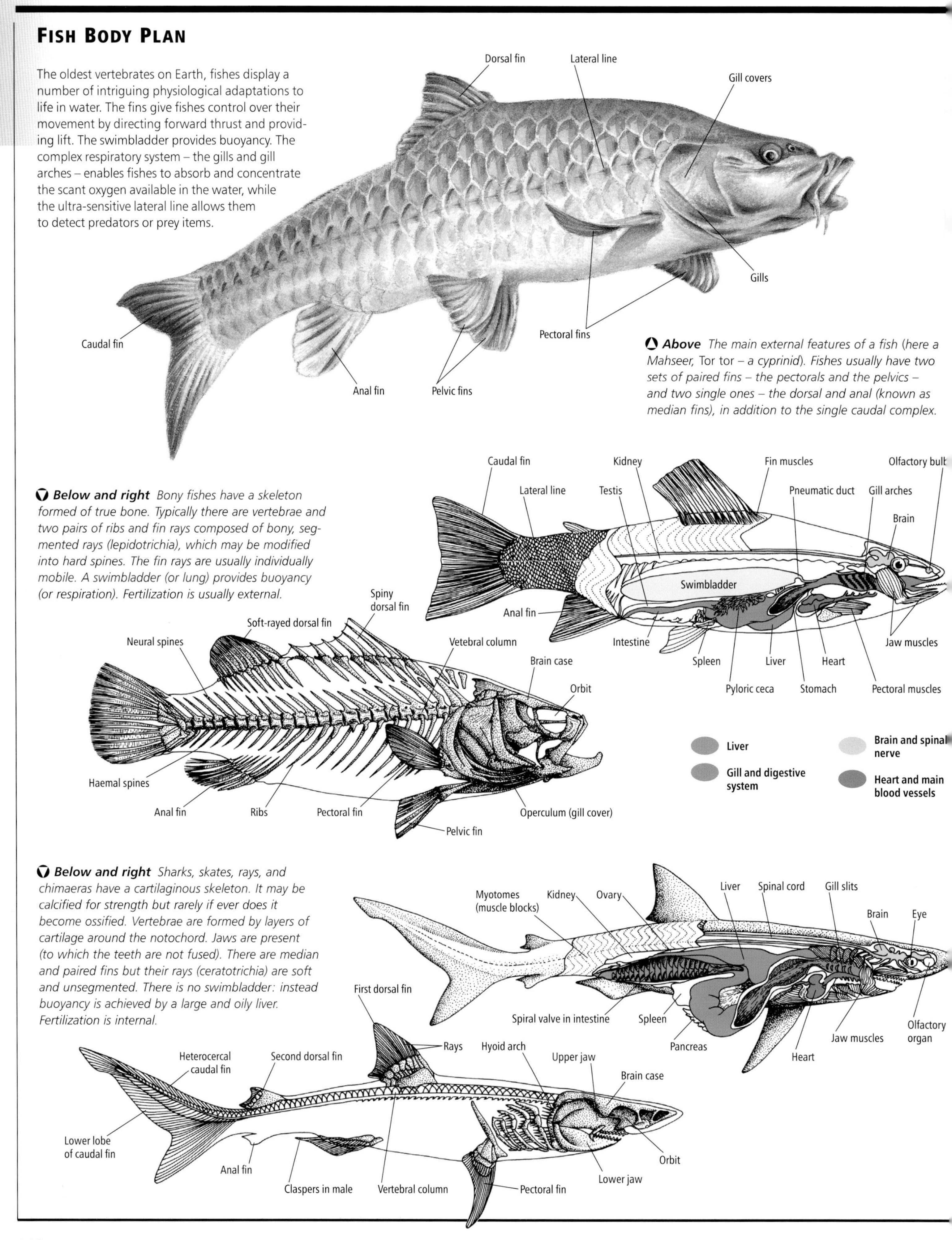

Above *The main external features of a fish (here a Mahseer,* Tor tor *– a cyprinid). Fishes usually have two sets of paired fins – the pectorals and the pelvics – and two single ones – the dorsal and anal (known as median fins), in addition to the single caudal complex.*

Below and right *Bony fishes have a skeleton formed of true bone. Typically there are vertebrae and two pairs of ribs and fin rays composed of bony, segmented rays (lepidotrichia), which may be modified into hard spines. The fin rays are usually individually mobile. A swimbladder (or lung) provides buoyancy (or respiration). Fertilization is usually external.*

Below and right *Sharks, skates, rays, and chimaeras have a cartilaginous skeleton. It may be calcified for strength but rarely if ever does it become ossified. Vertebrae are formed by layers of cartilage around the notochord. Jaws are present (to which the teeth are not fused). There are median and paired fins but their rays (ceratotrichia) are soft and unsegmented. There is no swimbladder: instead buoyancy is achieved by a large and oily liver. Fertilization is internal.*

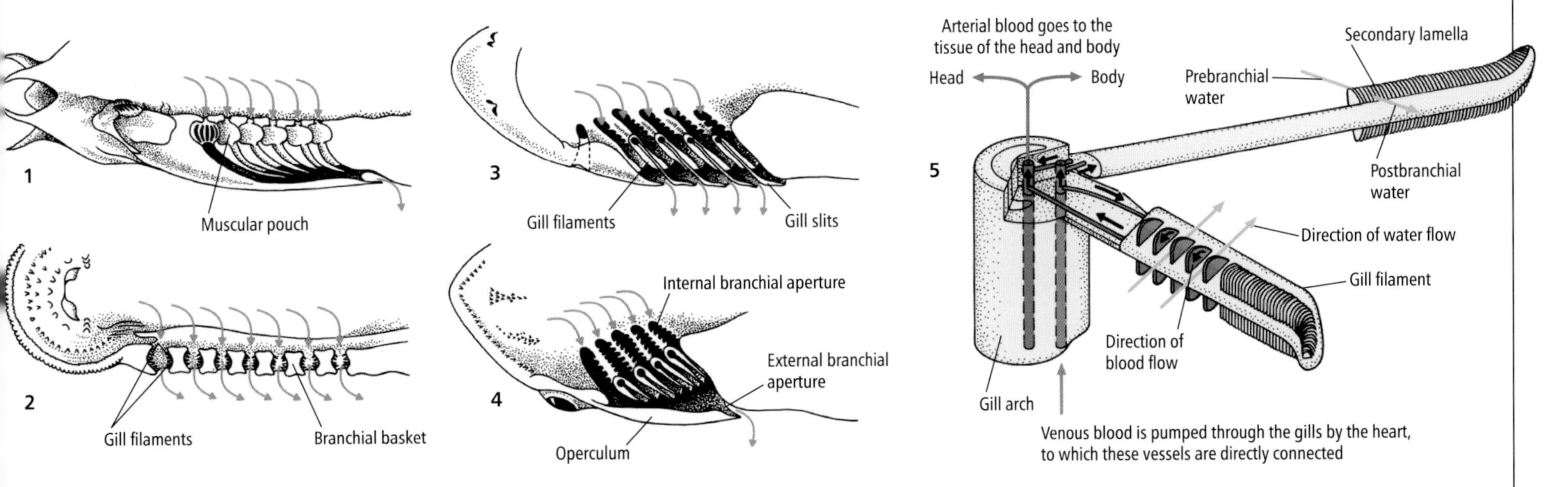

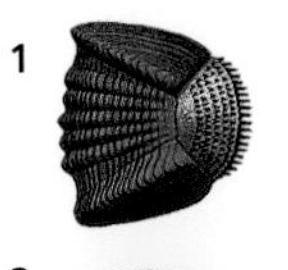

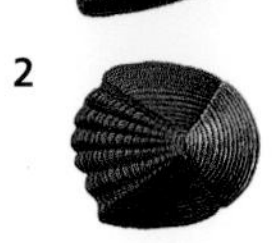

Right *The four basic types of fish scales:* ***1*** *Ctenoid scales, found in most modern bony fishes, overlap one another in the manner of roof tiles, and have a comblike posterior margin;* ***2*** *Cycloid scales – found, for example in salmon – have a rounded appearance and a smooth posterior margin ;* ***3*** *Rhomboid-shaped ganoid scales are found in certain "primitive" modern fishes, such as bichirs, gars, and bowfins;* ***4*** *Placoid scales, or dermal denticles, present in sharks, rays, and other cartilaginous fishes.*

Above *Gill structures in fishes:* ***1*** *In the hagfishes water passes through a series of muscular pouches before it leaves through a single common opening;* ***2*** *In the lampreys each gill has a separate opening to the outside and the gills are supported by an elaborate cartilaginous structure called the branchial basket;* ***3*** *In the living Chondrichthyes (namely, the sharks, skates, and chimaeras), the gills (except in the chimaeras) primitively open directly to the outside via five slits. (In the chimaeras an operculum or cover is developed);* ***4*** *In bony fishes the gills are protected externally by a bony operculum (cover);* ***5*** *Detailed diagram of blood flow through the filaments and the lamellae of fishes' gills. Note that the blood flows through the lamellae in the direction opposite to that of the water flowing over it to maximize the efficiency of exchange – the so-called "countercurrent system."*

EXTREME FISH FACTS

- **Smallest** The Dwarf pygmy goby (*Pandaka pygmaea*), length (mature male) 9mm (0.4in).
- **Largest** The Whale shark (*Rhincodon typus*), a cartilaginous fish, length 12.5m (41ft).
- **Fastest** A Sailfish (*Istiophorus platypterus*) swimming near the Florida coast was timed at 110km/h (68mph).
- **Fastest reflexes** It takes toadfishes only 6 milliseconds to swallow a passing fish – so fast that other fishes in the shoal will not even have noticed.
- **Commonest** Deep-sea bristlemouths (*Cyclothone* spp.) are found in great abundance throughout the world's oceans, at depths of over 300m.
- **Largest number of eggs** The Oceanic sunfish (*Mola mola*) lays some 250 million eggs in a single spawning.
- **Lowest reproduction rate** The Sand tiger shark (*Carcharias taurus*) gives birth to only one or two large young every two years.
- **Shortest lifespan** The Turquoise killlifish (*Nothobranchius furzeri*), which lives in seasonal pools, has a 12-week lifespan, the shortest of any vertebrate.
- **Longest lifespan** The Lake sturgeon (*Acipenser fulvescens*) can reach 80 years of age.
- **Most venomous** The Estuarine stonefish (*Synanceia horrida*) delivers venom that can kill a person.

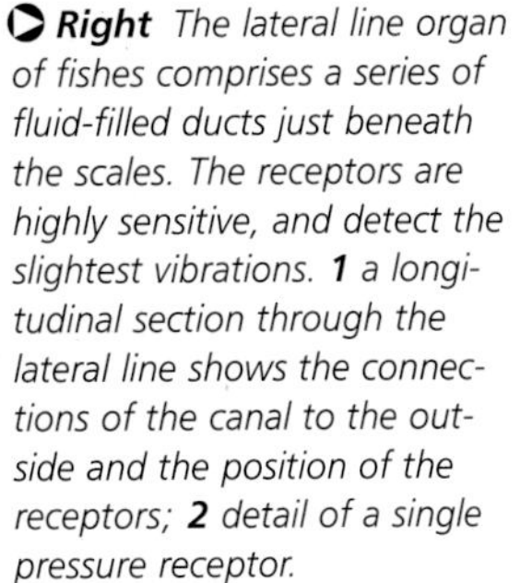

Right *The lateral line organ of fishes comprises a series of fluid-filled ducts just beneath the scales. The receptors are highly sensitive, and detect the slightest vibrations.* ***1*** *a longitudinal section through the lateral line shows the connections of the canal to the outside and the position of the receptors;* ***2*** *detail of a single pressure receptor.*

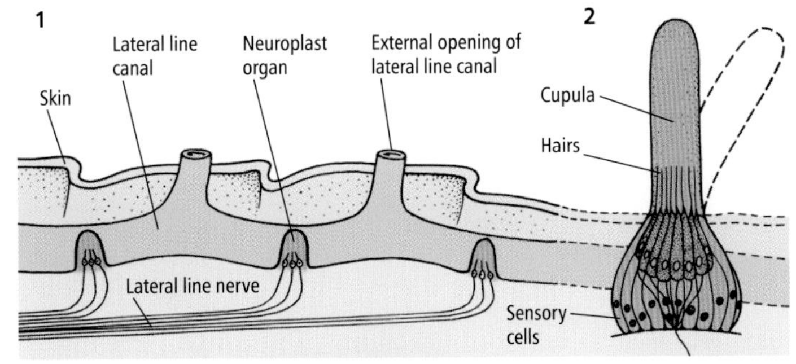

Right *The mouths of fishes display an enormous variety of forms, adapted to the ecology and feeding strategy of individual species:* ***1*** *The air-breathing Siamese fighting fish* (Betta splendens) *has an upturned mouth, ideal for catching the larvae of mosquitoes and other insects;* ***2*** *The Angel squeaker* (Synodontis angelicus), *a catfish, displays the long mouth barbels characteristic of this group. The fleshy protuberances help these suctorial bottom-feeders locate food;* ***3*** *The Moorish idol* (Zanclus cornutus) *has a small, protruding mouth equipped with many elongate, bristle-like teeth, with which it scrapes encrusting animals off rocks;* ***4*** *The Gulper eel* (Eurypharynx pelecanoides), *which feeds on crustaceans, has a huge backward extension of the jaws.*

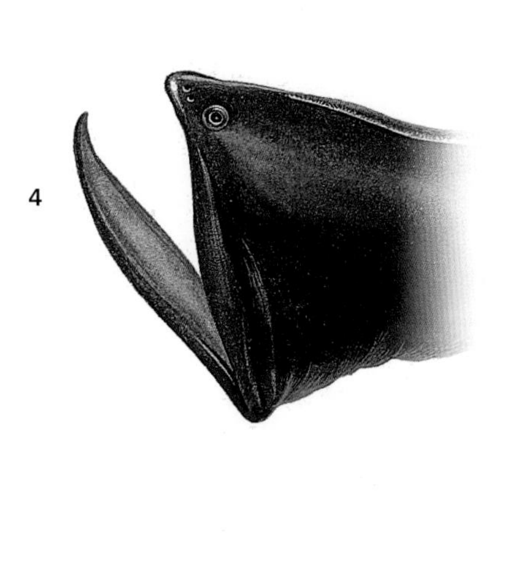

THE FASCINATION OF FISHES

Four thousand years of myth and mystery

HUMANS HAVE AN ENIGMATIC RELATIONSHIP with fishes. To start with, fishes live in a medium that people cannot enter without relying on technology that allows divers to take another medium (air) with them. Yet many people's awareness of fishes extends only as far as their use as food items, or as decorative aquarium or pond inhabitants.

It was not always so. Fishes were sacred to the ancient Egyptians – one form of the major deity Isis has a fish head. Moreover, a giant native fish, the Nile perch, was deified and the town of Esueh renamed Latopolis, from the Greek word for a fish, after several thousand of them were found embalmed there during the Hellenic period.

In Ancient and Medieval times, fishes were often associated with miraculous occurrences. A legend common to several cultures tells how a ring, accidentally dropped or deliberately thrown into a body of water, later returns to its rightful owner in the belly of a fish served up for dinner. This story recurs in such diverse sources as the Greek myth of the tyrant Polycrates of Samos, *The Arabian Nights,* and the Life of St Kentigern.

The sheer size of certain fishes has always held a special fascination – as witness the perennial angler's tale of "the one that got away." In 1558, the Swiss naturalist Conrad Gesner recounted how a man took his mule to a lake to drink, whereupon a pike bit the mule's lips and clung on. It was supposedly only after a fierce struggle that the mule prevailed and pulled the pike from the water. Somewhat better attested is the same author's account of the so-called "Emperor's pike," of which several paintings exist. Apparently, in 1497 in Germany, a pike was caught, around the "neck" of which was a copper ring bearing the inscription that it had been put there by Frederick II in 1230. This huge fish was 5.8m (19ft) long and weighed about a quarter of a tonne. To verify the catch and preserve it for posterity, the pike's skeleton was kept for centuries in Mannheim cathedral. Yet when it was examined by scientists in the 19th century, it was found to be a hoax – the skull came from one pike, but the body was made up of several pike bodies joined together.

Later fish forgeries were even more imaginative, feeding the popular superstition of grotesque hybrid creatures under the sea. From the 16th to the 19th century, a trade arose in "Jenny Hanivers," diabolical-looking sea monsters with humanoid heads. They were made either by stitching together fish and monkey carcasses to resemble a mermaid, or manipulating a ray or guitar fish to create a winged "sea devil." Mummified by drying in the sun, these morbid curios were brought back from their travels by European and American sailors and voyagers and exhibited in circuses or sideshows.

Other extraordinary aquatic phenomena have their basis in fact rather than fraud. Showers of fishes have been witnessed many times in the last thousand years, causing great alarm and consternation. Yet we now know that they are the result of freak weather conditions. A tornado passing over the sea and turning into a waterspout has the ability to suck up fishes along with the water so that, when the wind drops, the fishes fall from the sky. Most fishes that are caught up in waterspouts are small, but in India, a rain of fishes is reported to have contained individual specimens weighing up to 3kg (6.6lb). Occasionally, small fishes may be carried up to the height at which hail forms, and so come down embedded in ice.

Mariners' tales once abounded of monsters of the deep and sea serpents. Fanciful artists' impressions of these terrifying beasts appear on countless Medieval maps and nautical charts. Such creatures are thought to have had their origins in misinterpretations of rarely seen fishes such as the Oar fish (*Regalecus glesne*), sunfishes (*Mola* spp.), the Whale shark (*Rhincodon typus*), and the Megamouth shark (*Megachasma pelagios*). Likewise, a school of dolphins or porpoises swimming in line and surfacing periodically to breathe might well resemble a sea serpent if seen in poor conditions or from afar. Even so, stories of sea serpents persisted into modern times. In 1892, Dr A.C. Oudemans, Director of the Zoological Gardens in The Hague, recorded over 160 cases of what he considered

Above *This 2nd-century BC mosaic from the House of the Faun in Pompeii, gives evidence of the keen appreciation (and observation) of fishes by the Ancient Romans. Red and gray mullets, electric rays, dogfishes, and wrasses are all accurately depicted, as is a Moray eel, much prized by the Roman aristocracy.*

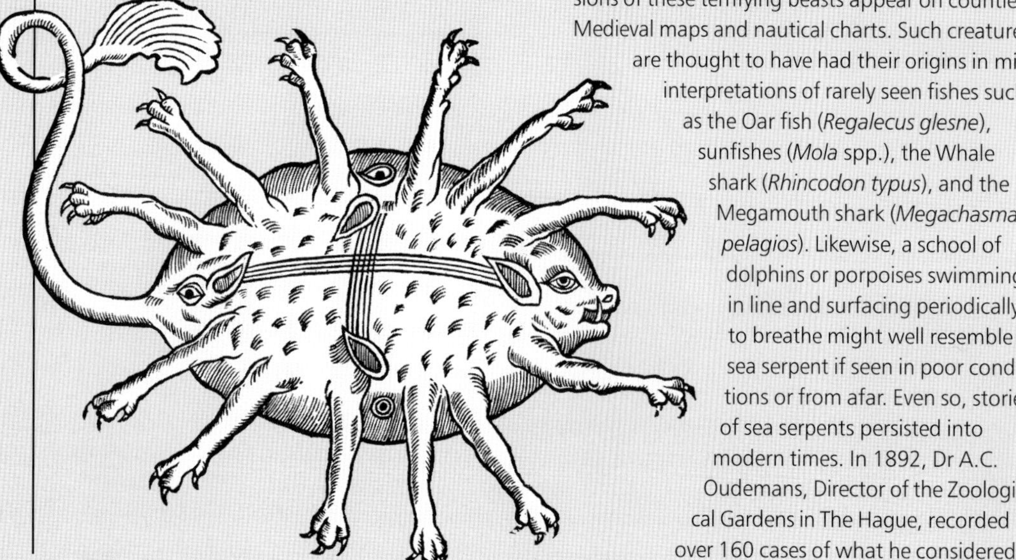

Left *Conrad Gesner's* Historia Animalium, *published in 1551–58, contained many accurate engravings of real animals alongside completely fanciful illustrations, such as this unidentified aquatic vertebrate.*

genuine recent sightings in his book *The Great Sea-Serpent*. Some of the reports came from the captains of Royal Navy vessels, who could be relied upon for their sobriety and, moreover, were equipped with telescopes.

Shark attacks, however infrequent, generate powerful myths. In a celebrated Internet hoax of 2001, a manipulated photo showed a hapless diver, suspended on a ladder beneath a helicopter, about to be eaten by a huge Great White shark leaping from the water. KEB/JD

group of placoderms, the ptyctodontids, the pelvic fins show sexual dimorphism – that is, they are different in males and females, those of the males being enlarged into sharklike claspers. From this arrangement, it is assumed that fertilization of eggs was internal.

Most of the early, and a few of the later, placoderms lived in freshwater; the rest, including the fascinating antiarchs, were marine. This enigmatic group appeared about 400 million years ago and became extinct some 70 million years later.

History of Fishes in Captivity

FOOD AND AQUARIA

Some 70 percent of the earth's surface is covered with water, which, unlike the land, offers a three-dimensional living space, very little of which is devoid of fish life. The overall contribution that fish make to the total vertebrate biomass is therefore considerable. Yet, because they are not as obvious a part of the environment as birds and mammals, fish have not been as generally appreciated as they deserve, although people have been interested in them for at least 4,000 years.

It is not known when fishes were first kept in captivity; nor is the reason for doing this known, although it seems likely that it was to provide a ready source of fresh food rather than for esthetic purposes. Around 4,000 years ago, the Middle East in general, and the fertile crescent of the rivers Tigris and Euphrates in particular, were far wetter and more fertile than in modern times. It was in this period that the first identifiable fish ponds were built by the Sumerians in their temples. The Assyrians and others followed later. It is conceivable that the idea of fish ponds came from the sight of fishes that had been left behind after a flood and survived for some time in natural, water-filled depressions. It is not, however, known which types of fishes were kept in these ponds. The Assyrians depicted fishes on their coins, but not accurately enough for them to be identified.

The story of fishes kept by the Egyptians is quite different. Their high standard of representational art has enabled the fishes in their ponds to be identified. Even more conveniently, they mummified some of their important species, allowing the accuracy of the drawings to be checked. Various species of *Tilapia* (still a highly valued food fish in the region), Nile perch, and *Mormyrus* (Elephant nose) are among those depicted. The Egyptians added a new dimension to the functional aspect of fish ponds, that of recreation. Murals depict fishing with a rod and line, which must have been for fun because it is not as efficient as netting for catching fish in commercial quantities. The Egyptians also worshiped their fish.

The Roman Marcus Terentius Varro (116–27 BC) wrote in his book *De Re Rustica* of two kinds of fish ponds: freshwater ponds (*dulces*) kept by the peasantry for food and profit, and saltwater ponds (*maritimae* or *sales*), which were owned only by wealthy aristocrats who used them for entertainment. Red mullet were especially favored because the dramatic color changes of the dying fish were admired by guests before the fish was cooked and eaten. Large moray eels were also kept and, in the most extreme examples, were decorated with jewelry and fed on unwanted or errant slaves (See Tarpons, Bonefishes, and Eels).

Although the Romans possessed glass technology, there is no record of any form of aquarium having been constructed. The Romans' involvement with fish was not totally for show, however.

They explored fish culture methods and were known to have transported fertilized fish eggs, which they may well have fertilized externally by stripping male and female fish.

Records of fishes are few in the western world after the fall of the Roman Empire and with the onset of the Dark Ages. The historian Cassiodorus (c.AD490–c.AD585) notes that live carp were taken from the Danube to the Goth king Theodoric at Ravenna in Italy, while Charlemagne marketed the live fishes he kept in ponds.

The tradition had doubtless been kept alive, however, by clergy and nobles. It is, for example, stated in the Domesday Book (1086) that the Abbot of St Edmund's had fish ponds providing fresh food for the monastery table and that Robert Malet of Yorkshire had 20 ponds taxed to the value of 20 eels. Stewponds were common in monasteries during the Middle Ages, and indeed were regarded as essential, since the church forbade the eating of meat on Fridays. The word "stew" in this context comes from the Old French *estui*, meaning "to confine," and is not connected – as is widely believed – with the means of preparation of the fish for consumption.

Modern-style aquaria were developed in the first half of the 19th century. At a meeting of the British Association for the Advancement of Science in 1833, it was shown that aquatic plants absorbed carbon dioxide and emitted oxygen, thereby benefiting fishes. The first attempt at keeping marine fish alive and the water healthy by using plants was, however, made by Mrs Thynne in 1846. Only six years later came the first sizeable public aquarium, built in the Zoological Gardens in London. In late Victorian times, as now, many homes had aquaria and the invention of the heater and thermostat later allowed more exotic fishes to be kept.

The best known of all aquarium fishes is the Goldfish. This species is native to China and has been bred for its beauty for over 4,500 years. In 475BC Fan Lai wrote that carp culture had been associated with silkworm culture (silkworms are the caterpillars of the Silk Moth - *Bombyx mori*) – the fish feeding on the feces produced by the caterpillars – since 2,689BC. About 2,000BC, the Chinese were, according to fishery experts, artificially hatching fish eggs. Red Goldfish were noted in AD350 and during the T'ang dynasty (about AD650) gold-colored fish-shaped badges were a symbol of high office. By the 10th century, basic medicines for fishes, such as poplar bark for removing fish lice from Goldfish, were available.

In the wild state, goldfish are coppery brown, but when first imported into Britain (probably in 1691) gold, red, white and mottled varieties were already available. By 1728, trade in the goldfish had grown considerably, with the merchant and economist, Sir Matthew Dekker, importing the fishes in large numbers and distributing them to many country houses. Goldfishes reached America in the 18th century and shortly thereafter became one of the most familiar fishes. The Goldfish is now believed to be not only the most popular fish kept in aquaria, but also the most popular of all pet animals.

Above left Kaiyukan Aquarium in Osaka, Japan, has the world's largest indoor tank, some 30ft (9m) deep. It contains fauna of the Pacific Ocean, including two huge Whale sharks.

Above right Some varieties of the universally popular Goldfish, created by selective breeding: **1** Bristol shubunkin; **2** Common goldfish; **3** Ranchu; **4** Veiltail; **5** Bubble-eye; **6** "Pom-pom."

Endangered Fishes

ENVIRONMENT AND CONSERVATION

In 1982, the Fish and Wildlife Service of the USA proposed removing the Blue pike (*Stizostedion vitreum glaucum*) and the Longjaw cisco (*Coregonus alpenae*) from the American List of Endangered and Threatened Wildlife – not, however, because numbers of both species had recovered to their former abundance but, on the contrary, because the two species were deemed extinct. The Blue pike, which formerly lived in the Niagara River and Lakes Erie and Ontario, had not been seen since the early 1960s. The Longjaw cisco, from Lakes Michigan, Huron, and Erie, was last reported in 1967. What caused their terminal disappearance after thousands of years of successful survival? Both species, directly and via their food chain, were severely depleted by pollution; in addition, the Longjaw cisco, in particular, suffered from predation by parasitic Sea lampreys that gained increasing access to its habitat after the Welland Canal was built, linking Lake Ontario with Lake Erie.

In South Africa a small minnow, called *Oreodaimon quathlambae*, was described in 1938. A few years later, it was extinct in its original locality in Natal. No canals had been dug, pollution was minimal, but an exotic species, the Brown trout (*Salmo trutta*), had been introduced to provide familiar sport for British expatriates. Small trout ate the same food as *Oreodaimon,* while large trout ate the *Oreodaimon* themselves. Fortunately, in the late 1970s, a small relict population was discovered living above a waterfall on the Tsoelikana River in the Drakensberg mountains, Lesotho. The fall had prevented the spread of trout, but they had recently been transplanted above the falls. Although this population is protected as far as possible from predation, a more serious threat to its survival is the overgrazing of the land adjoining the river, causing the silting up of the river and changes in the river's flow.

In parts of Malaysia, Sri Lanka, and in Lake Malawi some of the more brightly colored freshwater fishes are becoming ever harder to find. In the past, on top of any other pressures that may have existed, numbers were thought to have fallen because of collecting for the aquarium trade and concomitant environmental damage. Large-scale commercial breeding programs now meet most of the global demand for these fishes. However, while captive breeding may relieve some of the pressure on wild stocks, others – for example direct and indirect damage to natural habitats – remain as serious threats.

Formerly, large parts of the US southwest were covered with extensive lakes. With post-Pleistocene

SPECIAL FEATURE

STEALTH FISHES

Underwater camouflage strategies

FISHES HAVE A REMARKABLE RANGE OF methods for concealing themselves from predators or prey – or for mimicking the food items of other species, the better to hunt them.

Perhaps the simplest form of camouflage is countershading, as exhibited by sharks. A dark upper surface and a light underside enable them to approach prey unseen, since the light tone, when seen from below, blends in with the sky.

The ability to merge with the background, or cryptic coloration, is common to several species. Active cryptic coloration is used by flatfishes like the Turbot or the Peacock flounder, as they lie in wait on the seabed for their prey. Chromatophores (skin cells that can alter their pigmentation) enable these voracious hunters to change color rapidly to blend in with their surroundings.

The Leafy seadragon uses passive cryptic body shape and coloration, its fragmented outline disguising it perfectly in its weedy habitat. So successful is this strategy that neither predators nor prey recognize the seadragon as a fish. Less spectacular, though just as effective, are those species that burrow into the substrate, showing only their well-camouflaged faces, or that tone in almost imperceptibly with rocks. In the latter category is the Stonefish (*Synanceia verrucosa*), whose body may even, like the surrounding rocks, be mottled with algae. As divers can attest, it is all too easy to step inadvertently on a Stonefish and be injected with its fearsomely powerful venom.

Certain fishes have taken mimicry to a high level of sophistication. Angler fishes (*Antennarius* spp.) are so called for their habit of fishing with a "rod and line" – a stalk and a lure resembling a favored prey item (e.g. a worm) of their target species. This amazing mechanism, which folds away when not in use, is a development of the spine of the first dorsal fin.

Above *Hidden in the gravel of the sea floor, an Atlantic stargazer* (Uranoscopus scaber) *is barely visible. This species has venomous dorsal spines.*

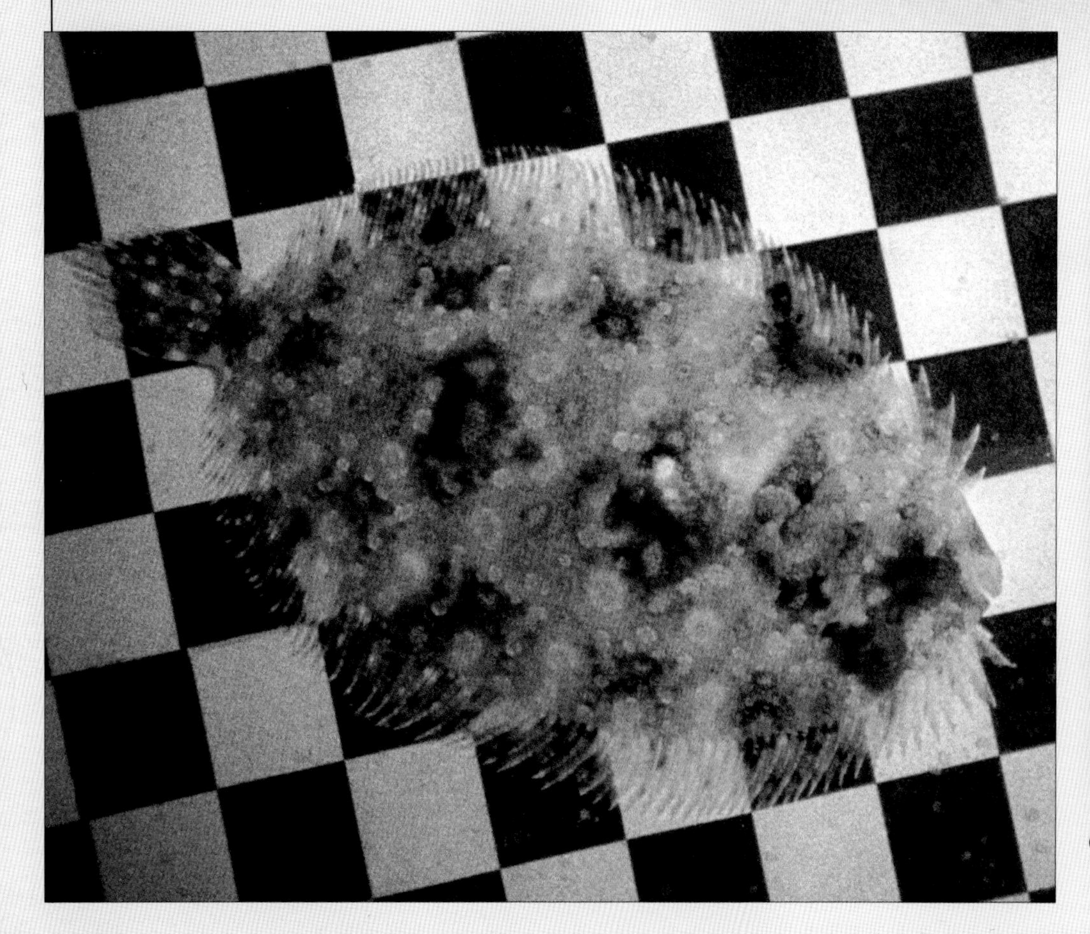

Below *The Peacock flounder* (Bothus lunatus) *is a predatorial fish that can change its coloration to match the substrate, so concealing itself from its prey. In experiments, this remarkable mimic has even replicated the pattern of a checkerboard.*

Above *The kelplike body shape of the Leafy seadragon* (Phycodurus eques) *camouflages it effectively in the reef environment that is its habitat.*

Right *A sturgeon catch on the Volga River, Russia, in the 1990s. The construction of dams, together with a growth of illegal fishing for these fishes' valuable caviar, has seen sturgeon stocks plummet.*

desiccation (i.e. about 100,000 years ago) the lakes, with their associated fishes, dwindled, and now some pupfish and related species survive only in minute environments. For example, the Devil's Hole pupfish (*Cyprinodon diabolis*) lives only in a pool 3 x 17m (10 x 55ft) located 18m (60ft) below the desert floor in southern Nevada; it is believed to be the vertebrate with the smallest natural distribution in the world. The species depends for its food on the invertebrates living in the algae on a rock shelf 3 x 5.5m (10 x 18ft) wide, close to the surface of the water. Although this habitat is situated in a protected area in the Death Valley National Monument, distant pumping of subterranean water lowered the water table, threatening to expose the shelf and deprive the fish of its only source of food. Attempts to transplant some of the fish to other localities failed. Hastily, a lower, artificial shelf was installed. Three specially built ponds were also stocked with some pupfishes, and the case was brought up before the US Supreme Court. The court ruled that the pumping should cease, whereupon the water level stabilized and the Devil's Hole pupfish was saved.

The examples given above illustrate what is happening and what can be done. All the fishes mentioned live in freshwaters, where the area occupied is small enough for low-level population changes to be monitored. The same detailed knowledge about marine fishes is lacking because they can occupy much greater territories, making the collection of such information very difficult.

For various reasons, then, a disproportionate number of species of fishes are in danger of extinction. Several measures have been put in place to attempt to secure their future. Many countries are signatories to CITES (Convention on International Trade in Endangered Species of Wild Fauna and Flora). Member countries produce lists of their threatened animals and plants and all agree that there shall be no unlicensed trade in certain of the species. While mammals and birds occupy the great majority of the pages in the latest editions of the IUCN (World Conservation Union) Red List, some 750 species of fishes are also included in the three highest categories of threat: Vulnerable, Endangered, and Critically Endangered. However, enforcing the regulations is a no easy matter, particularly for Customs officers, who are largely responsible for the implementation of the rules by ascertaining that the species in a particular shipment are what they purport to be. They are ill-equipped to identify accurately the 5,000 species they might be confronted with – something that even many professional fish specialists find difficult, since the diagnostic characters may be internal. Moreover, if fish are found to be illegal and are impounded by Customs, there remains the question of what to do with them. Even if it were possible to ship them back to their native land before they died, there is no guarantee they would be released into their original habitat. Public aquaria and zoological institutions are able to accept only a small proportion of illegal imports. As for the rest, the authorities are confronted with the harsh option of destroying the fish or the expensive alternative of allowing illegal shipments into the country, but placing a sale embargo on them until the correct documentation is presented.

Some countries follow the spirit of the laws of conservation better than others. For example, in the United States, dam construction on the Colorado River prevented many of the endemic species from breeding and many species quickly became very rare and were in danger of extinction. After prompting from scientists, the Dexter National Fish Hatchery was constructed in New

Mexico to which endangered species were taken and allowed to breed. The breeding program proved successful and many young were returned to suitable sites each year. But such ventures can provide only partial answers; the Amistad gambusia (*Gambusia amistadensis*), for example, lived and bred in a reed-fringed pool at Dexter but still went extinct in 1996. It could not be reintroduced in the wild because the Goodenough Spring, in which it lived, was lost forever at the bottom of a reservoir. Other species, such as the Gold sawfin goodeid (*Skiffia francesae*), now survive only in captivity and cannot be reintroduced because their home waters may now be populated by introduced species or may be highly polluted.

A cynic might question the point of trying to save small fishes from extinction. Altruistically, the response would be that they have as much right to life as humans have; more selfishly, it could be argued that they might ultimately be of use to humans. Living in a cave in Oman, for example, is a small, eyeless fish (*Garra* sp.) whose total population is probably 1,000. It is quite a remarkable fish because it can regrow about one-third of its brain (the optic lobes) – the only vertebrate known to do so. Such a discovery is potentially of great significance in human neurosurgery. Luckily, this fish was discovered. Yet how many more useful attributes were there in fishes that have died out?

The status of some species is more critical than others, but the fate of them all is in human hands. If endangered species are to survive, the right decisions must be taken – and the sooner, the better.

The Future of Fishes

ECOLOGY

For a few years prior to 1882 there was an extensive fishery for a tile fish (*Lopholatilus chamaeleonticeps*) off the eastern seaboard of the USA, particularly from Nantucket to Delaware Bay. This nutritious, tasty species reached about 90cm (3ft) in length and 18kg (40lb) in weight. Tile fishes lived in warm water near the edge of the continental shelf from 90 to 275m (300–900ft) down. During March and April 1882, many millions were found floating, dead, on the surface. One ship reported that it steamed for two whole days through tile fish corpses. For over 20 years no more were caught, but by 1915 the numbers had recovered enough to sustain a small fishery.

Despite the long gap in records, enough individuals must have survived to be able to rebuild the population, albeit to a lower level. The cause of this carnage is uncertain, but it is believed that ocean currents changed and the warm water in which the tile fish lived was rapidly replaced by an upwelling of deep, cold water in which they died. Although many fish species undergo natural variations in abundance, for no matter what reason, a major natural disaster can affect them all. Should the population be at a low level at such a time, the chance of extinction, and the loss of a resource,

Herrings and History

The humble herring, for long a staple food fish in many counties in northern Europe, has played a major role in the history of the region. The Hanseatic cities on the southern coast of the Baltic prospered greatly during the Middle Ages not only because of the rich supply of herring on their doorsteps, but also because of their commercial expertise in preserving them and exporting them throughout Europe.

Then the herring suddenly left the Baltic for the North Sea. The Hanseatic states collapsed economically and politically, and the Netherlands now had access to the herring. The efficient Dutch fleet proceeded to catch large quantities of herring, often in British waters. Charles I of England realized that a good way of boosting his exchequer would be to tax the Dutch for this privilege. To enforce payment of this unpopular tax and to keep British herring for the British, Charles had to build up the declining Royal Navy. The Dutch resolved to protect their fleet, and before long the two countries were at war.

An earlier "battle of herrings" occurred in 1429, when Sir John Fastolf routed the French, who tried to prevent him taking herrings to the Duke of Suffolk, who was besieging the city of Orléans.

Although these series of skirmishes were finally resolved, the dispute over the ownership of a "moveable" resource as represented by many fish species, continues. For instance, there is still no international agreement on the exploitation of deep-sea fishes in the North Atlantic. The fierce 1980s wrangles over herring quotas between member states of the European Union were merely modern, politely political equivalents of firing cannon balls at one another. KEB/JD

could occur. Fortunately, the tile fish population was high when the currents changed.

It is largely a self-evident truth that where industries are few, rivers are clean. This is certainly true of parts of Scotland, at least, where communities depend for financial viability on the spawning runs of the Atlantic salmon. For over a century the mainstay of the tourist and hotel business has been the anglers who stay in the hope of hooking a salmon. Nearer the coast are the netsmen, who catch salmon as they begin their upstream migration and sell their luxury food in the markets further south. These communities remained stable and thrived, until an accidental observation was made in the late 1950s.

Before that time, nobody knew where the salmon passed their growing period in the sea, but a US nuclear submarine spotted vast shoals of salmon under the ice in the Davis Strait (between Greenland and Baffin Island). When this information became generally available, commercial fishing boats went after this valuable catch. The stocks were rapidly decimated, leaving progressively fewer fish to return to their home rivers to spawn. There were fewer fish for the netsmen to catch and their livelihood suffered; there were fewer still for the angler to catch and even fewer to spawn and replenish the stocks.

On two famous pools on the River Tay, anglers normally landed about 500 fish before the end of May. In 1983, only 36 were caught. Anglers deserted the area, and hotels that used to open shortly after the New Year did not bother to open until Easter. This resulted in fewer jobs and faster economic decline in an already hard-pressed area that had few other employment opportunities.

Two further factors emphasize the ridiculous artificiality of the situation regarding salmon. First, there is a slow climatic change that tends to reduce the number of fishes involved in the spring spawning runs upriver. Second, salmon farming has reached the point where it can now provide such a large proportion of the market's demands that there is little necessity to continue cropping wild stocks at sea. Biologically, of course, it is also better to conserve vigorous and healthy wild populations as a genetic reserve, in case inbreeding in farmed fish weakens their genetic makeup. This is not such a hypothetical argument as it might at first appear; for example, the number of young farm-bred salmon that are born with foreshortened jaws and hence have difficulty in feeding and therefore a slow growth rate, is higher (at least, in some stocks) than in the wild.

The examples given above show how valuable fish stocks are. Fish should not be just a short-term source of profit for a few, but a continuing self-renewing resource; and they would be if used properly. When herring were caught by drift nets, before the Second World War, there were over 1,600km (1,000 miles) of net out in the North Sea each night. Yet, because the shoals moved up and down in the water, and the nets occupied only the top 6m (20ft), enough herring escaped to spawn and so maintain the stocks. Changing fishing techniques, using echo sounders, and adjusting the depth of the net resulted in a much better yield-to-effort ratio, but also brought a rapid destruction of the stocks. It was only in the early 1980s that drastic measures were taken (such as a fishing ban) to conserve stocks. Stringent, though not total, fishing restrictions are now being implemented to protect the dwindling stocks of another species, the Atlantic cod, in European waters. The lessons learned during the herring crisis may have prompted earlier action and stocks may consequently recover more quickly in this case.

It is not only prime food fish that are under threat. Species living in deeper waters, such as rattails and argentines, are now caught; their flesh is shredded, breaded, molded, and served as "fish fingers." When it becomes uneconomical to catch these species, what happens? To all these pressures must be added unnatural, that is, human-induced, catastrophes such as the dumping of industrial waste or oil spills. These events poison fish locally, but, more importantly, the toxic substances become concentrated up the food chain and can reach harmful levels in the fish that humans eat from the top of the chain. Oil spillages also have other less obvious effects, such as preventing eggs near the surface from hatching. However, for a convenient way of life, humans want waste disposed of and oil transported. Then there are the nets used to catch fish for fertilizer, which have a small mesh, so they also catch the young of larger species before they can spawn. Trawls can also destroy spawning grounds.

Are all these factors, of which many people are unaware, necessarily a recipe for disaster? The answer is not cut-and-dried, but is probably "yes," unless the international compromises that are regularly discussed can be agreed and adhered to. The seas can provide a sustainable crop if treated with care and thoughtfulness. The future of fishes is a human responsibility and is linked intimately with the quality of human life. It is up to people to realize this and act accordingly to ensure that the worst scenario never materializes. KEB/JD

Left *A Swordfish* (Xiphias gladius) *caught in a net off Sardinia, Italy. This species, much sought after for its firm flesh, is also sometimes taken as bycatch in other fisheries, and some populations are threatened.*

Below *Off Gloucester, Massachusetts, a fisherman empties cod from a dragnet. In some parts of the world, overfishing has depleted numbers of this once abundant food fish to below sustainable levels.*

Lampreys and Hagfishes

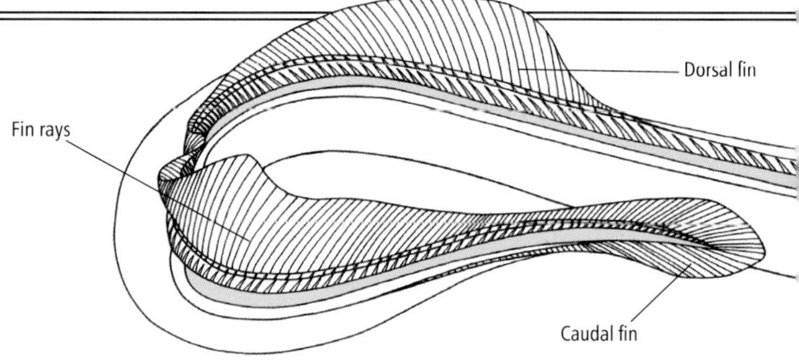

During the Devonian era (410–354 million years ago) the world's rivers and seas were occupied by heavily armored jawless vertebrates. Their bony plates and head-shields are now common as fossils, testifying to the success of their radiation. However, with the arrival of the jawed fishes, these jawless forms became less common and are now represented by just two groups, lampreys (class Cephalaspidomorphi) and hagfishes (class Myxini).

Neither group bears much resemblance to the Devonian forms; the living forms are eel-like and lack bony plates in the skin. Both are at a similar basal level of organization – at a pre-jaw stage – but both have had long, separate histories and there are many important internal differences. The relationship of the two living groups to each other and to the fossil forms was the subject of much controversy until the discovery of the first hagfish fossil in 1991 signaled that no close relationship exists between the two groups. The association of the two groups here is purely one of convenience.

Lampreys

FAMILY PETROMYZONTIDAE (PETROMYZONIDAE)

Lampreys live in the cooler waters of both hemispheres. All have a distinct larval stage (ammocoete), during which they hardly resemble the adults in structure or lifestyle. The adult is eel-like, with one or two dorsal fins, a simple caudal fin and no paired fins. The mouth is a disk adapted for sucking, bearing a complex arrangement of horny rasping teeth, the exact arrangement of which is peculiar to each particular species and heavily relied upon in classification. There are seven gills, each of which has a separate opening.

Water enters each side through an opening near the mouth and then flows into any of seven ducts, each of which leads to one of seven external gill openings. Each duct has a muscular pouch, which encourages water flow. Between the eyes, on the top of the head and just behind the single median (centrally located) nostril, is a small patch of translucent skin covering the pineal organ. In lampreys this organ is light sensitive and the light levels it receives control hormone levels. (In higher vertebrates this sensitivity to light is lost, but hormonal control is still exercised by the remaining part of the organ.) The lamprey's skin has glands that secrete a mucus with toxic properties that are thought to discourage larger fish from eating the lamprey. (The mucus should be removed before using lampreys for human consumption because it can cause severe stomach upsets.)

The inner ear of bony fish and the higher vertebrates consists of three semicircular canals at right angles to each other; they maintain balance as well as being involved in the hearing process. In lampreys only the two vertical canals are present; the horizontal canal is absent.

From a biological point of view there are two interconnected groups of lampreys: those that spend most of their adult life at sea and enter freshwater to breed (compared with those that spend their entire life in freshwater), and those that feed parasitically on fish as adults (compared

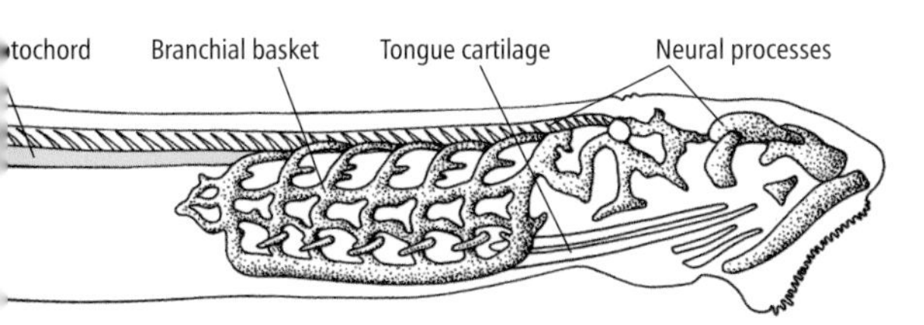

◀ ▼ ***Left and below*** *Lamprey anatomy. In lampreys the notochord does not have developed centra (i.e. the main part of vertebrae), but there is a series of paired arch cartilages. Dorsal and caudal fins are present. Formerly there was bone, but this is lost in the living forms.*

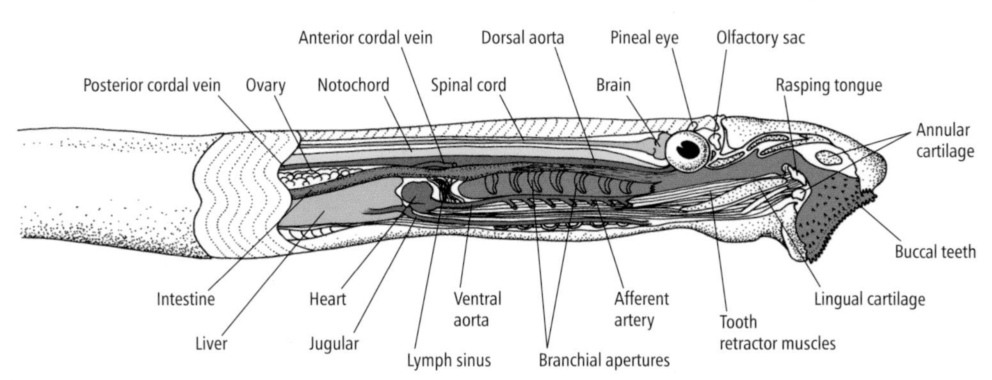

with those that feed on small invertebrates).

To some extent, all lampreys move upstream to spawn. The marine species, which are larger than their freshwater relatives, are anadromous – that is, they leave the sea to migrate up rivers. A problem immediately encountered by these species is the need to adjust the concentration of salts in their blood and body fluids as they move from salt water to freshwater.

The timing of the spawning run varies with species and locality. In northwest Europe, for example, the Brook lamprey commences its run in September or October, but, in the Adriatic region, the peak comes in February and March, while in northwest Russia there are spring and fall runs. Lampreys in the southern hemisphere spend much longer on the spawning run and may not spawn until a year or more after they enter freshwater. The spawning grounds may be several hundred kilometers from the estuary through which

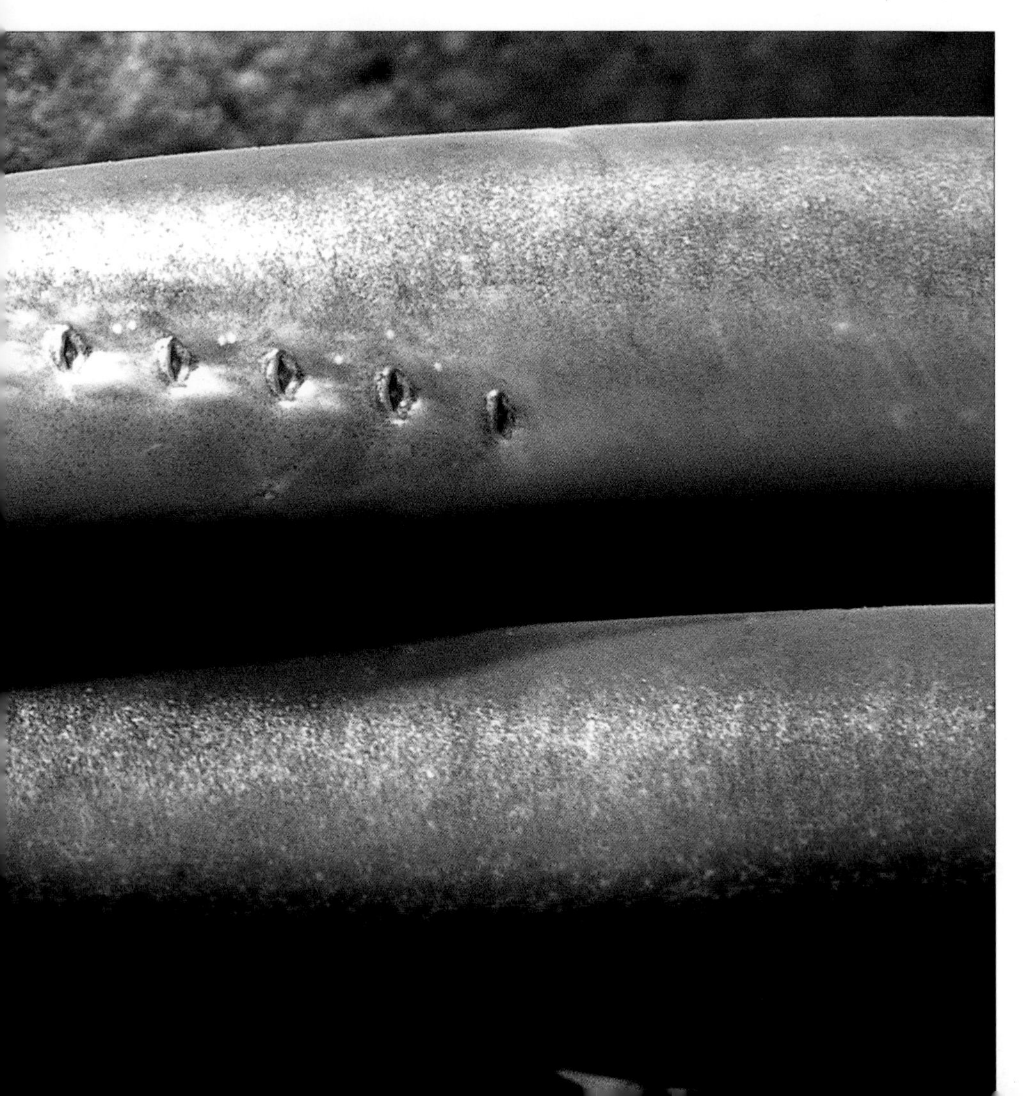

FACTFILE

LAMPREYS AND HAGFISHES

Superclass: Agnatha

Classes: Cephalaspidomorphi, Myxini

About 90 species in 13 genera and 2 families.

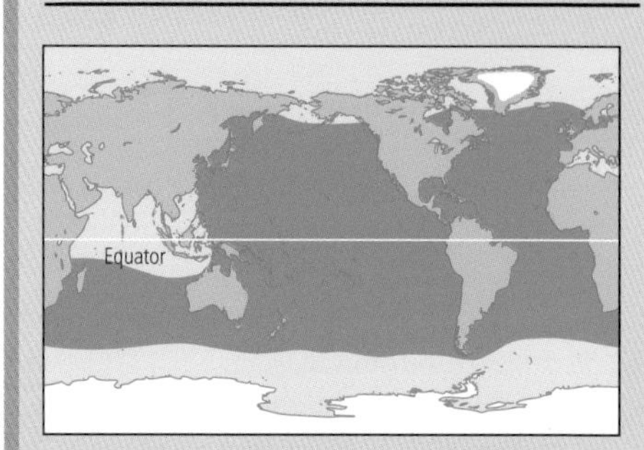

Distribution Worldwide in cool marine and freshwaters.

Size Length 20–90cm (8–35in).

ORDER PETROMYZONTIFORMES

LAMPREYS

Family Petromyzontidae or Petromyzonidae

About 40 species in 6 genera and 3 subfamilies.

Subfamily Petromyzontinae

About 36 species in 4 genera. Northern hemisphere. Species include: **Brook lamprey** (*Lampetra planeri*), **European river lamprey** (*L. fluviatilis*), **Northern brook lamprey** (*Ichthyomyzon fossor*), **Sea lamprey** (*Petromyzon marinus*). **Conservation status**: The Lombardy brook lamprey (*Lampetra [Lenthenteron] zanandreai*) is classed as Endangered, and the Greek brook lamprey (*Lampetra [Eudontomyzon] hellenicus*) is Vulnerable.

Subfamily Geotriinae

Sole member of subfamily **Pouched lamprey** (*Geotria australis*). Southern Australia, Tasmania, New Zealand, Chile, Argentina, Falkland Islands and South Georgia, possibly Uruguay and southern Brazil.

Subfamily Mordacinae

3 species of the genus *Mordacia*, including **Short-headed lamprey** (*M. mordax*). Chile, SE Australia, Tasmania. **Conservation status**: The Nonparasitic lamprey (*M. praecox*) is Vulnerable.

ORDER MYXINIFORMES

HAGFISHES

Family Myxinidae

43–50 species in 6 genera and 2 subfamilies.

Subfamily Myxininae

15–20 species in 4 genera. Atlantic, Pacific, including coasts of Argentina and New Zealand. Genera and species include: **North Atlantic hagfish** (*Myxine glutinosa*), *Notomyxine*, *Neomyxine*, and *Nemamyxine*.

Subfamily: Eptatretinae

28–30 species in 2 genera. Atlantic, Indian, and Pacific oceans. Genera: *Eptatretus*, including **Pacific hagfish** (*E. stouti*) and *Paramyxine*.

◀ ***Left*** *Clearly displayed on these sea lampreys are not only the tooth-studded sucking mouthparts but also the seven separate gill openings. When a lamprey latches onto a fish to feed, it is still able to pass water over its gills by drawing it in through the spiracle (the first gill opening).*

they enter the rivers from the sea. In slow-flowing stretches lampreys may be able to travel at 3km (about 2 miles) each day, overcoming weirs and rapids on the way. When exhausted by these efforts (during which they do not feed), lampreys cope with weariness by temporarily holding on to rocks with their suckers.

During the upstream migration and the subsequent spawning, some changes in body shape occur. These changes may be as minor as the changes in the relative positions of the two dorsal fins and the anal fin, or as substantial as the enlargement of the disk (sucker) in males of the southern species and genera, such as the Pouched lamprey and the two Australian *Mordacia* species. The males of the Pouched lamprey and the South American species *Mordacia lapicida* (but not of other species of the genus *Mordacia*) also develop a large, spectacular, saclike extension of the throat, whose function is, at the moment, unknown.

The spawning grounds, often used year after year, are chosen for particular characteristics, the most important of which appears to be gravel of the right size for the larvae to live in. Other features are less vital, although water about 1m (3ft) deep and of moderate current is favored. The males of the Sea lamprey arrive first and start building a nest by moving stones around to make an oval, shallow depression with a gravelly bottom. Large stones are removed and the rest graded, with the biggest being placed on the upstream part of the nest. All species of lamprey build similar nests, but in different species, the sex of the nest-builder differs and may not be consistent.

Spawning is usually a group activity, involving 10–30 individuals in the Brook lamprey, and smaller numbers in the Sea lamprey, often a pair to a nest. In the European river lamprey there is courtship: while the male is building the nest, the female swims overhead and passes the posterior part of her body close to the male's head. It is suggested that stimulation by smell may be involved in this behavior. Fertilization occurs externally, the male and female entwining themselves and shedding sperm and eggs, respectively, into the water. Using her sucker, the female attaches herself to a stone and the male then sucks onto her to remain in position during spawning. In the Brook lamprey two or three males may attach to the same female. Only a few eggs are extruded at a time, so mating takes place at frequent intervals over several days. The eggs are sticky and adhere to the sand grains, and the continued spawning activity of the adults covers the eggs with more sand. After spawning, the adults die.

The eggs hatch into burrowing larvae (called ammocoetes), which are structurally unlike the adults. Their eyes are small and hidden beneath the skin, and light is detected by a photosensitive region near the tail. The sucking mouth with its horny teeth is not yet developed. Instead, there is an oral hood, rather like a cowl or a greatly expanded upper lip, at the base of which is the mouth surrounded by a ring of filaments (cirri), which act as strainers. Water is drawn through the oral hood by the action of the gill pouches and of a valvelike structure behind the mouth called the velum. On the inner surface of the hood, there are rows of minute hairs (cilia) and large quantities of sticky mucus. Particles in the water are caught on this mucus and the action of the cilia channels the food-rich mixture through the mouth into a complex glandular duct (the endostyle) at the base of the pharynx. The ability of the ammocoetes to secrete mucus is important, for it enables them not just to feed but also to cement the walls of their burrows and stop them from collapsing.

Ammocoetes rarely leave their burrows, and then only at night, but they often change position. They lie partially on their backs inside their burrows, tail down, with the oral hood facing into the current. In this position the importance of the photosensitive tail region can now be appreciated: it helps the ammocoete to orient itself correctly. The longest phase in the life history of lampreys is spent as ammocoete larvae (often called "prides"). The Sea lamprey, for example, can spend seven years of its life in this stage, the Northern brook lamprey six, the Brook lamprey three to six years, and the Short-headed lamprey three years.

Although lampreys live successfully as ammocoetes and as adults, the change from one state to the other is a dangerous one, resulting in high mortality. The changes involved are profound, since the entire mouth and the feeding and digestive systems have to be restructured, eyes have to be developed and the burrowing habit abandoned for a free-swimming one. During this time, which may last for eight months, lampreys do not, and cannot, feed. It was discovered in Russia, for example, that large numbers of metamorphosing lampreys found dead during early spring had been effectively suffocated by the breakdown of the velum after the skin of the gills and the mouth had become blocked by mucus.

Below *Representative species of lampreys and hagfishes:* **1** *Juvenile – ammocoete larvae – European river lampreys* (Lampetra fluviatilis) *on the bottom, filtering detritus from water (8–10cm/ 3–4in);* **2** *Head of a juvenile European river lamprey showing the oral hood;* **3** *Mouth of an adult European river lamprey showing horny teeth. During the parasitic phase, the many teeth on the rim of the oral disk provide much-needed traction to help latch onto the side of the fish host. Once attached, lampreys use the teeth on the tongue in the center of the disk to penetrate the host;* **4** *Brook lampreys* (Lampetra planeri) *feeding on a sea trout (50–60cm/20–24in);* **5** *A Sea lamprey* (Petromyzon marinus) *building a nest (1m/3.3ft);* **6** *Head of a hagfish (a species of the genus* Myxine) *showing the mouth and nostril surrounded by tentacles;* **7** *A hagfish of the genus* Myxine, *showing its eel-like form (60cm/24in);* **8** *A hagfish (genus* Myxine) *about to enter the body of its prey. By twisting into a knot it can gain extra leverage for thrusting itself into the fish.*

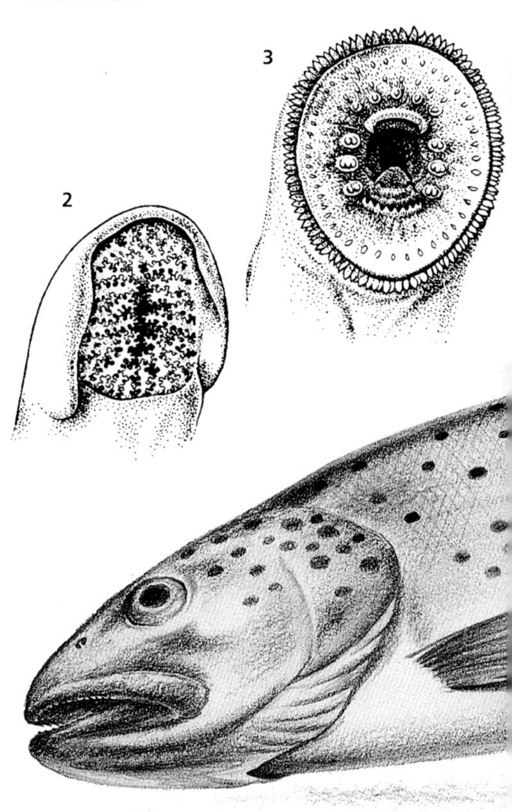

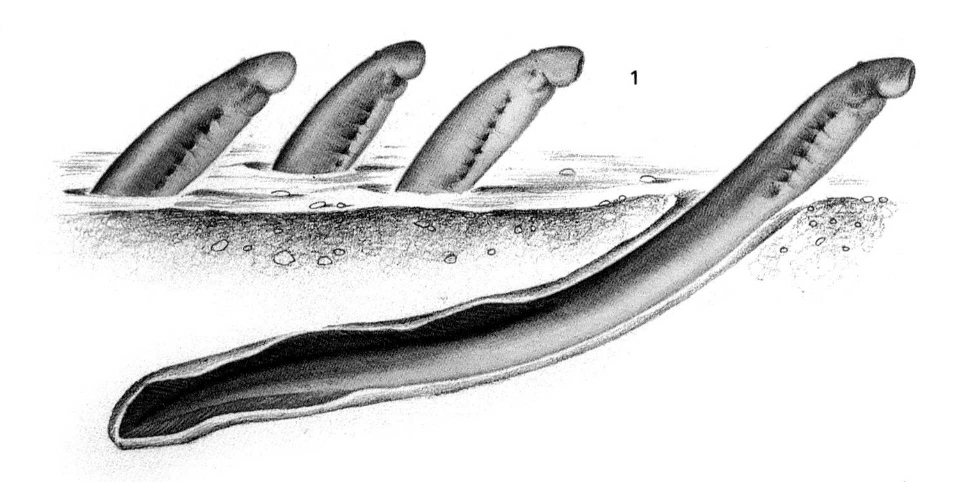

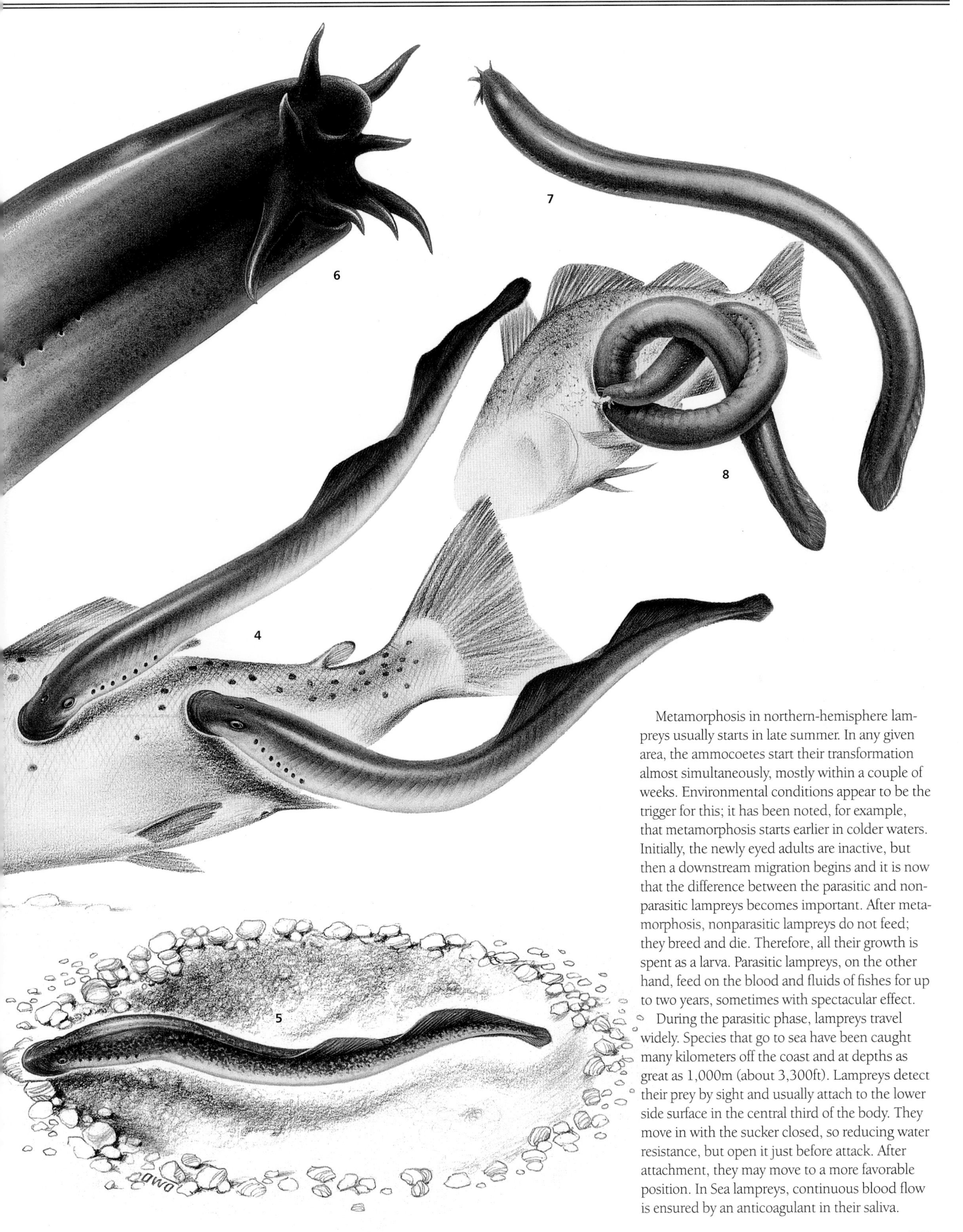

Metamorphosis in northern-hemisphere lampreys usually starts in late summer. In any given area, the ammocoetes start their transformation almost simultaneously, mostly within a couple of weeks. Environmental conditions appear to be the trigger for this; it has been noted, for example, that metamorphosis starts earlier in colder waters. Initially, the newly eyed adults are inactive, but then a downstream migration begins and it is now that the difference between the parasitic and nonparasitic lampreys becomes important. After metamorphosis, nonparasitic lampreys do not feed; they breed and die. Therefore, all their growth is spent as a larva. Parasitic lampreys, on the other hand, feed on the blood and fluids of fishes for up to two years, sometimes with spectacular effect.

During the parasitic phase, lampreys travel widely. Species that go to sea have been caught many kilometers off the coast and at depths as great as 1,000m (about 3,300ft). Lampreys detect their prey by sight and usually attach to the lower side surface in the central third of the body. They move in with the sucker closed, so reducing water resistance, but open it just before attack. After attachment, they may move to a more favorable position. In Sea lampreys, continuous blood flow is ensured by an anticoagulant in their saliva.

Above *Anchored against the flow, a European river lamprey uses its sucker to hold itself fast to a rock. The lamprey mouth is efficient and effective; as well as being used for feeding it is also used in carrying stones for the nest and in mating.*

Right *The Pacific hagfish is normally sedentary, living in burrows in mud and eating rarely; it has a low metabolic rate and stores a lot of fat. When it is hungry, its sedentary habits cease and it swims around tasting odors by lifting its head and spreading its barbels.*

The opening of the disk, which is essential for attachment, brings the teeth into play. The attachment is so strong that only rarely do fish manage to free themselves of lampreys, doing so by coming to the surface and turning over so that the lamprey's head is in the air. During the course of the parasitic phase, it has been calculated that 1.4kg (3lb) of fish blood will suffice to feed the lamprey from metamorphosis to spawning. Scarcely any fish is immune from attack, not even garfishes (family Lepisosteidae), which are covered with tough, diamond-shaped scutes (plates). Several lampreys may attack the same fish and newly metamorphosed individuals are the most voracious.

Lampreys are extremely adaptable and any change in the environment may have unforeseen consequences for other species. For instance, there used to be an extensive and profitable fishery for trout in the Great Lakes of North America in the early 19th century. When the Welland Canal was built to bypass Niagara Falls and thus allow ships to sail from Lake Ontario into Lake Erie (completed in 1829), it gave Sea lampreys access to the Lakes. Even so, they were not noted in Lake Erie for almost a century (1921), but thereafter spread rapidly from Lake Erie into Lakes Huron, Michigan, and Superior. By the mid-1950s, Sea lampreys were well established and had taken a terrible toll on the trout. Considerable efforts were made to control the lampreys and restock the trout, but these attempts were expensive and ultimately futile. The only real hope now is that a biological balance will be achieved before pollution wipes out all fish life in the Great Lakes.

Hagfishes

FAMILY MYXINIDAE

Hagfishes are superficially undistinguished and unprepossessing animals. They are eel like, white to pale brown, with a fleshy median fin on the flattened caudal (tail) region, and have four or six tentacles around the mouth. However, their apparent simplicity conceals a number of extraordinary, even unique, features. Hagfishes have neither jaws nor stomach, yet they are parasites of larger fishes. They have few predators but defend themselves by exuding large amounts of slime, and when their nostrils clog up with this slime they "sneeze." They can even tie their bodies into knots, and also have several hearts.

There are about 50 species, some of which can grow to 90cm (35in) or more in length and live in cold oceanic waters at depths from 20–300m (65–1,000ft) where the sediment is soft. The exact number of species is not known for certain because of the lack of agreed, tangible characters and the unknown degree of individual variability among the species. (The two best-known genera are *Eptatretus* and *Myxine*.) The relationship of the hagfishes to the other extant (living) jawless fishes, the lampreys, is uncertain and the subject of much debate. An important (and so far unique) discovery of a fossil hagfish took place in northeastern Illinois in 1991. Dating to the Late Carboniferous (Pennsylvanian) period, 300 million years ago, *Myxinikela siroka* shows remarkable similarities to extant forms, indicating that little evolutionary change has occurred in this group.

The mouth is surrounded by four or six tentacles according to species. There are no eyes; instead, there are two pigmented depressions on the head and other unpigmented regions on the head and around the cloaca (genital and anal chamber) that can detect the presence of light. The senses of touch and smell are well developed. There is a single median nostril above the mouth, which leads into the pharynx; a blind olfactory sac, which detects odors, is also present.

During breathing, water enters the nostril, from where it is pumped by the velum to the 14 or so pairs of gill pouches leading from the pharynx to

Right *Hagfishes have changed little in over 330 million years of evolution and although they have a partial skull, they have no backbone, only an undifferentiated and pliable notochord. What skeleton hagfishes do have is made of cartilage.*

the outside. In the gill pouches, oxygen is taken from the water by the blood, which simultaneously releases its carbon dioxide. It is thought that the skin also absorbs oxygen and excretes waste carbon dioxide. Externally, the gill pouches are visible as a series of small pores, increasing in size towards the rear. This row is continued by small openings of the slime glands. The only teeth of hagfish are on the tongue.

The main prey of hagfishes are dying or dead fish, although they can also be predatory, actively hunting for worms and other invertebrates. When they detect a victim, hagfishes swim rapidly upcurrent to reach it. Now the tongue and flexible body come into play, with the toothed tongue quickly rasping a hole in the side of the fish. In the case of a large fish, extra leverage is needed, so hagfishes loop their bodies into a half-hitch knot and use the fore part of this loop as a fulcrum to help thrust their head into the body of the fish. Hagfishes feed quickly and may soon be completely inside their victim, voraciously eating the flesh. It is quite common for a trawler to catch the grisly and almost-hollow remains of a fish with a sated hagfish inside it.

The hagfish's ability to knot the body has other uses. It can, for instance, be used to evade capture, especially when this strategy is combined with slime secretion. Knotting is also used to clear a hagfish of secreted slime that might otherwise clog gill openings and cause suffocation. In this case, it simply wipes itself free of slime by sliding through the knot. The nostril cannot be cleared in this maneuver, however, so a hagfish also has the ability to produce a powerful "sneeze" to clear its nose.

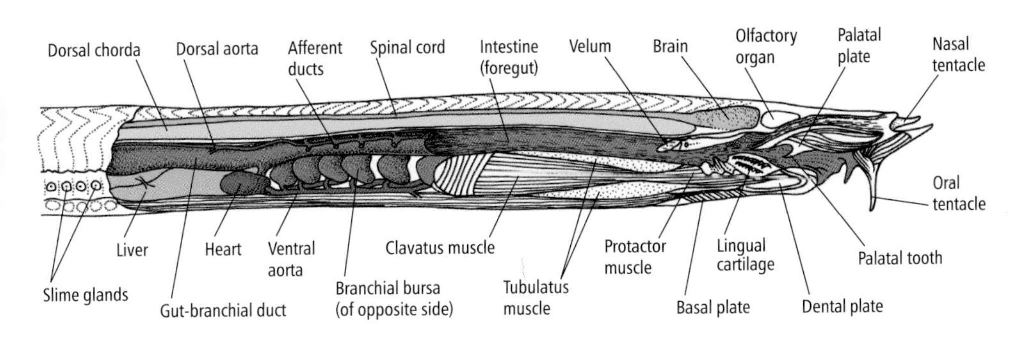

Slime is secreted into water as cottonlike cells, which, on contact with seawater, rapidly expand, coagulate, and form a tenacious mucous covering. A single North Atlantic hagfish measuring 45cm (18in) in length can, when placed in a bucket of seawater, turn the entire contents of the bucket into slime in a few minutes.

Like most fishes, hagfishes have a heart to pump the blood through the fine vessels of the gills. However, they also have a pair of hearts to speed the blood after it has passed through the gills, while yet another heart pumps the blood into the liver; there is also a pair of small hearts, like reciprocating pumps, near the tail. These accessory hearts are necessary because, in hagfishes, the blood is not always contained in restraining blood vessels. Instead, there is a series of open spaces or "blood-lakes," called sinuses, in which there is a great drop in blood pressure. The accessory heart that is found after each of these sinuses increases the pressure and pushes the blood on to the next part of the circulatory system. The tail hearts, however, are something of a mystery. They are minute, so they cannot pump much blood, and can be stopped by stimulating the skin. (They are worked by a plungerlike rod of cartilage activated from outside the heart by the movement of the body.) Furthermore, hagfishes are not greatly inconvenienced by their removal. It has also been found that at least some of the hearts double up as endocrine organs (ductless glands) that secrete hormones.

Little is known about the reproduction of hagfishes. Only a few (probably fewer than 30) eggs are laid, which measure some 2.5cm (1in) long, and are oval in shape, with adherent tufts at each end that attach the eggs to each other and to the sea bed. A zone of weakness near the end splits when a young hagfish is ready to hatch about two months later. Hagfish lack copulatory (mating) organs, so it is assumed that the eggs are fertilized externally, but how this is achieved remains a mystery. An adult hagfish has only one gonad (sexual organ), which develops into either an ovary or a testis as the fish matures. Despite few eggs being laid, hagfishes are abundant in some areas; local populations of up to 15,000 have been recorded. They have no larval stage (in marked contrast to lampreys) and their lifespan is unknown, but it may be long, thanks to their amazing resistance to infection of wounds, thought to be engendered by the slime. KEB/JD

Sturgeons and Paddlefishes

THE ONLY SURVIVORS OF AN ANCIENT GROUP of fishes known from the Upper Cretaceous period (95–65 million years ago), sturgeons and paddlefishes, form the subclass Chondrostei (together with five extinct orders). They are now confined to the northern hemisphere, within which there are two distributions, one centered on the Pacific Ocean, the other on the Atlantic. Sturgeons are the largest of all freshwater fishes, the longest-lived, and produce a roe – caviar – that is beloved of the world's gastronomes.

The sturgeons and paddlefishes are under a greater degree of threat than any other group of fishes. Overfishing and human despoliation of habitat have brought about a serious decline in many populations, with the result that 23 of the 25 species are classed by the IUCN as Vulnerable, Endangered, or Critically Endangered. Overfishing is especially problematic owing to these fishes' late sexual maturity. It is estimated that the number of sturgeon in the world's major river basins has fallen by some 70 percent since 1900. There is particular concern over the plight of sturgeon stocks in the Caspian Sea, victims both of pollution and of poaching for the highly lucrative illegal trade in roe.

Sturgeons

FAMILY ACIPENSERIDAE

Sturgeons have heavy, almost cylindrical, bodies that bear rows of large ivorylike nodules in the skin (scutes or bucklers), a ventral mouth surrounded by barbels (whiskers), a heterocercal tail (where the upper lobe is longer than the lower one), and a cartilaginous skeleton.

Some sturgeons live in the sea but breed in freshwater; others live entirely in freshwater. Little is known of the life at sea of anadromous sturgeons (those fish that migrate from the sea into rivers). They appear to take a wide variety of food including mollusks, polychaete (bristle) worms, shrimps, and fishes. Adults have few enemies, although they are known to be attacked and even killed by the Sea lamprey (*Petromyzon marinus*). The Baltic or Atlantic sturgeon has been found at depths of over 100m (330ft) in submarine canyons off the continental shelf. Although it was known that the Kaluga sturgeon inhabited fishing grounds off Sakhalin Island in the Russian Far East, it was not until 1975 that the first specimens were caught in the well-fished grounds off Hokkaido, northern Japan. The White sturgeon is known to travel over 1,000km (625 miles) during its time at sea. Freshwater sturgeons usually remain in shoal areas of large lakes and rivers feeding on crayfish, mollusks, insect larvae, and various other invertebrates, but rarely on fishes. Seasonal movements are from shallow to deeper waters in the summer and a return to the shallows in winter. In the River Volga, sturgeons overwinter along a 430-km (270-mile) stretch, aggregating in bottom depressions.

Anadromous sturgeons spawn during spring and summer months, although in some species there are "spring" and "winter" forms that ascend the rivers in their respective seasons. The spring form spawns soon after going up the river, while the winter form spawns the following spring. Additionally, some adults spawn every year, while others do so intermittently. Freshwater sturgeons make their way from their home streams and lakes into the upper or middle reaches of large rivers for spawning. The North American Lake sturgeon spawn over rocks in wave conditions when more suitable quiet areas are unavailable. Courtship in this species involves the fish leaping and rolling near the bottom.

Both anadromous and freshwater species cease feeding during the spawning period. Eggs (caviar) are produced in millions – over 3 million in a large female Baltic sturgeon 2.65m (8.7ft) long. The eggs are adhesive, attaching to vegetation and stones. Hatching takes about a week. Few data are available on the development of the young, but growth is generally rapid in the first five years: approximately 50cm (20in).

The size and age of sturgeons are impressive. The White sturgeon of North America and the Russian Beluga are the world's largest freshwater fishes. An 800kg (1,800lb) White sturgeon caught in Oregon in 1892 was exhibited at the Chicago World Fair, but the only great White sturgeon that has been weighed and measured came from the Columbia River (Canada and USA). Caught in 1912, it was 3.8m (12.5ft) long and weighed 580kg (1,285lb). When 1.8m (6ft) long, the White sturgeon is between 15 and 20 years old. The largest Lake sturgeon, caught in 1922, weighed in at 140kg (310lb), and the greatest recorded age was 154 years for a specimen caught in 1953. A Beluga caught in 1926 weighed over 1,000kg (2,200lb) and yielded 180kg (396lb) of caviar and 688kg (1,500lb) of flesh; it was at least 75 years old.

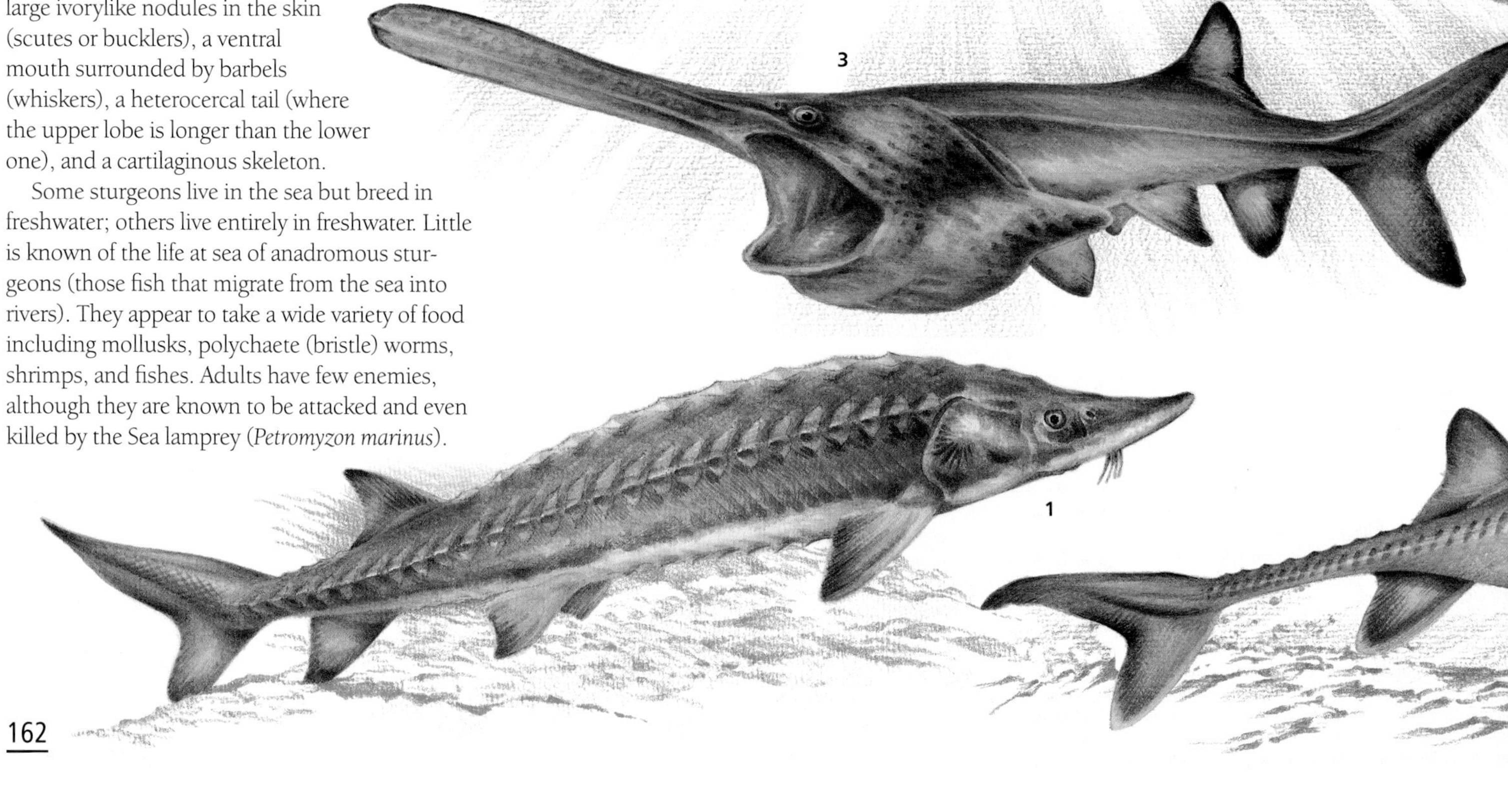

STURGEONS AND MAN

In Longfellow's epic poem *Hiawatha* (1855) the Lake Sturgeon was called "Mishe-Nahma, King of Fishes." This mighty fish also made an impression on the early explorers of North America, as shown by the proliferation of place-names such as "Sturgeon river." In ancient America and Russia, sturgeons were a valuable resource. Scutes (the bony plates on the body) were used as scrapers, oils as medicine, flesh as food, and eggs as caviar.

Steamboats in North America could trawl large numbers of sturgeons and used their oil as fuel. Thousands more were butchered by 1885, when over 2 million kg (5 million lb) of smoked sturgeon were sold at Sandusky, Ohio.

The eggs or roe provide the sturgeon's most prized product: the expensive delicacy caviar. The center of the caviar trade is the Russian Caspian Sea basin. Here, sturgeons were fished intensively for 200 years, and even by 1900 stocks had declined dramatically. The First World War and internal strife allowed some recovery of stock, but by 1930 feeding grounds were intensively overfished and dam construction further depleted the fishery by precluding breeding migrations. Another product that has contributed to the reduction in sturgeon stocks is isinglass. Derived from the sturgeon's swimbladder and vertebrae, it was used as a clarifier of wines and as a setting agent for jams and jellies. In 1885 about 1,350kg (3,000lb) of isinglass were exported (procured from around 30,000 sturgeons).

The development of sturgeon hatcheries in the 1950s conserved and increased populations; these hatcheries released millions of young (45 million in 1965) into the rivers. Sturgeon fishing is subject to strict quotas, although illegal harvesting has grown apace since the breakup of the Soviet Union. In both Russia and the USA the sturgeon's plight is now serious: the Aral Sea populations of the Ship and Syr-Darya shovelnose sturgeons are near extinction.

Below *It took five fishmongers from London, UK, to hold up this Beluga, which was caught in the English Channel in 1947 and weighed 181kg (400lb). Belugas produce the best known caviar.*

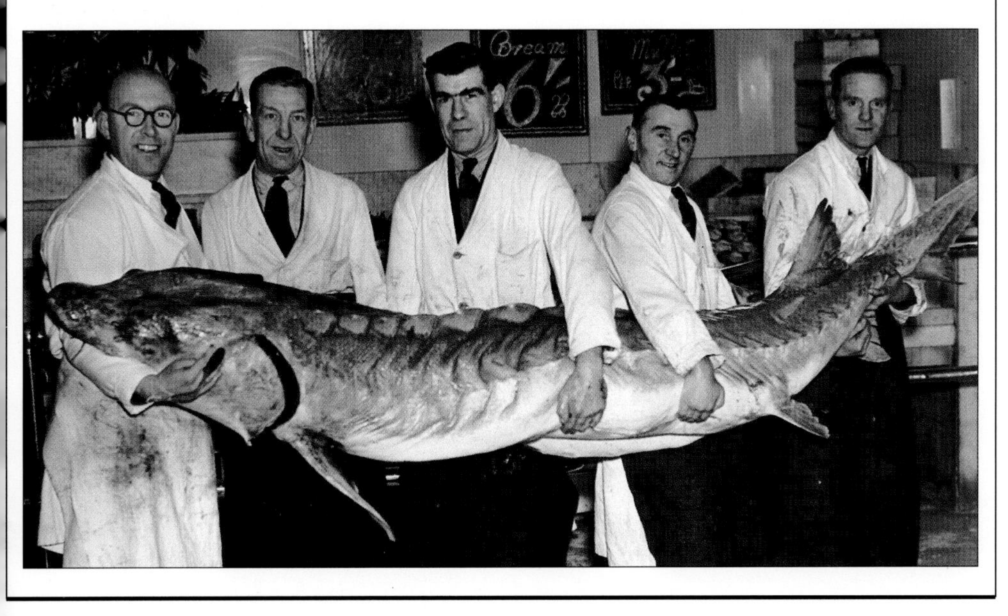

Left *Representative species of sturgeons and paddlefishes: **1** The Critically Endangered Common or Baltic sturgeon* (Acipenser sturio) *reaches sexual maturity at 7–9 years old; **2** The Pallid shovelnose* (Scaphirhynchus albus) *is cited on the IUCN Red List as Endangered; **3** Pollution, dams, and intensive fishing have contributed to the Vulnerable status of the American paddlefish* (Polyodon spathula).

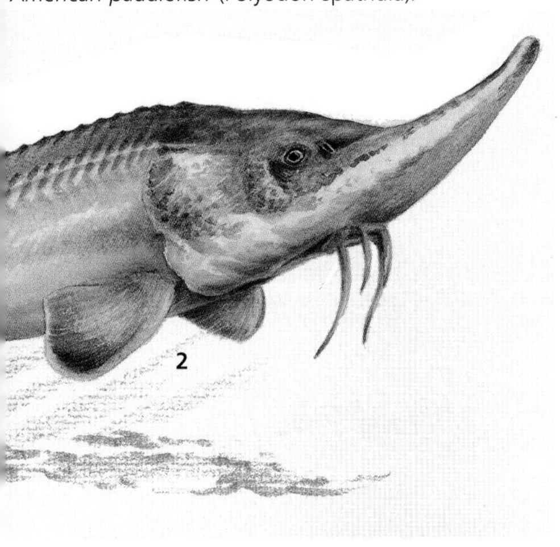

Paddlefishes

FAMILY POLYODONTIDAE

There are two paddlefish species: the American paddlefish of the Mississippi and the Chinese paddlefish of the Yangtze, but fossils are known from North America from the Cretaceous and Eocene periods (135–38 million years ago).

Paddlefishes are identified by their extended upper jaw, which forms a long, flat, broad snout. The huge mouth is sacklike and opens as the fish swims, scooping up crustaceans and other plankton. The American paddlefish occurs in silty reservoirs and rivers; it exceeds 1.5m (5ft) in length, weighs up to 80kg (175lb) and yields its own type of caviar. This species is largely nocturnal, resting at the bottom of deep pools during the day.

Spawning was first observed in the American paddlefish only in 1961. When water temperature reaches 10°C (50°F), adults are stimulated to move upstream to shallows, where spawning occurs in April and May in temperatures of around 13°C (55°F). About 7,500 eggs are produced per kilo of body weight by females, and they take about one week to hatch.

The function of the paddlelike upper jaw is uncertain. It has been suggested that it is an electrical sensory device for detecting plankton swarms; a stabilizer to balance the head against the downward pressure that is created in the huge mouth; a scoop; a mud-digger (although this is no longer believed to be the case); or even a beater to release small organisms from aquatic plants (though individuals that have lost their paddles in accidents are known to feed perfectly, from the evidence of their full stomachs).

The biology of the Chinese paddlefish is poorly known. It has, however, been established that – in contrast to its American plankton-eating relative – this species feeds predominantly on fish. It therefore possesses far fewer gill rakers – the "sieving" structures that filter out suspended food particles from the water. It is also reputed to grow to 7m (23ft); the largest recorded specimen, however, measured 3m (around 9.8ft). GJH/JD

FACTFILE

STURGEONS AND PADDLEFISHES

Order: Acipenseriformes

27 species in 6 genera and 2 families.

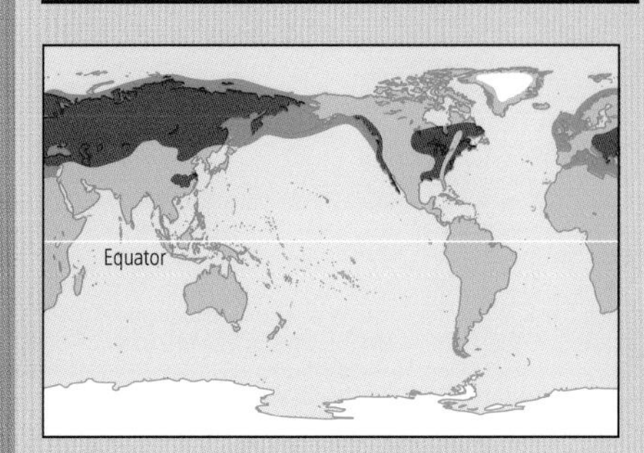

STURGEONS

Family Acipenseridae

25 species in 4 genera. North Atlantic, North Pacific, and Arctic oceans and associated feeder rivers. **Length:** (adult) 0.9–9m (3–20ft). Species include: **Common** or **Baltic sturgeon** (*Acipenser sturio*), **Lake sturgeon** (*A. fulvescens*), **Sterlet** (*A. ruthenus*), **White sturgeon** (*A. transmontanus*), **Beluga** (*Huso huso*), **Kaluga sturgeon** (*H. dauricus*), **Shovelnose sturgeon** (*Scaphirhynchus platorynchus*). **Conservation status:** The Syr-Darya shovelnose sturgeon (*Pseudoscaphirhynchus fedtshenkoi*) is Critically Endangered and the Ship sturgeon (*Acipenser nudiventris*) is Endangered.

PADDLEFISHES

Family Polyodontidae

2 species in 2 genera. Mississippi and Yangtze rivers. **Length:** (adult maximum) c. 3m (10ft). Species: **Chinese paddlefish** (*Psephurus gladius*), **American paddlefish** (*Polyodon spathula*). **Conservation status:** Both species are seriously threatened, the Chinese paddlefish of the Yangtze river being classed as Critically Endangered.

Garfishes and Bowfin

1

The garfishes and bowfin are living relicts of several widely distributed groups of extinct fishes. For example, garfishes of the extinct genus Obaichthys *inhabited South America during the Cretaceous period (142–65 million years ago), while fossils of the two living garfish genera have been found in the Cretaceous and Tertiary deposits of Europe, India, and North America. The ancestral group of the bowfin (the Halecomorphi) once contained 50 genera, with a lineage stretching back to the Jurassic (206–142 m.y.a.). During the Tertiary (65–1.8 m.y.a.), bowfins were widespread in Eurasia and North America.*

Nowadays, distribution of the two families is far more limited. Only one species of bowfin survives, in central and eastern North America. *Lepisosteus* gars exist only in North America and Mexico; members of the genus *Atractosteus*, however, are more southern in their distribution, extending from the southern USA into Central America (Costa Rica and Cuba).

FACTFILE

GARFISHES AND BOWFIN

Orders: Semionotiformes, Amiiformes

Families: Lepisosteidae, Amiidae

8 species in 3 genera.

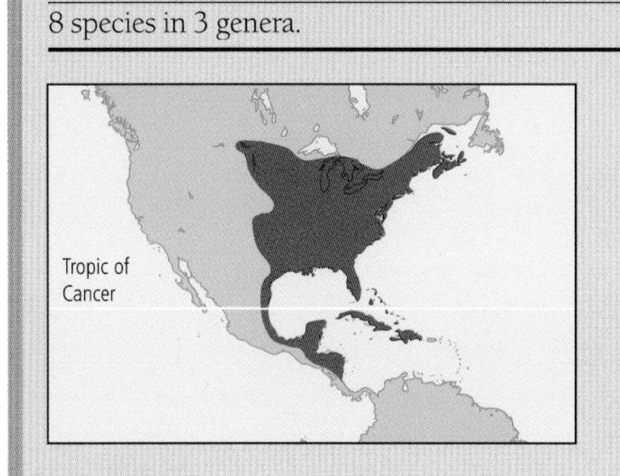

GARFISHES
Family Lepisosteidae
7 species in 2 genera. C America, Cuba, N America as far N as the Great Lakes. **Length:** 75cm–3m (2.5–10ft), **weight:** 7–135kg (15–300lb). Species: **Alligator gar** (*Atractosteus spatula*), **Cuban gar** (*A. tristoechus*), **Tropical gar** (*A. tropicus*), **Florida Spotted gar** (*Lepisosteus oculatus*), **Longnose gar** (*L. osseus*), Shortnose gar (*L. platostomus*), **Spotted gar** (*L. platyrhincus*).

BOWFIN *Amia calva*
Family Amiidae
Sole member of family.
N America from the Great Lakes S to Florida and the Mississippi Valley. **Length:** 45–100cm (18–39in), **weight:** maximum 4kg (9lb).

Garfishes

FAMILY LEPISOSTEIDAE

Garfishes are long-bodied predators with the habit of lurking in wait for prey alongside submerged branches. They are characterized by their long jaws with numerous pointed teeth and by their heavy, diamond-shaped, armorlike scales. Their swimbladder is connected to the esophagus or gullet, namely the tube that runs from the back of the mouth to the stomach. This arrangement performs like a lung, enabling garfishes to breathe atmospheric air.

The Alligator gar (*Atractostes spatula*) is one of the largest North American freshwater fishes. One specimen weighing over 135kg (300lb) and 3m (10ft) long was caught in Louisiana. This species is, however, now scarce throughout its range (an arc along the Gulf coast plain from Veracruz to the Ohio and Missouri rivers).

Sport fishermen widely hold the Alligator gar responsible for eating game fishes and waterfowl and, thus, for depleting angling stocks and other forms of wildlife. However, careful studies have shown that this species rarely feeds on these animals and preys mostly on forage fishes and crabs, although it will also take some game fishes and waterfowl. It has even been reported as eating other gars and – according to one account – of

Below *The characteristic habitat of the Longnose gar is slow-flowing streams, backwaters, and lakes with plenty of underwater vegetation. Small fishes and crustaceans are its habitual prey. The Longnose gar is rarely caught for eating, and its roe is poisonous.*

Left Garfish and bowfin species: **1** Alligator gar (Atractosteus spatula) – this species can grow extremely large, and there are unconfirmed reports of it attacking humans; **2** Longnose gar (Lepisosteus osseus), with Least killifish prey (Heterandria formosa); **3** Bowfin (Amia calva).

being capable of cutting a small alligator in half. The other two *Atractosteus* species are the Tropical gar (*A. tropicus*) of Nicaragua and the Cuban gar (*A. tristoechus*). Little is known of their biology; the Tropical gar inhabits the shallow, protected areas of Lake Nicaragua, where specimens can grow to over 1.1m (3.6ft) and weigh over 9kg (20lb); the Cuban gar can grow to around 2m (c.6.6ft).

The genus *Lepisosteus* has a wider distribution, from the northern Great Lakes to Florida and the Mississippi basin. The Spotted gar (*L. oculatus*) occurs throughout the Mississippi drainage and grows to over 1.1m (3.6ft) and a weight of 3kg (6.6lb). The distribution of the Longnose gar (*L. osseus*) is wider than that of the Spotted gar and it, too, inhabits brackish water in its coastal range.

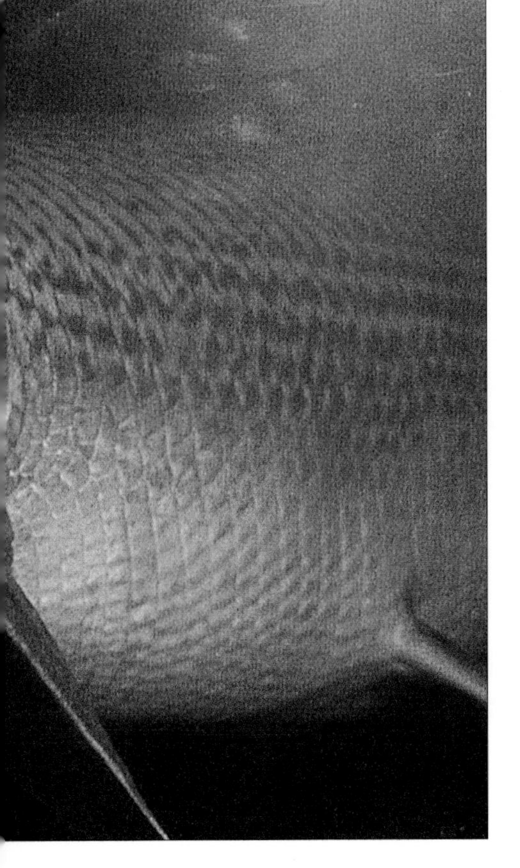

The Shortnose gar (*L. platostomus*) covers an area encompassing northeastern Texas, Montana, southern Ohio and the Mississippi, while the Florida Spotted gar (*L. platyrhincus*) occurs there and in the lowlands of Georgia.

The Longnose gar can grow to over 1.8m (5.9ft) and weigh 30kg (66lb). The females are longer than the males, the difference being as much as 18cm (7in) in their tenth or eleventh year. Males rarely survive beyond this period, but females may live for up to 22 years. Group spawning, in which a single female may breed with several males, occurs from March to August, according to locality, in shallow warm water over vegetation or in a female-excavated depression. The adhesive eggs, usually about 27,000 – but as many as 77,000 – per female, are deposited and hatch within 6–9 days. As in the bowfin, the young cling to vegetation by means of an adhesive pad on their snout. Growth is rapid, at 2.5–3.9mm (0.1–0.15in) per day.

Bowfin

FAMILY AMIIDAE

The bowfin's position within the historical classification of fishes has been hotly debated for many years, and among bony fishes its anatomy is probably the most thoroughly described. The structure of the skull alone forms the subject of a 300-page monograph (by Edward Allis) published in 1897. Current scientific opinion is that the bowfin is most closely related to the Teleostei, the so-called "perfectly boned" fishes.

The common name "bowfin" alludes to the species' long, undulating dorsal fin. In the Great Lakes it is known as the dogfish, and in the southern states of the USA as the grindle, while elsewhere it is variously referred to as the mudfish, choupique, cottonfish, or – intriguingly – the lawyer. Other distinctive features of the bowfin include its massive blunt head and cylindrical body. At the upper base of the tail is a dark spot, edged in orange or yellow in males (probably a sign for recognition). Females lack the edging and sometimes even the spot itself. Most bowfins grow to 45–60cm long (18–24 in) but a few reach 110cm (43in) and weigh 4kg (9lb).

Bowfins can use their air bladder as a "lung" and can survive for up to a day out of water, as long as the body is kept moist. This ability enables them to live in swampy, stagnant, oxygen-poor waters that are unsuitable for other predatory fish. There is also evidence that, during periods of drought, individuals may bury themselves in the bottom mud and enter a state of torpor or dormancy known as estivation.

Bowfins spawn in spring. Males move to shallow water and each prepares a saucer-shaped nest, 30–60cm (12–24in) across, by biting away plants growing in the bed of the river or lake. The males vigorously defend their nesting sites against other males and often spawn with several females. Females lay about 30,000 adhesive eggs, which the male guards and fans. In 8–10 days the eggs hatch and the young fish cling to vegetation by an adhesive snout pad. The male continues to guard the brood until the young are about 10cm (4in) long. Growth of these young is relatively slow; it takes them some 3–5 years to reach maturity.

Bowfins are predators that feed mostly on other fish (sunfish, bass, perch, pike, catfish, and minnows), frogs, crayfish, shrimps, various water insects, turtles, snakes – even leeches and rodents. Bowfin flesh itself is unsavory, and its plundering of sport fishes and comparative abundance do not endear it to anglers and conservationists. Some people, however, argue that, without the presence of bowfins, many stretches of angling water would become overcrowded, thus stunting the growth of the game fish, which – in turn – would make them far less attractive quarry for anglers. GJH/JD

Tarpons, Bonefishes, and Eels

EELS HAVE LONG BEEN PRIZED AND IN THE Middle Ages were a staple food. From the earliest days, eels were believed to be different, since, unlike other freshwater fish, they did not produce eggs and sperm at the start of the breeding season. This apparent enigma gave rise to some ingenious explanations – all of them fanciful – until the mystery was solved at the end of the 19th century. Now, even though people know a great deal about the biology of eels, their behavior still holds many secrets.

The subdivision of tarpons, bonefishes, eels, and notacanths comprises four orders, the members of each of which seem to be unlike the others, apart from certain anatomical similarities. They are united in having a larva that is quite unlike the adult: it is transparent, the shape of a willow leaf or ribbon, and is graced with the name of "leptocephalus." The four orders are the Elopiformes, containing the tarpons and tenpounders or ladyfishes; the Albuliformes, containing the bonefishes, halosaurs, and spiny eels; the Anguilliformes with about 15 families of eels; and the Saccopharyngiformes, containing the gulpers, swallowers, and their allies.

Tarpons and Allies

ORDER ELOPIFORMES

The un-eel-like tarpons and allies (Elopiformes) represent an ancient lineage. Fossils belonging to this group occur in the Upper Cretaceous deposits of Europe, Asia, and Africa (96–65 million years ago). In addition to developing from a leptocephalus larva, they all have gular plates – a pair of superficial bones in the skin of the throat – between each side of the lower jaw. Gular plates were far more common in ancient fishes than they are in living fishes. All members of this order are mainly marine fish, although they are also known to enter brackish water and even freshwater.

The Tarpon (*Megalops atlanticus*) is the largest fish in the group, weighing up to 160kg (350lb). It is a popular game fish because it gives the angler a fight, leaping into the air, twisting and turning to dislodge the hook. This species lives in tropical and subtropical waters on both sides of the Atlantic. The adults breed at sea and the larvae make their way inshore, where they metamorphose. The young live and grow in lagoons and mangrove swamps. These swampy regions can often be low in oxygen, but the tarpon can breathe atmospheric oxygen and does so at such times.

The Indo-Pacific tarpon (*M. cyprinoides*) is similar in appearance to its Atlantic relative and even has the last ray of the dorsal fin similarly prolonged. Its life cycle is also like that of the Tarpon. Apart from some minor anatomical differences, for example, in the way that processes from the swimbladder lie closely against the region of the inner ear in the skull, or in numbers of fin rays and vertebrae, the most obvious difference is that the

Below A young European eel (*Anguilla anguilla*) in freshwater; eels spend between 6 and 12 years (males) or 9 and 20 years (females) in such habitats before migrating to the sea to spawn.

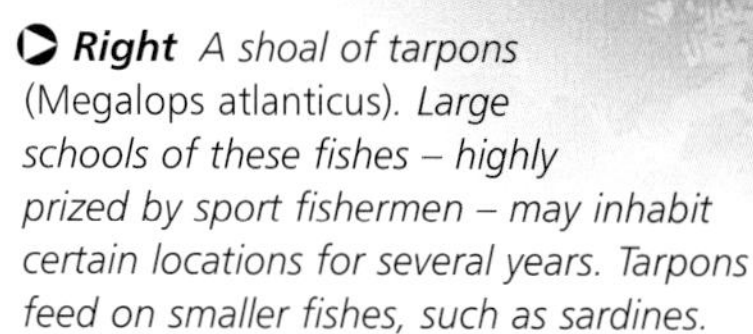

Right A shoal of tarpons (*Megalops atlanticus*). Large schools of these fishes – highly prized by sport fishermen – may inhabit certain locations for several years. Tarpons feed on smaller fishes, such as sardines.

Indo-Pacific tarpon is smaller, rarely reaching 50kg (110lb).

The tenpounders (*Elops* spp.) are also warm-water species, and occur in the tropical and subtropical Atlantic. Despite their common name, they can grow up to 6.8kg (15lb) in weight.

Bonefishes, Spiny Eels, and Allies

ORDER ALBULIFORMES

The bonefishes are widespread in shallow tropical marine waters. They are shoaling fishes and feed on bottom-living invertebrates, often in water so shallow that the dorsal fin and upper lobe of the caudal fin stick out of the water. Bonefishes rarely exceed 9kg (20lb) in weight and 1.1m (3.5ft) in length. Despite their poor flavor, they are much sought-after angling fish because of their fighting qualities.

The halosaurs (Halosauridae) and the spiny eels or notacanths (Notacanthidae) are scaled, deep-sea fishes rarely exceeding 2m (6.5ft) in length. They have a snout that extends in front of the mouth and which may be sharply pointed. The head is usually the deepest part of the body; the latter is partially eel-like and tapers away to a rat tail. In some species, a minute caudal fin is present, while in others, the anal and caudal fins are confluent., i.e. joined up. The dorsal fin, if present, is short and placed well forward. The notacanths have a series of isolated spines on the back and in front of the anal fin, hence their other common name of spiny eels; they are also stockier. Notacanths are distributed worldwide and feed on echinoderms (see urchins and starfishes), sponges, and sea anemones from the sea floor.

The halosaurs also live worldwide, mostly

FACTFILE

TARPONS, BONEFISHES, AND EELS

Subdivision: Elopomorpha

Orders: Elopiformes, Albuliformes, Anguilliformes, Saccopharyngiformes

About 800 species in about 156 genera, about 24 families, and 4 orders

Distribution Worldwide in all oceans; tropical and temperate waters.

See Tarpon, Bonefish, and Eel Orders ▷

at depths down to 1,800m (5,900ft), but one species, *Aldrovandia rostrata*, was caught in 5,200m (17,000ft) of water in the North Atlantic. They feed on invertebrates, mostly deep-dwelling forms, which at least some species are thought to dislodge from the seafloor with their snouts. Other species prey on small squid, and one species, the Abyssal halosaur (*Halosauropsis macrochir*) from the Atlantic, has a row of what are thought to be taste buds across the top of its head. Halosaurs are mostly dark-hued fishes, but *Halosaurus ovenii* is pinkish with silvery sheens. This species is also one of several in which the roof of the mouth has alternate light and dark stripes.

The Suckermouth spiny eel (*Lipogenys gillii*) from the western North Atlantic, has a toothless mouth that functions like a vacuum cleaner and sucks up vast quantities of ooze. The amount of organic material in ooze is probably small, but as this species has a long intestine, permitting maximum absorption of nutrients, it apparently survives on such an unpromising diet.

During a research voyage of 1928–30 , the research vessel *Dana* caught a leptocephalus larva 1.84m (6ft) long off South Africa. Much speculation appeared in the popular and pseudoscientific press along the following lines: "If a 10cm (4in) Conger eel leptocephalus produces an adult 2m (6.5ft) long, then this larva will produce an adult over 30m (100ft) long; hence it is a baby sea serpent and the sea serpent is an eel." Other giant leptocephali were caught and named *Leptocephalus giganteus*. In the mid-1960s, luck came to the aid of science, however, when another giant leptocephalus was caught. This time, however, it was in mid-metamorphosis and it could be established that the adult form was a notacanth. Reexamination of the other giant larvae showed they, too, could be referred to notacanths, thereby establishing the relationships of the notacanths with the eels and dispelling one set of rumours about sea serpents. Unlike other eels, however, notacanths hardly grow after metamorphosis and the 30m (100ft) adult – the sea serpent – is merely a product of imaginative extrapolation.

Eels

ORDER ANGUILLIFORMES

The exact number of families of eels is uncertain. Differences of opinion arise because some groups are poorly known – sometimes on the basis of just one or a few specimens. In addition, there is the problem of matching the known leptocephali with the known adults.

The larvae of European eels (*Anguilla anguilla*) hatch in the Sargasso Sea and then drift back in the Gulf Stream and take about three years to reach the colder, shallower, and fresher European coastal waters. Here they shrink slightly, metamorphose into elvers and move upstream, where they grow and feed until, some years later, the urge to migrate comes upon them. The American eel (*A. rostrata*) breeds in the western part of the Sargasso and its leptocephali take only one or two years to reach freshwaters.

The breeding grounds have not been found for all anguillid eels, however. The commercially significant Japanese eel (*Anguilla japonica*), for example, has an unknown spawning ground. Eels from the eastern part of southern Africa, for example the Mottled eel (*A. nebulosa*), the Northern eel (*A. bicolor bicolor*), and the East African or Longfin freshwater eel (*A. mossambica*) all breed at depth to the east of Madagascar.

As adults, the European, American, and Japanese eels are widespread. A fourth species of eel in eastern Africa, the Madagascar mottled eel (*A. marmorata*) also occurs on many Pacific islands and as far east and north as Hong Kong and southern Japan. In complete contrast, *Anguilla malgumora* lives in the river Bo in Borneo, *A. anterioris* in mountain streams in New Guinea, and the Celebes longfin eel (*A. celebensis*) only in the northeast of Celebes. Other species have intermediate ranges and all have to find their way to their breeding grounds. How, and why? No one knows the answers, but a few interesting possibilities can be suggested.

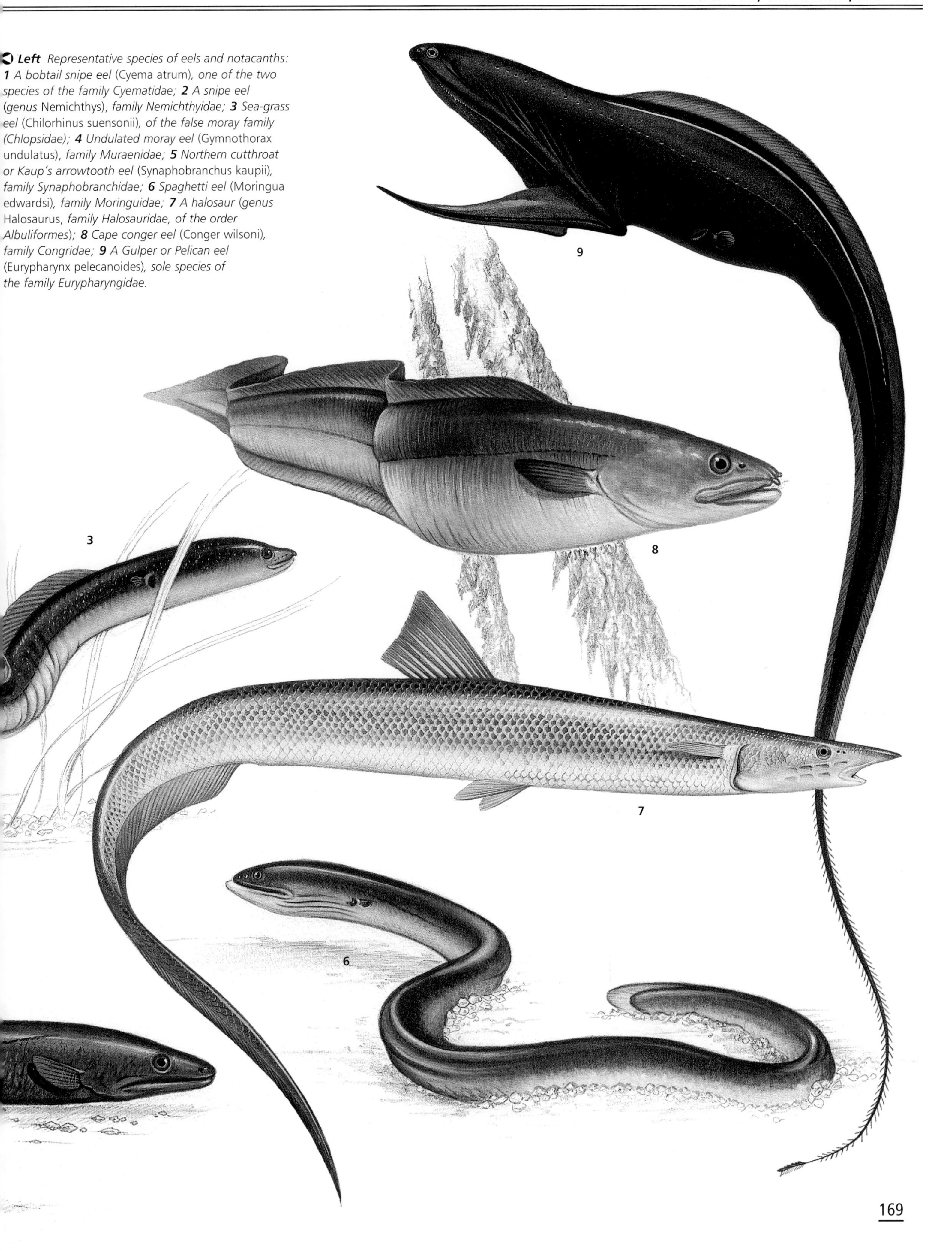

Left *Representative species of eels and notacanths:* **1** *A bobtail snipe eel* (Cyema atrum), *one of the two species of the family Cyematidae;* **2** *A snipe eel (genus* Nemichthys), *family Nemichthyidae;* **3** *Sea-grass eel* (Chilorhinus suensonii), *of the false moray family (Chlopsidae);* **4** *Undulated moray eel* (Gymnothorax undulatus), *family Muraenidae;* **5** *Northern cutthroat or Kaup's arrowtooth eel* (Synaphobranchus kaupii), *family Synaphobranchidae;* **6** *Spaghetti eel* (Moringua edwardsi), *family Moringuidae;* **7** *A halosaur (genus* Halosaurus, *family Halosauridae, of the order Albuliformes);* **8** *Cape conger eel* (Conger wilsoni), *family Congridae;* **9** *A Gulper or Pelican eel* (Eurypharynx pelecanoides), *sole species of the family Eurypharyngidae.*

FEATURE

THE SARGASSO SEA MYSTERY

Solving the puzzle of where eels come from

THE BEST-KNOWN EEL IS THE EUROPEAN EEL (family Anguillidae). Although anguillid eels are the only family that spend most of their life in freshwater, the European eel's life history will serve as an illustration for all the other families, not least because the eel's familiarity has, over the centuries, occasioned so much curiosity about eel reproduction.

From before the days of the Greeks and Romans, this eel has been an important source of food. Aristotle and Pliny wrote about large eels going down to the sea and small eels coming back from the sea. Other freshwater fish had been noticed to have eggs and sperm at the start of the breeding season, but not the eel; it was therefore concluded that the eel was "different" and two millennia of speculation were born. Aristotle was certain that baby eels sprang out from "the entrails of the earth," while Pliny, on the other hand, concluded that young eels grew from pieces of the adult's skin scraped off on rocks. Subsequent suggestions continued to be as unrealistic. In the 18th century, for example, the notion of the hairs from horses' tails giving rise to eels was in vogue and, in the 19th, a small beetle was advocated as the natural mother of eels. The truth, when finally resolved, like the denouement of a detective story, was hardly more credible.

Countless millions of eels had been caught and gutted for eating over the centuries, but not until 1777 were developing ovaries identified by Professor Mondini of Bologna. In 1788, his finding was challenged by Spallanzani, who pointed out that none of the 152 million eels from Lake Commachio ever showed such structures. Nonetheless, Spallanzani was struck with the determination of the eel to reach the sea, even traveling overland on damp nights. In 1874, indisputable testes were found in an eel in Poland, but it was not until 1897 that an indisputably sexually mature female was caught in the Straits of Messina. The beetle myth was laid to rest; eels must lay eggs in the sea – but where? When the eels reappear on the coasts they are about 15cm (6in) long. Why were smaller specimens never caught?

The answer had been available, but unrealized, since 1763. In that year, the zoologist Theodore Gronovius illustrated and described a transparent fish like a willow leaf, which he called leptocephalus. One hundred and thirty-three years later (1896) Grassi and Calandruccio (the two biologists who found the sexually mature female eel) caught two leptocephali and kept them alive in aquaria. The leptocephali, being caught near the coast, were on the point of metamorphosing; their transformation in the aquaria, therefore, tied up, at least, a part of the story.

Above *Swimming leptocephali of the common eel. Adult eels of various species arrive in the Sargasso Sea from Europe and North America, sometimes travelling from thousands of miles away, to mate and spawn (only with others from their own native region, it has been found). The larvae then embark on a long journey back to where their parents came from; the map charts the odyssey of the European eel on the Gulf Stream.*

Below *The transparent leptocephalus larva of an eel, which is shaped like a willow leaf, shrinks before metamorphosing into its adult form.*

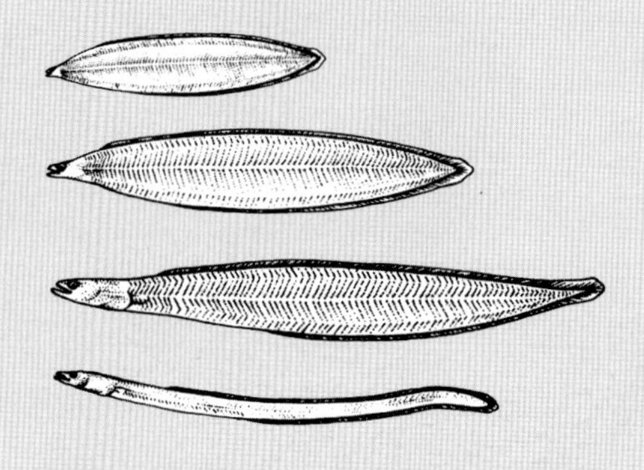

The hunt was now on for leptocephali and the eel breeding grounds. Johannes Schmidt took leptocephali samples and followed their decreasing size. Finally, he found the smallest of all, 1cm (0.4in) long, in an area of the west Atlantic between 20 and 30° N, 48 and 65° W: the Sargasso Sea.

On the basis of much subsequent research, scientists now know that the European Eel probably breeds from late February, probably to May or June, at moderate depths of around 180m (600ft) – the eyes of the adults enlarge on their 4,000 mile journey – and at a temperature of about 20°C (68°F). The Sargasso Sea is one of the few areas where such a high temperature extends to such a depth. There are, throughout the world, 16 species of the genus *Anguilla* and all are thought to breed in deep warm waters, although only two, the European and the American eel (*Anguilla anguilla* and *A. rostrata*), in the Sargasso Sea. No adult eels, however, have been captured in the Sargasso – and no eggs have ever been collected in the area.

In the European eel, as the larvae are carried by the Gulf Stream, might not the adults also follow it back to the breeding grounds? However, if the breeding grounds are characterized by three parameters – temperature, salinity, and pressure – could they be guides? However, these latter possibilities present difficulties. For example, an adult eel leaving northern Europe could follow a temperature gradient and reach the Sargasso, but an eel leaving Italy would not (the Mediterranean is warmer than the adjacent parts of the Atlantic). The appropriate pressure (as a measure of depth) could be reached very much closer to European shores than the other side of the Atlantic. Also, it should not surprise anyone that adult eels are not particularly sensitive to salinity changes, as they have already had to cope in their earlier migration with change from freshwater to salt water.

Scientists' ignorance is considerable, but one cannot leave the problems without some potentially exciting speculations. The first is that many animals are known to be able to navigate using Earth's magnetic field, and although no magnetic-field detectors have yet been found in the eel, the possibility remains. A second possibility involves the movements of continents – continental drift.

Most of Europe and North America were once joined together as one continent, as were Africa and South America. At one stage in the shaping of the present continents, there was a narrow sea between the two. Is it possible that when eels first appeared, they bred in this sea and still need its physical conditions to ensure successful reproduction, even though it entails a journey of thousands of kilometers by an inefficient swimmer?

From tagged migrating eels it has been extrapolated that they take 4–7 months to make the journey and, during that time, they do not eat. Further enigmas occur with the allegedly passive journey of the leptocephali in the water currents that bring them back to the feeding grounds. It appears that they do not eat; yet, it takes them two to three years to migrate across the ocean. No food has been found in the guts and they develop peculiar forward-pointing teeth, mostly on the outside of their jaws, which are ill-adapted to catching food. However, they grow as they migrate, until, at metamorphosis, they shrink and lose their teeth.

A few clues are available that hint at partial solutions to these conundra. A Conger eel (*Conger conger*) larva that was studied was shown to be able to absorb nutrients and vital minerals from the seawater. The mouth of this species is lined with minute projections (villi), which are thought to absorb the nutrients. However, this has not yet been shown to be the case in anguillid eels. The metamorphosis is rapid, and a great deal of calcium must be made available very quickly to mineralize the bones and convert them from the juvenile to the adult state. The forward-pointing teeth of the leptocephalus, which disappear at metamorphosis, contain a great deal of calcium, and so it is possible that these may act as the necessary calcium reservoirs.

As the leptocephalus metamorphoses, it modifies its compressed body into the adult's shape and develops pigment, pectoral fins, and scales. Before pigmentation develops, the young fish is called a glass eel and latterly an elver. Depending upon local conditions, it is either the glass eels or the elvers that ascend the rivers in vast numbers. In the 19th century, before estuarine pollution had its effect, millions of elvers were caught and

Below *Between leptocephalus and adulthood, eels pass initially through a phase where the body form is developed but still transparent. These "glass eels" and the next phase, the elver, are culinary delicacies.*

eel fairs were held. Elvers caught on the River Severn in England were particularly famous. There, elvers were used principally for food, but in Europe, they were also used to make preservatives and glue. Elvers are still caught but their fate is to be taken to eel farms and grown to maturity in less time than it would take in the wild. Adult eels are very nutritious and are popular in many parts of the world, smoked or stewed; jellied eels were once widely eaten in the poor East End of London.

The popularity of eels remains high, even though their blood contains an ichthyotoxin (fish toxin), which can be dangerous. It is particularly important not to let eel blood come into contact with the eyes or any other mucous membrane. However, the poison is quickly destroyed by cooking, so most eel eaters remain blissfully unaware of their toxicity.

Moray eels were favorite pets of the ruling class in Ancient Rome. The scientific name of the Mediterranean moray (*Muraena helena*) commemorates a rich and powerful Roman citizen, Licinius Muraena, who lived toward the end of the 2nd century BC. According to the writer Pliny the Elder, Licinius kept moray eels in captivity as an ostentatious public display of his wealth. Later, in the time of Julius Caesar, one Gaius Herrius is said to have constructed a special pond for his morays, which so impressed the emperor that he offered to buy the eels. Gaius refused but, presumably to retain favor with Caesar, did lend him 6,000 morays to display at a banquet. Other stories recount how wealthy owners bedecked their pet eels with jewelry.

Perhaps the most notorious account of eels in the decadent society of Imperial Rome concerns a certain Vedius Pollo, who would entertain his dinner guests by feeding recalcitrant or superfluous slaves to his morays. This and similar stories about the voracity of these eels may be the source of some pervasive myths, principally that morays find skin divers irresistible. Tales abound of moray eels attacking divers, delivering a poisonous bite, and refusing to loosen their grip once they have clamped their teeth onto their victims. Little credence can be given to such rumors. First, a moray's bite is not poisonous, and any infection of the wound is secondary. Second, Moray eels' teeth are adapted for holding small prey – they are long and thin, and the larger teeth are hinged to permit the smooth passage of prey into the stomach. This does not, however, mean that they will not release after biting, especially if the quarry is too large to swallow.

Most morays have teeth like those of the Mediterranean moray, but species of the tropical

Right *Nestling amid* Tubastraea *sun corals, two Spotted moray eels* (Gymnothorax moringa) *strike a characteristic pose, with their heads just protruding from their lair. This species is common around reefs.*

genus *Echidna* have blunt, rounded teeth adapted for crushing crabs, which are their major food item. The ribbon morays (genus *Rhinomuraena*) are characterized by having the anterior nostrils expanded into tall, leaflike structures. The Blue ribbon eel (*Rhinomuraena quaesita*) is brilliantly colored and much prized by aquarists. In males, the body is turquoise blue, the dorsal fin bright yellow with a white margin, the front and underside of the head yellow, and the rest blue; females are yellow with a black anal fin.

Moray eels are found in all tropical seas, usually in shallow waters. They will also occasionally enter freshwater.

The false morays (Chlopsidae; also sometimes referred to as Xenocongridae) are probably close relatives of the morays. Like them, they lack scales and have thick skin covering the continuous median (central) fin. The pectoral fins are reduced or, in at least one case, lost. The coloration is mottled like that of many morays. The pores on the head are much smaller than in morays and the body is much thinner, this combination being deemed sufficient justification to consider them a separate family. Representatives are found worldwide, but little is known of them because they are secretive fishes.

The leptocephalus of the Bi-colored false moray (*Chlopsis bicolor*) is extremely common in the Mediterranean, but until recently, the adults were poorly known. This species was described over a hundred years ago, but only a very few specimens were in museum collections until divers began using underwater narcotics to drug fish and succeeded in catching large numbers. This species, which also lives off Florida, is a small fish (shorter than 25cm, 10in) that lacks pectoral fins and is dark brown above and paler below.

Chlopsis olokun was known from just a single specimen captured by the research vessel *Pillsbury* off the estuary of the St. Andrews River, Ivory Coast, at a depth of 50m (165ft). (The specific name is derived from Olokun, a sea deity in the Yoruba religion of the region.) Fryer's false moray (*Xenoconger fryeri*) is widespread in the Pacific, but *C. olokun* was the first member of this genus to be found in the Atlantic.

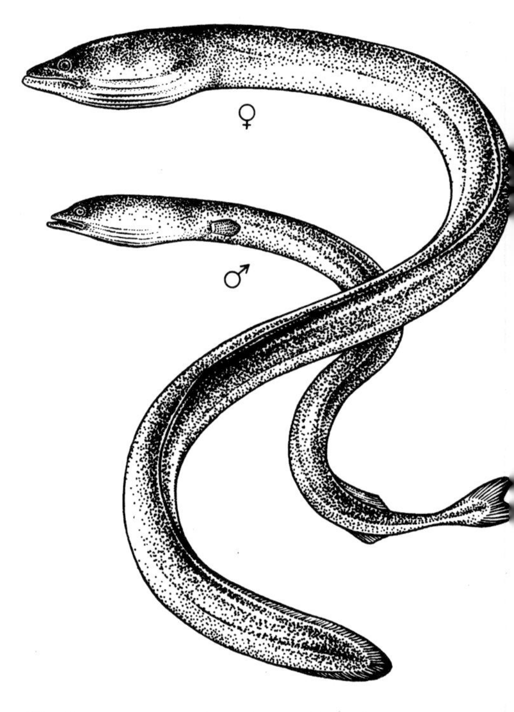

Above *Some eel species, such as the Spaghetti eel* (Moringua edwardsi), *exhibit pronounced sexual dimorphism; here the size and the configuration of the fins are very different.*

Below *The Blue ribbon moray* (Rhinomuraena quaesita; *here a male, as indicated by the yellow dorsal fin) is the only moray species that can undergo abrupt changes in sex and coloration.*

COLONIES OF GARDEN EELS

There are about 25 species of garden eels grouped into two genera (*Gorgasia* and *Heteroconger*), and it seems that the daily cycle and life histories of all are very similar. Colonies usually live in shallow water where there is a reasonable current and enough light for the eels to see their food. There, the eels give a passable imitation of prehensile walking sticks facing into the current. The densest colonies are deepest, with perhaps one eel every 50cm (20in). The distance between individuals depends upon the length of the eel, with each fish occupying a hemisphere of water with a radius similar to the eel's body length, so that it is just separated from its neighbor.

A colony of *Gorgasia sillneri* in the Red Sea has been studied intensively and has produced some interesting information. The eels' burrows themselves are twisted and lined with mucus secreted by the skin (during the period of study, no eel ever left its burrow). The day starts about half an hour before sunrise when the eels emerge and, by sunrise, all are out eating. Any disturbance causes the whole colony to disappear. For a period either side of noon, the eels rest in their burrows, but from midafternoon to sunset, they are out feeding again. An interesting phenomenon is that while the eels are fully retracted in their burrows, their feeding space is taken over by another, totally unrelated fish – a species of sand diver (*Trichonotus nikii*) from the family Trichonotidae, which feeds at the level of where the eels' heads would be. When the eels are out feeding, the sand diver, as its name suggests, hides in the sand.

Even mating is conducted from the security of the burrow with a male stretching out and rippling his iridescent fins at an adjacent female. If she is receptive, she stretches out of her burrow and they intertwine. A few eggs are shed and fertilized at a time and a couple may mate 20 times a day. Detailed study of a plot within a colony showed that females never moved their burrows but some males did, to be near females. However, as this movement was never observed, it was presumed to take place at night. The males also moved away from the females after breeding.

After going through a leptacephalus stage, small eels settle down together until they are about 25cm (10in) long, when the colony breaks up and individual eels try to find a place in an adult colony. The adults resent this intrusion and, unless the juvenile can stand up to aggression, it has to move away to a new area or the fringes of the established colony.

A major predator of garden eels is the ray, whose effect as it swims over a colony has been likened to a mower cutting grass. Whether the eels retreat in time or are caught, has not, however, been satisfactorily ascertained.

The common name "spaghetti eels" appositely describes the six or so members of the family Moringuidae. These long, thin, and scaleless fishes, usually have poorly developed dorsal and anal fins; they are also sometimes referred to as "worm eels," although this common name is probably best reserved for the members of the family Ophichthidae, otherwise also known as snake eels. They are commonest in the Indo-Pacific, but a few species live in the Atlantic, mostly in the west. Most unusually for eels, they have a distinct caudal fin, which can vary from having distinct forks (with two lobes) to being trilobed.

In some species of spaghetti eels, there are marked differences between the sexes. In the West Indian worm eel (*Moringua edwardsi*), for example, the females are nearly twice as long as the males, have more vertebrae, and the heart lies much further back in the body. These differences engendered several misidentifications in the past.

Spaghetti eels are burrowing fishes but, unlike most others, they burrow head-first, preferring a bed of sand or fine gravel. Their life history is exemplified by the Pacific species *Moringua macrochir*. As the wormlike leptocephalus with its minute eyes and reduced fins metamorphoses, it starts burrowing. It subsequently rarely leaves the burrow and then only at night, until the onset of sexual maturity. At that point, the eyes and fins enlarge and the metamorphosing forms swim around at night to find a mate.

Garden eels (family Congridae, subfamily Heterocongrinae) have acquired their common name because they live in underwater colonies (gardens), rooted in the sea bottom where they wave around like plants. They are closely related to conger eels and are characterized by having a long thin body, small gill openings, and a hard fleshy point at the rear of the body, with the caudal fin hidden below the skin. They burrow tail-first, their modified tail regions being used to penetrate the sand. Their distribution is similar to that of spaghetti eels.

The cutthroat eel famiy (Synaphobranchidae) is subdivided into three groups: the arrowtooth or mustard eels (subfamily Ilyophinae or Dysommatinae) – with about 16 species, the cutthroat eels proper (subfammily Synaphobranchinae) – with around 9 species, and the monotypic Simenchelyinae, containing the Snubnose parasitic eel (*Simenchelys parasitica*). All have leptocephali with an unusual characteristic: telescopic eyes, indicating a very close relationship.

Dysomma species are characterized by enlarged teeth in the roof of the mouth (vomerine teeth). There is also a tendency towards the loss of the pectoral fin. These species tend to be laterally compressed and to taper evenly towards the tail. The swimbladder is very long and the nostrils are tubular. The main reason for including this otherwise unremarkable Indo-Pacific subfamily (Ilyophinae or Dysommatinae) is because the following history of the genus *Meadia* sums up some of the problems faced by ichthyologists.

A fish was bought in a fish market near Kochi in Japan. No one knew what it was and it became the type, that is, the specimen on which the scientific description was based, of the new species and genus – the Abyssal cutthroat eel (*Meadia abyssalis*). During the Second World War, this specimen was destroyed and no further specimens were caught until 1950, off Japan and Hawaii. Although the search for this brownish, scaleless eel was not especially assiduous, it does show how fate can so easily eradicate the only representatives available to the scientific world of a species that is probably common in its own particular habitat.

The cutthroat or synaphobranchid eels are characterized by their gill slits being almost joined up ventrally, a feature that gives them their common name. Rare in collections, 12 specimens were caught in the course of one expedition by the deep-sea research vessel, *Galathea*. They were sent, in three separate parcels for safety, to a researcher in New Zealand. Unfortunately, the second parcel, containing five specimens, was destroyed by fire in a mail-storage warehouse in Wellington in July 1961.

Cutthroat eels live in cold, deep waters, where temperatures average 5°C (41°F), and where they feed on crustaceans and fishes. (One specimen of *Synaphobranchus kaupii*, trawled at a depth of 1,000m (3,300ft), had eaten octopus eggs.) They are thick-set, scaled eels with large jaws and small teeth. The scale pattern is like that of the Snub-nosed parasitic eel, and several species have a characteristic nick in the ventral outline of the body just below the pectoral fin. The genus *Haptenchelys* is the odd one out in this subfamily because it lacks scales. The distribution of these eels is worldwide and all come from waters that are 400m to 2,000m deep (1,300–6,600ft). The adults are bottom dwellers and the larvae have telescopic eyes, but beyond that, little is known about their life history.

The third species in this assemblage (the Snub-nose parasitic eel) is believed to be the only member of its subfamily. However, as is often the case, there is no universal agreement on this matter. This deep-sea eel, which grows to around 60cm (24in), has been caught off the eastern coast of Canada, in the western and northern Pacific, off South Africa, and off New Zealand. Whether these discrete collecting localities reflect different local species or just the collecting effort is impossible to say. It seems to be plentiful because it is caught in large numbers, which could mean that it is a gregarious species or that it may need a particular habitat. It has a blunt head with a small transverse mouth. Jaws and jaw muscles are strong. The gill slits are short and lie horizontally below the pectoral fins. The dorsal fin starts well in advance of the anal fin. The body is cylindrical until the anus and compressed thereafter. The small scales are grouped and angled acutely to the lateral line.

The young eat small crustaceans, but the adult is probably at least partly parasitic or a scavenger. The stomachs of the first specimens found were full of fish flesh, as are those of most of the subsequent specimens. Evidence from the North Atlantic suggests that these eels feed on the flesh of halibut and other large fish, but probably only injured or moribund specimens.

The remaining families in the order Anguilliformes are all deep-sea eels and are modified for their environment in spectacular, if different, ways.

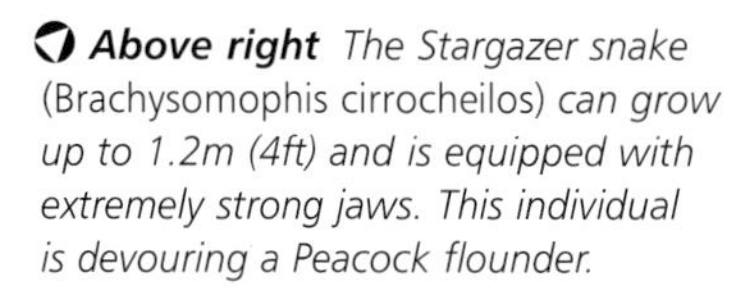

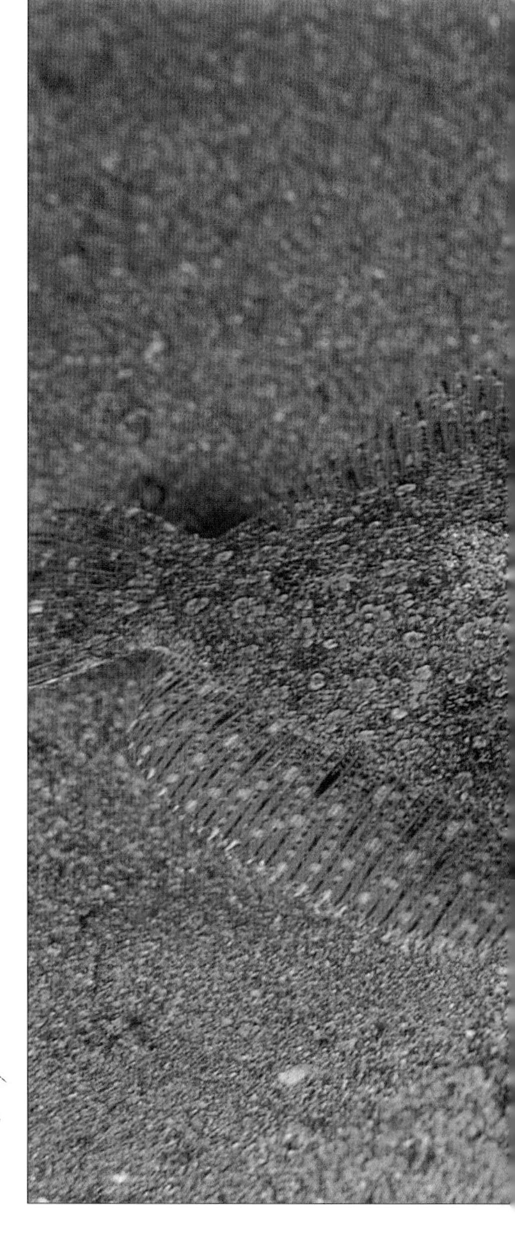

Above *Cutthroat eels have a single ventral gill slit; found in most oceans, some species of this family (Synaphobranchidae) are parasitic.*

Above right *The Stargazer snake* (Brachysomophis cirrocheilos) *can grow up to 1.2m (4ft) and is equipped with extremely strong jaws. This individual is devouring a Peacock flounder.*

Far right *A Gold-spotted snake eel* (Myrichthys ocellatus). *Members of this family (Ophichthidae), which have stiffened tails that allow them to burrow into sand, lie partially buried waiting to attack small fish and crustaceans.*

Left *The large-eye conger eel* (Ariosoma marginatum) *is a Pacific species. It hunts at night, with its head protruding obliquely from the substrate.*

Members of the family of snipe eels (Nemichthyidae) are large-eyed, extremely elongated animals. The rear of the body is little more than a skin-covered continuation of the spine. Here, the vertebrae are weak and poorly ossified, i.e. contain small quantities of bone matter; hence, few complete specimens are known. Undamaged specimens can have over 750 vertebrae, the highest number known. This family's common name alludes to its long and widely flaring jaws, fancifully believed to resemble the bill of the snipe. It had been thought that there were two groups of snipe eels, those with extremely long divergent jaws and those with short divergent jaws. However, enough specimens have been caught by now to show that mature males are short jawed, and mature females and juveniles of both sexes are long jawed. Before maturity, both the inside and outside of snipe eels' jaws are

covered with small, backward-pointing teeth, presenting a passable imitation of sandpaper. It seems likely that because adults have lost many teeth, they eat little; it is also possible that they may die after they have reproduced.

Scientists are still not sure how snipe eels feed because their jaws can close only posteriorly, i.e. the two tips of the snout cannot be brought together. A few observations of living eels from deep-sea research submarine vessels have shown that they spend their time hanging vertically, mouth down, in the water, either still or with bodies gently undulating. A few specimens caught with food still in their stomachs revealed that the major food source is deep-sea shrimps. Typical of these creatures are very long antennae and legs. It is therefore suggested that snipe eels feed by entanglement, i.e. that once the long antennae or legs of the crustacean become caught up in the teeth inside or outside the jaws, they are followed down to the body, which is then consumed.

There are three recognized genera of snipe eels (*Avocettina*, *Nemichthys*, and *Labichthys*). They are widely distributed in warmer parts of the oceans at depths down to 2,000m (6,600ft). Their leptocephali are easily identifed by their thinness and the long caudal filament, a precursor of the adult's prolongation of the caudal region. These larvae .metamorphose when they are 30cm (12in) long.

Gulper Eels and Allies

ORDER SACCOPHARYNGIFORMES

The remaining four families are all members of the deep-sea gulper or swallower eels. The two monotypic genera of bobtail snipe eels (Cyematidae) are especially distinctive. These fishes look rather like darts, with a long, thin point divided into diverging dorsal and ventral parts. These eels occur in all tropical and subtropical oceans at depths of 500–5,000m (1,600–16,500ft), but rarely to the greatest depth. They are laterally compressed, small fish that grow up to 15cm (6in) long. *Cyema* has a dark velvety skin and minute, but functional, eyes; *Neocyema* is bright red. Their biology is poorly known, although what is known of their breeding has some interesting aspects. Many eels breed in clearly circumscribed areas, where the physical conditions meet the stringent requirements of their physiology. By contrast, in the Atlantic, the bobtail snipe eels spawn over large stretches of the warmer parts of the north section of this ocean.

The leptocephalus is quite different to that of other eels. While the typical leptocephalus is a willow-leaf shape (save for minor variations), the leptocephalus of the curtailed snipe eels can be nearly as deep as long. The little evidence available suggests that these species spend at least two years in the larval phase.

On account of their mouths, three families of deep-sea eels: the single-jawed eels (Monognathidae), the swallowers (Saccopharyngidae), and the Gulper or Pelican eel (Eurypharyngidae) are sometimes grouped under the name "gulper eels."

Monognathids are known from about 14 species found in the Atlantic and Pacific oceans. Their scientific name, which means "one jawed," alludes to the fact that they lack upper jaw bones and have a conspicuous lower jaw that can be longer than the head. All known specimens are small, the largest individual being just 16cm (6.3in) long. Only one species, *Monognathus isaacsi*, has developed any pigment. It has been suggested that single-jawed eels are in fact juvenile swallowers (saccopharyngids), but adult saccopharyngids have many more vertebrae than the monognathids. However, the largest monognathids have the most vertebrae, so perhaps vertebral formation continues as the fish grows. This puzzle will be solved only when either a sexually mature monognathid is found or when a clear monognathid–saccopharyngid series of stages can be built up, i n other words, specimens showing a spectrum of changes from one type of eel to the other.

◐ ***Left*** *The elongated jaws of a snipe eel* (Nemichthys sp). *They bear filelike teeth that entangle with the long legs and antennae of deep-sea crustaceans.*

The swallowers are archetypal deep-sea fishes. They have huge mouths, elastic stomachs, toothed jaws, and a luminous organ on the tail. One major enigma is that only nine species are generally recognized (based on fewer than 100 specimens), while about 14 species of single-jawed eels are usually admitted. Thus, if single-jawed eels are juvenile swallowers, then there are at least five more swallowers awaiting discovery, or five single-jawed eel species have been wrongly described.

Only a few specimens have been seen alive. One fortunate catch involved a specimen of *Saccopharynx harrisoni* from the western Atlantic. This fish, although trawled in 1,700m (5,600ft) of water, was not in the net but entangled by its teeth in the mouth of the trawl, and so escaped being crushed by the weight of fish during the haul to the surface. Also, lacking a swimbladder, it did not suffer the fate of many deep-water fishes – the massive expansion of swimbladder gas as the water pressure reduces during the ascent to the surface.

For deep-sea fishes, the swallowers are large, growing to over 2m (6.5ft) long, but most of the body is tail. The huge mouth has, unsurprisingly, necessitated some morphological changes. The gill arches, for example, are a long way behind the skull and are dissociated into two separate lateral halves. The opercular (gill cover) bones are not developed and the gill chambers are incompletely covered by skin. These modifications mean that the respiratory mechanism of these eels is unlike that of other fishes. Another peculiarity of these fishes, apart from the absence of a pelvic girdle, is that the lateral line organs, instead of lying in a subcutaneous canal, stand out from the body on separate papillae or protuberances. It is surmised that this adaptation makes the fish more sensitive to vibrations in the water and enhances its chance of finding a suitable fish to cram into the distensible stomach. The escape of the prey, once in the mouth, is prevented by two rows of conical and curved teeth on the upper jaw and a single row of alternating large and small teeth on the lower jaw.

At the end of the long, tapering tail of all swallowers there is a complex luminous organ whose function is unknown. Indeed, the whole arrangement of the luminous organs is unusual. On top of the head are two grooves that run backward towards the tail. These contain a white, luminous tissue that glows with a pale light. The grooves separate to pass either side of the dorsal fin, each ray of which has two small, angled grooves containing a similar white tissue. The tail organ is confined to about the last 15cm (6in) of the body. Where the body is shallowest, there is a single, pink, club-shaped tentacle on the ventral surface. Further back, where the body is more rounded, there are six dorsal and seven ventral scarlet projections (papillae) on pigmentless mounds. The main part of the organ lies behind them and is a transparent leaflike structure with an ample supply of blood vessels. Its dorsal and ventral edges are prolonged and scarlet, whereas the organ is pink because of the blood vessels. The main organ is split into two zones by a band of black pigment with red spots. There are further fingerlike papillae even nearer the end of the body where the tail narrows and the black of the rest of the body is replaced by red and purple pigments. The small papillae produce a steady pink light, while the leaf-like tail organ can produce flashes on top of a constant reddish glow.

The organ may be a lure but this seems unlikely, since the contortions required to place the organ where it will act as a lure, even for such a long, thin, and presumably prehensile fish, would leave the body in a position where it would be unable to surge forward to grasp the prey.

Equally bizarre is the last member of this group of deep-sea eels, the Gulper or Pelican eel (*Eurypharynx pelecanoides*), the only member of its family (Eurypharyngidae). The mouth is bigger than that of the saccopharyngids, and the jaws can be up to 25 percent of the body length. The jaws are joined by a black elastic membrane. The eyes and brain are minute and confined to a very small area above the front of the mouth.

Almost nothing is known of the biology of the Gulper eel. The teeth are tiny, so it probably feeds on minute organisms. A small complex organ is present near the tail, but it is not known if it is luminous. There is no swimbladder, but there is an extensive liquid-filled system of vessels (the lymphatic system), which may aid buoyancy. The lateral line organs are external and show up as two or three papillae emerging from a small bump.

Gulper eel larvae metamorphose when less than 4cm (1.5in), but even at this diminutive size have already developed the huge mouth. Interestingly, the larvae live much nearer the surface of the sea (100–200m; 330–660ft) than the adults, which live at great depths. KEB/JD

Tarpon, Bonefish, and Eel Orders

Tarpons and Allies
Order Elopiformes

Most tropical seas. Rarely brackish or freshwater. Length maximum about 2m (6.5ft), weight maximum 160kg (350lb). About 8 species in 2 genera and 2 families including: **Indo-Pacific tarpon** (*Megalops cyprinoides*), **Tarpon** (*M. atlanticus*), **Tenpounder** or **Ladyfish** (*Elops saurus*).

Bonefishes, Spiny Eels, and Allies
Order Albuliformes

Most tropical seas (bonefishes) and deep seas worldwide (halosaurs and spiny eels). Length maximum about 2m (6.5ft). About 29 species in 8 genera and 3 families including: **Bonefish** (*Albula vulpes*), **Suckermouth spiny eel** (*Lipogenys gillii*).

Eels
Order Anguilliformes

All oceans, N America, Europe, E Africa, Madagascar, S India, Sri Lanka, SE Asia, Malay Archipelago, N and E Australia, New Zealand. Length maximum about 3m (10ft). About 738 species in c. 141 genera and c. 15 families including: **anguillid** or **freshwater eels** (family Anguillidae), including **American eel** (*Anguilla rostrata*), **European eel** (*A. anguilla*), **Japanese eel** (*A. japonica*); **snipe eels** (family Nemichthyidae); **conger eels** (family Congridae); **moray eels** (family Muraenidae), including **moray eels** (genus *Muraena*); **cutthroat** or **synaphobranchid eels** (family Synaphobranchidae); **worm** or **spaghetti eels** (family Moringuidae); **false morays** or **xenocongrid eels** (family Chlopsidae); **Snub-nosed parasitic eel** (subfamily Simenchelyinae, sole species *Simenchelys parasitica*).

Gulper Eel & Allies
Order Saccopharyngiformes

Mostly deep waters in tropical and temperate Atlantic, Indian, and Pacific oceans. Length maximum c. 2m (6.5ft). 26 species in 5 genera and 4 families including: **Gulper eel** (*Eurypharynx pelecanoides*).

Tarpons and allies
Tarpons Megalopidae
Tenpounders Elopidae
Bonefishes, spiny eels and allies
Bonefishes Albulidae
Halosaurs Halosauridae
Lipogenyidae
Notacanthids Notacanthidae
Eels
Anguillid or freshwater eels Anguillidae
Duckbill eels Nettastomatidae
Heterenchelyidae
Pike congers Muraenesocidae
Snipe eels Nemicthyidae
Worm or spaghetti eels Moringuidae
False morays or xenocongrid eels Chlopsidae
Gulper eel and allies
Curtailed or bobtail snipe eels Cyematidae
Gulper eels Eurpharyngidae
Saccopharyngids Saccopharyngidae

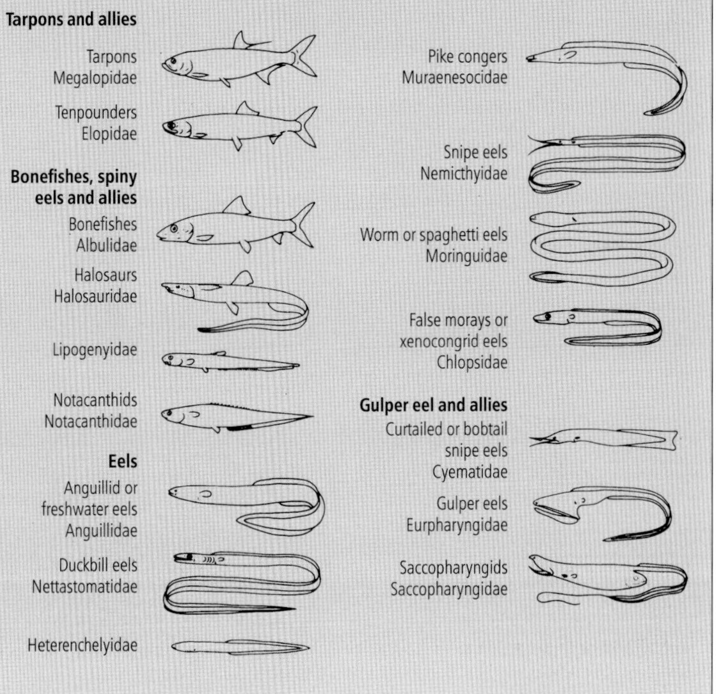

FISH OUT OF WATER

How fishes are adapted for life on land

THAT HUMANS EXIST ON EARTH REFLECTS THE fact that, in the Devonian period, more than 350 million years ago, some ancient fishes emerged from the water, adapted progressively to a terrestrial environment, evolved, and . . . one of the results is humans. Some bony fishes now leave the water for various lengths of time, although it is not suggested that the end point of their excursions will be as dramatic as that of their ancestors.

It should be pointed out that the ability of a fish to breathe air does not equate with its ability to leave water. The spectacular attributes of the lungfishes, among others, are dealt with elsewhere . This section concerns fishes that actively travel out of water. Naturally, there are degrees of extra-aquatic activity, ranging from small fishes that skitter along the water's surface momentarily to escape predation, to the Pacific moray eel which may spend up to ten minutes out of water, and others that spend far longer on land.

What problems do they face? Several, including (not in order of severity): respiration, temperature control, vision, desiccation, and locomotion.

An initial problem facing fishes leaving water is that the surface area of the delicate gill filaments is reduced since they clamp together because they are no longer kept separate by the buoyancy of water. Drying out of the gills also occurs. Consequently, either other means of respiration have to be present or ways of protecting the gills must exist. This usually involves the development of a sac into which the air is sucked; such sacs or cavities have a moist lining that is richly endowed with blood vessels (vascularized). As is always the case with fishes, there are, of course, exceptions. The Chilean cling fish (*Sicyaces sanguineus*), for example, which spends a substantial part of its life out of water, has a vascularized layer of skin in front of the sucker formed from its ventral fins. When it needs oxygen, it raises the front of its body off the rocks and exposes this skin patch to the air. More "orthodox" are the anabantoids or labyrinth fishes of the suborder Anabantoidei, which include the Climbing perch (*Anabas testudineus)* and the totally unrelated Walking catfish (*Clarias batrachus*) of the family Clariidae. Both these types of fishes have pouches above the gill chamber, with linings expanded into convoluted shapes that increase the surface area. The Electric eel (*Electrophorus electricus*), which spends only a small part of its life out of water, can use its gills for aerial respiration, while the mudskippers (*Periophthalmus* spp.), which spend a lot of their life out of water, can use their skin.

Fishes on land have a problem in keeping cool, a difficulty that is compounded because most

Above *On their migration back to the sea, mature European eels* (Anguilla anguilla) *will cross wet ground if necessary.*

Left *Mudskippers move by "crutching" – the fish swings its pectoral fins forward while supporting its body weight on the pelvic fins. Then, by pressing downward and backward with the pectorals, the body is both lifted and drawn forward.*

"semiterrestrial" fish live in tropical areas. Although living in a coolish climate, one technique for keeping cool is displayed, once more, by the Chilean cling fish: it does not produce heat by muscular activity. At its best, this species has been described, when out of water, as "inactive and difficult to tell when dead." It also lives in a shore zone where waves wash over it from time to time.

The length of time it can stay on land depends very much on the weather. When cloudy, it can stay out for about two days and lose only 10 percent of its body water, while in captivity it has been shown that it takes about a 25 percent loss of body water to cause death. Chilean cling fish that die in the wild have lost little water, but the sun has driven up their body temperatures to some 24°C (75°F), which is fatal for the species. Adult Chilean cling fish do not breathe continuously through the vascularized skin in front of the sucker, but alternate this with holding air in the mouth and pharynx and exchanging gases through their moist lining.

When on land, fish have to see and, unless modifications occur, they would all be shortsighted. Two physical responses are possible: the first is to change the shape of the lens, and/or the second, to keep the lens spherical and change the shape of the cornea. Mudskippers have done both and have such excellent vision that they can catch insects on the wing.

Eyes are delicate and must, obviously, be protected on land. Mudskippers achieve this by means of a thick layer of clear skin that protects the eye against physical damage. To keep their eyes lubricated (since fish lack tear ducts), mudskippers roll them back into the moist sockets. The walking catfishes (*Clarias* spp.), which probably cannot see farther than 2m (6.5ft) on land, protect their eyes by a thick layer of clear skin and by confining them to within the body contour – that is, they do not protrude from the side of the head. Since conditions are brighter on land than in water, to cope with excess light hitting the retina, some fishes – notably those of the blenny genus *Mnierpes* – have developed a layer of pigment – the fishes' equivalent of sunglasses.

For eels and eel-like fishes, locomotion on land is just like swimming. *Clarias* species, for instance, employ waving or swimming movements of the body, but supplement this with the use of the pectoral fin spines as bracers. An African species, the Sharptooth or North African catfish (*Clarias gariepinus*), burrows in drying pools and comes out at night to feed. Migrations of up to a thousand individuals have been recorded and it is suggested that the barbels (whiskers) are used to keep in touch. The Walking catfish originally from Southeast Asia, was introduced into Florida, where it has flourished and its ambulatory habits have, not surprisingly, caused much disquiet among many residents of that state, who occasionally find them "walking" across roads. KEB/JD

Right *The Climbing perch of southern Asia can survive for days or even weeks out of water if its auxiliary breathing mechanism – the "labyrinth organ" – is kept moist.*

Herrings and Anchovies

1

2

SOME OF THE MOST FAMILIAR FISH IN THE world belong to an order comprising just four living families: herrings and anchovies, plus the denticle and wolf herrings. Their influence has been enormous. The economically highly valuable North Atlantic herring has been the subject of wars, and its migrations have brought down governments and caused the dissolution of states.

Herrings and their relatives (Clupeidae) are a cosmopolitan, largely marine group of fishes. Compared with many groups, they are beautifully coherent and easy to diagnose. Freshwater representatives of the herrings and shads occur in the eastern USA, the Amazon basin, West and Central Africa, eastern Australia, and occasionally and sporadically elsewhere. The anchovies are coastal forms in all temperate and tropical regions, with freshwater species in the Amazon and Southeast Asia. The wolf herrings are solely marine, and the single species in the Denticipitidae lives only in freshwater, occurring in a few rivers in West Africa.

Herrings and Shads

FAMILY CLUPEIDAE

Within this group, the herrings and shads family contains the largest number of species (about 214). They are of great commercial importance. Thanks to their former abundance, highly nutritious flesh, and shoaling habits, they became a prime target for fishing fleets. In 1936–37 members of this family made up 37.3 percent by weight of all the fish caught in the world. About half of this weight was from one species, the Pacific Sardine.

In the North Atlantic, the Herring has long been exploited. In 709 AD salted herring were exported from East Anglia in England to the Frisian Islands; the fisheries even appear in Domesday Book (1086). The great advantage of this herring (and others) was that it could be preserved in a variety of ways: pickled in brine, salted, or hot and cold smoked (either salted first and split or not, resulting variously in kippers and red herrings). These preservation techniques were devised to keep fish edible before the advent of freezing and canning.

Herrings spawn in shoals in the warmer months, laying a mat of sticky eggs on the sea floor. After hatching, the young become pelagic (free-swimming in the upper and middle layers). Their food, at all stages in their life history, consists of plankton, especially small crustaceans and the larval stages of larger crustaceans. Herrings can swim at a maximum speed of 5.8km/h (3.6mph).

Sprats are small relatives of herrings. The fish often marketed as sardines are the young of the Pilchard, another herring species. Whitebait are the young of both the North Atlantic herring and the Sprat.

Shads (*Alosa* spp.) are larger members of the herring clan. An American Atlantic species (*A. sapidissima*) reaches nearly 80cm (31in) in length. In 1871, they were introduced into rivers draining into the Pacific and are now widespread along the Pacific coast of the USA, where they grow to 90cm (35in). In the European North Atlantic there are two species, both now scarce, the Allis shad and the Twaite shad. The latter species has some nonmigratory dwarf populations in the Lakes of Killarney (Eire) and some Italian lakes. Freshwater populations of the American alewife are also known.

The external appearance of members of the herring family is very similar. Most are silvery (darker on the back), with no scales on the head. They lack spines in the fins and have scales that are very easily shed (deciduous scales).

Anchovies

FAMILY ENGRAULIDAE

Species in the anchovy family are longer and thinner than many clupeids. They have a large mouth, a pointed, overhanging snout, and a round belly without the scute-covered keel that is present in the clupeids.

Engraulis mordax from the eastern Pacific grows to about 18cm (7in). It is in great demand, more as a source of oil and as meal than as a delicacy. Only a very small proportion of the catch is canned or made into paste. It was once far commoner than it is now. There is, for example, a record of a single set of purse-seine nets catching over 200 tonnes in November 1933.

Further south, the Peruvian anchovetta, which is abundant in the food-rich Humboldt Current, is one of the major resources of South American countries on the Pacific seaboard. It is caught in vast shoals off Ecuador, Chile, and especially Peru, but its numbers decline drastically during occurrences of the *El Niño* Southern Oscillation climatic event. The Anchovy is found in the warmer parts of the North Atlantic and the Mediterranean. Before being sold, in whatever form, anchovies are packed into barrels with salt and kept at 30°C (86°F) for three months until the flesh has turned red. Anchovies rarely grow to more than 20cm (8in) long, and may live for seven years, but are usually sexually mature at the end of the first or second year. They are plankton feeders and eat the edible contents of the seawater filtered out by their long, thin gill rakers.

Wolf and Denticle Herrings

FAMILIES CHIROCENTRIDAE, DENTICIPITIDAE

The family Chirocentridae houses only one genus and two species, the wolf herrings. These giants among the herrings grow to over 1m (39in) long and live in the tropical parts of the Indo-Pacific Ocean (west to South Africa and the Red Sea and from Japan to New South Wales). They are elongate fish, strongly compressed, and with a sharp belly keel. They have large, fanglike teeth on their jaws and smaller teeth on the tongue and roof of the mouth. Avid hunters, wolf herrings make prodigious leaps. These species are not used for food because they have numerous small intermuscular bones and also, if caught, struggle violently and snap at anything, including fishermen. Within the intestine, these fish have a spiral valve, the function of which is to increase the absorbtive surface; spiral valves are very rare in bony fishes.

The Denticle herring is the only representative of the Denticipitidae. It measures 8cm (3in) long and is found only in the fastest flowing parts of a few rivers near the Nigerian border with Dahomey. It is silvery with a dark stripe along the side. A seemingly inconspicuous and insignificant fish, it is noteworthy for the presence of a large number of toothlike denticles over the head and front part of the body, hence its common name. Their significance and function are unknown. Fossils, hardly distinguishable from the living species, have been found in former lake deposits in Tanzania – about 20–25 million years old – and named *Palaeodenticeps tanganikae*. The living Denticle herring is believed to be the most primitive member of the clupeomorphs. KEB/JD

Right *Atlantic herrings schooling in a cave in Scotland. Schooling, along with their silvery sides, excellent hearing, and fast escape responses help these fish to fend off predators. Atlantic herrings seek out food by using only their visual sense.*

Left *Representative species of herrings and anchovies:* ***1*** *The Sprat* (Sprattus sprattus) *feeds on planktonic crustaceans;* ***2*** *The North Atlantic herring* (Clupea harengus)*;* ***3*** *Pilchards* (Sardina pilchardus) *spawn in open sea or near the coast, producing 50,000–60,000 eggs;* ***4*** *During spawning, anchovies* (Engraulis encrasiocolus) *in some areas venture into lakes, estuaries, and lagoons;* ***5*** *A pair of Wolf herrings* (Chirocentrus dorab).

FACTFILE

HERRINGS AND ANCHOVIES

Subdivision: Clupeomorpha

Order: Clupeiformes

About 357 species in 83 genera and 4 extant (living) families.

HERRINGS
Family Clupeidae
About 214 species in about 56 genera. Oceans worldwide; about 50 species occur in freshwaters in Africa. **Length:** maximum 90cm (35in). Species include: **Allis shad** (*Alosa alosa*), **Alewife** (*A. pseudoharengus*), **Twaite shad** (*A. fallax*), *A. sapidissima*, **North Atlantic herring** (*Clupea harengus*), **Pilchard** (*Sardina pilchardus*), **Sprat** (*Sprattus sprattus*), **Pacific sardine** (*Sardinops sagax*). **Conservation status**: the Alabama shad (*Alosa alabamae*) is Endangered; the Liparia (*A. macedonica*) and Venezuelan herring (*Jenkinsia parvula*) are Vulnerable.

ANCHOVIES
Family Engraulidae
About 140 species in 16 genera. Oceans worldwide; some 17 species occur in freshwater or brackish habitats. **Length:** maximum about 50cm (20in). Species include: **Peruvian anchovetta** (*Engraulis ringens*) and **Anchovy** (*E. encrasicolus*).

WOLF HERRINGS
Family Chirocentridae
2 species in 1 genus: *Chirocentrus dorab* and *C. nudus*. Indian and W Pacific oceans. **Length:** maximum 1m (39in).

DENTICLE HERRING
Denticeps clupeoides
Family Denticipitidae. Sole member of family. Rivers in W Africa (near border of Nigeria with Benin). **Length:** maximum 8cm (3in).

Bonytongues and Allies

1

2

4

ESSENTIALLY TROPICAL FRESHWATER FISHES *bonytongues, butterflyfish, mooneyes, featherbacks, and elephantnoses make up the diverse group known as osteoglossomorphs. Some featherfins, however, may also occur in brackish water. Although they possess toothed jaws, the main bite is exerted by teeth on bones in the tongue pressing against teeth on the roof of the mouth. This feature is responsible for the general name sometimes applied to the whole group – the bonytongues.*

Osteoglossomorphs share a number of other structural characteristics, including complex ornamentation or markings of the scales and, unusually, an intestinal arrangement in which part of the gut is arranged on the left side of the gullet and stomach (it passes on the right in most fishes).

A Wide Spectrum

OSTEOGLOSSOMORPH FAMILIES

The **bonytongues** are moderate to very large fishes with prominent eyes and scales and dorsal and anal fins placed well back on their long bodies. The giant Amazonian arapaima is reputed to grow to lengths of 5m (16ft) and weights of 170kg (375lb). These estimates have not been verified, but if they were authenticated, then this species would rank as the giant of freshwater fishes. There is no doubt that it grows to 3m (10ft) and weighs up to 100kg (220lb) in the wild – still a huge size.

In addition to using its gills in the normal way to breathe, the Arapaima is also an air-breather, absorbing oxygen via its swimbladder, a structure that is joined to the throat and has a lunglike lining. A similar bladder condition prevails in the African or Nile arowana (*Heterotis niloticus*), which also has an accessory respiratory organ above the gills. How much these fish use air is uncertain, but both can penetrate inhospitable, oxygen-poor swamps for spawning. Both fishes build nests and guard their young. The Nile arowana constructs a walled nest of broken vegetation measuring about 1.2m (4ft) in diameter. By contrast, the two South American arowanas (*Osteoglossum bicirrhosum* and *O. ferreirae*), the Asian Dragon fish, and the Australian saratogas brood eggs and young in their mouths. In the South American arowanas and Dragon fish, the brooding is carried out by the male. In *S. leichardti*, it is the female that is reported as being the brooder; no details are currently available for *S. jardinii*.

Bonytongues are carnivores that eat insects, other fish, amphibians, rodents, and even birds, snakes, and bats. *Heterotis*, however, consumes mud, plankton, and plant detritus.

At the other end of the size spectrum of osteoglossomorphs lies the West African **Butterflyfish,** which is only about 6–10cm (2.4–4in) long. This species, also an air breather, inhabits grassy swamps where it swims close to the surface, trailing below the amazingly long and separated rays of its pelvic fins as it feeds on floating insects and fish. With its ability to leap high out of the water and to skit over the surface, the butterflyfish can also feed on flying insects.

Mooneyes are herringlike fishes of modest size which extend the distribution of osteoglossomorphs into North America. There are only two species in this family: the Mooneye itself and the Goldeye. Both are freshwater species with strongly compressed bodies (that is, flattened from side to side), and silvery scales; they also possess a keel-like belly. Mooneyes are predatory fishes that feed on other fishes and insects. Both are fished commercially, especially the Goldeye.

Featherbacks are laterally compressed fishes that swim by undulating a very long anal fin, extending from the tiny pelvic fins to the tip of the tail. The Clown knifefish, one of two Asian species, can grow to 1m (3.3ft) in length and

◐ **Left** *Representative species of bonytongues and allies:* ***1*** *A young specimen of an arowana (genus* Osteoglossum*) at the water's surface;* ***2*** *the* Butterflyfish (Pantodon buchholzi) *was given its name after leaping into a butterfly collector's net;* ***3*** *the* Arapaima (Arapaima gigas) *builds a nest about 0.5m (1.6ft) across in the sand for its young;* ***4*** *Bronze* featherback (Notopterus notopterus)*;* ***5*** *the Churchill* (Petrocephalus catastoma) *is a species of snoutless elephantnose and* ***6*** Campylomormyrus rhynchophorus *is an example of an elephantnose with a long snout.*

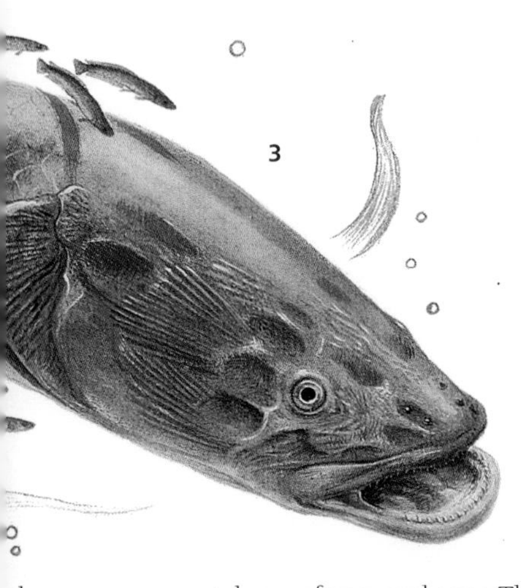

FACTFILE

BONYTONGUES AND ALLIES

Subdivision: Osteoglossomorpha

Order: Osteoglossiformes

About 217 species in 29 genera and 6 families.

Distribution N and S America, Africa, SE Asia, Australia in rivers, swamps, lakes.
Size Length 6cm–3m (2.4in–10ft); **weight** up 100kg (220lb).
Diet Varied, including insects, fish, plankton.

BONYTONGUES
Family Osteoglossidae
7 species in 4 genera. S America, Africa, Malaysia, New Guinea, Australia. Species: **Arapaima** (*Arapaima gigas*), **Nile arowana** (*Heterotis niloticus*), **Dragon fish** (*Scleropages formosus*), **Australian saratogas** (*S. jardinii, S. leichardti*), **South Anerican arowanas** (*Osteoglossum* spp.). **Conservation status**: Dragon fish Endangered.

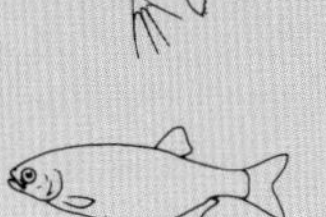

BUTTERFLYFISH *Pantodon buchholzi*
Sole member of family Pantodontidae. W Africa.

MOONEYES Family Hiodontidae
2 species of the genus *Hiodon* – the Mooneye (*H. tergisus*) and Goldeye (*H. alosoides*). N America.

FEATHERBACKS
Family Notopteridae
8 species in 4 genera. Africa, SE Asia. Species include: **Bronze featherback** (*Notopterus notopterus*), **African knifefish** (*Xenomystus nigri*).

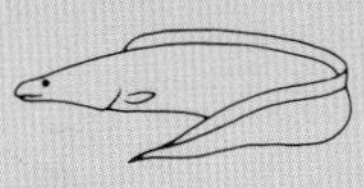

ELEPHANTNOSES
Family Mormyridae
Nearly 200 species in 18 genera. Africa. Species include: **Peter's elephantnose** (*Gnathonemus petersii*), **Elephantnose** (*Campylomormyrus tamandua*).

ABA ABA *Gymnarchus niloticus*
Sole member of family Gymnarchidae. Africa.

shows some parental care of spawned eggs. This care consists of the male fanning the eggs (which are laid on a sunken branch or other solid surface) and defending them during their 5–6 day development. Other featherbacks are smaller than the Clown knifefish. The African knifefishes, *Papyrocranus* and *Xenomystus*, also have accessory respiratory structures above the gills and can inhabit swampy pools. All of these fishes have large mouths and predatory habits, feeding largely on aquatic invertebrates and other smaller fishes.

The **elephantnoses** or mormyrids make significant contributions to the fish stocks of many African lakes, rivers, and floodpools. Most are bottom dwellers feeding on worms, insects, and mollusks. Some species are long-snouted, but, although there is a tendency in others for a forward and downward prolongation of the snout region, within the family head shapes are highly variable. Some species (many of which are known as "whales") have no snout at all. Small to medium in size, all elephantnoses have small mouths, eyes, gill openings, and scales. Dorsal and anal fins are set well back and the deeply forked caudal fin has an exceptionally narrow peduncle. The muscles are modified to form electric organs that emit a continuous field of weak discharges at varying frequencies around the fish. Moreover, there are electroreception centers in part of the enlarged cerebellum of the brain (giving mormyrids the largest brains to be found among lower vertebrate animals). The cerebellum is so large that it extends over the surface of the forebrain. The electrosensitive system acts as a sort of radar, detecting distortions in the electrical field from objects coming within it. This would seem to be an ideal adaptation for nocturnal activities, including social and breeding interactions, in murky waters.

Much early research into electrogenic activity was conducted on the **Aba Aba**. Recently separated by systematists from the mormyrids and assigned to its own family (Gymnarchidae), it is a large, predatory osteoglossomorph that is reported as growing to 1.6m (65in). The Aba Aba has a remarkable shape in that it lacks anal and caudal fins and moves by undulating a dorsal fin that extends almost the entire length of its eel-like body. The swimbladder serves as a lung, and *Gymnarchus* constructs a 1-m (39-in) flask-shaped, often floating, nest of grass which it is reputed to defend with vigor. Around 1,000 eggs are laid, which take about five days to hatch.

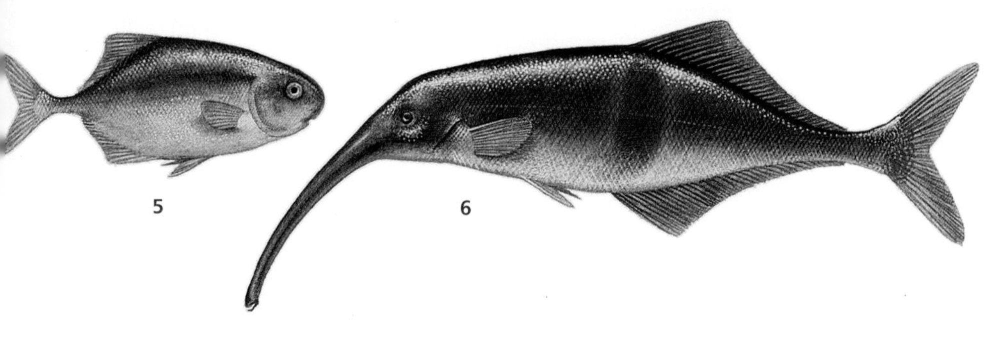

Popular but Problematic

CONSERVATION AND ENVIRONMENT

Apart from their scientific interest, humans are involved with osteoglossomorphs at various levels. There are capture fisheries for Arapaima and arowanas in South America, and for the Nile arowana and the Aba Aba in West Africa and the upper White Nile. Larger mormyrids are more widely fished in Africa, although they are not universally acceptable as food. For example, many women in East Africa avoid eating them due to the superstition that their children might be born with elephantine snouts. Bonytongues, mooneyes, and *Gymnarchus* are all rated as good sport fish by anglers. In pond culture, good growth rates have been achieved, but in the case of the Nile arowana, harvesting may be thwarted by their superb ability to leap the seine net, behavior also exhibited by its South American, Asian, and Australian cousins.

The Butterflyfish, featherbacks, and a number of small mormyrids are particularly interesting for aquarists because of their specialized features, but they are not popular, since breeding them in captivity is difficult. Some of the larger species, most notably the Arapaima, are popular in public aquaria. When first introduced into an aquarium, however, large arapaimas can cause problems by trying to swim through the walls of tanks; they too are hard to breed in captivity. However, pond-breeding of several species, including the Arapaima, is now being achieved on a regular basis. In the Far East, the Dragon fish has been bred in captivity for many years and the best, large fish command astounding prices because of the symbolism attached to them. RGB/JD

Pike, Salmon, Argentines, and Allies

MEMBERS OF THE ORDERS INCLUDING pike, smelts, salmon, and their allies are of great interest to many people. They contain prize angling fishes, important food fishes, and fishes of great interest to biologists for their migratory habits. The last include many examples of diadromy – the phenomenon of migration between freshwater and salt water. Diadromous species that spawn in freshwaters are said to be anadromous, while those spawning in the sea are said to be catadromous.

The superorder Protacanthopterygii was originally created to contain primitive teleostean fishes such as pike, salmon, lantern fishes, whale fishes, and galaxiids. However, research has shown that this is an artificial grouping united largely on primitive characters that do not indicate true relationships. Consequently the superorder and its orders and families have often been revised and continue to be unstable. The classification currently most widely accepted comprises three orders, the Esociformes, Osmeriformes, and Salmoniformes (see Salmon, Pike, and Related Families).

Pike

FAMILY ESOCIDAE

The pike are renowned sport fishes, some of which grow to a very large size and are known for their fighting abilities once hooked. They are powerful and aggressive predators, mostly on other fish, and generally lead solitary lives. In many fish communities their feeding has a major impact on the abundance and behavior of smaller species.

The distribution of pike is basically circumpolar. Of the five species, the Northern pike is widespread in North America, Europe, and Asia, but the others are more local in range, with one in Siberia and three in North America. The biggest of them is the muskellunge, or muskie, which may reach more than 30kg (66lb) and 1.5m (5ft) in length, while the Northern pike may exceed 20kg (44lb) and 1m (3.3ft).

The North American pickerels are now regarded by most experts as two species – the Chain pickerel, and two subsubspecies of *E. americanus* known as the Redfin pickerel and the Grass pickerel. All are small fishes, the Redfin and Grass pickerels rarely growing to more than 30cm (1ft) long. The Chain pickerel is somewhat larger and, in areas where it cohabits with the others, may hybridize and produce fishes that are often claimed to be Redfin or Chain pickerels of record size. It is not always easy to distinguish the two or three species, and even more difficult to establish the true nature of the hybrids.

All pike species are similar in appearance, being slender, elongate fish somewhat laterally compressed with a long, flattened, almost alligator-like snout. The mouth itself is also long and has large pointed teeth. Perhaps the most distinctive feature of pike is the clustering of their dorsal and anal fins. This concentration of finnage at the rear enables them to accelerate rapidly and has endowed them, and other fish with similar fin arrangements, with the name of "lurking" or "ambushing predators." They skulk among vegetation around the margins of lakes and rivers and surge out to catch passing prey.

Pike are mainly freshwater fish, but a few venture into mildly saline waters in Canadian lakes and the Baltic Sea. Spawning takes place among vegetation in still or gently flowing marginal shallows during the early spring. The female pairs with a male and the two spend several hours together releasing and fertilizing small batches of quite large eggs (2.3–3.0mm, about 0.1in, in diameter). A large female may lay several hundred thousand eggs. Factors known to influence the success of pike reproduction include the degree of inundation by high water levels of terrestrial vegetation, which acts as excellent spawning habitat, and the warmth of the late summer. Young pike are also subject to considerable cannibalism from both their peers and their elders. From the moment they hatch, young pike are predators, initially eating insects and small crustaceans but very soon becoming fish eaters like the adults. Large pike may also occasionally take small mammals and birds.

Pike are highly prized by anglers, especially in Europe. In North America, where salmon species are more diverse, abundant, and freely available, the popularity of pike is not as great, although many fishermen have the aspiration of catching a large muskellunge.

Right *The Redfin pickerel* (Esox americanus americanus) *is widespread throughout North America and inhabits swamps, lakes, and backwaters. Its snout is shorter than that of many other* Esox *species.*

Above *Along with many other members of the order Salmoniformes, the European Brown trout* (Salmo trutta trutta) *is greatly sought after as a food fish.*

Mudminnows

FAMILY UMBRIDAE

The mudminnows are closely related to pike. They were once included in the families Daliidae and Novumbridae, but are now united in one family. Their present disjunct distribution is a relict one, and fossils found in Europe and North America show that their former distribution was much like that of the extant pike. They now occur in eastern Europe, in the Danube and Dniester river systems (European mudminnow, *Umbra krameri*); in eastern North America (Southern mudminnow, *Umbra pygmaea*, which has been introduced to Europe); in the Chehalis River, Washington State, USA (*Novumbra hubbsi*); in the Great Lakes and Mississippi drainage (*Umbra limi*); and Alaska and Eastern Siberia (Alaska blackfish, *Dallia pectoralis*).

Mudminnows are small fishes, rarely exceeding 15cm (6in) in length. The caudal fin is rounded and the dorsal and anal fins are set far back, as in the pike. Their mottled dark brown or olivaceous coloring is cryptic. All are carnivorous and feed on small invertebrates and larval fishes, which they seize by making rapid lunges. They are sluggish, retiring fishes, hiding among vegetation waiting for the prey to come within striking distance.

Mudminnows are capable of living in high densities in poorly oxygenated swampy areas as they can utilize atmospheric oxygen. They are tolerant of drought, which they escape by burrowing into soft mud and ooze, and also tolerate cold, especially the Alaska blackfish, much as one might expect from where it lives. In many books there are accounts of the Alaska blackfish being able to withstand freezing. This frequently repeated untruth seems to originate from a book by L. M. Turner in 1886. In his *Contributions to the Natural History of Alaska,* he wrote "The vitality of this fish is astonishing. They will remain in . . . grass buckets for weeks, and when brought in the house and thawed out they will be as lively as ever. The pieces which are thrown to the ravenous dogs are eagerly swallowed; the animal heat of the dog's stomach thaws the fish out, whereupon its movements cause the dog to vomit it up alive. This I have *seen* . . ." Sadly for sensation, properly controlled experiments have shown that, although the Alaska blackfish is capable of surviving at very low temperatures, it cannot withstand freezing or being icebound.

FACTFILE

PIKES, SMELTS, SALMONS...

Superorder: Protacanthopterygii

Orders: Esociformes, Osmeriformes, Salmoniformes

Over 300 species in 89 genera and 16 families.

Distribution Worldwide in seas and freshwater.

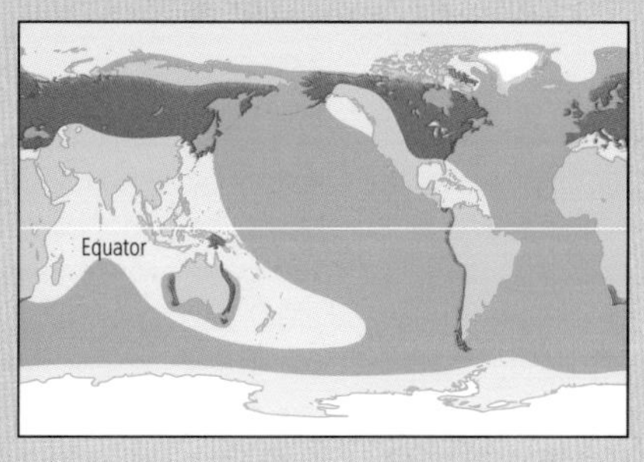

See Salmon, Pike, and Related Families ▷

Argentines

FAMILY ARGENTINIDAE

The argentines take their common name from their silvery sheen, and comprise the two genera, *Argentina* and *Glossanodon*. They are exclusively marine species. They are also called herring smelts and reveal their osmeroid relationships by having an adipose fin and by superficially resembling the anadromous freshwater osmerid smelts. They are mostly small, usually less than 30cm (12in). They are elongate, slender, silvery fish, darker on the back, usually lacking distinctive markings or coloration. They have scales and a well-developed, rather flaglike dorsal fin high on the back, usually in front of the ventral fins, which are in the abdominal region. The head is longish with a pointed snout, the mouth small and terminal. The eyes are very large, a common feature of fish living at some depth in the sea. Although not well known, argentines are widespread in most oceans of the world; they are found down to about 1,000m (3,300ft), mostly a few hundred meters down, where they probably live in aggregations, if not in coordinated schools. From their teeth and stomach contents it is known that they are carnivores, living on small crustaceans, worms, and other prey. Their small size and the depths at which they occur mean that they are not of prime importance to commercial fisheries, but they are taken for processing. They also act as a forage fish for larger and more significant food fishes of deep waters.

Although the adults are deepwater fish, the eggs and young are found in the surface waters of the ocean, usually over the continental shelf. Their eggs are 3–3.5mm (0.1in) in diameter. They are slow-growing fish and are long lived, one estimate being that they may live for 20 years or more.

Microstomatids

FAMILY MICROSTOMATIDAE

Despite its wide distribution, the family Microstomatidae has been only poorly studied. The Slender argentine is a greatly elongated, mesopelagic species that probably feeds on zooplankton and is generally solitary. It spawns throughout the year in the Mediterranean Sea. The Stout argentine is also greatly elongated and found in continental slope regions, with planktonic eggs and larvae. *Xenophthalmichthys danae* is bizarre, with tubular eyes that look forward like a pair of car headlights.

Deepsea Smelts

FAMILY BATHYLAGIDAE

The family Bathylagidae consists of small, dark, large-eyed fishes found worldwide, with many species showing daily vertical migrations to depths of about 3,500m (11 ,500ft). Their diets

Right *A river predator par excellence, the Northern pike* (Esox lucius) *has a streamlined body and an array of sharp teeth with which to devour its prey. It is the most widespread freshwater species in the world.*

are usually dominated by plankton, although euphausiids (luminescent shrimplike crustaceans) are taken by some species. Both eggs and larvae are typically planktonic.

Barreleyes

FAMILY OPISTHOPROCTIDAE

The family Opisthoproctidae contains six genera and ten species in tropical and temperate seas down to about 1,000m (3,300ft). All species have tubular eyes. In the deep-bodied species of the genera *Opisthoproctus*, *Macropinna,* and *Winteria* the eyes point directly upwards. In contrast, in the fragile, slender-bodied forms of the genera *Dolichopteryx* and *Bathylynchnops* the eyes point forwards. The remaining species, *Rhynchnohyalus natalensis*, which is known from only three examples, is apparently intermediate.

The Mirrorbelly (*Opisthoproctus grimaldii*) grows to about 10cm (4in) long and lives in the North Atlantic. Its body is silvery, with dark spots on the back. The sides of the body are covered with very deep scales. A swimbladder is present. The skull is so transparent that in live or freshly dead specimens, the brain can be seen clearly behind the eyes. The spherical lenses in the tubular eyes are pale green. The ventral edge of the body is flattened and expanded into a shallow trough known as the sole. The base of the sole is silvery but covered with large thin scales and a dark pigment. The sole is believed to act as a reflector for the light produced by bacteria in a gland near the anus. The light from the gland passes through a lens and is then reflected downwards by a light-guide chamber just above the flattened part.

The Barrel-eye (*Opisthoproctus soleatus*), a more widespread species, has a pigmentation pattern on its sole different from that of its only congener *O. grimaldii*, so it is thought that the sole enables species recognition in the areas where the two species live together. The upward pointing, tubular eyes, which afford excellent binocular vision, would easily be able to perceive the light directed downwards. The main food of the North Atlantic species (*O. grimaldii*) seems to be small, jellyfish-like organisms.

The Brownsnout spookfish (*Dolichopteryx longipes*) is slender and very fragile. It is also very rare. The fins are elongated like a filament and there is no swimbladder. The muscles are poorly developed; indeed, it has lost so much of its ventral musculature that the gut is enclosed only by transparent skin. It is, then, probably a very poor swimmer, and the tubular eyes may be advantageous in avoiding predators. Unlike *Opisthoproctus*, there is a light-producing organ associated with the eye. The species has been caught, infrequently, in all tropical and subtropical oceans between 350 and 2,700m (1,150–8,860ft) deep.

Slickheads, Leptochilichthyids

FAMILIES ALEPOCEPHALIDAE, LEPTOCHILICHTHYIDAE

The family Alepocephalidae comprises the two subfamilies Bathylaconinae and Alepocephalinae, which together contain at least 63 species in approximately 24 genera. *Bathyprion danae* is a pikelike fish living at depths of some 2,500m (8,200ft) in the South Indian Ocean and the North and Southeast Atlantic. It probably hovers waiting for very small fishes or crustaceans to come close when, so its fin positions would suggest, it surges forward to grab the prey.

Above Rhynchohyalus natalensis, *a barreleye, emits bacterial bioluminescence The uncanny appearance of the barreleyes explains their other name – spookfishes.*

Right *The capelin* (Mallotus villosus), *a species of smelt, congregates in large shoals, which move inshore to spawn in the spring. In the process, many become stranded, as on this beach in Newfoundland, Canada.*

The family Leptochilichthyidae is represented only by the genus *Leptochilichthys,* which contains just three rare species. Their biology is unknown.

Tubeshoulders

FAMILY PLATYTROCTIDAE (SEARSIIDAE)

The common name of slickheads originates from the fact that the head is covered with a smooth skin, whereas the body has large scales. Most species are dark brown, violet, or black. Light organs are rare among slickheads, but one genus, *Xenodermichthys*, is distinguished by the presence of tiny, raised light organs on the underside of the head and body.

The members of the family Platytroctidae, which was until recently known as the Searsiidae, are deepsea fishes with large, extremely light-sensitive eyes. The lateral-line canals on the head are greatly enlarged and expanded. They are found in all except polar waters and the family is characterized by a unique light organ on the shoulder above the pectoral fin. Light-producing cells are contained in a dark sac, which opens to the outside by a backwards-pointing pore. When the fish is alarmed, a bright cloud is squirted out, which lasts a few seconds and enables the fish to escape into the darkness.

A living example of the genus *Searsia* "was seen to discharge a bright luminous cloud into the water on being handled. The light appeared as multitudinous bright points, blue-green in color." There are also series of stripelike or rounded luminous organs underneath the body.

Smelts

FAMILY OSMERIDAE

The osmerids, or smelts, may well have acquired their alternative common name from the fact that the European smelt and several other members of the family have a strong smell like that of a cucumber, owing to the presence of the compound trans-2-cis-6-nonadienal. Mostly small, silvery fish, smelts live in coastal and brackish cool waters in the northern hemisphere and undertake an anadromous migration to spawn in rivers. Numerous landlocked populations are also known. Smelts are carnivores, feeding on small invertebrates seized with their sharp conical teeth.

Their importance to subsistence fisheries in the far north is significant. They can be numerous and have a high fat content. When not eaten by the original inhabitants of the British Columbian coast, they were dried, and because of their fat content could be set alight and used as a natural candle. Hence *Thaleichthys pacificus* now has the common name of Candle fish.

The ayu (sometimes considered as being within its own monotypic family, Plecoglossidae) is a particularly remarkable member of this family and lives in Japan and adjacent parts of Asia, where it is of great economic importance. Its body is olive brown with a pale yellow blotch on the side. The dorsal fin is expanded and, like the other fins, has a reddish tint. When these colors, especially the reds, are enhanced in the breeding season, the Japanese name for the fish changes from *ayu* to *sabi*, which means rusty. Both sexes become covered in warty nuptial tubercules at the onset of breeding. The upper jaw of the male shortens and the female's anal fin expands. These changes start during the summer and the fish breed in the fall.

The fish mature in the upper reaches of rivers and move downstream towards the sea to breed. Spawning is carried out at night after a 10cm (4in) pit has been excavated. Each female produces

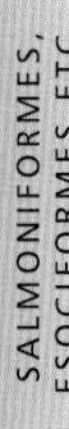

some 20,000 adhesive eggs, which hatch in about three weeks, depending on temperature. The larvae stay in the river until they are about 2.5cm (1 in) long when they move into the sea.

This seaward migration of the larvae is part of an interesting survival strategy. If the young stayed in the river, having spawned in the fall, they would have to endure the cold and potentially compete with larger young of the species that spawned in spring (when most fishes spawn). However, during winter the temperature of the sea is more stable than that of the river and at sea food is more abundant. On the other hand, the young ayu have to have developed a physiological (osmotic) mechanism to enable their small bodies to cope with the shock of transferring from freshwater to salt water. During winter they feed on zooplankton and small crustaceans and, by spring, when they return to freshwater, they have grown to about 8cm (3in) long. Then they migrate upstream in huge shoals, when thousands are caught and taken to capture ponds to facilitate a rapid growth rate and to provide an easily accessible source of food. The fish that escape continue up to the fast-flowing upper reaches, where each individual establishes a territory for itself among rocks and stones. Here they feed on diatoms and algae until summer or fall, when they move downstream to spawn. The ayu is an annual fish as almost all adults die after spawning. The very small percentage that survive spawning spend the winter at sea and repeat the cycle.

Concomitant with the change of diet from young to adult and the move from salt water to freshwater, the teeth change drastically. While at sea the ayu's diet is carnivorous and it uses its conical teeth to catch small crustaceans and other invertebrates. Adults, by contrast, feed on algae and have a whole series of groups of teeth forming comblike structures. Even more unusual is the fact that the teeth lie outside the mouth.

The comb-teeth develop under the skin of the jaw and erupt, shedding the conical teeth, when the fish enter freshwater. Each comb-tooth consists of 20–30 individual teeth, each one shaped like a crescent on a stick: narrow in the plane of the fish but very broad transversely. The gutter of the crescent faces inwards and, because of the different lengths of the arms of the crescent on each tooth, forms a sinuous gutter across the width of the comb-teeth. The combs of the upper and lower jaws juxtapose outside the mouth when it is closed. At the front of each lower jaw is a bony, pointed process that fits into a corresponding recess in the upper jaw. On the midline of the floor of the mouth is a flange of tissue, which is low at the front but higher at the back, where it branches into two. Each branch bends back on itself to run forwards, parallel to the sides of the jaw and decreasing in height towards the front. Muscles link this device with a median bone in the branchial series.

It is known that adult ayus eat algae within closely guarded territories, but exactly how they do so has not yet been explained. Grazing marks have frequently been seen on stones covered by algae in ayus' territories and it is usually stated that these are formed by the comblike teeth. However, as these teeth are outside the jaw, any algae so scraped off would be liable to be washed away by the current and lost.

It is suggested here that the ayu is a filter-feeder in a manner analogous to that of the baleen whales. The ayu's snout is fleshy and slightly overhangs the front of the upper jaw. Behind the snout, at the front of the upper jaw, is a row of about eight small conical teeth. Hanging down from the palate is a complex series of curtainlike structures, which have a relationship to the various flanges on the floor of the mouth. It is possible that if the ayu were to face into the current and scrape the algae off the rocks with the conical teeth, the algae so scraped would wash into the mouth. The mouth could then be closed tightly

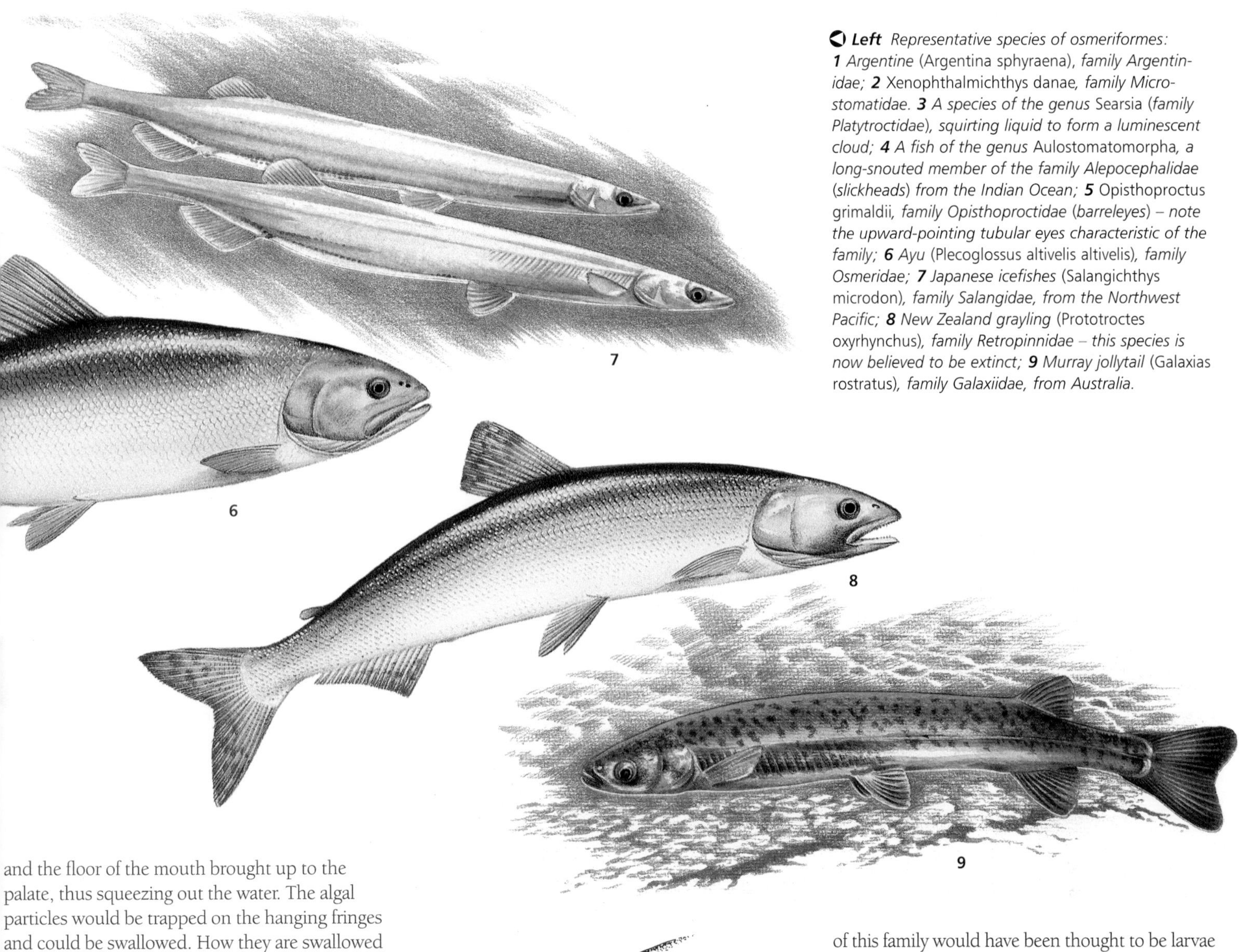

Left *Representative species of osmeriformes:* ***1*** *Argentine* (Argentina sphyraena), *family Argentinidae;* ***2*** Xenophthalmichthys danae, *family Microstomatidae.* ***3*** *A species of the genus* Searsia *(family Platytroctidae), squirting liquid to form a luminescent cloud;* ***4*** *A fish of the genus* Aulostomatomorpha, *a long-snouted member of the family Alepocephalidae (slickheads) from the Indian Ocean;* ***5*** Opisthoproctus grimaldii, *family Opisthoproctidae (barreleyes) – note the upward-pointing tubular eyes characteristic of the family;* ***6*** *Ayu* (Plecoglossus altivelis altivelis), *family Osmeridae;* ***7*** *Japanese icefishes* (Salangichthys microdon), *family Salangidae, from the Northwest Pacific;* ***8*** *New Zealand grayling* (Prototroctes oxyrhynchus), *family Retropinnidae – this species is now believed to be extinct;* ***9*** *Murray jollytail* (Galaxias rostratus), *family Galaxiidae, from Australia.*

and the floor of the mouth brought up to the palate, thus squeezing out the water. The algal particles would be trapped on the hanging fringes and could be swallowed. How they are swallowed is unknown. If the comb-teeth are a part of the filtration system, they can act only as a long-stop, but if so, how the entrapped particles are removed is a mystery. Clearly, much work remains to be done on this enigmatic but commercially important species.

Icefishes

FAMILY SALANGIDAE

Noodle fishes of the family Salangidae are very small, slender, transparent fishes from the western Pacific. Although marine, they move into estuaries to spawn. Their head is tiny and pointed and the deepest part of the body is just in front of the dorsal fin. The only coloration on the 10cm (4in) long Japanese icefish (*Salangichthys microdon*) is two rows of small black spots on the belly. Although small, they are sometimes so abundant that they can easily be caught for food. To the Japanese they are a delicacy called *shirauwo*.

The shape and transparency of the adult is very similar to that of the larvae of some other fish species. Indeed, if it were not for the fact that sexually mature examples are known, adult specimens of this family would have been thought to be larvae of some unknown species. In lower vertebrates, it is possible for a species to evolve by a process called neoteny or paedomorphosis in which the body development is curtailed while sexual development continues as normal. With time, the ability to attain the adult form is lost and a new species is formed. Both the Salangidae and the closely related Sundasalangidae are considered to have evolved in this way and so are said to be neotenic.

Above *The ayu has remarkable dentition:* ***1*** *juxtaposing comb teeth on the outside of the jaw;* ***2*** *a fleshy snout with a few canine teeth behind it, and* ***3*** *elaborate skin folds within the mouth.*

Sundaland Noodlefishes

FAMILY SUNDASALANGIDAE

The first two of the very small freshwater species that form the family Sundasalangidae from Thailand and Borneo were described as recently as 1981, with another five species discovered in the late 1990s. The body is scaleless and transparent, there is no adipose fin, and several bones in the skull are lacking. Superficially, they resemble noodlefishes but have particular features of the paired fin girdles and gill arches that make them unique. They are among the smallest of all fish species, one of them becoming mature at 1.5cm (0.5in) long.

New Zealand Smelts

FAMILY RETROPINNIDAE

The family Retropinnidae comprises the two subfamilies Prototroctinae and Retropinninae, commonly known as the southern graylings and southern smelts, respectively, which together contain 5 species in 3 genera.

The scaled, cylindrical southern graylings of the Prototroctinae are represented by just two species: the New Zealand southern grayling (*Prototroctes oxyrhynchus*) and the Australian southern grayling (*P. maraena*). The New Zealand species was first described and named in 1870 when it was abundant and regarded as a good source of food, but the last known specimens were caught by accident in a Maori fish trap in 1923 and it is now presumed extinct. There are only a few records describing live *P. oxyrhynchus*, although like its Australian congener it is known that it had only one ovary or testis. Early reports of its body color are inconsistent. One suggests that it was slaty on the back, merging to silvery on the sides and belly with patches of azure. The fins were orange, slaty dark at the tips and the cheeks had a golden tinge. It apparently migrated regularly from the sea to freshwater, where it spawned. Although they once existed in their millions, there are now less than 40 bodies preserved in the great national museums of the world and a stuffed specimen in the Rotoiti Lodge of the New Zealand Deer Stalkers' Association, Nelson.

P. maraena appears to be following the same path towards extinction as its New Zealand cousin and is now severely endangered. Why these two remarkable fish species should have declined so dramatically is unknown for sure, although the extinction of *P. oxyrhynchus* has been attributed to a combination of habitat degradation resulting from deforestation plus the impacts of introduced salmonids. However, neither explanation is regarded as conclusive.

The Retropinninae, or southern smelts, are slender, small-mouthed fishes from southeastern Australia, Tasmania, and New Zealand. Some populations are migratory, while others are landlocked. Because they are very variable in form, their taxonomy is far from clear. However, there appears to be two Australian species, including the Tasmanian smelt, and two New Zealand species, including the Cucumberfish.

Galaxiids

FAMILY GALAXIIDAE

The family Galaxiidae comprises the three subfamilies Lovettiinae, Aplochitoninae, and Galaxiinae, which together contain about 40 species in 8 genera. The galaxliids are distinctive small fishes found in all the major southern land masses (Australia, New Zealand, South America, and South Africa) and also on some of the more remote southern islands such as New Caledonia, Auckland, Campbell, and the Falklands. The first galaxiid was collected by naturalists with Captain Cook in New Zealand during 1777. The generic name *Galaxias* was given to the fish because it was a dark, black-olive in coloration and covered with a profusion of small gold spots resembling a galaxy of stars. As a result of these markings, an absence of scales, and the presence of smooth leathery skins, galaxiids are fish of distinctive appearance. Unlike some northern relatives, they have no adipose fin, but their single dorsal fin is placed over the anal fin well back towards the tail. Most are small, 10–25cm (4–l0in) in length, but one species reaches 58cm (23in), while there are several tiny species, 3–5cm (1–2in) long. Most are tubular, cigar-shaped fish with blunt heads, thick fleshy fins, and a truncated tail. A few are stocky, thick-bodied fish and most of them are secretive species that hide among boulders, logs, or debris in streams or lakes. Many are solitary, but a few show shoaling behavior. All but about six species have a freshwater distribution, and a few are marine migratory fish in which the larval and juvenile phases are spent at sea.

Migratory species spawn mostly in freshwater and rarely in estuaries. After their eggs hatch, the resulting larvae, about 1cm (0.4in) long, are swept to sea. Their ability to cope with a sudden transition from freshwater to seawater shows remarkable flexibility. They spend 5–6 months at sea before migrating back into freshwaters in spring as elongate transparent juveniles. It will be months before these fish reach maturity, at about a year in some species, two to three years in others.

What is known about the spawning of these fish is equally remarkable. The Inanga (*Galaxias maculatus*) is known to spawn in synchrony with the lunar or tidal cycle, spawning over vegetated estuary margins during high spring tides. When these high tides retreat, the eggs are stranded in the vegetation and protected from dehydration only by humid air. Development takes place out of water, the eggs not being reimmersed until the next set of spring tides two weeks later. Then the eggs hatch, releasing the larvae, which are swept quickly out to sea. The Banded kokopu (*G. fasciatus*), from small heavily forested streams, spawns during floods and deposits its eggs in leaf litter along stream margins. When the floods subside, the eggs are left stranded among rotting leaves, where they develop. Hatching cannot occur until there is another flood, and then the larvae are swept downstream and out to sea. It is clear that both these modes of spawning involve substantial risks. The breeding habits described above are exceptional and as far as is known, most galaxiid species lay their eggs in clusters between rocks and boulders. A few very small species pair up for spawning and lay their eggs on the leaves of aquatic plants.

Although the regions where these fish live tend to have moist climates, some species have become adapted to surviving droughts and therefore aestivate. Some live in pools that lie on the floor of wet podocarp (southern hemisphere conifer) forests in water usually only a few centimeters deep, which covers leaf litter shed from the towering forests above. In the late summer and fall, these pools frequently dry up and the fishes disappear into natural hollows around the buttresses of trees. They survive for several weeks in these damp pockets until rainfall restores water to the forest floor. The fish wait for the return of water before spawning. As the pools are increasingly replenished, larvae are enabled to disperse around the forest floor and invade available habitats.

That such small fish should have attained importance for commercial fisheries may seem remarkable. However, when European settlers

Left *Brown galaxias* (Galaxias fuscus) *at Woods Point in Victoria, Australia. This species lives in clear, mountain streams above the snowline that have gravelly or rocky substrates.*

Above *The Australian smelt* (Retropinna semoni) *lives most commonly in freshwater, but can also tolerate brackish environments. This species is widely distributed throughout southeastern Australia.*

arrived in New Zealand in the mid-19th century they found that vast populations of the sea-living juveniles of several *Galaxias* species migrated into rivers during spring. The Maoris of New Zealand exploited huge quantities of them, and not surprisingly the Europeans followed their example, calling the tiny fish whitebait on account of their similarity to unrelated fish they knew at home in England. Quantities caught now do not compare with those of the early years, but a good fishery persists in some rivers. Colossal numbers of fishes are caught, with each individual weighing only about 0.5g and so requiring about 1,800 fishes to make up a kilogram. The catches of individual fishermen vary from just a few to many kilograms, with exceptional catches reaching several hundred kilograms in a day. These fishes make a delicious gourmet seafood and command a correspondingly high price in the shops.

Aplochiton zebra is a fairly widespread fish, occurring in rivers of Patagonia and the Falkland Islands where it is called trout. This handsome fish has dark vertical stripes over the back and sides of its scaleless body. Its persistence in the Falkland Islands has unfortunately been placed in jeopardy by the introduction of the Brown trout from Europe, a sad predicament for a species first collected by Charles Darwin when he visited the Falklands. Recent collections suggest that this species may enter coastal waters; otherwise very little is known of its biology.

Historically the very broad distribution of the galaxiid fishes has attracted intense interest. Long ago zoologists asked how a group of freshwater fishes could be so widely distributed around southern lands. Not only is the family as a whole widely dispersed, but the Inanga is found in Australia, Tasmania, Lord Howe Island, New Zealand, Chatham Islands, Chile, Argentina, and the Falkland Islands. With such a broad range, this species is one of the naturally most widely distributed freshwater fishes known. Noting the remarkable range of this species, zoologists of the late 19th century suggested that there must have been former land connections between the areas where the fish are present, and so proposed a vicariance biogeography. Some thought that Antarctica or the ancient continent of Gondwana might have been involved. Another explanation is that these fishes dispersed by migrating through the sea (a dispersal biogeography).

This debate has continued into modern times and it is now generally accepted that arguments in favor of a Gondwana-based vicariance biogeography or a dispersal biogeography are not necessarily mutually exclusive. However, a recent extensive consideration of the distribution patterns of galaxiids and other fishes in the southern hemisphere concluded that evidence from areas including genetics, morphology, recent dispersal events, and parasitology all supports, or is consistent with, a dispersal biogeography. The galaxiids are indeed sufficiently ancient to have formerly inhabited Gondwana, but no compelling evidence indicates that their present distribution reflects a former broad Gondwana-based range.

A PIKE IN THE SOUTHERN HEMISPHERE?

If the galaxiids are the salmonoids of the southern hemisphere, and salmonoids are related to pike, is there a southern hemisphere pike? The answer is probably yes.

Described as recently as 1961, the Salamanderfish (*Lepidogalaxias salamandroides*) is a small fish about 4 cm (1.5in) long, and is the sole member of its family (Lepidogalaxiidae). It is found in numerous localities in western Australia. Originally thought to be a galaxiid, unlike those fishes it has scales and dorsal and anal fins set much further forward, while the anal fin of the mature male is highly modified, with gnarled and hooked rays and unusual dermal flaps in order to function as an intromittent organ during internal fertilization. The Salamanderfish feeds mainly on aquatic insect larvae and small crustaceans.

Little is known of the biology of this unique species, but it is apparently capable of surviving drought by burrowing in mud or under damp leaves. Its distribution appears to be confined to small, temporary acidic pools and ditches, principally in the sand-plain area between the Blackwood and Kent rivers in Western Australia.

Salmon, Trout, and Allies

FAMILY SALMONIDAE

The family Salmonidae comprises the three subfamilies Coregoninae, Thymallinae, and Salmoninae, which together contain about 66 species in 11 genera. Included in this grouping are the salmon, trout, charrs, and whitefishes, which consitute some of the most famous and important fishes of the northern hemisphere. Unsurprisingly, their great commercial importance as food and sport fishes has ensured that they have been particularly well studied by science.

Members of the subfamily Coregoninae are commonly known as the whitefishes, although this common name may also be applied specifically to at least two species within the subfamily and is also used for a group of completely unrelated species of marine fish. The whitefishes are relatively plain, silvery fish found predominantly in the cold deep lakes of Asia, Europe, and North America. There are three genera, although species are highly variable morphologically and genetically leading to continuing debate over classification at this level even when modern molecular techniques are used. A few species are anadromous, but most have exclusively freshwater distributions. Many species, such as the vendace and the European whitefish, are important for both commercial and recreational fisheries.

The subfamily Thymallinae contains the single genus *Thymallus*, whose name is derived from the supposedly thymelike smell of their flesh. These fishes inhabit cool, swift-flowing rivers, but will sometimes enter brackish water. The grayling is widespread in Asia and Europe, while the slightly more colorful Arctic grayling occurs at higher latitudes in North America.

The subfamily Salmoninae is one of the most important groups of fish in the world, with the Atlantic salmon being its best known species. Many people pay high prices for its flesh and even higher prices for the pleasure of trying to catch it by angling in often breathtakingly beautiful surroundings. This species is also much sought after by commercial fishermen, although such activities are now strictly controlled throughout the distribution range of this magnificent fish. Although salmon is now a luxury item, in the 19th century apprentices in London protested about being fed salmon six days a week.

Above *Male and female River or Brown trout* (Salmo trutta fario) *attend their nest – a shallow scrape in the substrate – in a fast-flowing, clear stream.*

Below *A twelve-day-old Brown trout. Larval fish of the family Salmonidae such as this, which have hatched but not yet absorbed their yolk-sacs, are known as alevin. Seen suspended below its body, the yolk-sac forms the alevin's vital first source of energy.*

The significance of the only salmon species native to the Atlantic is emphasized by the range of different names applied to each stage of its life history. On hatching, the individual is called an alevin and quickly grows to become a fry. Later, when it is a few centimeters long and has developed dark blotches or parr marks on the body, it becomes a parr. When it later migrates to the sea, the parr marks become covered by a silvery pigment and the fish becomes a smolt. The fish may then remain at sea for one winter and return to freshwater to spawn as a grilse or one-sea-winter fish, or it may stay at sea for one or more additional winters and return to spawn as a larger spring-run or multi-sea-winter fish. After spawning it becomes a kelt, at which stage most fish die, although some survive to go back to sea and return again on future spawning migrations.

After adult salmon have spawned in their home streams, the eggs remain in the gravel for a considerable time. Hatching time can be predicted from a knowledge of the local water temperature conditions and is usually between April and May in northern Europe.

The young are about 2cm (0.8in) long on hatching and for the first six weeks or so live in the gravel, feeding on their yolk sacs. With the exhaustion of the yolk supply, they emerge from the gravel and start feeding as fry on insect larvae and other invertebrates. As they grow, they develop into parr and their markings give them camouflage for hunting actively. The length of time spent in freshwater before becoming a smolt varies from five years in the north of the range to one year in the south.

Not all members of the year class migrate to the sea. A few males stay in freshwater to become precociously mature and have been seen shedding sperm in the company of mating adults. The migrating young spend some time in estuaries, acclimatizing themselves for coping with salt water. In the sea they grow rapidly and can reach 14kg (30lb) in three years. They feed on fish and spend up to four years in the sea building up strength for the rigors of the spawning migration. When returning to the natal stream for spawning they can travel 115km (70mi) a day.

Above *Representative species of salmoniformes:* ***1*** *Atlantic salmon* (Salmo salar)*;* ***2*** *Rainbow trout* (Oncorhynchus mykiss)*;* ***3*** *Grayling* (Thymallus thymallus)*;* ***4*** *Sockeye salmon* (Oncorhynchus nerka), *family Ophidiidae;* ***5*** *Charr (genus* Salvelinus)*;* ***6*** *Sea, Brook, or Brown trout* (Salmo trutta trutta)*;* ***7*** *Shortjaw cisco* (Coregonus zenithicus).

Above *Large-scale salmon farming in sea pens, pioneered in Norway in the 1960s, has grown into a lucrative business in many countries. However, disease is a danger with such high concentrations of fishes.*

Right *A spawning pair of Sockeye salmon* (Oncorhynchus nerka) *– on the right, the male with its distinctive hump and hooked jaw. This species is fished extensively on the Pacific coast of North America.*

Some populations of landlocked salmon exist in lakes in the far north of America and Europe. They never grow as large as the others, but still run up streams to spawn. It is thought that their access to the sea was blocked after the last ice age.

There are probably seven species of Pacific salmon, whose generic name *Oncorhynchus* means "hooked snout." Their life history is generally very similar to that of the Atlantic salmon, although two North American species – the Sockeye salmon and the Rainbow trout – and the somewhat doubtful species amago from Lake Biwa in Japan have landlocked forms. In the Atlantic Ocean, the Atlantic salmon grows to about 32kg (70lb) in weight, but in the Pacific Ocean the Chinook salmon has been recorded at 57kg (126lb).

The Eurasian genus *Hucho* includes the Huchen, a slender species from the Danube, and other species in Central Asia. Attempts were made to translocate the huchen to the River Thames in England in the 19th century. Despite rumors of it surviving into the early 20th century, there is no reliable evidence that the introduction succeeded. Such an introduction would not even be attempted by today's more environmentally aware fisheries managers.

The charrs of the genus *Salvelinus* come from the cold deep lakes and rivers of Europe and North America. Only in the very north of the Atlantic are they migratory. The sole European species, the Arctic charr, is very variable in appearance and until relatively recently a different species was named from almost each lake. In the breeding season the males sport a spectacular deep red on their underside, from which the name charr is derived from the Gaelic *tarr,* meaning "belly." If anything their flesh exceeds that of the salmon in quality and those lucky enough to have the chance to sample it should have charr steaks, lightly boiled with bay leaves, cold on toast. Despite its common name of Brook trout, the eastern American species *Salvelinus fontinalis* is a charr. In its native haunts, the migratory form can grow to nearly 90cm (3ft) long, but the European introductions rarely reach half that length. In Europe, the Brook or Brown trout hybridizes with both the native River or "other" Brown trout and the introduced Rainbow trout. The offspring of both mismatches are a striped fish called the Tiger or Zebra trout, which are sterile.

The Brook trout of Europe has generated much confusion regarding its nomenclature. Being a very variable species in both form and behavior, it has been given a variety of common names, such as Brown trout, Sea trout, Lake trout, and Salmon trout. Brown trout living in lakes may become very large cannibals and are often called Lake trout or ferox. Brown trout that migrate to the sea to feed become very silvery and are called Sea trout or Salmon trout (not a hybrid between a salmon and a Brown trout, although these do occur). As a result of the richer feeding opportunities in the sea, migratory Brown trout reach nearly twice the size of their nonmigratory siblings.

A similar situation occurs on the western coast of North America with the Rainbow trout. This commonly introduced denizen of fish farms in Europe has an extensive natural range from southern California to Alaska. In the northern part of its range it is migratory, and much larger and more intensively colored than in the south. The Canadians call the migratory form the Steelhead trout, whereas the nonmigratory form is referred to as the Rainbow trout. Some years ago a fishery in the United Kingdom bought from North America a large number of Rainbow trout to enhance the trout fishing in their waters. Unfortunately they were provided with the migratory form and very little of their investment ever returned.

Salmon and trout have been introduced into many countries as game and food fishes. The Brown trout is now found in the North American west, almost all southern hemisphere countries, and even in some tropical ones where there are cool streams at high altitudes in which they can thrive. In many cases, the present distribution of the subspecies corresponds to the former extent of the British Empire, where the fish was introduced by keen sport fishermen in colonial service.

To complete this important family, two more poorly known species of uncertain affinities must be mentioned. In the lakes of the Ohrid region on the borders of Albania, the former Yugoslav Republic of Macedonia, and Greece live fish placed in the genus *Salmothymus,* which may be southern landlocked salmon. Finally, the genus *Brachymystax* comes from rivers of Mongolia, China, and Korea. No satisfactory conclusions have been reached on its taxonomic relationships because very few specimens have ever been available for study. IW/RMcD/KEB

Salmon, Pike, and Related Families

ORDER ESOCIFORMES

Pikes
Family Esocidae

Freshwater; Northern Hemisphere. Length maximum about 1.5m (5ft). 5 species in 1 genus: **Northern pike** (*Esox lucius*), **Amur pike** (*E. reicherti*), **muskellunge** or **muskie** (*E. masquinongy*), **Chain pickerel** (*E. niger*), *E. americanus* with subspecies **Redfin pickerel** and **Grass pickerel**.

Mudminnows
Family Umbridae

Freshwater; parts of Northern Hemisphere. Length maximum about 15cm (6in). 5 species in 3 genera: **European mudminnow** (*Umbra krameri*), *U. limi, U. pygmaea*, **Alaska blackfish** (*Dallia pectoralis*), *Novumbra hubbsi*. Conservation status: The European mudminnow is classed as Vulnerable.

ORDER OSMERIFORMES

SUBORDER ARGENTINOIDEI
SUPERFAMILY ARGENTINOIDEA

Argentines or Herring smelts
Family Argentinidae

Marine; Atlantic, Indian, and Pacific oceans. Length maximum about 60cm (24in). 19 species in 2 genera, including: **Alice argentine** (*Argentina aliceae*), **Greater argentine** (*A. silus*), and **Pygmy argentine** (*Glossanodon pygmaeus*).

Microstomatids
Family Microstomatidae

Marine; tropical to temperate seas, Atlantic, Indian, and Pacific oceans. Length maximum about 21cm (8in). 17 species in 3 genera, including: **Slender argentine** (*Microstoma microstoma*), **Stout argentine** (*Nansenia crassa*), *Nansenia oblita*, and *Xenophthalmichthys danae*.

Deepsea smelts
Family Bathylagidae

Marine; Atlantic, Indian, and Pacific oceans. Length maximum about 25cm (10in). 15 species in 1 genus including: **Longsnout blacksmelt** (*Dolicholagus longirostris*), **Goiter blacksmelt** (*Bathylagus euryops*), and **Eared blacksmelt** (*Lipolagus ochotensis*).

Barreleyes or spookfishes
Family Opisthoproctidae

Marine; tropical to temperate, Atlantic, Indian, and Pacific oceans. Length maximum about 25cm (10in). 10 species in 6 genera, including: *Opisthoproctus grimaldii, O. soleatus, Dolichopteryx longipes, Rhynchohyalus natalensis*.

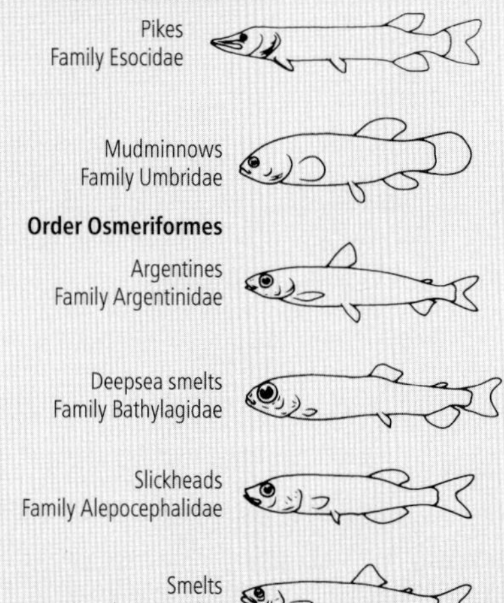

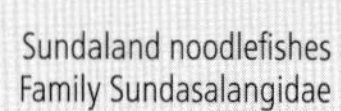

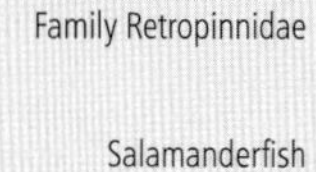

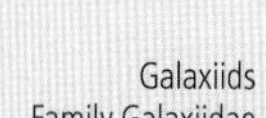

SUPERFAMILY ALEPOCEPHALOIDEA

Slickheads
Family Alepocephalidae

Deep sea; all oceans. Length maximum about 76cm (30in). At least 63 species in about 24 genera, including: **Small-scaled brown slickhead** (*Alepocephalus australis*), **Fangtooth smooth-head** (*Bathyprion danae*), the **Black warrior** (*Bathylaco nigricans*), and the **Blackhead salmon** (*Narcetes stomias*).

Leptochilichthyids
Family Leptochilichthyidae

Deep sea; eastern Atlantic, western Indian, and eastern and western Pacific oceans. Length maximum about 30cm (12in). 3 species in 1 genus, including: *Leptochilichthys microlepis*.

Tubeshoulders
Family Platytroctidae (Searsiidae)

Marine; all oceans (absent from Mediterranean). Length maximum about 38cm (15in). 37 species in 13 genera, including: *Barbantus curvifrons, Mentodus crassus*.

SUBORDER OSMEROIDEI
SUPERFAMILY OSMEROIDEA

Smelts
Family Osmeridae

Marine, anadromous and coastal freshwater; Northern Hemisphere in Arctic, Atlantic and Pacific oceans. Length maximum about 45cm (18in). 13 species in 7 genera, including: **European smelt** (*Osmerus eperlanus*), **Pond smelt** (*Hypomesus olidus*), **Capelin** (*Mallotus villosus*), **Candle fish** (*Thaleichthys pacificus*), **ayu** (*Plecoglossus altivelis*). Conservation status: The **Delta smelt** (*Hypomesus transpacificus*) is Endangered.

Icefishes or noodlefishes
Family Salangidae

Anadromous and freshwater; Sakhalin, Japan, Korea, China to northern Vietnam. Length maximum about 18cm (7in). 11 species in 4 genera, including: **noodlefish** (*Salanx cuvieri*), **Japanese icefish** (*Salangichthys microdon*). Conservation status: *Neosalanx regani* is classed as Vulnerable.

Sundaland noodlefishes
Family Sundasalangidae

Freshwater; Borneo and southern Thailand. Length maximum about 3cm (1in). 7 species in 1 genus, including: *Sundasalanx microps, S. mekongensis*.

SUPERFAMILY GALAXIOIDEA

New Zealand smelts
Family Retropinnidae

Freshwater and brackish water (some partially marine); New Zealand, Chatham Islands, southeastern Australia, and Tasmania. Length maximum about 15cm (6in). 6 species in 3 genera: **cucumberfish** (*Retropinna retropinna*), **Tasmanian smelt** (*R. tasmanica*), **Australian smelt** (*R. semoni*),**New Zealand southern grayling** (*Prototroctes oxyrhynchus*), **Australian southern grayling** (*P. maraena*), **Stokell's smelt** (*Stokellia anisodon*). Conservation status: The New Zealand southern grayling is now classed as Extinct, while the Australian southern grayling is Vulnerable.

Salamanderfish
Lepidogalaxias salamandroides
Family Lepidogalaxiidae

Sole member of family. Freshwater; southwestern Australia. Length maximum about 4cm (1.5in). Conservation status: Lower Risk/Near Threatened.

Galaxiids
Family Galaxiidae

Freshwater and diadromous; Australia, New Zealand, New Caledonia, southernmost Africa and southern South America. Length maximum about 58cm (23in). About 40 species in 8 genera, including: **Flathead galaxiid** (*Galaxius rostratus*), **Golden galaxiid** (*G. auratus*), *G. maculatu G. fasciatus, Aplochiton zebra*. Conservatio status: 4 species are Critically Endangered including the **Swan galaxias** (*Galaxias fontanus*) and **Barred galaxias** (*G. fuscus*); species are Vulnerable, including the **Gian kokopu** (*Galaxias argenteus*) and **Canterbury mudfish** (*Neochanna burrowsius*).

ORDER SALMONIFORMES

Salmon, trout, charrs, whitefishes, and allies
Family Salmonidae

Freshwater and anadromous; Northern Hemisphere. Length maximum about 1.5m (5ft). About 66 species in 11 genera, including: **Bear Lake whitefish** (*Prosopiun abyssicola*), **vendace** (*Coregonus albula*), **European whitefish** (*C. lavaretus*), **graylin** (*Thymallus thymallus*), **Arctic grayling** (*T. arcticus*), **Arctic charr** (*Salvelinus alpinus*), **Brook trout** (*S. fontinalis*), **Atlantic salmo** (*Salmo salar*), **Brown trout** (*S. trutta*), **Chinook salmon** (*Oncorhynchus tshawytscha*), **Humpback** or **Pink salmon** (*O. gorbuscha*), **Sockeye** or **Red salmon** (*O. nerka*), **amago** (*O. rhodiurus*), **Rainbow trout** (*O. mykiss*), **huchen** (*Hucho hucho*). Conservation status: 19 species are endangered, 4 of which are Critically Endangered, including the **Apache trout** (*Oncorhynchus apache*) of th Colorado river system in the United States and the **Ala Balik** (*Salmo platycephalus*) of Turkey and Central Asia. 5 species are classed as Endangered, including the huchen, and 10 species are classed as Vulnerable, including the **bloater** (*Coregonus hoyi*) and the Shortjaw cisco (*C. zenithucus*

Right *The Northern pike (*Esox lucius*) is an aggressive, solitary species that lurks in wait for its prey, often poised with its body in an "S" shape, ready to strike. This opportunistic hunter utilizes all kinds of cover for its ambushes; here, a pike emerges from the cockpit of a sunken aircraft. Its marbled green and brown coloration give it excellent camouflage in a range of environments.*

RETURN TO BASE

The life cycle of the Sockeye salmon

THE LIFE HISTORY OF THE SOCKEYE SALMON of the Pacific Northwest exemplifies that of other anadromous species – that is, those that live most of their lives at sea but return to freshwater streams in order to spawn and eventually die. In the case of the sockeye, which travels further than any other salmon, this remarkable migration may see the fish travel up to 1,600km (1,000 miles).

From spring until late summer, sockeyes run in large schools upstream in an effort to return to the streams in which they were hatched. As they journey up the rivers of Alaska (the Kasilof, Kenai, and Russian) and the Canadian province of British Columbia (the Fraser, Skeena, Nass, and Nootka), they are faced with numerous obstacles and hazards such as rapids and falls. The key to their remarkable "homing" ability is memory and a sense of smell. It has been demonstrated that adult salmon remember and later trace the smell of their birth stream resulting from its combination of chemicals in the water contributed by the surrounding rocks, soils, vegetation, and other factors. Under natural conditions, occasional errors of navigation are made by the fish, which has the advantage of allowing the species to extend its range if more suitable habitat becomes available.

The males usually arrive back first but the females normally select the spawning site, where they are aggressively courted by the males. During maturation, hormonal changes dramatically alter the color of sockeyes – their heads turn green and their backs to a deep red – and the male's jaws elongate to form a hook, called the kype. (All seven Pacific salmon species belong to the genus *Oncorhynchus*, meaning "hooked jaw," and the sockeye develops this feature most pronouncedly.)

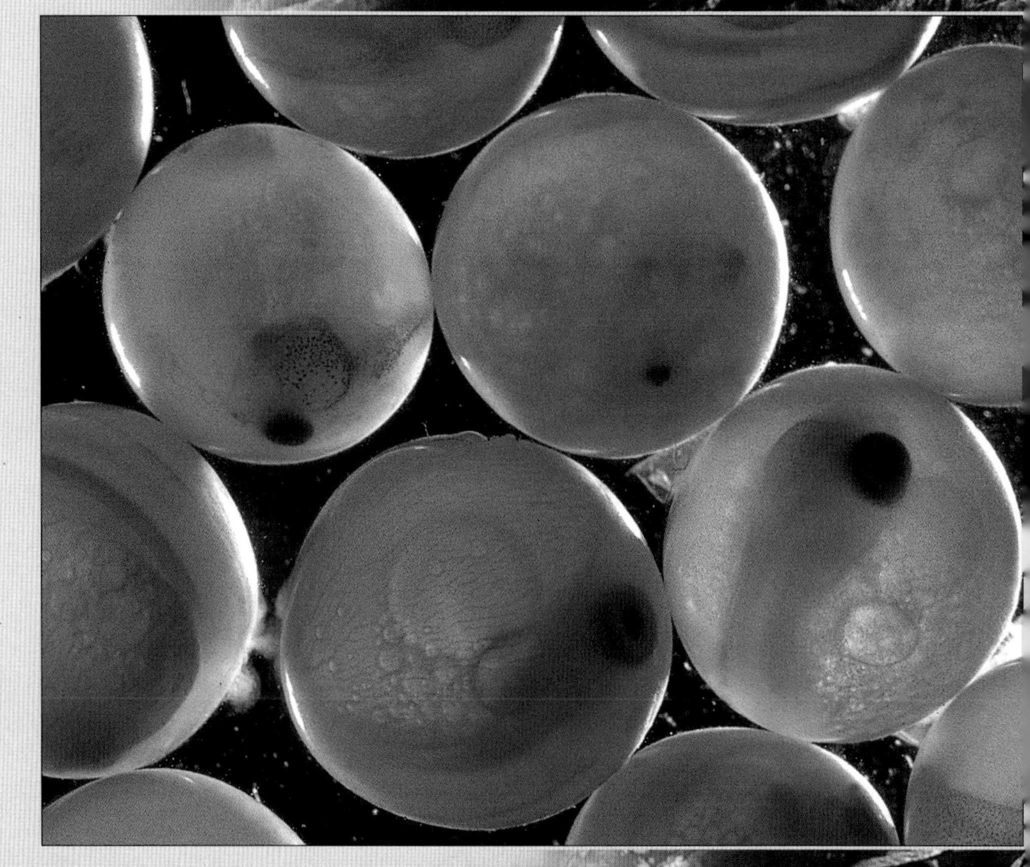

Above *Laid in the fall, the eggs incubate over winter – protected beneath the gravel, and often several feet of snow and ice. Around one month after laying, eyes begin to appear in the eggs. The fish are most vulnerable in the egg-to-fry stage.*

Below left *Many salmon die on their upstream run, killed by exhaustion, predators, or pollution. Where hydroelectric dams bar their way, "fish ladders" have been built to help them reach hatcheries.*

At the spawning site, a nest or redd is excavated in suitable substrate by vigorous movements of the female's tail until it is up to 3m (10ft) long and 30cm (12in) deep. Depending on the size of the adult fish, female sockeyes can lay between 2,500 and 7,500 eggs. The pair lie alongside one another for spawning and accompany the act with much trembling and jaw opening. The male ejects a milky fluid (milt) containing his sperm onto the eggs to fertilize them. The female then covers the fertilized eggs with gravel for protection until they hatch. Each act of spawning lasts for about five minutes, and the whole process may continue for a fortnight. In between, the adults rest in deep holes in the river bed. After each spawning session, the redd is filled in and another excavated. Exhausted by their arduous upstream journey and the rigors of digging and guarding the redds, the adults tend to die about a week after spawning is complete.

After hatching and developing into fry over a period of around a year in the freshwater streams of their birth or in nearby lakes, the young salmon head downstream to the sea. They are known at this developmental stage as smolts or fingerlings. Once they have entered the Pacific, they range far out into mid-ocean, south of the Aleutian Islands. Here, they will spend 2–4 years maturing and developing their characteristic orange-red flesh that makes them so highly prized by Northwest coast fishermen. Then, in the summer of their fourth year, they head inland to the mouths of the great rivers to begin the cycle anew.

Sockeyes are the most important commercially of all Pacific species. They are caught in purse-seine and gill nets by First Nation and other fishermen. The high fat content (stored for the long migration) makes their flesh particularly rich and flavorful.

Above *With their distinctive green heads and red bodies, adult Sockeye salmon in prime spawning condition mass in the Adams River in British Columbia on their annual migration run. Runs in odd-numbered years are generally larger than those in even-numbered years.*

Left *In the late winter, the eggs hatch into alevins, tiny fishes with large attached yolk sacs, from which they gain their nutrition. The orange yolk sacs contain a balanced diet of protein, carbohydrates, vitamins, and minerals.*

Bristlemouths and Allies

Bristlelike teeth, luminous organs, eyes on stalks – these are some of the characteristics of the order Stomiiformes, a worldwide group of deepsea fishes contained in four families and several subfamilies. The classification of the order is very uncertain and the subject of active research, hence some of the families and subfamilies included here may well change over time.

Practically all stomiiform species have luminous organs and many also have a luminous chin barbel, thought to act as a lure. They are flesh-eating fishes with a large, widegaping, toothed mouth. Most are scaleless and black or dark brown in color, but one family in particular, the bristlemouths (Gonostomatidae), are mostly silvery, midwater fishes. Typically an adipose fin is present, but this and the pectoral and dorsal fins have been lost in some lineages.

Bristlemouths

FAMILY GONOSTOMATIDAE

The Gonostomatidae are known as bristlemouths because of their fine, bristlelike teeth. The genus *Cyclothone*, with about 12 species, occurs in all oceans. It is probably – along with *Vinciguerria* – the most common genus in the world regarding numbers of individuals: trawls can haul up tens of thousands of these small fish at a time. They feed on small crustaceans and other invertebrates and, in turn, are a most important source of food for larger fishes, including some of their relatives.

The maximum size of a stomiiform species depends upon the richness of the environment. In areas of abundance, such as the Bay of Bengal and the Arabian Sea, a species may reach 6cm (2.5in) long, but in a polluted area, like the depths of the Mediterranean, or in an area poor in food, they are much smaller. The Mediterranean species *Cyclothone pygmaea* grows to only 2.5cm (1in).

Some species occur worldwide, while others have a very limited distribution. As well as a two-dimensional distribution that can be shown on a map, there is also a three-dimensional distribution because species are separated vertically. Generally, silvery or transparent species live nearer the surface than their dark-colored relations. It is also generally true that the species living deeper have weaker and fewer light organs than those living above them. The swimbladder is poorly developed in bristlemouths, which may explain why they do not undertake the extensive daily vertical migrations common in many deepsea fishes.

Cyclothone species show differences between the sexes, not with different light-organ patterns, as in the lanternfishes, but in the nature of the nasal complex. In males, as they mature, the olfactory organs grow out through the nostrils; this does not occur in females. In some species, such as the Veiled anglemouth, the hydrodynamic disadvantage of the protruding nasal plates (lamellae) is thought to be compensated for by the development of an elongate snout, or rostrum.

Barbeled Dragonfishes

FAMILY STOMIIDAE

There are currently six subfamilies in the family Stomiidae; however, this may be revised in the light of new research and information.

One of these subfamilies, the scaled dragonfishes (subfamily Stomiinae) includes the single genus, *Stomias*, with about 11 species. Elongated predators lacking an adipose fin, the scaly dragonfishes are found in the Indo-Pacific and Atlantic Oceans. They have large, easily shed, hexagonal scales that produce a honeycomb pattern on their dark bodies. *Stomias* species lack a swimbladder, so they can easily undertake extensive daily vertical migrations. There are usually two rows of small light organs along the ventral margin of the body. At their longest, the scaled dragonfishes rarely exceed 30cm (12in).

Members of the snaggletooths (subfamily Astronesthinae) have a dorsal fin that begins slightly behind the midpoint of the body but well ahead of the anal fin; all genera have an adipose fin except for *Rhadinesthes*. They are midwater predators with fairly compressed, elongate bodies that are usually black.

The black dragonfishes (subfamily Idiacanthinae) are elongated, scaleless fishes also lacking a

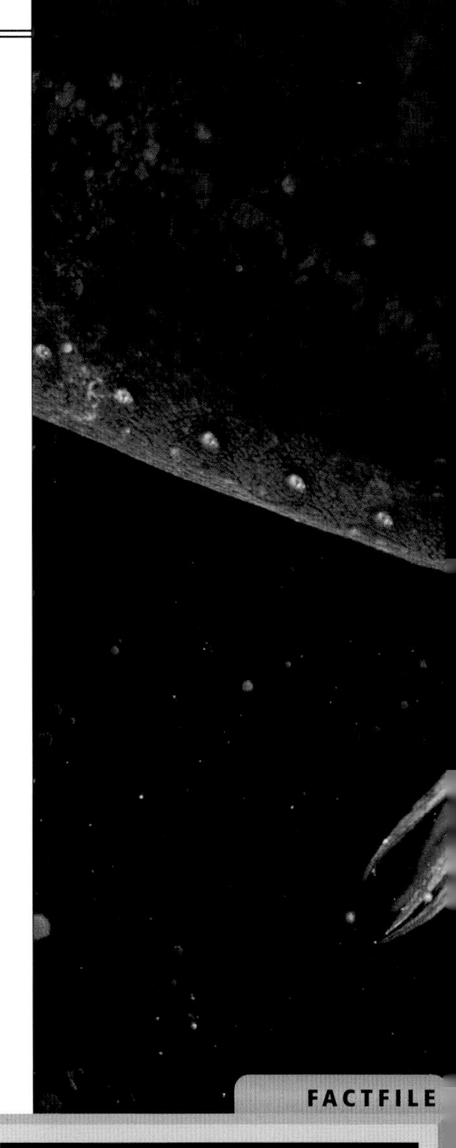

FACTFILE

BRISTLEMOUTHS AND ALLIES

Class: Actinopterygii

Order: Stomiiformes

About 320 species in over 50 genera and about 4 families.

Distribution All oceans within the temperate zones but not uniformly distributed within these limits.

Size **Length** maximum adult 35cm (14in): most are much smaller.

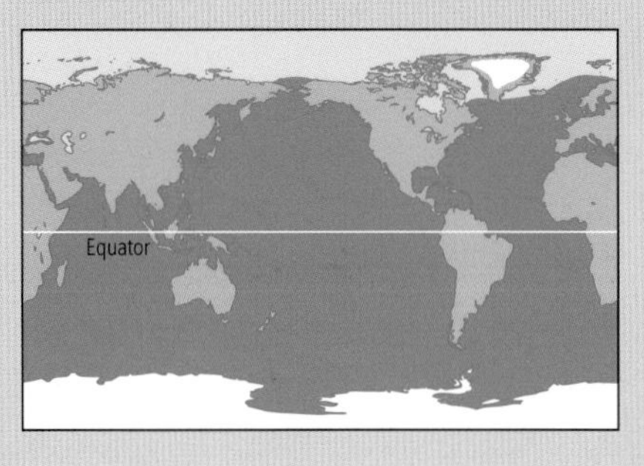

See families table ▷

◐ ***Left*** *Juveniles and adults of the Atlantic fangjaw* (Gonostoma atlanticum) *stay mainly at a depth of 300–600m (980–1,970ft) during the day but at night they go up to between 50–200m (160–660ft). This species is oviparous, with planktonic eggs and larvae.*

Above Echiostoma barbatum *belongs to the stomiid family and the subfamily Melanostominae, commonly referred to as the scaleless black dragonfishes. Note the large photophore behind its eye and the tiny photophores all over its body and fins.*

swimbladder. Worldwide in distribution, the number of species is uncertain; there may be fewer than six, or just one variable species.

The North Atlantic species known as the Ribbon sawtail fish is remarkable for extreme differences between the sexes and its most peculiar larvae. The sex of the larvae cannot be determined until they are about 4cm (1.5in) long. They are, however, so peculiar that they were assigned to a separate genus (*Stylophthalmus*) until it was realized that they were the young of *Idiacanthus*. They are stalkeyed – the eyes are at the end of cartilaginous rods that extend up to one-third of the body length. The body is transparent, the intestine extends beyond the tail, and the pelvic fins are not developed. However, the pectoral fins (lost in the adult) are well developed. The larva also lacks the luminous organs of the adults.

During metamorphosis (the change from larval to adult form), the eye stalks gradually shorten until the eyes rest in an orthodox (normal) position in the orbit of the skull. The pectoral fins are lost and only the female develops pelvic fins. She also grows a luminous chin barbel, develops rows of small luminous organs on the body, and strong jaws with thin, hooked teeth. The male never has a barbel nor teeth, but develops a large luminous organ just below the eye. The female is black, while the male is brown. The male does not grow after metamorphosis and therefore remains less than 5cm (2in) long, whereas the female feeds actively on prey of suitable size and can grow to over 30cm (12in) long.

The general biology of *Idiacanthus* is poorly known. The smallest larvae are caught at the greatest depths and metamorphosing larvae at about 300m (980ft), so they probably spawn at considerable depths. As the catches of larvae are sporadic, it has been thought that they may shoal or otherwise agglomerate. The adults undergo a daily vertical migration from 1,800m (6,000ft) deep during the day to reach the surface at night.

The viperfishes (subfamily Chauliodontinae) contains about six species in the one genus, *Chauliodus*. These are midwater fishes distributed worldwide in oceans between 60° N and 40°S. Sloane's viperfish has been recorded in all oceans, but distinct and discrete populations have been recognized by some authors. "In all oceans" does not necessarily mean that the species is equally and universally distributed within the stated range. Oceans consist of distinct water masses varying in temperature, salinity, current, and food supplies, and the way that fish are distributed within these water masses is exemplified by the distribution of *Chauliodus* species.

Two small species of *Chauliodus*, the Dana viperfish and *C. minimus* (no common name), live respectively in the central water masses of the North and South Atlantic. The larger Sloane's viperfish, lives in the richer waters that flow around the poorer central water masses. It can reportedly grow to over 30cm (12in) long, more than twice the length of the central-water species. Even the oxygen content of water can limit a distribution: *Chauliodus pammelas*, for example, lives only in the deep waters off Arabia and the Maldives, both of which have a low oxygen content. To cope with these conditions, it has gill filaments much longer than those of its relatives.

Chauliodus species are highly specialized predators. The second ray of their dorsal fin is elongated, highly mobile, and has a luminous lure at the end. Their teeth vary in shape, but this variation is remarkably consistent throughout each species. The front teeth on the upper jaw have four sharp ridges near the tip and are used for stabbing. The longest teeth (which imply a remarkable gape) are the front two on the lower jaw. Normally, when the mouth is closed, they lie outside the upper jaw, but when they impale the prey, their natural curvature tends to push the prey into the roof of the mouth. At the base of both the second and third upper jaw teeth there is a small tooth sticking out sideways, which is thought to protect the large luminous organ below the eye.

These specialized, predatory modifications do not end with the teeth. The heart, ventral aorta (main blood vessel), and gill filaments are all much further forwards than is usual; they lie between the sides of the lower jaw, the gill filaments extending almost to the front of it. Bearing in mind the fragility and importance of the gill filaments, how does large prey pass through the mouth without damaging them? The answer lies in the backbone. In almost all fishes the backbone consists of a series of firmly articulated bones that allow normal flexibility. In these viperfishes, however, the front vertebrae generally do not develop and the spinal column remains a flexible rod of cartilage (softer than bone). Although this is normal in embryonic and juvenile states (where it is known as the notochord), bony vertebrae generally replace it in adults.

Its retention in *Chauliodus* enables the head to enjoy a remarkable freedom of movement. Firstly, as the back muscles pull the head upwards, the hinge between the upper and lower jaws is pushed forwards. At the same time, the mouth is opened and the shoulder girdle, to which the heart is attached, is pulled backwards and downwards. Special muscles then pull the gill arches and their filaments downwards, that is, away from the path of the prey. Finally, movable teeth in the throat clutch the prey and slowly transfer it to the elastic stomach. With the prey stowed away, these organs return to normal.

Viperfishes have been seen alive from a deepsea submersible. They hang still in the water, head lower than tail, with the long dorsal fin ray curved forwards to lie just in front of the mouth. The body is covered with a thick, watery sheath enclosed by a thin epidermis or "skin." This gelatinous layer is thickest dorsally and ventrally and thinnest laterally. It contains nerves, blood vessels, and many small luminous organs.

In addition to the lure, *Chauliodus* species have various kinds of luminescent organs (photophores). Along the ventral part of the fish are complex organs with two kinds of secretory cells, a pigment layer and a reflector. This type of organ is specialized below the eye, protected by teeth and transparent bones, with pigment layers and reflectors so arranged that the light shines into the eye. It is believed that this arrangement makes the eye more sensitive to light. Small light organs above and in front of the eye are thought to illuminate possible prey, suggesting that sight is an important factor in feeding.

Scattered throughout the gelatinous sheath, as well as inside the mouth, are small, simple photophores, which, in life, emit a bluish light. They are spherical and their bioluminous product is secreted into the hollow center of the organ. They are controlled by nerves, unlike the ventral luminous organs, and their function is unknown, but some interesting observations have been made.

When a viperfish is relaxed, its ventral organs produce a bluish light. When touched, however, pulses of light illuminate the whole body. In addition to this, it has been demonstrated in an experiment that the intensity of the ventral photophores can be adjusted to match the amount of light received by the fish from above. In the clearest parts of the oceans, all traces of sunlight have been absorbed at about 900m (3,000ft). Interestingly, fishes below that depth lack ventral photophores. *Chauliodus*, however, lives higher and is therefore affected by low light levels. The photophore near the eye varies its intensity with that of the ventral photophores and presumably, by balancing internal and external light levels, viperfishes can produce the correct level of light from their ventral photophores to match the background illumination, thus making themselves less liable to predation from below.

Species of *Malacosteus*, a black-skinned genus of the subfamily Malacosteinae, are known as loosejaws. They have no floor to the mouth and the stark jaws with their long teeth are reminiscent of the cruel efficiency of gin traps (used – illegally – on land to snare mammals). Some species in this family are unusual in that they have a cheek light organ producing a red light (most photophores produce a blue-green light).

The species of the subfamily Melanostominae are called scaleless black dragonfishes, live in all oceans, and are predators. Some species are elongated, while others are squatter, but all have dorsal and anal fins set far back and large, fanglike teeth and rows of small, ventral, luminous organs. Almost all species have chin barbels, which can vary from the very small, through multibranched versions, to ones six times the length of the body.

Deepsea Hatchetfishes and Allies

FAMILY STERNOPTYCHIDAE

In the family Sternoptychidae species are laterally compressed – flattened from side to side – and deepchested and are therefore known as deepsea hatchetfishes. Some have tubular eyes and, in many, the mouth is vertical. Species in the genus *Argyropelecus* have upwardly directed tubular eyes with a yellow, spherical lens. They feed on very small crustaceans. Some species undergo a small daily vertical migration. The genus *Polyipnus* has over 30 species, mostly in the western Pacific – perhaps the best known are Nutting's hatchetfish

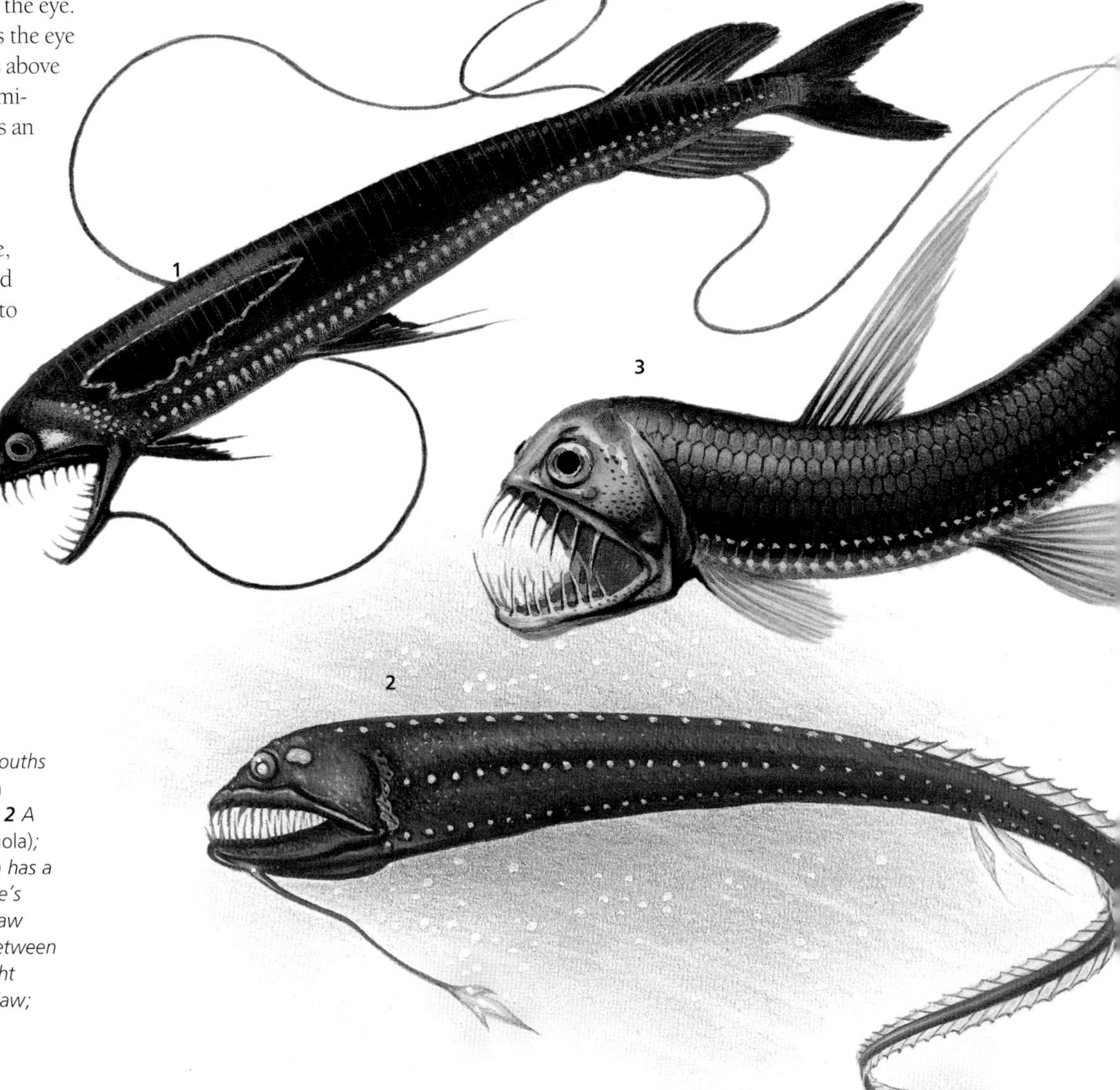

Right *Representative species of bristlemouths and allies:* **1** Grammatostomias flagellibarba *belongs to the scaleless black dragonfishes;* **2** *A female Ribbon sawtail fish* (Idiacanthus fasciola)*;* **3** *The Pacific viperfish* (Chauliodus macouni) *has a row of photophores along its belly;* **4** *Sloane's viperfish* (C. sloani)*;* **5a** *The Stoplight loosejaw* (Malacosteus niger) *has a depth range of between 0–2,500m (0–8,200ft);* **5b** *Here the Stoplight loosejaw displays its remarkably distensible jaw;* **6** *Pacific hatchetfish* (Argyropelecus affinis).

and the Three-spined hatchetfish. All species stay close to land at depths of between 45–450m (150–1,500ft). Like all members of this family, these hatchetfishes have large, elaborate, downwardly pointing light organs.

The photophores of four species in the genus *Sternoptyx*, which includes the Highlight hatchetfish, have been studied intensively. They have two elliptical patches on the roof of the mouth, which lack pigment, reflectors, or colored filters. They luminesce independently of the ventral organs and can glow for about half an hour before gently fading. Apart from, presumably, attracting prey, there is a sort of light guide that lets some of this light to be led close to the eye, where it may be used to balance the light production from the ventral organs so that it matches the background daylight.

The deepsea hatchetfishes are exceedingly beautiful silvery fishes, although what advantage their hatchet shape gives them is unknown. The Pearlsides, a small, spratlike fish often found at night at the surface of the North Atlantic, is thought to be a primitive relative of the deepsea hatchetfishes. It resembles them in many internal details but has a more normal fish shape and is placed in a separate subfamily, the Maurolicinae, along with 13 or so other species belonging to seven genera. The three genera mentioned above have their own subfamly, the Sternoptychinae.

Lightfishes

FAMILY PHOTICHTHYIDAE

Lightfishes are found in the Atlantic, Indian and Pacific oceans, mainly at depths of 200–400m (660–1,300ft) during the day and 100m (330ft) at night. Their diet consists mainly of copepods and they usually feed intensively during the afternoon to early evening. Lightfishes have a similar body shape to the Gonostomatidae and have well developed gill rakers. The lower jaw barbel is absent throughout the family, while the adipose fin is present in all genera except *Yarrella*. KEB/JD

Bristlemouths and Allies

Bristlemouths
Family Gonostomatidae

26 species in 7 genera. Genera and species include: *Cyclothone pygmaea*, Veiled anglemouth (*C. microdon*), *Diplophos orientalis*, *Manducus greyae*, *Vinciguerria*.

Barbeled Dragonfishes
Family Stomiidae

228 species in 27 genera and 6 subfamilies.
Snaggletooths Subfamily Astronesthinae. 30 species in 5 genera. Species include: Snaggletooth (*Astronesthes gemmifer*), Panama snaggletooth (*Borostomias panamensis*), *Heterophotus ophistoma*, *Neonesthes capensis*, *Rhadinesthes decimus*.
Scaled dragonfishes Subfamily Stomiinae. Single genus *Stomias* containing about 11 species. Species include: Alcock's boafish (*S. nebulosus*), Blackbelly dragonfish (*S. atriventer*).
Black dragonfishes Subfamily Idiacanthinae. Fewer than 6 species in one genus or just one variable species. Species include: Ribbon sawtail fish (*Idiacanthus fasciola*), Pacific blackdragon (*I. antrostomus*).
Viperfish Subfamily Chauliodontinae. About 6 species in one genus (*Chauliodus*). Species include: Sloane's viperfish (*C. sloani*), Dana viperfish (*C. danae*), *C. minimus*, *C. pammelas*.
Loosejaws Subfamily Malacosteinae. 15 species in 3 genera. Species include: *Aristostomias lunifer*, Shiny loosejaw (*A. scintillans*), *Malacosteus indicus*, Stoplight loosejaw (*M. niger*), *Photostomias guernei*, *P. mirabilis*.
Scaleless black dragonfishes Subfamily Melanostominae. About 160 species in 16 genera. Species include: Highfin dragonfish (*Bathophilus flemingi*), *Flagellostomias boureei*, *Grammatostomias flagellibarba*, *Melanostomias melanops*, *Odontostomias micropogon*, Longfin dragonfish (*Tactostoma macropus*).

Deepsea Hatchetfishes and Allies
Family Sternoptychidae

49 species in 10 genera and 2 subfamilies.
Subfamily Sternoptychinae c.35 species in 3 genera. Species include: Lovely hatchetfish (*Argyropelecus aculeatus*), Nutting's hatchetfish (*Polypipnus nuttingi*), Three-spined hatchetfish (*P. tridentifer*), Highlight hatchetfish (*Sternoptyx pseudobscura*).
Subfamily Maurolicinae c. 14 species in 7 genera. Species include: Pearlsides (*Maurolicus muelleri*), *Sonoda megalophthalma*, *Thorophos nexilis*, Constellation fish (*Valenciennellus tripunctulatus*).

Lightfishes
Family Photichthyidae

About 18 species in 7 genera. Species include: Slim lightfish (*Ichthyococcus elongatus*), Stareye lightfish (*Pollichthys mauli*), *Yarrella blackfordi*.

Bristlemouths
Gonostomatidae

Barbeled Dragonfishes
Stomiidae

Snaggletooths
Astronesthinae

Scaled dragonfishes
Stomiinae

Black dragonfishes
Idiacanthinae

Viperfishes
Chauliodontinae

Loosejaws
Malacosteinae

Scaleless black dragonfishes
Melanostominae

Deepsea Hatchetfishes and allies
Sternoptychidae

Light From Living Fishes

How fishes produce and use illumination

AS TERRESTRIAL CREATURES WHO, IN THEIR waking hours, are accustomed to light, humans have little appreciation of lifestyles that must cope with continual darkness. Yet that is precisely what life would be like in the depths of the world's oceans were it not for bioluminescence, the production of light by living organisms. Not to be confused with phosphorescence or fluorescence – that is, light produced by nonliving things through "excitement" or "stimulation" of crystals – bioluminescence occurs in living organisms It is the result of the chemical reaction of a substance, usually a compound called luciferin, and an enzyme, referred to as a luciferase. Bioluminescence occurs on land, the best examples probably being fireflies that glow in the evening sky or fungi that glow on the forest floor. It does not, however, occur in freshwater fishes. Its greatest display, however – both in variety and intensity – occurs in the sea.

Of the more than 20,000 living species of fishes, perhaps 1,000–1,500 bioluminesce. None of the lampreys and hagfishes or lungfishes are known to, but six genera of midwater and benthic (bottom-living) sharks and species representing nearly 190 genera of marine bony fishes are luminescent. This is best seen in the lanternfishes (Myctophidae), the bristlemouths (Gonostomatidae), several of the subfamilies of barbeled dragonfishes (Stomiidae) – for example Melanostominae or scaleless black dragonfishes; Malacosteinae or loosejaws; Chauliodontinae or viperfishes – the slickheads (Alepocephalidae), tubeshoulders (Platytroctidae), and the anglerfishes (various families). Several shallow-water and bottom-living fish families also contain luminescent species, and they are better understood behaviorally and physiologically in that they are easier to capture and study. Among these families are the ponyfishes, slimys or slipmouths (Leiognathidae), flashlight fishes (Anomalopidae), pinecone fishes (Monocentridae), *Porichthys* species of toadfishes (Batrachoididae), and several of the cardinal fishes (Apogonidae).

Above *This leftvent species (genus* Linophryne*) of the anglerfish family Linophrynidae is displaying its glowing lure. Species of the family have a barbel hanging from their throat that also generates light.*

Below *A complex luminous organ (photophore) from a deepsea hatchetfish (genus* Argyropelecus*). A common pattern is shown here. The light-producing organs are partly screened by pigment cells and backed by a reflective layer that directs the light to the lens, in some cases through filters that change the color of the light emission.*

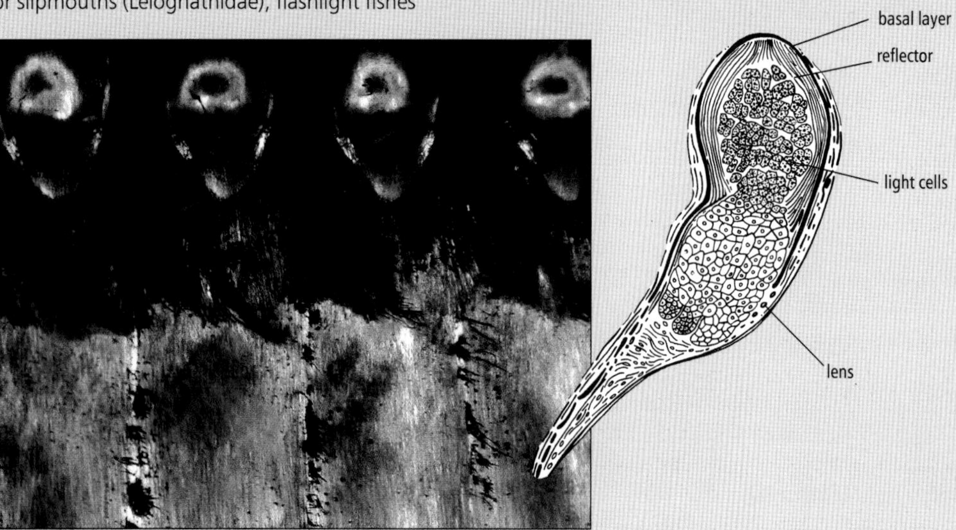

The origin of the light and its associated chemistry may be divided conveniently into two categories. The first are those with self-luminous photophores – specialized structures usually arranged in rows and consisting of highly complex lenses, reflectors, and pigmented screens. The skin photophores produce light via the photogenic (light-generating) cells and reflect it through the lens and cornea-like epidermis (transparent surface tissue). The more than 840 photophores in the belly of the Plainfin midshipman (*Porichthys notatus*) produce a gentle, even glow, the intensity of which can be slowly modulated (modified) to match the downwelling moonlight upon the sandy bottom.

The second category involves luminous bacterial symbionts – bacteria that live in harmony with their host fish. They are maintained in complex organs and are nurtured by the host fish in exchange for a more brilliant level of light. This extrinsic (externally generated) form of illumination does not permit the fish to control the intensity or duration of the bacterial light that is produced. In response to this the hosts have evolved fascinating mechanisms, like lids, that allow them to turn off their light when it

Above *The snaggletooth* Astronesthes niger *develops pale luminous patches with growth on its shoulder and lower sides. This particular species was photographed under ultraviolet light to indicate the fluorescence of photophores on its body and barbel.*

Right *The Flashlight fish has a large light organ below the eye that flashes on and off. The light is produced by bacteria housed in cells well provided with blood vessels. The role of the organ is uncertain, but in parts of the eastern Indian Ocean it is used by native fishermen as a bait.*

would be a hindrance or when it is not needed.

What, then, is the function of bioluminescence in fishes? In the case of most deepsea fishes, it is one of camouflage through counter-illumination. Even at 1,000m (3,300ft) in clear tropical waters, downwelling light would silhouette a lanternfish against the light, thus making it visible to an upward-searching predator were it not for the weak glow of the rows of photophores on its underside, which cancel out the shadowy form of its silhouette and make it "disappear." Other photophores on its body can also be used to advertise its species and its sex. The large suborbital light organs (organs located under the eyes) of *Aristosfomias*, *Pachystomias*, and *Malacosteus* emit red light, which, when coupled with its red-sensitive retina, might act like a "snooperscope" to hunt prey that can see only blue-green shades. Other uses in the deep sea include luring, as has been achieved by many scaleless black dragonfishes with elongate luminous chin barbels, or anglerfishes with luminous escae (bait) at the end of their modified first dorsal spine (ilicium). Concealment is also a function whereby slickheads and certain grenadiers or rattails (Macrouridae) presumably behave like squids and octopuses, leaving a predator snapping at a luminous ink cloud.

The behavior and bioluminescent function of shallow-water fishes is becoming better understood as a result of nocturnal observers with scuba gear and the improvement in aquarium collecting and husbandry. The pinecone fishes (*Monocentris japonicus* and *Cleidopus gloriamaris*) live in shallow water and apparently lure nocturnal crustacean prey to their jaws with the light organs located in their mouth and on their jaws. Bioluminescence is, however, best perfected in the anomalopid flashlight fishes. These small, black, reef-associated fishes possess an immense light organ directly under each eye, capable of emitting enough light to be seen from 30m (100ft) away. In evening twilight they migrate from deep water to feed along the reef edge and return before daylight to the recesses of the deep reef. They use the light for many purposes, including finding food, attracting food, communication, and avoiding predators. Flashlight fishes continually blink, either by raising a black eyelidlike structure over the light organ or by rotating the entire organ into a dark pocket. The living light produced by the Flashlight fish (*Photoblepharon palpebratus*) is the most intense yet discovered.

This brief summary of fish bioluminescence reflects the meager knowledge we have of life in the nocturnal sea. As our ability to descend into the deep ocean at night improves, many more forms of extraordinary behavior associated with bioluminescence are coming to light and our body of knowledge is expanding rapidly. JEM/JD

Lizardfishes and Lanternfishes

FACTFILE

LIZARDFISHES AND LANTERNFISHES

Orders: Aulopiformes, Myctophiformes

About 475 species in about 75 genera and 17 families.

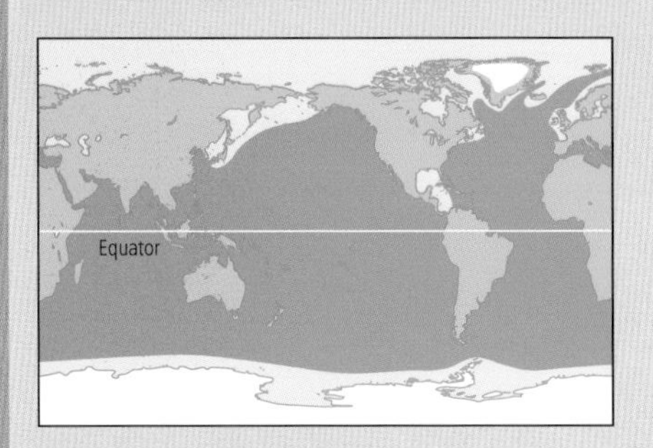

LIZARDFISHES AND ALLIES Order Aulopiformes

About 225 species in c. 40 genera and 15 families. Shallow to deep water of all oceans. **Length:** maximum up to 2m (6.5ft). Families include: **flagfins** (Aulopodidae); **lizardfishes** (Synodontidae); **greeneyes** (Chlorophthalmidae); **tripod fishes** and **grideyes** (Ipnopidae); **barracudinas** (Paralepididae) and **daggertooths** (Anotopteridae), including the **Daggertooth** (*Anotopterus pharao*); **lancetfishes** (family Alepisauridae), including **Longnose lancetfish** (*Alepisaurus ferox*); **sabertooths** (family Evermannellidae); **pearleyes** (family Scopelarchidae); **telescope fishes** (family Giganturidae).

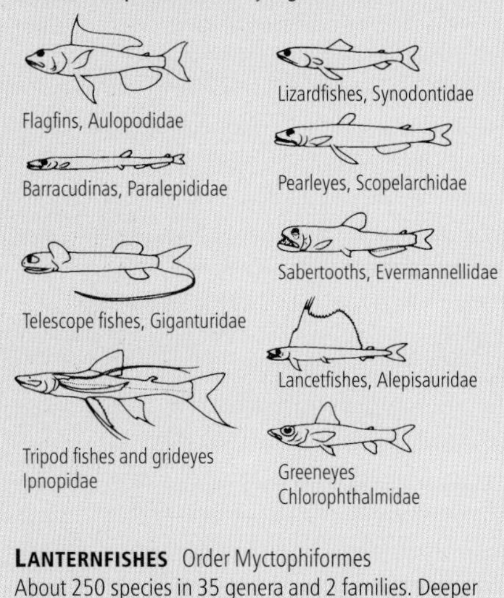

LANTERNFISHES Order Myctophiformes

About 250 species in 35 genera and 2 families. Deeper water of all oceans. **Length:** maximum 30cm (12in). Families: Neoscopelidae; **lanternfishes** (Myctophidae), including the **Blue lanternfish** (*Tarletonbeania crenularis*), **Jewel lanternfish** (*Lampanyctus crocodilus*).

Lanternfishes, Myctophidae   Neoscopelidae

VARIOUS EXTRAORDINARY FISHES BELONG to the two orders of lizardfishes and lanternfishes: tripod fishes, which perch on the bottom in very deep water on their stiffened pelvic fins and lower tail lobe; grideyes of the genus Ipnops, *which have greatly flattened eyes covering the top of the skull; that famous delicacy the Bombay duck; the large predatory lancetfish; the bizarre, highly modified telescope fish; the lanternfishes, speckled with luminous organs; and others.*

The 17 or so families in these orders were formerly considered by some authorities to be more realistically included in one order. The arrangement here follows the latest classifications in which just two families, Myctophidae and Neoscopelidae, comprise the Myctophiformes; all the others are Aulopiformes. The myctophiforms are now hypothesized to be more closely related to more advanced teleosts than to aulopiforms.

Lizardfishes and Allies

ORDER AULOPIFORMES

The anatomically diverse fishes included in the order Aulopiformes are united as a group by a peculiar arrangement of some of the small bones in the skeleton of the gills. Although the precise function of the gill modification is unknown, it is significant because it has not been observed in any non-aulopiform fishes. Within the order numerous arrangements of genera and families have been proposed over the past several decades, but the two most recent comprehensive studies of the group divide aulopiforms into four subgroups.

Only one of those groups, the synodontoids, includes members that primarily inhabit warm shallow waters. Flagfin fishes of the family Aulopodidae, the namesake for the order, are among the most primitive aulopiforms. They live in the Atlantic, parts of the Pacific, and around southern coasts. They are scaled, bottom-living fishes with slender bodies and large heads. The second ray of the dorsal fin is characteristically elongate. An adipose fin is present. The teeth are small and lie in closely packed rows. Small, bottom-living invertebrates are the main food.

Flagfins are surprisingly colorful for fishes that may live as far down as 915m (3,000ft); browns, reds, and pinks are well represented among the dozen or so species. *Aulopus purpurissatus* is a highly colored, edible species from Australia. The edges of the scales are crimson against a purple or scarlet background. The fins are yellow with rows of red spots. Its common name of "Sergeant Baker" apocryphally alludes to the name of the soldier who first caught this fish in New South Wales. There is no evidence for this, but many red-colored fishes have a military common name in allusion to the red coats of British troops.

Most other members of this shallow or moderately shallow warm-water fish group belong to the

Right *One characteristic of the genus* Synodus, *to which this Twospot lizardfish (*S. binotatus*) belongs, is the single band of palatine teeth that occur on each side of the mouth.*

Above *Generally bottom-dwelling predators, this pair of Lighthouse lizardfish* (Synodus jaculum) *from Wakatobi, Indonesia, will hardly leave the substrate except when in flight. This species can be identified by the big black blotch on the caudal peduncle.*

Right *By far the most common reef-dwelling lizardfish, the Variegated lizardfish* (Synodus variegatus) *is often found hiding in the sands of deep lagoons to depths of over 40m (131ft). It preys on small fishes by seizing them from passing schools.*

lizardfish family Synodontidae. Lizardfishes are bottom dwellers, spending time propped up on their pelvic (ventral) fins waiting for their prey to swim by. They get their common name not just from the very lizardlike shape of the head but also from their rapid feeding movements. The most famous family member is the Bombay duck, popular in Indian restaurants as the crispy, salty delicacy eaten before the meal. This dish is the sundried, salted fillet of a large-mouthed, large-toothed fish from the Ganges estuary. Its slender, cylindrical body, with a soft dorsal fin and an adipose fin, is typical of many lizardfishes. Almost all lizardfishes are predatory, using their curved, needlelike teeth to seize fish and invertebrates.

The next subgroup, the chlorophthalmoids, is made up of a diversity of fishes including the moderately shallow greeneyes and the very deep ipnopids. Greeneyes, or chlorophthlamids, are the orthodox members of the group. They are laterally compressed, silvery fishes with large eyes. They grow up to 30cm (12in) long, have dorsal and ventral fins well forward, and an adipose fin directly above the anal fin. The common name for the family comes from the green light reflected by the tapetum lucidum (a reflective layer in the retina of the eye) in some species. Yellow eye lenses are common; the yellow coloring is believed to act as a selective filter enabling the fish to "see through" the downwardly directed luminous camouflage emanating from the ventral light organs of the small crustaceans that form its prey.

Greeneyes are widespread and fairly abundant in the North Atlantic, where they shoal at depths of 200–750m (660–2,460ft). In all species the lateral line system is well developed and greatly expanded into special organs on the snout, head, and gill covers. These organs enable the fish to detect the approach of small prey.

The dim light from the perianal organ (an organ around the anus) is produced by bacteria. The presence of this light is believed to enable a fish to maintain contact with its fellows and perhaps to facilitate mating among those in breeding condition. Ten species previously thought to belong to the greeneye family, including the Cucumber fish (*Paraulopus nigripinnis*), lack a perianal organ.

Right *Representative species of lizardfishes and lanternfishes:* ***1*** *The Indo-Pacific Gracile or Graceful lizardfish* (Saurida gracilis) *is common in shallow lagoons and reef flats;* ***2*** *Antarctic jonasfish* (Notolepis coatsi)*;* ***3*** *The Longnose lancetfish* (Alepisaurus ferox) *is oviparous, with planktonic larvae;* ***4*** *The Metallic lanternfish* (Myctophum affine) *feeds on plankton.*

These ten species are now classified in the genus *Paraulopus* in a new aulopiform family, Paraulopidae, and are believed to be more closely related to the lizardfishes than to the greeneyes. Comprehensive studies of anatomical details often result in major changes in fish classifications.

Tripod fishes of the chlorophthalmoid family Ipnopidae occur as deep as 6,000m (19,700ft). They have been photographed on the deepsea floor resting on their stiffened pelvic fins and the lower lobe of the tail, facing into the current, with their batlike pectoral fins raised over the head in the manner of forward-pointing elk horns. The 18 or so species of tripod fishes are found worldwide in deep water. However, they are found only in a particular type of oceanic water mass called central oceanic water. Each species of tripod fish lives only in a particular area, defined by subtle parameters of temperature and salinity. *Bathypterois atricolor* may occur in waters as shallow as 300m (1,000ft), while *B. longicauda* lives as deep as 6,000m (19,700ft); *B. filiferus* lives only off South Africa, while *B. longipes* is circumglobal. All, however, can live only where the seafloor is composed of ooze or very fine sand, which permits a firm "foothold" for their fins. One puzzle is that there are large areas of deep sea, for example the North Pacific, with ideal conditions (as far as can be seen) in which they do not live. For deepsea fish they are common. Off the Bahamas, intensive surveys have shown there can be almost 90 fish per sq km (233 per sq mile). The batlike pectoral fins have an elaborate nerve supply; thus, they are sensory, but for what use is not known. The eyes are extremely small. The fish feed on small crustaceans that one can only presume are detected by the fins as they drift past in the current. As with deepsea organisms in general, there is much that remains unknown, including their life history.

Tripod fishes are grouped with several other genera, including *Ipnops*, in the family Ipnopidae. *Ipnops* comprises 3 species, one of which is called the grideye, which look somewhat like flattened tripod fish without the tripod. They appear to be eyeless. They have achieved fame because of the enigma of the two large, flat, pale yellow plates on the top of the head. For about half a century it was thought that they were luminous organs that directed light upwards. The advent of deepsea photography revealed that these plates are highly reflective. The flash from a deepsea camera that photographed one was clearly reflected from the plates. Deepsea collecting has made more specimens available to scientists, and the puzzle of the plates has been solved: they are a mixture of modified eyes and skull bones. Each plate, which covers half of the head width, is a transparent skull bone. The reflective layer below is a highly modified eye retina. Ordinary eye structures, such as the lens, have been lost; only the light-sensitive retina remains, spread out over the top of the head and protected by the skull. This is remarkable, but it is difficult to understand what it means to the fish. It can detect light coming down from above, yet cannot focus on an object. Moreover, *Ipnops* eats marine worms, which live below it.

The alepisauroid group of aulopiforms represents another diverse assemblage of fishes, all of which inhabit deep oceanic waters. The barracudinas of the family Paralepididae derive their common name from their superficial resemblance to the predatory barracudas of coral reefs, to which they are not related. Barracudinas are slender fishes with large jaws, many small, sharp, pointed teeth on the upper jaw, and a mixture of larger stabbing teeth on the lower. A small adipose fin is present and the single, soft-rayed dorsal fin is in the rear half of the body. Scales, when present, are fragile and easily shed. This is thought to be an adaptation for swallowing large prey by permitting easier body expansion. In many species there is a fleshy keel between the anus and the anal fin.

The family is widespread but the range of individual species is more limited. *Notolepis coatsi*, a species growing to more than 10cm (4in) long, is confined to Antarctic waters. The Ribbon barracudina (*Arctozenus risso*) inhabits the eastern North Pacific.

Above *The Spinycheek lanternfish* (Benthosema fibulatum) *grows to a maximum length of 10cm (4in) and at night occurs in the upper 200m of the ocean. The light organs (photophores) are visible here as small shiny dots.*

The barracudinas are unusual among aulopiforms in that they have light organs. In the genus *Lestidium* these organs consist of ducts that extend from the head to the ventral fins. The bacteria therein produce a bright, pale-yellow light. The function of the luminosity is unknown. It is probably not for camouflage because the species with luminous organs have an iridescent skin as well as being translucent. Perhaps the organs function as lighthouses, enabling individuals of the same species to recognize one another.

The eyes of barracudinas are arranged in such a way that their best field of binocular vision is directly downward. However, barracudinas seen from submersibles oriented themselves head-up, or nearly so, in the water so that they were looking along a horizontal plane (straight ahead). It is also conjectured that the head-up posture presents a smaller silhouette for predators beneath to spot.

The Daggertooth may grow to 1.5m (4.9ft) and lives in cool polar waters. It was formerly considered part of the barracudina family, but most experts now place it and two congeners in their own family, the Anotopteridae. The Daggertooth can take prey of up to half of its own length; an intact specimen 75cm (30in) long and taken from a whale's stomach in the Antarctic was found to contain two barracudinas, 27 and 18cm (11 and 7in) long, both engorged with krill.

Lancetfishes are large, predatory, midwater alepisauroids. None have luminous organs but all have large stabbing teeth and distensible stomachs. Species in the genus *Alepisaurus* can reach 2.2m (7.2ft) in length, but their bodies are so slender that a fish of this size will only weigh 1.8–2.3kg (4–5lb). They are similar in shape to the Daggertooth but have a long, very high dorsal fin that can be folded down into a deep groove along the back and become invisible. The function of this large dorsal fin is unknown but it has been suggested that it might be used like the similarly large dorsal fin of the sail fishes in helping to "round up" shoals of small fishes. The Longnose lancetfish feeds on deepsea hatchet fish, barracudinas, squids, octopuses, and almost anything else available. Most of the fishes eaten do not undertake daytime migrations. The very abundant lanternfishes, which do undertake daily migrations, are rarely eaten by lancetfish. Lancetfishes are, however, eagerly eaten by tunas and other surface predators when the opportunity arises.

Other alepisauroids include the closely related mesopelagic sabertooths of the family Evermannellidae and pearleyes of the family Scopelarchidae, some of which may enter more shallow areas at night. As the name suggests, sabertooths have long fanglike teeth typical of many mesopelagic fishes. Pearleyes obtained their name from a white spot, or pearl organ, on the eye. This organ is believed to help the fish detect light from a wide peripheral area. Pearleyes have slightly telescopic eyes that are pointed up and forward.

The last subgroup of aulopiforms contains the telescope fishes of the family Giganturidae. They are cylindrical, silvery fishes with forward-pointing tubular eyes set back on the snout. Scales and luminous organs are absent. The pectoral fin is set high up on the body; the caudal fin has an elongated lower lobe; adipose and pelvic fins are missing. Also missing are a large number of bones. Indeed, the adult fish has been subjected to a loss of many features typically present in its relatives. Many of the remaining bones are still cartilaginous. The teeth are large, sharp, and depressible, thus easing the passage of large prey. The inside of the mouth and stomach are lined with dense, black pigment, which, it has been suggested, blacks out the luminous organs of its last meal. The abdominal region is elastic; thus, telescope fish are capable of swallowing food much larger than themselves. The pectoral fins are inserted above the level of the gills and are thought to help ventilate them during the slow passage of a large fish down the throat. Certainly, some such device would have to be present because the normal water currents cannot flow through the blocked mouth to the gills.

Historically, telescope fishes have been classified in their own order. The gill skeleton of adults is so highly modified that some of the structures involved in the gill modifications of other aulopiforms are absent. However, in very young telescope fishes, when the gill skeleton is still cartilaginous, the aulopiform arrangement of the gill structures is evident. It is ironic that the young stages provide the best evidence that telescope fishes are aulopiforms: they are so different from adults that until recently they were thought to represent a separate family. Telescope fishes are widespread in tropical and semitropical oceans at depths down to about 3,350m (11,000ft). All are small fishes, rarely exceeding 15cm (6in).

Lanternfishes and Allies

ORDER MYCTOPHIFORMES

The 300 or so species in the order Myctophiformes are commonly known as lanternfishes because of the impression created by their extensive speckling of luminous organs. More precise common names for species within this order include Jewel lanternfish and Blue lanternfish, conveying the conspicuous and ornamental nature of the photophores. These carnivorous fishes live in the middle depths of all oceans and rarely exceed 30cm (12in) long. The overall body shape is very similar in all species, but the pattern of light organs differs from species to species and is the basis for species definition. As well as the small photophores there are larger organs – upper and lower glands – mostly near the tail, which indicate the sex of the fish. Usually the female lacks the upper glands and the lower ones are less conspicuous or even absent. The fish react strongly to light signals: one aquarium fish appeared to be most interested in the researcher's luminous wristwatch. Brighter light sources have little effect.

Lanternfishes live 300–700m (1,000–2,300ft) down. There are both silvery bodied and dark-bodied forms. Many display an upward migration at night, occasionally to as little as 50m (165ft). Those with functional swimbladders have less fat than those without. Fat, being lighter than water, helps the latter maintain neutral buoyancy. Contrary to this, the Blue lanternfish may have no gas in its swimbladder, almost no body fat, and is negatively buoyant – denser than water. KEB/CCB

Characins, Catfishes, Carps, and Allies

CARPS, CATFISHES, CHARACINS, SUCKERS, loaches, and their allies are the dominant freshwater fishes in Eurasia and North America and arguably so in Africa and South America. (Only the catfishes are native to Australia.) The approximately 6,500 species are predominantly freshwater fish. Just two families of catfishes and one species of cyprinid are found at sea, although several genera may spend time in brackish water.

The major groups are well defined (although their relationships remain the subject of much controversy). However, one species – *Ellopostoma megalomycter* from Borneo – is a puzzle, since it does not quite fit in anywhere, although some authorities believe that it belongs within the river loach family (Balitoridae).

Diverse and Numerous

CLASSIFICATION AND MORPHOLOGY

The superorder Ostariophysi as a whole is divided into two series, the otophysi and the anotophysi, the former containing two hundred times more species than the latter. The otophysi have two main unifying characters. First, the presence of an "alarm substance," or pheromone, secreted from glands in the skin when a fish is threatened and which causes a fright reaction in other otophysans. Perhaps understandably, the alarm substance is not present in families of heavily armored catfishes, but less comprehensible is its absence in certain species of the cave-dwelling characins and cyprinids.

The second diagnostic character is the Weberian apparatus. This is an elaborate modification of the first few vertebrae (individual backbones) into a series of levers, known as ossicles, that transmit compression waves of high-frequency sound received by the swimbladder to the inner ear. Consequently, otophysans have acute hearing. No one is certain how this complex "hearing aid" developed, but a clue may come from the anotophysi, which possess "head ribs" that may represent a prototypic form of such a mechanism.

The anotophysi are a diverse and somewhat incongruent group. The milkfish (*Chanos chanos*, sole member of the family Chanidae) is a food fish from the region of Southeast Asia. It looks rather like a large herring with small, silvery scales but lacking ventral scutes ("plates"). Milkfish are cultured intensively in fish ponds in many areas and can grow to well over 1m (3.3ft) in length. They can also tolerate a wide range of salinity.

The Beaked sandfish (*Gonorhynchus gonorhynchus*) is the only member of its family (Gonorhynchidae). A shallow-water species from the temperate and tropical Indo-Pacific, it has an elongate body, long snout, and a ventral mouth. There is no swimbladder. What could be fossil relatives of this species have been found in Alberta, Canada.

Unlike *Gonorhynchus*, the Hingemouth or Snake mudhead (*Phractolaemus ansorgii*) from West African freshwaters has a dorsal mouth that extends like a short periscope. Its swimbladder is divided into small units and can be used for breathing atmospheric air. Growing to little more than 15cm (6in) long, it is confined to the Niger and parts of the Zaïre basin.

The remaining anotophysan family, made up of the knerias and their closest relatives (Kneriidae), consists of small freshwater fishes that feed on algae in tropical and nilotic African freshwaters. They exhibit marked sexual dimorphism whereby the male has a peculiar rosette, of unknown function, on its operculum or gill cover. The genera *Cromeria* and *Grasseichthys* are neotenic, that is, they become sexually mature while still having a larval body form. These two genera are considered by some authorities to be kneriids; others place them in a separate family. Both of these small, transparent fishes from West African rivers – unlike their fellow family members – lack scales and a lateral line.

The characins, catfishes, carps, and New World knifefishes are a highly successful group and display a remarkable mixture of evolutionary conservatism and extreme radicalism, which, when coupled with the plasticity or variability at the species level, makes their taxonomy very difficult. For such common fish, they are an enigmatic group. KEB/JD

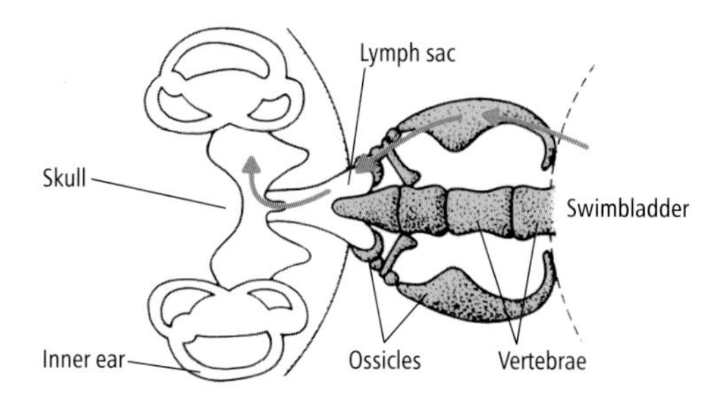

Above *Weberian apparatus seen from above, a characteristic feature of many fishes belonging to the series Otophysi. It transmits vibrations from the swimbladder to the inner ear, giving the fish greatly enhanced hearing.*

Right *With their short, triangular teeth equipped with razorlike edges for stripping flesh, piranhas (here, a Black-eared Piranha,* Serrasalmus notatus*) are supremely well adapted for feeding on other fishes and carrion. Observations in captivity suggest that the alleged "feeding frenzy" does not occur unless about 20 fishes are gathered together.*

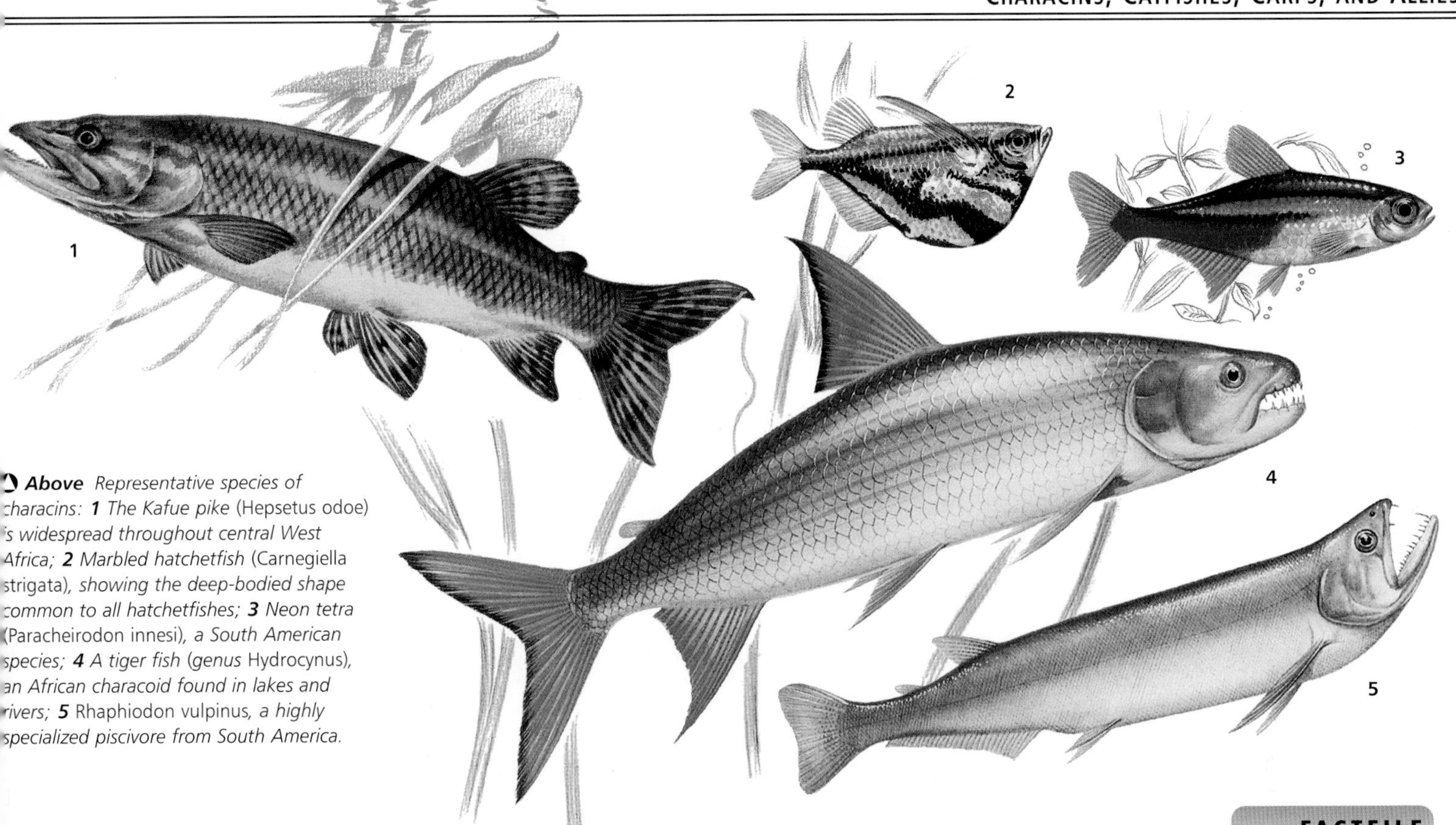

Above Representative species of characins: **1** The Kafue pike (Hepsetus odoe) is widespread throughout central West Africa; **2** Marbled hatchetfish (Carnegiella strigata), showing the deep-bodied shape common to all hatchetfishes; **3** Neon tetra (Paracheirodon innesi), a South American species; **4** A tiger fish (genus Hydrocynus), an African characoid found in lakes and rivers; **5** Rhaphiodon vulpinus, a highly specialized piscivore from South America.

Characins and Relatives

ORDER CHARACIFORMES

There are over 1,340 living species of characiforms, about 210 of which are found in Africa and the remainder in Central and South America. This discontinuous distribution implies that some 100 million years ago, characiforms were widespread in the area of the landmass known as Gondwanaland, which later split to form Africa and South America, Antarctica, and Australia.

Superficially, the characiforms resemble members of the carp family (cyprinids) but they usually have a fleshy adipose ("second" dorsal) fin between the caudal and true dorsal fins and have teeth on the jaws but not in the pharynx or throat. In addition to their complete functional set of teeth, characiforms also have a replacement set behind those in current use. In some species all the "old" teeth on one side of the upper and lower jaws fall out and the replacements take their place. Once they are firmly in position, the teeth on the other side of the jaws are replaced. In predatory characiforms, all the old teeth drop out and are rapidly replaced in one go. As soon as the replacement teeth are functional, a new set of teeth begins to grow in the tooth-replacement trenches.

There are three families (with several subfamilies) of characiforms in Africa. They are both carnivorous and omnivorous, but are less varied than the Neotropical species. The most primitive African characoid is the Kafue or African pike (*Hepsetus odoe*), the sole member of the family Hepsetidae. It is a fish eater of the lurking-predator type, with a large mouth equipped with strong, conical teeth that prevent the prey's escape. Unusually for characiforms, it lays and guards its several thousand eggs in a floating foam nest (considered a great gastronomic delicacy). On hatching, the larvae hang from the water surface by special adhesive organs on their heads.

The fin eaters, or ichthyborids, are elongate members of the citharinids (family Citharinidae) that subsist by biting mouthfuls of scales and nipping notches out of other fishes' fins. Their close relatives are two groups of relatively harmless fishes: the grass eaters or distichontids, with which they share the subfamily Distichodontinae, and the moon fishes, or citharinids (subfamily Citharininae). Their distinctive feature is scales with a serrated edge (ctenoid scales), unlike all other characins, which have smooth (cycloid) scales.

The final African characiform family is the Alestidae – the African tetras – containing some 18 genera and 100 species. The genus *Alestes* is one of the best-known groups of characoids, thanks to their popularity with aquarists. They are fairly colorful fishes, often with a single lateral stripe and red, orange, or yellow fins. The body can be short and deep or elongate (fusiform). All show sexual dimorphism in anal fin shape; in females the margin is completely straight, while in males it is convex. These fishes also exhibit a curious sexual dimorphism in the caudal vertebrae. Yet how they recognize it, and what benefit it confers on them, is unknown. The African tetras have very strong, multicusped teeth, ideal for crushing and grinding their food of insects, fish and insect larvae, plankton, and assorted vegetation.

The tigerfishes or water dogs, a genus containing several highly predatory species, have gained notoriety in their native lands. Their generic common name derives from their long, conical teeth, which overlap the outside of the jaws when the mouth is shut, and from their black body stripes (although these are horizontal rather than vertical). The Goliath tigerfish (*Hydrocynus goliath*) from the Congo region grows to over 1.5m (5ft) and weighs more than 45kg (100lb) and is so predatory that there are even unsubstantiated reports of them attacking people. Tigerfishes and water dogs are excellent sport fish. Oddly, tigerfish lose all their old teeth at once. As a result, toothless individuals have occasionally been caught, but it only takes a matter of days for the replacement teeth to emerge and become functional.

FACTFILE

CHARACINS, CATFISHES, & ALLIES

Superorder: Ostariophysi

Orders: Characiformes, Siluriformes, Cypriniformes, Gymnotiformes, Gonorhynchiformes

About 6,500 species in at least 960 genera and around 60 families.

Distribution Worldwide, largely freshwater.

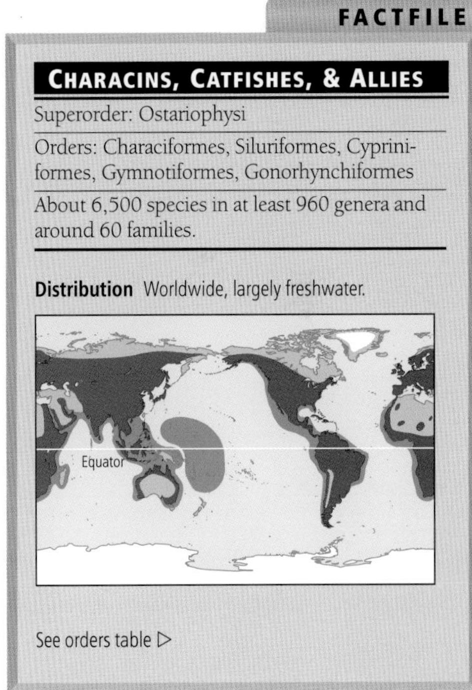

See orders table ▷

Characins, Catfishes, Carps, New World Knifefishes, and Milkfishes

SERIES: OTOPHYSI

Characins and relatives
Order Characiformes

Freshwaters of Africa, S and C America, and southern N America. Length maximum 1.5m (5ft). Over 1,340 species in over 250 genera and probably 15 families. Families, genera, and species include: Characidae, including Cardinal tetra (*Paracheirodon axelrodi*); African tetras (family Alestiidae), including tiger fishes (genus *Hydrocynus*); freshwater hatchet fishes (family Gasteropelecidae); croaking tetras or glandulocaudines (subfamily Glandulocaudinae); Lebiasinidae, including Splashing tetra (*Copella arnoldi*); Serrasalminae, including piranhas (e.g. genera *Pygocentrus* and *Serrasalmus*), silver dollars (e.g. genera *Metynnis, Myleus*), wimple piranhas (genus *Catoprion*). The Naked characin (*Gymnocharacinus bergi*) of Argentina is classed as Endangered.

Note: Characiform classification underwent a major revision in the late 1990s and early 2000. As a result, families and subfamilies changed; the new systematics are reflected in the following pages. A large group of species – including many of the small, colorful, best-known tetras – were also earmarked for further study and placed in a group designated as *incertae sedis*, i.e. of uncertain identity.

Catfishes
Order: Siluriformes

Most habitable freshwaters; members of two families (Ariidae, Plotosidae) can inhabit tropical and subtropical seas. Length maximum 3m (10ft). Over 2,400 species in over c. 410 genera and 34 families. Families, genera, and species include: loach catfishes or amphiliids (family Amphiliidae); callichthyid armored catfishes (family Callichthyidae); walking catfishes (family Clariidae), including the Walking catfish (*Clarias batrachus*); crucifix fish (family Ariidae); North American freshwater catfishes (family Ictaluridae), including Blue and Channel catfishes and bullheads (genera *Ictalurus* and *Ameiurus*), flatheads (genus *Pylodictis*) and madtoms (genus *Noturus*); suckermouth armored catfishes or loricariids (family Loricariidae); parasitic catfishes (family Trichomycteridae), including candirús (genera *Vandellia, Branchioica*); long-whiskered, antenna catfishes or pimelodids (family Pimelodidae); schilbeids (family Schilbeidae); sheatfishes (family Siluridae), including European wels (*Silurus glanis*); Asian hillstream catfishes or sisorids (family Sisoridae). Thirty-seven species of catfishes are currently classed as threatened to some degree, including the Critically Endangered Smoky madtom (*Noturus baileyi*), Barnard's rock catfish (*Australoglanis barnardii*), and Cave catfish (*Clarias cavernicola*).

Carps and Allies, or Cyprinoids
Order: Cypriniformes

N America, Europe, Africa, Asia almost exclusively in freshwater. Length maximum 3m (10ft). About 2,660 species in around 279 genera and 5 families. Families, genera, and species include: cyprinids (family Cyprinidae), including bighead and silver carps (genus *Hypophthalmichthys*), bream (genus *Abramis*), Common carp (*Cyprinus carpio carpio*), Grass carp (*Ctenopharyngodon idella*), mahseers (genus *Tor*), Roach (*Rutilus rutilus*), snow trouts (genus *Schizothorax*), squawfish (genus *Ptychocheilus*), Tench (*Tinca tinca*), Yellow cheek (*Elopichthys bambusa*); algae eaters or gyrinocheilids (family Gyrinocheilidae); river loaches (family Balitoridae or Homalopteridae); "true" loaches (family Cobitidae), including weather fish (*Misgurnus fossilis* and *M. anguillicaudatus*); suckers or catostomids (family Catostomidae). One hundred and ninety-one species of cyprinids are currently classed as threatened in some measure.

New World Knifefishes
Order Gymnotiformes

C and S America, exclusively in freshwate[r]. Length maximum 2.3m (7.5ft). About 62 species in 23 genera and 6 families. Families, genera, and species include: Ghost knifefishes (family Apteronotidae); Electri[c] knifefish or Electric eel (*Electrophorus electricus*); naked-back knifefishes (family Gymnotidae).

SERIES: ANOTOPHYSI

Milkfish and Allies
Order: Gonorhynchiformes

Indo-Pacific Ocean; freshwaters in tropica[l] Africa. Length maximum 2m (6.5ft). Abo[ut] 35 species in 7 genera and 4 families. Gen[era] and species include the Milkfish (*Chan[os] chanos*), Beaked sandfish (*Gonorhynchus gonorhynchus*), and Hingemouth or Snake mudhead (*Phractolaemus ansorgii*).

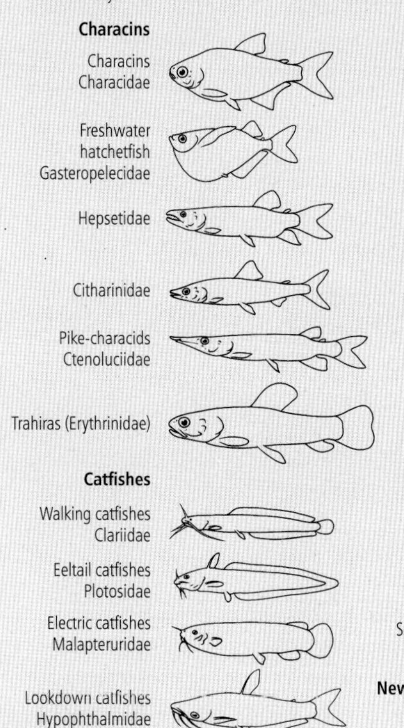

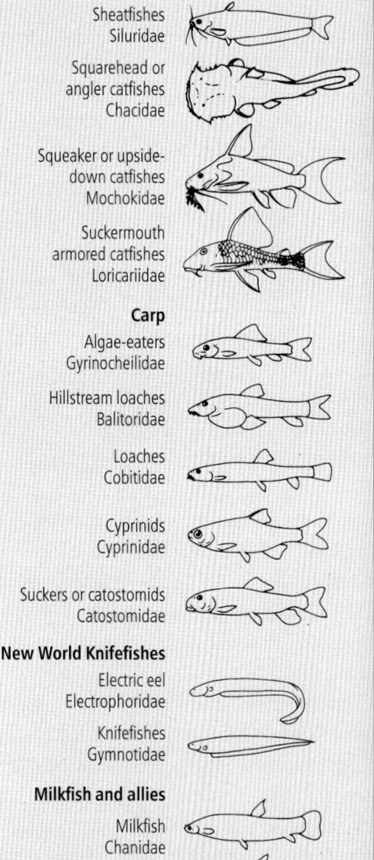

Below *The term "tetra" covers a multitude of small characins from Sout[h] America and Africa. Rummy-nosed tetras* (Hemigrammus rhodostomus) *inhabit the Orinoco and Amazon river basins in Brazil and Venezuela.*

The diversity in Neotropical characins ranges from voracious predators, through tiny vegetarians, to blind subterranean species. The piranhas are the most fabled of all predatory fishes. They are stocky and tough, having a deep head and short powerful jaws with triangular, interlocking, razor-sharp teeth. They are shoaling animals, and feed communally on smaller fish or, allegedly, on larger injured or supposedly healthy prey.

Each piranha makes a clean bite of about 16 cubic cm (1cu in) of flesh. Shoal size then dictates the speed with which the victim is despatched. In some areas, the feeding frenzy is triggered by blood in the water and it takes only a few minutes for a victim to be reduced to little more than a skeleton. People wading or bathing in rivers have also been attacked, but these are rare occurrences. One legend relates how a man and horse fell into the water and were later found with all the flesh picked off, yet the man's clothes were found to be undamaged. Ironically, in reality it is humans who customarily prey on piranhas – large numbers are fished and eaten (usually fried) by river people (*caboclos*) all along the Amazon and other South American rivers.

Piranhas rarely exceed 60cm (2ft) and are excellent sport fish. Anglers use a stout wire trace to catch them or, quite simply, a hand line with a baited hook. Such is the appetite of hungry piranhas that an excellent catch can be made in a remarkably short time.

The piranhas' close relatives include vegetarian or omnivorous genera known to aquarists as silver dollars and pacus. Although they are similar in appearance, their nature is quite unlike that of the piranhas. Pacus(*Colossoma macropomum*) in particular, have strong jaws and grinding teeth with which they can crack hard seeds and fruits. The silver dollars (*Metynnis* and allies) are more straightforward plant eaters.

The wimple-piranhas (*Catoprion* spp.) eat the scales of other fish. Their lower jaw is longer than the upper and their teeth are everted, that is, they point outward, which enables them to scrape the scales of their prey in a single upward swipe. In the presence of wary potential prey, they live on insects and other small invertebrates.

The genera of South American "salmons" have many features that suggest they are a primitive group. *Catabasis* is known only from a single preserved specimen. These fish are presumed to be extant but none have been seen since this specimen was caught in 1900. The trout predator tetras (*Salminus* spp.) are known as *dourados* in Brazil and are, apparently, the most primitive characiforms known. Despite its importance as a genus of food and sport fish, the classification of the four species of *Salminus* is uncertain.

The cachorros (*Acestrorhynchus* spp.) are the Neotropical genus equivalent of the African genus *Hepsetus*. They are streamlined pikelike predators with formidable teeth. They often hunt in open water and are capable of producing short, high-speed bursts; they can also leap out of the water.

The family of freshwater hatchetfish (Gasteropelecidae) is completely different. It consists of deep-bodied fish with long pectoral fins; they rarely exceed 10cm (4in) and are capable of powered flight. The deep chest houses powerful muscles, which are necessary for turning the pectoral fins into wings. Prior to flight, hatchetfish may "taxi" for distances of up to 12m (40ft), for most of which the tail and chest trail in the water. The flight phase is marked by a buzzing caused by the very rapid flapping of the fins. The flight distance rarely exceeds 1.5m (5ft) – but can be as long as 39m (100ft) in optimal conditions – and they have been seen at a height of 90cm (3ft) above the water, although a height of up to 10cm (4in) is most common. The energy cost of flight to the fish is unknown, but it is thought that flight is used when they are frightened by predators. Normally, hatchetfish are found near the water surface feeding on insects. Species of elongate hatchetfishes (*Triportheus* spp.) also have their chest developed into a sort of keel and have large winglike pectoral fins, similar to those of the other hatchetfish. These fish are able to use the breast muscles and pectoral fins to jump about 1m (3.3ft) above the water surface to escape predators, but this cannot be regarded as true flight.

Most characins are generalized egg scatterers but in some, there is specialized breeding behavior. The male croaking tetras or glandulocaudines, for example, have scales modified into special glands, known as "caudal glands," at the base of the caudal fin or tail. These glands secrete a pheromone – a chemical that the opposite sex are supposed to find irresistible. If the pheromone alone is insufficient to attract a mate, some males are also equipped with a wormlike lure – particularly well developed in the Swordtail characin (*Stevardia riisei*) to signal to the female of his choice. It is unknown whether this is intended to be a prey mimic or whether it is a visual cue to induce the female to approach, align herself at the right angle beside the male, and allow him to insert sperm into her genital (sexual) opening. These are the only characoids believed to employ internal egg fertilization.

The Splashing tetra (*Copella arnoldi*), a member of the family Lebiasinidae, lays its eggs on overhanging leaves or rocks, thus avoiding the high predation that eggs experience in water. The male courts a gravid female until the point of egg laying is reached. Leading her to the overhang of his choice about 3cm (1in) above the water, he makes trial jumps up to this spawning site. The female then follows, sticks briefly to its surface, using water surface tension, and lays a few eggs. The male then leaps and fertilizes them. Alternatively, both fish may jump together. This process continues until about 200 eggs are laid and fertilized. After spawning, the female goes on her way, but the male remains nearby to splash water over the eggs until they hatch. Hatching takes about three days, and once the fry fall into the water, the male's "nursemaid" activities are over.

One Panamanian characin, the Sábalo pipón (*Brycon petrosus*) also lays its eggs out of the water. This species indulges in terrestrial group spawning, with about 50 adults moving by lateral undulations or tail flips onto the banks and laying eggs on the damp ground. Males are distinguished by a convex anal fin and short, bony spicules (tiny needle-like structures) on the fin rays. There is no parental care and the eggs take about two days to hatch.

Most of the South American tetras, fishes beloved of aquarists, have received their common name from a contraction of their former scientific group (subfamily) name, Tetragonopterinae (most are now regarded as *incertae sedis*). Many of these highly successful fishes are brilliantly colored, which, apart from rendering them commercially important in the ornamental aquatic world, holds the members of shoals together in the wild. Tetras are omnivorous to the extent that they eat anything they can fit into their mouths, although most are predatory, feeding largely on small insects and aquatic invertebrates. Within this group is the Mexican Blind Cave Characin (*Astyanax jordani*). BB/JD

Catfishes

ORDER SILURIFORMES

There are about 2,400 species of catfishes, assigned to some 34 families. Most are tropical freshwater fishes, but some inhabit temperate regions (Ictaluridae, Siluridae, Diplomystidae, and Bagridae) and two families (Plotosidae and Ariidae) are marine.

Catfishes are named for their long barbels, which give them a bewhiskered appearance reminiscent of cats (though barbels are not present in all species and are not diagnostic for the group). Characters defining catfishes include the fusion of the first 4–8 vertebrae, often with modification of the chain of bones or ossicles connecting the swimbladder to the inner ear; lack of parietal bones, i.e. a paired bone on the roof of the skull; a unique arrangement of blood vessels in the head; an absence of typical body scales and strong dorsal and pectoral fin spines.

Although catfishes lack typical scales, the bodies of many are not entirely naked. The thorny, talking catfishes (doradids), the Asian hillstream catfishes (sisorids), and the loach catfishes (amphiliids) all have bony scutes or plates around the sensory pores of the lateral line and sometimes along the back. The loricariid and callichthyid armored catfishes may be completely encased in these scutes. Strong serrated pectoral and dorsal fin spines are widespread in catfishes. Locking mechanisms keep these spines erect and these, coupled with the bony armor, must deter potential predators. Catfish swimbladders may be partially or completely enclosed in a bony capsule. Why this should be is not clear, since the most obvious association of a reduction of the swimbladder and benthic (bottom-hugging) habits, does not hold. For example, reduced and encapsulated swimbladders occur in fast-swimming species of lookdown catfishes (Hypophthalmidae) and bottlenose or barbel-less catfishes (Ageniosidae), while the benthic electric catfishes (Malapteruridae) have the largest swimbladders of all.

There are more catfish species in South America than in the rest of their area of distribution. Both the world's smallest and largest catfishes occur here. The Spiny dwarf catfish (*Scoloplax dicra*, of the family Scoloplacidae) of Bolivia is a minute, partially armoured fish whose total adult length is less than 13mm (0.5in),whereas the Jáu (*Paulicea* sp.) of the family Pimelodidae from Amazonia, can grow to more than 3m (about 10ft). The European wels (*Silurus glanis*) of the family Siluridae can attain a similar size. Of the 16 families in South America, most live in the Amazon Basin, but four are endemic to the Andes.

The suckermouth armored catfishes (Loricariidae) are the largest of all catfish families, with over 600 species. As their name suggests, their mouths are suckerlike with thin, often comblike teeth adapted to scraping algal mats. Most species are active at night and hide during the day in crevices and logs. The males of some species bear long spines on their opercular apparatus (the bones forming the gill cover) and use them in cheek-to-cheek territoriality fights with other males. Twig catfish (*Farlowella* sp.) are especially slim loricariids – long, thin, and resembling dead twigs – hence their name.

Closely related to the loricariids are the climbing catfishes (family Astroblepidae). The family's 40 species inhabit Andean torrents. These specialized fishes are able to climb smooth, almost vertical rock faces by utilizing the muscles of the ventral surface and using their pelvic fins to form a sucker.

The callichthyid armored catfishes (Callichthyidae) have mail-like plates thought to be able to resist desiccation when ponds dry up. Several genera (i.e. *Callichthys*, *Lethoplosternum*, *Megalechis* and *Hoplosternum*) can withstand marked temperature changes and propel themselves over dry land with their pectoral spines. They also build floating bubble nests. The genus *Corydoras* – with nearly 200 species – is well-known to aquarists, who have recorded the breeding habits and development of many species. Unusually, it has been shown that the females of some species of *Corydoras* drink sperm when mating and channel it rapidly (and without any deterioration) through their gut, releasing it through their anus, onto a pouch created by their pelvic fins and in which they hold several eggs. Once these sticky eggs have been fertilized, the female places them on a suitable surface and abandons them.

The doradids vary greatly in size, from *Physopyxis lyra* (5cm, 2in) to the Soldier or Snail-eating cat (*Megalodoras irwini*) (1m, 3.3ft). Most species are bottom dwellers and, surprisingly, play an important part in seed dispersal by their fruit-eating habits (the seeds are not digested and pass through the fish unharmed).

The long-whiskered or antenna catfishes (Pimelodidae), with 300 species, is a family of great morphological diversity and includes some of the largest known catfishes, e.g. the Red-tailed cat (*Phractocephalus hemioliopterus*). Most species are omnivorous, but the larger ones are fish eaters or carnivores. One species is recorded as having eaten monkeys that fell in the river. Pimelodids are commercially important in parts of the Amazon Basin.

The parasitic catfishes (Trichomycteridae) contain some 155 species. Some are parasites, inhabiting and laying their eggs in the gill cavities of the larger pimelodid catfishes. The candirú is perhaps the most notorious representative of the family. Mammals (including humans) have had their urethra penetrated by this slender fish when urinating while immersed in a stream. The candirú probably mistakes the flow of urine for water being expelled from the gill chamber of a large catfish. It therefore swims "upstream" and enters the urethra, where it lodges itself with its powerful gill

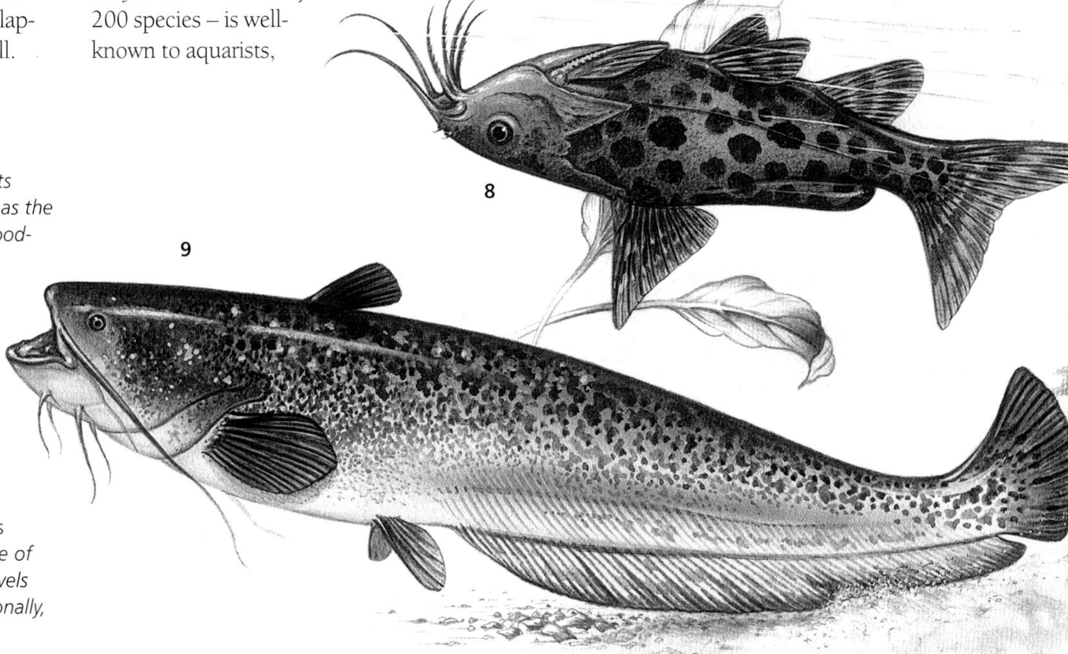

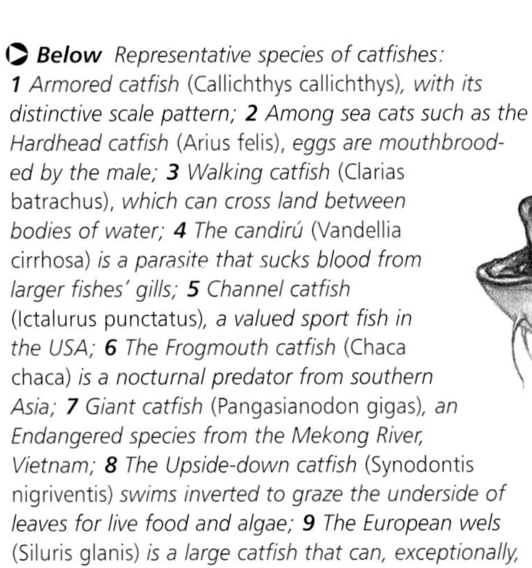

Below *Representative species of catfishes:* **1** *Armored catfish* (Callichthys callichthys), *with its distinctive scale pattern;* **2** *Among sea cats such as the Hardhead catfish* (Arius felis), *eggs are mouthbrooded by the male;* **3** *Walking catfish* (Clarias batrachus), *which can cross land between bodies of water;* **4** *The candirú* (Vandellia cirrhosa) *is a parasite that sucks blood from larger fishes' gills;* **5** *Channel catfish* (Ictalurus punctatus), *a valued sport fish in the USA;* **6** *The Frogmouth catfish* (Chaca chaca) *is a nocturnal predator from southern Asia;* **7** *Giant catfish* (Pangasianodon gigas), *an Endangered species from the Mekong River, Vietnam;* **8** *The Upside-down catfish* (Synodontis nigriventis) *swims inverted to graze the underside of leaves for live food and algae;* **9** *The European wels* (Siluris glanis) *is a large catfish that can, exceptionally, grow to 5m (16.4ft).*

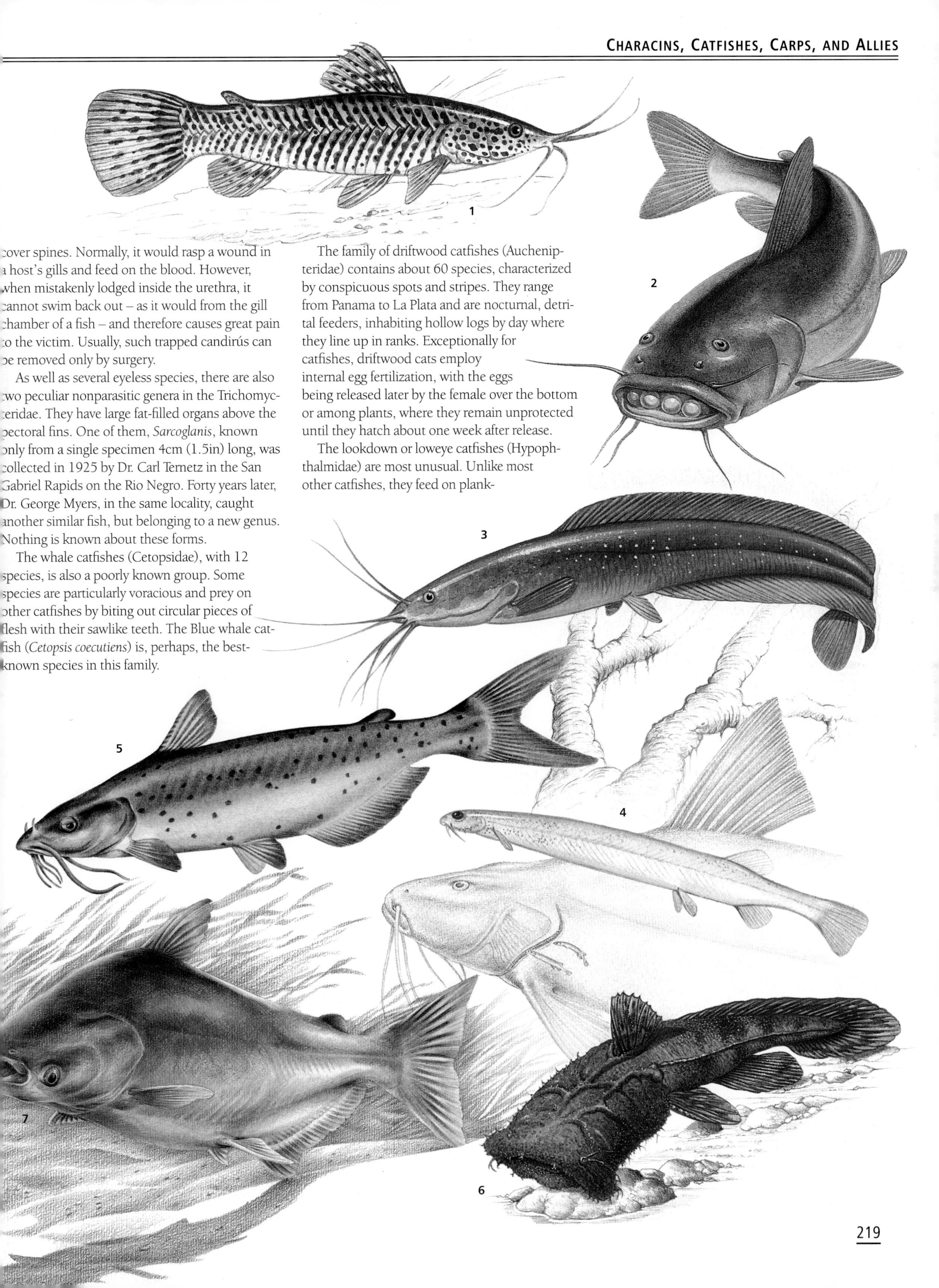

cover spines. Normally, it would rasp a wound in a host's gills and feed on the blood. However, when mistakenly lodged inside the urethra, it cannot swim back out – as it would from the gill chamber of a fish – and therefore causes great pain to the victim. Usually, such trapped candirús can be removed only by surgery.

As well as several eyeless species, there are also two peculiar nonparasitic genera in the Trichomycteridae. They have large fat-filled organs above the pectoral fins. One of them, *Sarcoglanis*, known only from a single specimen 4cm (1.5in) long, was collected in 1925 by Dr. Carl Ternetz in the San Gabriel Rapids on the Rio Negro. Forty years later, Dr. George Myers, in the same locality, caught another similar fish, but belonging to a new genus. Nothing is known about these forms.

The whale catfishes (Cetopsidae), with 12 species, is also a poorly known group. Some species are particularly voracious and prey on other catfishes by biting out circular pieces of flesh with their sawlike teeth. The Blue whale catfish (*Cetopsis coecutiens*) is, perhaps, the best-known species in this family.

The family of driftwood catfishes (Auchenipteridae) contains about 60 species, characterized by conspicuous spots and stripes. They range from Panama to La Plata and are nocturnal, detrital feeders, inhabiting hollow logs by day where they line up in ranks. Exceptionally for catfishes, driftwood cats employ internal egg fertilization, with the eggs being released later by the female over the bottom or among plants, where they remain unprotected until they hatch about one week after release.

The lookdown or loweye catfishes (Hypophthalmidae) are most unusual. Unlike most other catfishes, they feed on plankton, which they sieve from the water through fine gill rakers. Long barbels help to funnel the plankton into the mouth. As they are surface feeders, *Hypophthalmus* species have a high fat content and paper-thin bones that increase buoyancy.

Of the eight families of catfishes in Africa, only three are endemic, four are shared with Asia and one, the sea catfishes (Ariidae), also occurs in the coastal waters of Asia, Australia, and North and South America. Although there is less diversity in form among African catfishes than among South American ones, there are many unusual forms, some displaying remarkable parallels with those of South America. The richest diversity of species in Africa occurs in the principal equatorial river basin, the Zaïre, Africa's equivalent of the Amazon. A major difference between the continents, however, is the series of rift valley great lakes in Africa, some of which harbor endemic groups of catfishes.

The most widespread of these catfishes are the bagrids (family Bagridae), with over 200 species. Some species of *Bagrus*, a Nilotic catfish genus, weigh over 5kg (11lb). Small bagrids, Zaïrean endemics, live in torrents, while a "flock" of several *Chrysichthys* species, among them the giant of them all, *C. grandis*, which can weigh as much as 190kg (420lb), inhabits Lake Tanganyika. This generalized family also has members in Asia.

The walking catfishes (Clariidae) contain about 30 species and 10 genera. They are long bodied, with long dorsal and anal fins and broad, flat heads. Some species have an organ at the top of the gill chamber that enables them to breathe atmospheric air and so survive when water is deoxygenated. Clariids, like the South American callichthyids, can travel overland from one water body to another, hence their common name. The largest clariid is the Vundu (*Heterobranchus longifilis*), exceeding 50kg (110lb). Other genera are small, anguilliform (eel-like) fishes with burrowing habits; *Uegitglanis zammaranoi* from Somalia is, however, subterranean and eyeless; the Cave clarias (*Clarias cavernicola*) from Namibia is also eyeless. This family also occurs in Asia, but is represented by fewer species there.

The endemic family Malapteruridae, with its two species, is the only family of electric catfishes. All catfishes appear capable of detecting electrical activity, but only *Malapterurus* is actively electrogenic. The dense, fatty electric organ covers the flanks of the fish and gives it a cylindrical, sausage-shaped appearance. These catfishes are capable of generating strong electric impulses (up to 450 volts), which are used for both defense and stunning prey. Some *Malapterurus* specimens exceed 1m (3.3ft) in length and weigh up to 20kg (44lb).

The squeakers, or upside-down catfishes (family Mochokidae), are exclusively African. Over 100 of its 170 or so species belong to the genus *Synodontis*, a few species of which can swim upside-down and utilize food on the surface, as well as feeding normally on the bottom. This habit, along

Above *The young of the Saltwater or Coral catfish* (Plotosus lineatus) *form rolling "feeding-balls" that resemble giant sea urchins. Plotosids are notorious for the dangerous wounds they can inflict with their pectoral spines, which have venomous glands at the base.*

Right *The whiskerlike barbels from which the catfishes derive their name are prominently on display on this Black bullhead catfish* (Ictalurus melas), *a North American species that tends to inhabit turbid, silty waters.*

Below *The striking transparency of the Glass catfish* (Kryptopterus bicirrhis; *Siluridae) is a camouflage adaptation, though its physiology is unknown.*

with the ability of at least some species to produce sounds, explains the family's common names.

The endemic loach catfish family (Amphiliidae) contains about 47 small species. Some (e.g. the hillstream whiptails, *Phractura* and *Andersonia*) display remarkable parallelism with the South American twig catfishes in their elongate, plated bodies. Amphiliids inhabit the faster, cooler upper reaches of rivers, clinging to the cobbled substrate.

The schilbeid catfishes (Schilbeidae), a family shared with Asia, has about 20 species in Africa. They have a short dorsal fin and compressed, deep bodies. They are fast-swimming shoalers, which, in their large numbers, are both important predators on other fishes and, in their turn, form a large food source for fish-eating perches. Schilbeids are ubiquitous fishes that can change their diets easily and have quickly adapted to artificial situations, such as dammed lakes and reservoirs.

After carps and their allies, catfishes are the dominant element of Asia's fish fauna. Compared with African catfishes, Asian catfishes are not well-known, scattered as they are throughout the Indonesian islands and isolated rivers and lakes of China and high Asia. Twelve families occur in and around the continent, seven of which are endemic.

The Bagridae are widely distributed, with the species-rich genus *Mystus* (about 40 species) being typical. Most Mystus species are small - around 8-35cm (3.2-14in) in length; some, however, like the Asian Red-tailed Bagrid (*M. nemurus*), can grow much larger, around 60cm (24in) in this case. The accurately named lancers (*Bagrichthys* spp.) from Borneo and Sumatra are unusual in that the dorsal fin spine of the adults is nearly the length of the fish's body.

The sheathfishes (Siluridae) are a significant family with species in Eurasia, Japan, and offshore islands. The family contains the 2m (6.5ft) long voracious predator, the Helicopter catfish (*Wallago attu*) and the largest of all, the European wels (*Silurus glanis*). The *Wallago* follows shoals of carps in their upstream migrations and can leap clear of the water during a feeding frenzy.

Species of the endemic Asian hillstream catfishes (Sisoridae) with 100 species, and the torrent catfishes (Amblycipitidae) with about 10 species, are small mountain-stream dwellers, clinging to the substrate by the partial vacuum caused by corrugations of their undersides. The two species of the small family of frogmouth catfishes (Chacidae) from Borneo are well-camouflaged, flattened fishes with cavernous mouths and large heads. They resemble some anglerfishes and feed in a similar way, apparently being able to lure prey to within reach of their gape.

The shark catfishes (Pangasiidae), numbering about 21 species, are possibly the most economically important of the Southeast Asian catfish families. In Thailand, *Pangasius* species have been pond-reared on fruit and vegetables for over a century and are sold in vast numbers. One of the world's largest freshwater fishes, the Giant catfish (*Pangasianodon gigas*) lives in the Mekong river, growing to 3m (10ft). Despite the anecdotal evidence of travelers' tales, it too is totally vegetarian.

The tandan catfish family (Plotosidae) contains about 32 species, some of which live in the Indo-Pacific Ocean. However, two genera, *Tandanus* and *Neosilurus* (the eel-tailed catfishes), live in the freshwaters of Australia and New Guinea. Specimens of *Tandanus* can weigh up to 7kg (15lb). These fishes build circular nests, about 2m (6.5ft) in diameter, from pebbles, gravel, and sticks. The several thousand eggs are guarded by the male until they hatch after seven days.

Closely related to the walking catfishes (Clariidae) is the family Heteropneustidae, known as the stinging or airsac catfishes, species of which have long air sacs that extend backwards from each gill cavity. These catfishes can deliver a very potent sting, with enough venom to kill a human.

The sea cats (Ariidae) have a circumtropical, largely marine distribution. Males of this family are mouthbrooders; up to 50 large, fertilized eggs are carried for as long as two months, during which time the male does not feed.

Apart from a few ariid species reaching coastal regions, only one family, the North American freshwater catfishes (Ictaluridae), with its 7 genera comprising some 45 species, is present in North America. The flatheads (*Pylodictis* species) are the largest, growing to some 40kg (88lb). The blue and channel catfishes (*Ictalurus* spp.) form the basis of a large catfishery in the Great Lakes and the Mississippi valley. The madtoms (*Noturus* species) – named for the venomous glands at the base of the pectoral fins – comprise some 25 species. There are also three cave-dwelling, eyeless ictalurids, which appear to have evolved independently. The Mexican blindcat (*Prietella phreatophila*) is known only from a well in Coahuila, Mexico; the Widemouth blindcat (*Satan eurystomus*) and the Toothless blindcat (*Trogloglanis pattersoni*) have appeared only in 300m- (1,000ft) deep artesian wells near San Antonio in Texas. It is thought they live in deep water-bearing strata. GJH/JD

Carps and Allies

ORDER CYPRINIFORMES

The cyprinoids form a major lineage of largely freshwater, egg-laying fishes. All lack jaw teeth but most have a pair of enlarged bones in the pharynx or throat, the teeth of which work against the partner bone and a pad on the base of the skull. A less conspicuous unifying feature is a small bone (the kinethmoid) that enables the upper jaw to be protruded, i.e. extended. Most cyprinoids also have scaleless heads. Cyprinoids are indigenous to Eurasia, Africa, and North America. Unlike the other major otophysan lineages, they are not native to South America and Australia.

Of the five families, with over 2,660 species and around 280 genera, by far the largest is the family Cyprinidae, with over 2,000 species. The Cyprinidae – chubs, minnows, mahseers, carps, barbs, etc.– reflect the distribution of the cyprinoids and are well known to both freshwater anglers and aquarists. Even in areas where they never lived naturally, many are familiar to fishermen and pondkeepers as well as to gourmets.

The Common carp (*Cyprinus carpio carpio*) epitomizes the family. Probably native to Central Europe and Asia, it has been introduced to all continents capable of supporting fish life. The carp is extremely tolerant of a wide range of conditions, so much so, that, in Central Africa, where it was introduced to provide food for expatriates, it has colonized areas so successfully that it is now the commonest cyprinid. In the United Kingdom, to which it was probably introduced in Roman times, it is loved by anglers, for whom catching an 18kg (40lb) specimen is a lifetime's goal. In South Africa, where the carp was introduced in the early 20th century, a 38kg (83lb) specimen was caught.

The Japanese in particular have cultivated carp as objects of beauty; several hundred years of intensive breeding has released the potential colors and many "brocaded" carps (known as Koi)

Below *Representative species of cyprinoids, knifefishes, and milkfishes:* **1** *Mahseer* (Tor tor), *a native of the Indian subcontinent;* **2** *Shiner* (Notropis lutrensis); **3** *Rosy barbs* (Barbus conchonius) *show marked sexual dimorphism, the male being the more colorful;* **4** *The Common carp* (Cyprinus carpio) *is a popular game and food fish in central Europe;* **5** *Two examples of* Gastromyzon, *a hillstream cyprinoid that has paired fins forming suction disks, allowing it to cling to rocks;* **6** *Gudgeon* (Gobio gobio); **7** *Luciobrama, a pikelike cyprinoid predator;* **8** *Sucking loach or Chinese algae eater* (Gyrinocheilus aymonieri); **9** *The Flying fox* (Epalzeorhynchous kallopterus) *is a favored aquarium species;* **10** *The Razorback sucker* (Xyrauchen texanus) *has a very limited distribution in the western USA, and is classed as Endangered;* **11** *Ghost knifefish* (Apteronotus albifrons); **12** *Electric knifefish* (Electrophorus electricus); **13** *Milkfish* (Chanos chanos).

are sold for high prices to beautify ponds and aquaria. These colored fishes have become so popular that large breeding and export centers now exist, not just in Japan, but in China, Singapore, Malaysia, Sri Lanka, the USA, Israel, and several European countries. In its native Eurasia, the Carp is bred for food, but even there, modifications have been made. Careful selection first produced carp with few scales (Mirror carp), then with none (Leather carp). Further selective breeding has resulted in strains lacking the hairlike, intermuscular bones that create problems for diners.

The majority of cyprinids are smallish fishes, measuring only some 10–15cm (4–6in) long when adult. There are, however, some notable exceptions. The Golden Mahseer (*Tor putitora*) of Himalayan and Indian rivers can grow to 2.7m (9ft) long and can weigh 54kg (120lb). In North America, the Colorado Squawfish (*Ptychocheilus lucius*) of the Colorado and Sacramento rivers, which was an important food source to the native Americans of those areas (hence the common name), used to grow to over 1.8m (6ft) long. Now, this species is practically extinct in the Colorado (because of damming) and much smaller and rarer in the Sacramento (because of overfishing). Of similar length is the Yellow cheek (*Elopichthys bambusa*) from the Amur River in northern China. Both species are unusual among cyprinids (which are toothless) in that they are specialised fish eaters. Other, but smaller, examples of fish eaters are the Rhinofish (*Barbus mariae*), from a few rivers in East Africa; *Luciobrama macrocephalus* from southern China, a fish with a disproportionately elongate head; and the Dab Bao or Pla Pak Pra (*Macrochirichthys macrochirus*) from the Mekong River. The last of these is a strongly compressed fish with a large, angled mouth and a hook and notch on the lower jaw. It can also raise its head to increase the gape when lunging at its prey.

Most cyprinids, including the mahseers, eat almost anything: detritus, algae, mollusks, insects, crustacea – even cheese sandwiches. The bighead and silver carps (*Hypophthalmichthys* spp.) of China are specialized plankton feeders with gill rakers modified into an elaborate filtering organ. One species, the Grass carp (*H. molitrix*), eats plants and has been introduced to many countries to clear weeds from canals, rivers, and lakes.

The shape and distribution of the pharyngeal or "throat" teeth often indicate the diet. Mollusk eaters have crushing, molarlike teeth, which are closely packed; fish eaters have thin, hooked teeth; vegetarians have thin, knifelike teeth for shredding; omnivores come somewhere in between. Yet even within one species, the teeth may vary. In Africa, for example, fish of the species *Barbus altianalis* living in a lake with no snails have "middle-of-the-road-type" teeth, while those in a snail-rich lake a few kilometres away, have thicker, lower, more rounded teeth. The young, however, all start off with the same type of teeth.

The widespread and species-rich genus *Barbus* (used here to include, in addition to *Barbus* itself,

genera referred in some works as *Puntius*, *Capoeta*, and *Barbodes*) derives its name from the (usually) four barbels around the mouth; these barbels have taste buds with which the fish can taste the substrate before eating. A particular specialization of some African species is the varied shape and thickness of the mouth and lips according to diet. A broad-mouthed form with a wide, sharp-edged lower jaw feeds on epilithic algae, i.e. algae that grow as an encrusting layer on rocks. A narrow-mouthed form with thick, rubbery lips feeds by sucking up stones and their associated fauna. The fact that these two extremes are found in one and the same species explains why the fish that is now regarded as *B. intermedius* formerly had over 50 scientific names.

Most cyprinids do not exhibit sexual dimorphism, but some of the small *Barbus* species do. In Central Zaïre, among submerged tree roots, live small "butterfly barbs," 4cm (1.5in) long, including *Barbus hulstaerti* and *B. papilio*. In these strikingly marked species, the males and females have conspicuously different color patterns.

Although, generally, the cyprinids are not as brilliantly colored as the characins, some of the Southeast Asian *Rasbora* species come close. The small Malaysian species, Brittan's rasbora (*Rasbora brittani*) and Axelrod's rasbora (*R. axelrodi*) are the cyprinid equivalents of the popular Glow-light and Neon tetras (order Characiformes).

Totally without color, however, are the subterranean cyprinids. Cave-dwelling members of the genera *Barbus*, *Garra*, and *Caecocypris* have lost all pigment – and their eyes – as a result of living in lightless habitats underground.

The epigean (i.e. above ground) *Garra* species live in Africa, India, and Southeast Asia. They are bottom-dwelling fishes with a sucking and sensory disk on the underside of the head. The related "shark" genera (*Labeo* and *Epalzeorhynchos*), with a similar distribution, have an elaborate suctorial mouth and graze on algae. One African species of *Labeo* has specialized in grazing on the flanks of submerged hippopotamuses.

In the cold mountain streams of India and Tibet live the poorly known snow trout (*Schizothorax* spp.), cyprinids that imitate the lifestyle of trout and salmon. They are elongate fish, up to 30cm (1ft) long with very small scales (or none at all) except for a row of tilelike scales along the base of the anal fin. Nepalese fishermen capture snow trout by fashioning a worm-shaped wire surrounded by a loop, which is tightened when the fish strikes at the "worm."

While it is true that cyprinids are essentially freshwater fishes, some can nevertheless tolerate considerable degrees of salinity. In Japan, species of redfin (*Tribolodon*) have been found up to 5km (3 miles) out to sea. In Europe, the freshwater Roach (*Rutilus rutilus*) and Bream (*Abramis brama*) also live in the Baltic Sea at about 50 percent salinity. They are, however, among the few exceptions.

As if cyprinids were not already a compendium of eccentric habits, at least two species are known to get drunk. In Southeast Asia, both the cigar "sharks" (*Leptobarbus* spp) and the Silver-and-red barb (*Hampala macrolepidota*) gorge themselves on the fermented fruit of the chaulmoogra tree when it falls into the water. They even congregate before "opening time." When intoxicated, these fishes float helpless in the water, but are relatively safe as their flesh is made unpalatable by the alcohol. Counterbalancing such antisocial cyprinid behavior, the Eurasian tench (*Tinca tinca*) has gained a folk reputation as a doctor fish. It is reported that injured fish seek out tench and rub their wounds

in its slime. While it is undeniable that the tench has a copious coating of slime, its healing properties have never been scientifically attested.

The suckers (Catostomidae) have many species in North America and a handful in north Asia. For a long time, they were thought to be the most primitive cyprinoids, due to the shape of their pharyngeal (throat) bones and teeth. Suckers have also developed highly complex sacs from the upper gill-arch bones and it now seems likely, taking into account their distribution, that they are a highly specialized group of cyprinoids. They are generally innocuous, unspectacular fish, apart from two genera, each containing a single species: the Chinese sailfin sucker (*Myxocyprinus asiaticus*) and the Razorback sucker (*Xyrauchen texanus*) from the Colorado River. Both are deep bodied, with a triangular profile. The function of this shape is to force them close to the bottom of the river during flash floods (both species live in rivers susceptible to flash floods). These two suckers are therefore a good example of parallel evolution.

The "true" loaches are a family (Cobitidae) of (frequently) eel-like fishes with minute, embedded scales and a plethora of barbels around the mouth. Bony processes enclose most, or all, of the swimbladder, making its normal volume changes rather awkward. All the cyprinoids have a tube connecting the swimbladder to the pharynx, allowing the fish to swallow or expel air, but with their constricted swimbladder, loaches use this

Left *The beautifully banded Clown or Tiger loach* (Botia macracanthus) *is native to freshwater habitats on the Indonesian islands of Borneo and Sumatra. It feeds on crustaceans, worms, and plant matter.*

Above Familiar and widespread in Europe, where it is fished recreationally, the Roach (Rutilus rutilus) is a highly adaptable fish that can tolerate poor-quality water. So successful is the Roach that it can overrun an area where it has been introduced.

more often than many. One species, known as the Weather fish or Weather loach (*Misgurnus fossilis*), from eastern Europe, has been kept for centuries by peasants as a living barometer. Its agitation and continued "burping" as it expels air with the changes in atmospheric pressure that accompany the approach of thunderstorms made it one of the earliest weather forecasters.

Not all of this family are eel-like in appearance, however. One group, the Botinae, have shorter bodies, often compressed, and a pointed snout. There are only three genera in this group, the best known being *Botia* – which includes some popular aquarium fishes, most notably the striking Clown loach (*B. macracanthus*). This species also exhibits a most unusual behavior, resting on its side and giving the impression that it is dead.

Most loaches have small dorsal and anal fins symmetrically placed near the rear of the body. The poorly known but appropriately named long-finned loaches, *Vaillantella* spp., however, have a dorsal fin the length of the body. Loaches (in the broadest sense of the word) can be divided into two subgroups: those that have an erectile spine below the eye and those that do not. Many are secretive fish, liking to hide under stones during the day. It was an overdue discovery when the first cave-living species was found in Iran in 1976. Since then, two more have been found in southwestern China. Loaches live in Eurasia, not in Africa (apart from an arguably introduced species in European North Africa).

Despite their common name, the gyrinocheilids, or algae eaters (Gyrinocheilidae), feed largely on detritus. There are only four species, all from Southeast Asia. They have a ventral, protrusile mouth, like the hose of a vacuum cleaner. With this, they suck in the fine substrate (and scrape off encrusting algae) and filter out the edible material.

Gyrinocheilus species lack a pharyngeal tooth apparatus; whether this is lost or has never been developed is unknown. These fishes – often misleadingly called "sucking loaches" – are unique among cyprinoids in that the gill cover is sealed to the body for most of its length, leaving just top and bottom openings. The top opening is covered by a valve and takes in water to oxygenate the gills and expels it through the bottom opening. Thus, breathing, in sharp contrast to other fishes, is through the gill cover, rather than the mouth.

The river loaches, balitorids, or homalopterids (Balitoridae), consist of two subgroups: the flat loaches (Balitorinae), which live mostly in fast-flowing waters, even torrents, in Southeast Asia, and the nemacheilines (Nemacheilinae), which are almost exclusively Eurasian in distribution. Greatly flattened, the Balitorinae have both pectoral and pelvic fins fused into suckers. The mouth is ventral and, while the fish graze on the rich algal growth in such fertile waters the snout is protected by a remarkable development of bones that in other cyprinoids lie below the eye. In the hillstream species, these bones are curved forward in front of the snout, strengthened, and act like the bumpers of a dodgem car.

The nemacheilines are Eurasian eel-like fishes which are very similar in overall shape to the Weather loach and its closest relatives. One species, the enigmatic *Nemacheilus abyssinicus* is African. A Mr Degen was collecting for the British Museum in Ethiopia in 1900. His collection from the mouth of a stream feeding Lake Tsana purportedly contained the specimen described under that name some years later. No more specimens have ever been found; but then, no one has revisited the site and used his collecting techniques. Or perhaps one jar could have been misplaced on a museum shelf?

Equally enigmatic are the description and illustration of *Ellopostoma* from Borneo. The original specimens were lost by 1868 and no more were found for many years. It is now reported from Malaysia and Thailand, however, but is still poorly understood.

New World Knifefishes

ORDER GYMNOTIFORMES

The New World knifefishes consist of six families of the order Gymnotiformes. All are eel-like to varying degrees and all lack the pelvic girdle and fins, as well as the dorsal fin. The anal fin, in striking contrast, is extremely long based, possessing 140 or more rays, and forms the main means of propulsion, both forward and backward. The tail (caudal) fin is either lacking or highly reduced.

New World or South American knifefishes possess electric organs, which, in most species, consist of specially modified muscle cells. However, in the ghost knifefishes (family Apteronotidae) modified nerve cells perform this function. In most cases, the electric field is weak and is used mainly to help the fish navigate at night (most species are nocturnal), locate food, and communicate with each other. However, in the Electric knifefish (*Electrophorus electricus*), better known as the Electric eel, powerful impulses capable of (reportedly) stunning a horse can be generated. This species, which can grow to 2.3m (7.5ft), uses electricity to stun its prey and in self-defense, making it both a lethal hunter and a formidable adversary.

Milkfish and Allies

ORDER GONORHYNCHIFORMES

One of the most widespread and commercially significant species of the gonorhynchiformes is the Milkfish (*Chanos chanos*). This silvery, streamlined fish, about the size of a Gray mullet, congregates in large shoals in warm waters around island reefs and along continental shelves, and has long been fished and cultivated for food, especially in the Philippines, Taiwan, and Indonesia.

Another marine species, the elongate Beaked sandfish, or Beaked salmon (*Gonorhynchus gonorhynchus*) is found along the shorelines of the southern Pacific, Indian Ocean, and southeastern Atlantic off Namibia and South Africa. It is fished commercially throughout its range.

The freshwater Hingemouth (*Phractolaemus ansorgii*) takes its name from its ability to extend its mouth into a small trunk. Its swimbladder functions as a lung, enabling this species to survive in unoxygenated waters KEB/GJH/JD

FISHES UNDERGROUND

The locations, forms, and lives of cave fishes

THERE ARE AROUND FORTY SPECIES OF FISHES, belonging to some 13 different families, that spend their lives in lightless, underground waters. In some cases, it is uncertain whether an underground (hypogean) population represents a separate species or merely a highly modified population of a surface-living (epigean) species.

Cave fishes are colorless. However, they appear to be pink-colored. This coloration is caused by the blood that runs along tiny vessels (capillaries) that lie close to the skin surface. In a number of cases, the vessels – particularly those that run parallel to the ribs – are extremely distinct and can be seen easily with the naked eye.

Cave fishes also frequently have reduced scales, the degree of reduction ranging from species to species. At least one species, the Indian Blind Catfish (*Horaglanis krishnai*) is scaleless. Not all cave fishes live in caves, however. Some, for example, live in water-bearing strata or rock layers (known as aquifers) where the rock is honeycombed with water-filled channels.

Subterranean fishes live in tropical and warm temperate countries that have not been affected by recent glaciation. In Australia, two species, a sleeper goby (or eleotrid) and a swamp eel (or synbranchid), live in the Yardee Creek wells on the North West Cape. Madagascar has two sleeper gobies, while Africa has a barb in the Democratic Republic of Congo (DRC), a swamp eel in Mauritania, and a clariid, or walking catfish, in Namibia; three species: a catfish and two carps or cyprinids, live in Somalia.

Oman has two species of the cyprinid genus *Garra*. Iran has a subterranean barb and another *Garra*-type cyprinid; Iraq, yet another garrine, as well as the Iraq blind barb (*Typhlogarra widdowsoni*) and – sharing the same sinkhole at the Sheik Hadid Shrine – the much rarer *Caecocypris basimi*. Wells in Kerala, southern India, hold the Indian blind catfish (*Horaglanis krishnai*) a small species from the catfish family Clariidae. Three species (two loaches and yet another garrine) have recently been found in China. Cuba has two freshwater ophidioids – relatives of the marine cusk eels; Mexico has one and also has characoids and a swamp eel. The subterranean fauna of Brazil consists of catfishes and characoids. The USA has a rich cave fauna in many sites, from the Ozark and Cumberland plateaux to Texas, comprising catfish and amblyopsids. There may also be cave fish in New Guinea and Thailand, but Europe has no cave fish.

Some of the cave species now known belong to families or subfamilies that are, basically, marine. Yet the vast majority of cave fishes are found in freshwater habitats. Prime examples of this are the ophidoids of Cuba, Mexico, and the Bahamas, which are found in freshwater limestone wells or pools. *Lucifuga spelaeotus* from Mermaid's Pool in the Bahamas is such a fish; so is its closest relative, *L. dentatus*, from Cuban caves. Yet the family to which they belong (Bythitidae – a member of the order Ophidiiformes, hence the term "ophidioids"), contains about 90 species, all but a few of which are marine.

Quite how these fishes ended up living in freshwater is open to speculation. In the geological past, fishes living in subterranean saltwater-bearing strata could have been trapped in such underground waters as land masses moved apart through continental drift. They would then slowly adapt themselves to brackish or freshwater over a long period of time. Such explanations have been put forward to explain the presence of *Stygichthys* and *Lucifuga* in Cuba. Debate and doubts also surround the possibilities of how these fish came to live where they do. There is some evidence, in at least one instance – that of the Swampfish (*Chologaster cornuta*), a close relative of the American amblyopsids, or cavefishes – of preadaptation. *Chologaster*, for instance, has small eyes and actively shuns light. It is therefore possible that, in desert areas such fish may well have, where possible, followed the falling water table underground.

The evolution of cave fishes was a much-loved subject of Charles Darwin, but his views of isolation and natural selection were supplanted by the more recent theory of adaptive shift, which states that cave dwellers can adapt to the dark while still interbreeding with their surface kin – as demonstrated by the blind and eyed types of Mexican tetras, which meet when rivers flow into and out of caves and can be crossbred in the laboratory. However, new research suggests that Darwin's isolation theory may be right after all. Now DNA profiling of the Characiform genus *Astyanax* has clearly distinguished the cave fish from the surface-dwellers, even when living close together. Thus cave fishes cannot have descended from the eyed fish nearby but must have evolved inside the caves a long time ago.

So far as is known, cave fishes live longer than their surface-dwelling relatives. This may be a response to their irregular and sparse food supply, which comes into the cave during floods as detritus, or is provided by other cave animals, all of which are ultimately supported by "the outside."

Reproduction in cave fishes has never been observed in the wild, except in the Cuban ophidioids, which give birth to fully formed young. Other cavefish groups are probably egglayers. In captivity, the only species that is bred in aquaria

Above *Mexican blind cave tetras* (Astyanax jordani) *are endemic to Mexico. Once they have spawned, these fishes hide their eggs in the crevices of the rocky caves that they inhabit in the central San Luis Potosí region.*

(be it by aquarists or by commercial breeders) is the Mexican blind cave tetra, or characin (*Astyanax jordani*), which, until recently, was believed to be no more than a cave form of the widespread Mexican tetra (*Astyanax mexicanus*). The reproductive strategies of the surface-living and subterranean species are similar, so much so that in at least one case (*Cueva chica*), they interbreed, producing a whole range of forms, from fully eyed, fully colored specimens, to pink-bodied, completely blind ones.

It is important to be able to breed cave fishes because, first, wild populations have very restricted ranges and are low in numbers and, second, one cannot answer questions about them or improve their chances of survival until this has been done. In some cases, cave fishes have been brought into breeding condition, but some final, essential ingredient has been missing and the fishes have failed to spawn or have died. Thus, one needs to identify what these naturally occurring stimuli or triggers are. Perhaps the torrential flooding of the caves in DRC where the Blind cave barb (*Coecobarbus geertsi*) lives is vital in this case – perhaps not – but, at the moment, not enough is known about the secret lives of these fishes, most of which are facing threats to their survival in the wild. KEB/JD

Codfishes, Anglerfishes, and Allies

THE SUPERORDER PARACANTHOPTERYGII IS probably best known for its great commercial importance, the codfishes and their allies being responsible for a huge proportion of the annual marine harvest. The superorder contains a remarkable variety of forms, from species with a typical piscine body plan, like the Trout-perch, Codfish, Hake, and Haddock, to those with peculiar shapes and colors, fishes flattened like a pancake from top to bottom, highly camouflaged fishes, fishes that live inside various invertebrates, and others with bioluminescent lures and reproducing by sexual parasitism.

The superorder, which, following current systematics, comprises five orders, was originally created in 1966 (and somewhat redefined in 1969 and again in 1989), as a taxonomic repository for a number of groups excluded from the major grouping of spiny-finned fishes (Acanthopterygii) – namely, all those forms that were of a similar evolutionary grade but that lacked the characteristics of the spiny-finned fishes. Although the erection of the superorder enabled the acanthopterygians to be somewhat more succinctly defined, the validity of the Paracanthopterygii is still rather tenuous. Over the years, the group has been plagued with problems and confusion, the major difficulty being the lack of unique characters shared by all members of the group. Despite several attempts by some of the best contemporary ichthyologists, there is still no rigorous definition of the group; in other words, there is no satisfactory basis to believe that it constitutes a natural group.

Nearly all members of this superorder are marine, most inhabiting rather shallow water but some living at great depths within the mesopelagic, bathypelagic, and abyssal zones of the major oceans of the world. There are a few (only about 20) exceptional individual species of some otherwise marine families that live in freshwater, along

FACTFILE

CODFISHES, ANGLERFISHES ETC.

Superorder Paracanthopterygii

Orders: Percopsiformes, Ophidiiformes, Gadiformes, Batrachoidiformes, Lophiiformes

About 1,225 species in 267 genera, and 39 families.

Distribution Worldwide in all oceans; tropical and temperate waters.

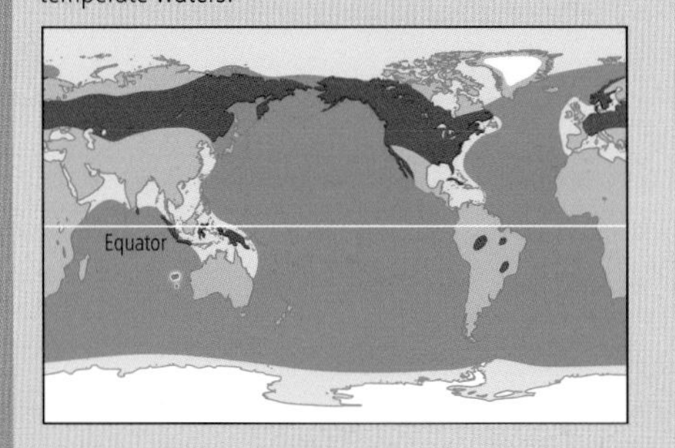

See The 5 Orders of Codfishes, Anglerfishes, & Allies. ▷

Above *Representative species of trout-perches, cuskeels, and codfishes:* ***1*** *Trout-perch* (Percopsis omiscomaycus), *family Percopsidae;* ***2*** *Pirate perch* (Aphredoderus sayanus), *sole member of the family Aphredoderidae. One extraordinary feature of this species* ***2a*** *is the movement of the relative position of its anus while the fish is growing – when still a fry, the anus is sited far back on the body, while it has migrated further forward by the time the fish reaches adulthood;* ***3*** *Pearlfish* (Echiodon drummondi), *family Carapidae, showing the adult inside a sea cucumber;* ***4*** *Cuskeel* (Abyssobrotula galatheae), *family Ophidiidae;* ***5*** *Threadtailed grenadier* (Coryphaenoides filicauda) *family Macrouridae;* ***6*** *Luminous hake* (Steindachneria argentea), *family Steindachneriidae;* ***7*** *Poutassou or Blue whiting* (Micromesistius poutassou), *family Gadidae;* ***8*** *Atlantic cod* (Gadus morhua), *family Gadidae;* ***9*** *Tadpole cod* (Salilota australis), *family Moridae.*

with the nine species belonging to the order of trout-perches and their allies that occur exclusively in freshwater habitats in North America.

Trout-perches and Allies

Order Percopsiformes

The trout-perches and their allies are small fishes, attaining a maximum length of just 20cm (8in). They are thought to be intermediate in structure between soft-rayed, primitive fishes, such as salmon, trout, and herrings, and spiny-rayed, derived forms, such as rockfishes and basses. They appear to represent remnants of a once larger, more widely distributed assemblage. Fossils clearly belonging to this group (the marine genus *Sphenocephalus*) are rather well known from the Upper Cretaceous period (95–65 million years ago) of Europe. There are, in addition, several fossil genera of freshwater forms from the Eocene (about 55–34 m.y.a.) of North America.

The two species of Trout-perches, family Percopsidae, derive their common name from the presence of a "troutlike" adipose fin combined with their vaguely "perchlike" first dorsal fin, which is spiny in the front, but made up of soft rays behind. The Sand roller (*Percopsis transmontana*) is restricted to slow-flowing, weedy parts of the Columbia River drainage in Washington, Oregon, and Idaho. It has scales with a comblike free margin (ctenoid scales) and a cryptic, greenish coloration with dark spots. Reaching a maximum length of about 10cm (4in), it is only half the size of its more widely distributed congener the Trout-perch (*Percopsis omiscomaycus*), which ranges from the west coast of Canada to the Great Lakes and the Mississippi–Missouri river system. Although two rows of spots are present, the body is translucent and the lining of the abdominal cavity can be seen through the sides. Both species feed on bottom-living invertebrates and are themselves eaten by a number of predatory fishes.

The Pirate perch (*Aphredoderus sayanus*), from still and slow-flowing waters of the eastern USA, is the only member of the family Aphredoderidae. A sluggish, dark-hued fish, it grows to a maximum length of 13cm (6in) and lacks the adipose fin of the trout-perches. It feeds on invertebrates and small fishes. Its most unusual feature is the strange development of the vent: in juveniles, the anus is located in the normal position, just ahead of the anal fin, but it moves forward as the fish grows so that in adults it lies beneath the throat.

The family Amblyopsidae contains five genera, three of which live only in caves in the limestone regions of Kentucky and adjacent states. The Swampfish (*Chologaster cornuta*) is an eyed, pigmented species found in sluggish and still waters from West Virginia to Georgia. Despite having functional eyes, it shuns light and hides under stones and logs during the day. The Spring cavefish (*Forbesichthys agassizii*) lives in subterranean waters of Kentucky and Tennessee. It lacks the dark stripe of its only congener but still has functional eyes. In both species, but especially the latter, there are series of raised sense organs on the skin.

The other four species of the family are blind. The Southern cavefish (*Typhlichthys subterraneus*) lacks not only eyes, but pigment and pelvic fins as well. Rows of papillae, sensitive to vibrations, are present on the body and on the tail fin. It is thought to have achieved its wide distribution, from Oklahoma to Tennessee and northern Alabama, by traveling through underground waterways. The Northern cavefish (*Ambylopsis spelaea*) is white, has minute eyes covered by skin, and tiny pelvic fins. Like its relatives, it has vertical rows of sensory papillae on the body. Its reproductive strategy is unusual: the female lays a few relatively large eggs that, once fertilized, are carried in the gill chamber of the mother for up to ten weeks until they hatch. The remaining genus of the family Amblyopsidae, *Speoplatyrhinus*, containing only the Alabama cavefish (*Speoplatyrhinus poulsoni*) is exceedingly rare, being classed by the IUCN as Critically Endangered.

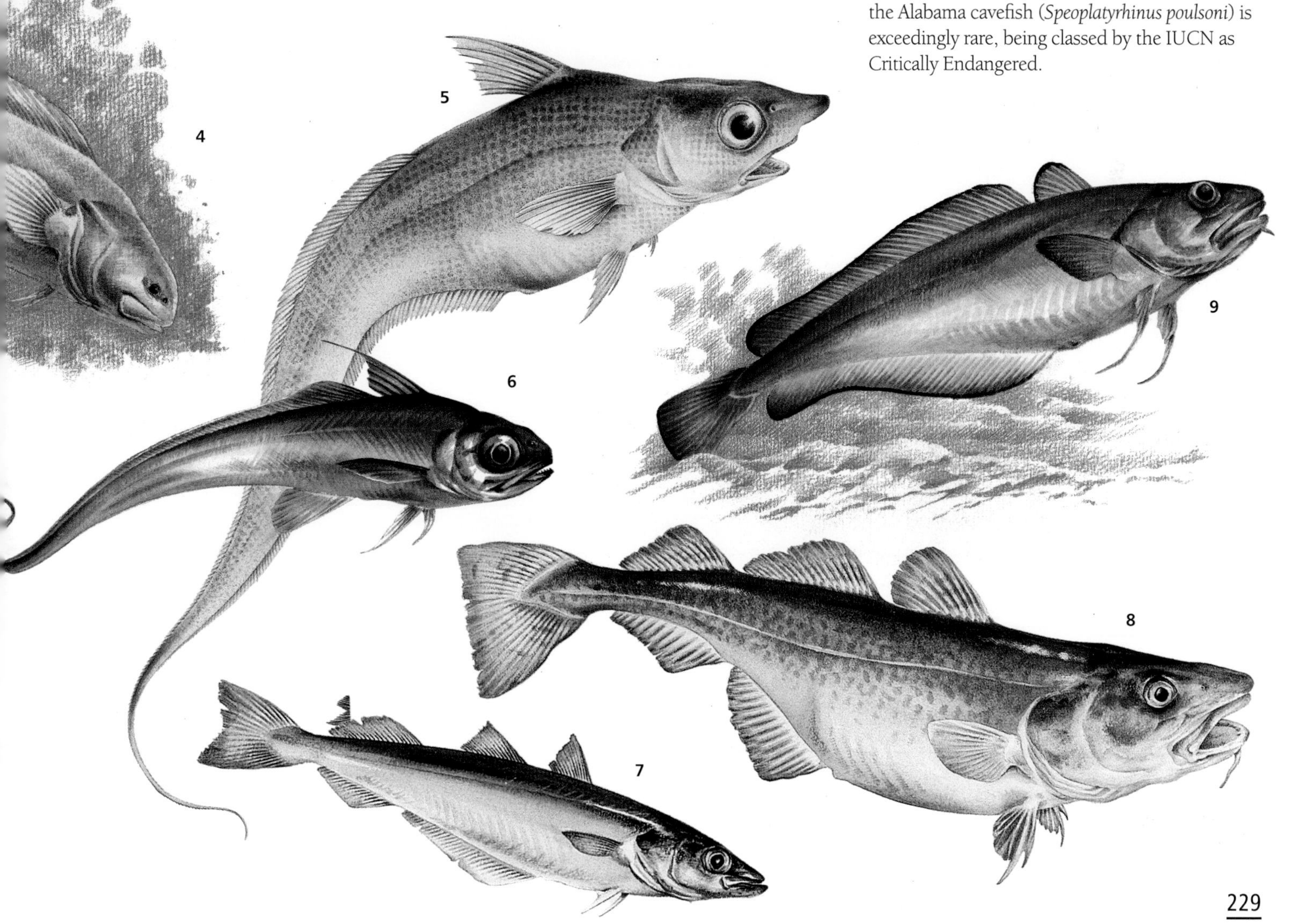

Cuskeels and Allies

ORDER OPHIDIIFORMES

The cuskeels and their allies constitute five families of rather similar looking forms, all having a rather small head and a long tapering body, with long-based dorsal and anal fins that extend far backwards and are usually attached to the tail. The pelvic fins, if present at all, are located far forward on the body, under the gill covers or even beneath the throat or on the chin.

The carapids or pearlfishes (family Carapidae) occur most commonly in warm tropical seas. All are elongate, slender fishes with long pointed tails. They have a complicated life history in which they pass through two dissimilar larval stages known as the *vexillifer* and *tenuis* stages. For a long time these two larval forms were thought to belong to two separate, relatively unrelated groups.

All carapids are secretive and live inside the bodies of various marine invertebrates, for example, sea cucumbers, clams, tunicates, sea urchins, or any other animal with a suitable body cavity. Some small species living inside oysters have been found entombed within the shell wall, hence the name "pearl fish." A common Mediterranean species, the Common pearlfish (*Carapus acus*), attains a length of about 20cm (8in). It lives within the body cavity of large sea cucumbers (typically the holothurian species *Holothuria tubulosa* and *Stichopus regalis*), which, like many of its relatives, it enters tail-first by way of the anus. Rather surprisingly, it has been known to feed on the internal organs of its host. The larvae are free living, and this semiparasitic habit is taken up only by the adult. Carapids usually eat bottom invertebrates or small fishes.

The family Ophidiidae contains little-known but interesting fishes commonly referred to as cuskeels or brotulas. All are elongate fishes with long-based dorsal and anal fins that are often fused with the tail fin. Brotulas are typically deeper-bodied and broader-headed than cuskeels, and have threadlike pelvic fins situated below the rear of the head. The similar-looking pelvic fins of cuskeels are situated beneath the throat. In some species the male has a penislike intromittent organ that passes packets of sperm to the genital duct of the female. Some species are egg layers, while in others the eggs hatch inside the mother and emerge as fully formed individuals.

Most of the cuskeels are small, secretive, burrowing fishes from warm seas. The Spotted cuskeel (*Chilara taylori*) from the eastern Pacific burrows tail-first into the sand or rocky crevices. When it emerges from its burrow, it aligns itself vertically, with only the last part of its body in the substrate. The Kingklip (*Genypterus capensis*) is found only off the coast of South Africa from Walvis Bay to Algoa Bay, where it inhabits a wide range of depths from 50 to about 450m (164–1.476ft). Growing to a maximum length of 1.5m (5ft), it is a giant among the cuskeels. Its flesh is excellent and the liver apparently has a quality much sought after by gourmets. It is not commercially important, however, as catches are irregular and sparse.

The brotulas are found worldwide, mostly in deep waters. The few species that live in shallow waters are shy and secretive, hiding away among rocks or in corals. This tendency to avoid well-lighted habitats may have led to the evolution of cave-dwelling forms in the Yucatán Peninsula of Mexico and in Cuba. The caves are all close to the sea and the water is brackish but variable in salinity. The Mexican blind cavefish (*Ogilbia pearsei*), found only in Balaam Canche Cave, in Yucatán, Mexico, is extremely rare, known from only a very few specimens. Its eyes are minute and covered by skin. Species of the genus *Lucifuga* have the best claim to being freshwater members of the family. Their coloration is remarkably variable, ranging from off-white to deep violet or dark brown. They give birth to fully formed young. The Nassau cavefish (*Lucifuga spelaeotes*) is known only from a single isolated population in the Bahamas. It was discovered in 1967 in Mermaid's Pool, a small freshwater sink in the limestone region near Nassau. Described as new to science in 1970, its future is now seriously threatened by commercial development.

Left *Atlantic cod has long been an extremely important commercial fish. Yet at the beginning of the 21st century, overfishing and climate change have seen stocks shrink dramatically, especially in the North Sea.*

Codfishes and Allies

ORDER GADIFORMES

The Gadiformes, containing the codfishes and their allies, is the largest of the five paracanthopterygian orders, with almost 500 species. It includes numerous commercially important species (together comprising over one-quarter of the world's marine catch), such as the Atlantic and Pacific codfishes, the Hakes, Haddock, Pollock, as well as many smaller and deepwater species of great biological interest but of little or no applied importance.

The group is widely distributed around the world in deep and shallow seas, from the tropics to the far northern and southern polar regions. They are mostly elongate fishes, with the pelvic fins situated far forward on the body, often in front of the pectoral fins. The dorsal and anal fins, sometimes contiguous with the tail fin, are long-based, the dorsal often divided into two or three separate units, the anal often divided into two units. No true spines are present in any of the fins.

Many of the commercially important species form shoals and the number of individuals they contained was enormous, until severe overfishing took its toll. For many years, an estimated 400 million Atlantic cod (*Gadus morhua*) were caught each year in the North Atlantic. At any one time the number of individuals of that species was enormous. A single female Atlantic cod typically produces over six million eggs; thus, if left alone, some fisheries biologists predict that stocks could regain their former abundance in a relatively short period of time.

Some of the 12 currently recognized families in this order are of little relative importance and only brief reference will be made to them.

The Macrouridae, containing the grenadiers, or rattails, is a large, widely distrbuted family of deepwater fishes with long, tapering bodies. The mouth is on the underside of a large head, nearly always equipped with an unusually long, pointed snout. In many species the males have drumming muscles attached to the swimbladder that they use to produce surprisingly loud sounds, apparently to attract members of the opposite sex for reproductive purposes. Many have a bioluminescent organ of unknown function, lying lengthwise beneath the skin of the abdomen, with an opening to the outside just in front of the anus.

The Moridae is a worldwide family of deepsea cods, found in all oceans almost from pole to pole. There are nearly 100 species grouped into 18 genera, nearly half of all species confined to genera *Physiculus* and *Laemonema*. The configuration of dorsal and anal fins is variable, with one to two, but rarely three dorsal fins, and one or two anal fins. Species in this family rarely exceed a length of 90cm (3ft).

Blue hake (*Antimora rostrata*), a morid cod despite the common name, has been found in the North Pacific, North and South Atlantic, and Indian Oceans at depths from about 500 to 1,300m (1,650–3,900ft). The first ray of the short-based first-dorsal fin is very long. The body color is dark violet to blackish brown. The Red codling (*Pseudophycis bachus*), again a morid, is common off South Australia, Tasmania, and New Zealand where it is used as a food fish. It lives in much shallower waters than many of its relatives, from as little as 50m (165ft) but it is most often trawled at depths of 200 to 300m (660–980ft). The first specimen was captured during Captain James Cook's second voyage (1772–75) and described by the famous team of Marcus Elieser Bloch and Johann Gottlob Schneider in 1801. The Japanese codling (*Physiculus japonicus*) has a light organ, as do many other members of this family. The organ is bulbous and has a canal that opens into the rectum near the anus. There is a reflector above the gland and the light shines out through a scaleless area of skin in front of the anus.

The so-called codlets of the family Bregmacerotidae, about 12 species all contained within a single genus (*Bregmaceros*), inhabit the surface waters of tropical and subtropical oceans; all are marine but a few venture into estuaries. Their anal and second dorsal fins are mirror images of each other; both have long bases and are higher at the front and back than in the middle. The first dorsal fin is nothing more than a long single ray, emerging from the dorsal surface of the head just behind the level of the eyes. All are small fishes, attaining a maximum length of only about 12cm (4.7in). The McClelland's codlet (*Bregmaceros mcclellandii*) is as widespread geographically as the rest of the family but has a considerably greater depth range, from the surface to 4,000m (13,000ft). Members of this family are difficult to characterize and identify, and almost nothing is known of their biology.

The fishes of the Hake family Merlucciidae have elongate bodies, a short first dorsal fin and a much longer second dorsal fin. Both the second dorsal and the anal fins are separate from the tail. In the Pacific hake (*Merluccius productus*), the anal and second dorsal fins possess a deep notch that nearly, but not quite, divides these fins in half. The European hake (*Merluccius merluccius*), the range of which extends into the Mediterranean, can grow to a maximum length of 140 cm (4.6 feet) and weigh more than 15kg (33lb). It has been an important food fish for the population of Western Europe throughout historic times. It is a nocturnal feeder, its large mouth eager for squids and small fishes, even for its own species. They are caught in midwater, but by day they live close to the bottom. They begin spawning in December, or even as late as April, in water deeper than 180m (600ft), but as the season moves along they migrate and continue to spawn in shallower water. The eggs float at the surface, and future hake stocks are very dependent on the weather. If the wind blows the eggs away from the rich inshore feeding grounds very few young survive, thereby causing the failure of the fishery a few years later.

The Deep-water Cape hake (*Merluccius paradoxus*) is found around South Africa, especially off the west coast where the water is richest. It is very similar to the European hake in appearance but is considerably smaller in size, with females (82cm,

Below *The Poor cod* (Trisopterus minutus), *a small gadoid, occurs in abundance in the Eastern Atlantic and the Mediterranean. It is eaten locally and also used in the production of fishmeal.*

The 5 Orders of Codfishes, Anglerfishes, and Allies

Trout-perches and Allies
Order: Percopsiformes

North America, from SE United States to Alaska and Quebec. All confined to freshwater. Length 5cm (2in) to a maximum of 20cm (8in). 9 species in 6 genera and 3 families including: Percopsidae, with two species, the **Trout-perch** (*Percopsis omiscomaycus*), and the **Sand roller** (*P. transmontana*); Aphredoderidae, containing a single species, the **Pirate perch** (*Aphredoderus sayanus*); and the **cavefishes** (family Amblyopsidae) 6 species in 5 genera, including the **Spring cavefish** (*Forbesichthys agassizii*) and **Southern cavefish** (*Typhlichthys subterraneus*). The **Alabama cavefish** (*Speleoplatyrhinus poulsoni*) is Critically Endangered, while three other species are classed as Vulnerable.

Cuskeels and Allies
Order: Ophidiiformes

Widely distributed in the Atlantic, Pacific, and Indian oceans. Nearly all marine, but some confined to fresh or weak brackish water. Length 5 cm (2in) to a maximum of 2m (6.5ft). About 355 species in 92 genera and 5 families. Families include: **carapids** or **pearlfishes** (family Carapidae) with c.32 species in 7 genera; **cuskeels** (family Ophidiidae) with c.209 species in 46 genera; **viviparous brotulas** (family Bythitidae) with at least 90 species in 31 genera; **aphyonids** (family Aphyonidae) 21 species in 6 genera; **false brotulas** (family Parabrotulidae) 3 species in 2 genera. Seven species are classed as Vulnerable, including the **New Providence cuskeel** (*Lucifuga spelaeotes*).

Trout-perches and allies
- Cavefishes — Amblyopsidae
- Pirate perch — Aphredoderidae
- Trout-perches — Percopsidae

Cuskeels and allies
- Carapids or pearlfishes — Carapidae

Codfishes and allies
- Codlets — Bregmacerotidae
- Codfishes — Gadidae
- Grenadiers or rattails — Macrouridae
- Merlucciid hakes — Merlucciidae
- Morid cods or moras — Moridae

Anglerfishes
- Fanfins — Caulophrynidae
- Linophrynidae
- Sea devils — Ceratiidae

Codfishes and Allies
Order: Gadiformes

Worldwide in all oceans. Almost exclusively marine; one species confined to freshwater, a second with some populations that are confined to freshwater. Length 10 cm (4in) to a maximum of 2m (6.5ft). About 482 species in 85 genera and 12 families including: Ranicipitidae, containing a single species, the **Tadpole cod** (*Raniceps raninus*); Euclichthyidae, containing a single species, the **Eucla cod** (*Euclichthys polynemus*); **grenadiers** or **rattails** (family Macrouridae) 285 species in 38 genera; Steindachneriidae, containing a single species, the **Luminous hake** (*Steindachneria argentea*); **morid cods** or **moras** (family Moridae) c.98 species in 18 genera; **pelagic cods** (family Melanonidae) containing a single genus, *Melanonus*, and 2 species; **Southern hakes** (family Macruronidae) 8 species in 2 genera; **codlets** (family Bregmacerotidae) containing a single genus, *Bregmaceros*, with c.12 species; **Eel cods** (family Muraenolepididae) containing a single genus, *Muraenolepis*, with 4 species; **phycid hakes** (family Phycidae) 27 species in 5 genera; **merluccid hakes** (family Merlucciidae) containing a single genus, *Merluccius*, with 13 species, including the **Atlantic hake** (*M. merluccius*) and **North Pacific hake** (*M. productus*); **codfishes** (family Gadidae) 30 species in 15 genera, including the **Atlantic cod** (*Gadus morhua*), **Burbot** (*Lota lota*), **European ling** (*Molva molva*), **Haddock** (*Melanogrammus aeglefinus*), **Pacific cod** (*Gadus macrocephalus*), **Pollock** (*Pollachius virens*), and **Alaska pollock** (*Theragra chalcogramma*). The **Skulpin** (*Physiculus helenaensis*) is Critically Endangered, while the Atlantic cod and the Haddock are Vulnerable.

Left *The Striped anglerfish* (Antennarius striatus) *uses a lure comprising a stalk, or illicium, and a bait, or esca. The esca's design, which varies between species, mimics the food of the prey fish, such as a worm. When not in use, the lure is held back against the head.*

Toadfishes
Order: Batrachoidiformes

Widely distributed in the Atlantic, Pacific, and Indian oceans. Primarily marine, rarel entering brackish water; a few species con fined to freshwater. Length 7.5cm (3in) to maximum of 57cm (22in). 69 species in 1 genera and 1 family, Batrachoididae including: the **Atlantic midshipman** (*Porichthys plectrodon*), **Oyster toadfish** (*Opsanus tau*), **Splendid toadfish** (*Sanopus splendidus*), and **Venomous or Cano toadfish** (*Thalassophryne maculosa*). Five species of toadfishe are classed as Vulnerable, including *S. splendidus*.

Anglerfishes
Order: Lophiiformes

Worldwide in all oceans. Marine, with rare incursions into brackish or freshwater. Length 6cm (2.5in) to a maximum of 1.5m (5ft). At least 310 species in 65 genera and 18 families including: **goosefishes** or **monkfishes** (family Lophiidae) 25 species in 4 genera, including the **Common goosefish** (*Lophius piscatorius*); **frogfishes** (family Antennariidae) 42 species in 12 genera, including the **Striated frogfish** (*Antennarius striatus*) and the **Sargassum frogfish** (*Histrio histrio*); Tetrabrachiidae, containing a single species, the **Four-armed frogfish** (*Tetrabrachium ocellatum*); Lophichthyidae containing a single species, **Boschma's frogfish** (*Lophichthys boschmai*); **handfishes** or **warty anglers** (family Brachionichthyidae), containing a single genus, *Brachionichthys*, with 4 species; **gapers, coffinfishes,** or **sea toads** (family Chaunacidae) 2 genera and up to 14 species, including the **Redeye gaper** (*Chaunax stigmaeus*) and **Rosy gaper** (*Bathychaunax roseus*); **batfishes** (family Ogcocephalidae) 9 genera and 62 species, including the **Atlantic batfish** (*Dibranchus atlanticus*) and the **Shortnose batfish** (*Ogcocephalus nasutus*); and the 11 families of the deepsea Ceratioidei, 157 valid species in 35 genera, including the footballfishes (family Himantolophidae) – a single genus, *Himantolophus*, with 18 species, including *Himantolophus groenlandicus*); the Oneirodidae – 62 species in 16 genera including the Short-rod anglerfish (*Microlophichthys microlophus*) and the Bulbous dreamer (*Oneirodes eschrichtii*); seadevils (family Ceratiidae) – 4 species in 2 genera, including Krøyer's deepsea angler (*Ceratias holboelli*) and the Triplewart seadevil (*Cryptopsaras couesii*), Gigantactinidae (21 species in 2 genera, including *Gigantactis vanhoeffeni*), and Linophrynidae (25 species in 5 genera, including *Borophryne apogon*, *Haplophryne mollis*, and *Linophryne lucifer*). The **Spotted handfish** (*Brachionichthys hirsutus*) of Australia is Critically Endangered.

Above *The burbot* (Lota lota) *is the only wholly freshwater species of the order Gadiformes. Characterized by the single barbel on its chin, it lives in wide, slow flowing rivers or in deep lakes, and hunts in the dusk or at night.*

2.7ft) and males (53cm, 1.7ft) reaching quite different maximum lengths. This species lives close to the bottom just over the continental slope at depths of 200–850m (655–2,790ft). In contrast to the European hake, spawning probably occurs from September to November. It is fished mainly by bottom trawl but also by longline. When fresh the flesh is delicious and well textured, but loses its flavor and texture with keeping.

Hakes also occur off New Zealand (the Southern hake, *Merluccius australis*), and off the Pacific and Atlantic coasts of South America (the South Pacific or Chilean hake, *Merluccius gayi*; and the Argentine hake, *Merluccius hubbsi*, respectively). Like the other members of this family, these species are all commercially important and when not used as prime food for humans are used as pet food, fishmeal, and fertilizer.

The Gadidae is by far the most commercially important family of the Paracanthopterygii. Most of the approximately 30 species are found on the continental shelves of the northern hemisphere, but members of one genus of Rocklings (*Gaidropsarus*) also lives off New Zealand, Kerguelen, and South Africa. They have two or three dorsal fins and one or two anal fins, but none of the fins has spines. Many species have a chin barbel and some have additional barbels on the snout.

Practically all species are marine, but an interesting exception is the Burbot (*Lota lota*), an eel-like fish with a mottled brown body. The first dorsal fin is short and just barely makes contact with the long, second dorsal fin. It is widespread throughout sluggish or still waters of cold northern parts of Eurasia and North America. During the last century it was common in rivers on the east coast of England, from Yorkshire south to East Anglia, but it is now probably extinct there. Certainly no reliable reports of its presence have appeared in recent years. A major factor contributing to its demise has been the dredging of drains and canals. This has not only removed the weeds that gave the young protection but also increased the speed of currents, to which sluggish species like the Burbot cannot adapt.

The Burbot is a winter and early spring spawner, from November to May, considering its whole area of distribution, but mainly from January to March in Canada. Large females are well known for producing exceedingly large numbers of eggs. Fecundity estimates in Canada range from 45,600 eggs in a 34-cm (13-in) female to 1,362,077 eggs in a 64-cm (25-in) individual. Burbot are largely nocturnal and feed on invertebrates and bottom-living fishes. Although their flesh is nutritious and tasty, and the liver contains a lot of vitamin A, it is not widely accepted as food for human consumption. However, it is fished commercially in Finland, Sweden, and the European part of Russia, and is of minor commercial importance in Alaska and Canada.

The Lings (genus *Molva*), of which there are only two species, look rather like large marine Burbot. The Blue ling (*Molva dypterygia*) is found in the Barents Sea and west to Greenland (including Iceland) and Newfoundland; around the British Isles and south to Morocco and into the Mediterranean Sea. Its sister species, called simply the Ling (*Molva molva*), has an almost identical distribution but a much smaller range in Greenland waters and is only rarely seen in the northwestern Mediterranean. The Ling is most common at depths of about 300m (1,000ft). A large female can live for 15 years and can weigh as much as 22kg (50lb). It is a commercially important fish, but its main claim to fame is its extraordinary reproductive capabilities. A female Ling 1.5 m (5 feet) long and weighing 24kg (54lb) was found to have 28,361,000 eggs in her ovaries, the largest number of eggs ever recorded for a vertebrate.

The Atlantic cod (*Gadus morhua*) and the Haddock (*Melanogrammus aeglefinus*) are highly prized food fish from the North Atlantic. During the 19th century, when fishing grounds in the far North Atlantic were first exploited, Atlantic cod measuring 2 m (6.6 feet) in length and weighing up to 90kg (200lb) were recorded. These days, because of intensive fishing, an 18kg (40lb) individual is regarded as a giant and most of those caught commercially average less than about 4.5kg (10lb). Throughout its range the Atlantic cod exists in fairly discrete populations (usually referred to as races), but the migration of individuals from one population to another has precluded the accordance of subspecific status to these groups.

The Atlantic cod spawns in late winter and early spring in depths of less than 180m (600ft). The eggs are pelagic and scattered widely by the currents. The young hatch at the surface and feed on small planktonic organisms, but when they are about two months old and about 2.5cm (1in) long, they descend to live close to the bottom.

Adult Atlantic cod feed on large invertebrates, but fish form a larger part of the diet as they grow. During the day, individuals form dense shoals off the bottom, but at night they separate and become more or less solitary. The northern populations migrate south to spawn. The lifestyle of the Haddock (*Melanogrammus aeglefinus*) is similar to that of the Atlantic Cod, but they are a much smaller fish hardly reaching 3.5kg (8lb) in weight.

There are many smaller and commercially relatively unimportant gadid genera in both the North Atlantic and North Pacific oceans. Common in the Atlantic are the Poor cod (*Trisopterus minutus*), locally abundant and said to be good eating, but not extensively fished; the Pouting (*Trisopterus luscus*), of which more than 22,000 metric tons have been harvested in good years; the Blue whiting (*Micromesistius poutassou*), with a 708,000 metric-ton annual harvest; the Pollack (*Pollachius pollachius*), not of great commercial importance, but often marketed fresh and frozen; and the Saithe (*Pollachius virens*), a commerically important species, similar to the Atlantic cod and Haddock. In the Pacific, the Pacific cod (*Gadus macrocephalus*) and the Alaska pollock (*Theragra chalcogramma*) are the most abundant and important members of the codfish family. The Pacific cod is now the dominant trawl-caught bottom fish in British Columbia. In Canada and the western North Pacific, the major types of gear used are trawls, but also longlines, troll, and handlines. Although this species appears to consist of a number of distinct populations, with different behavior patterns, it is overall a biological species quite distinct from the Atlantic cod. The Walleye pollock presently constitutes the largest demersal fish resource in the world. Composed of 12 major stocks distributed in different areas of the North Pacific, the catch in recent years has exceeded 6,700,000 metric tons.

Toadfishes

ORDER BATRACHOIDIFORMES

The toadfishes are a group of sluggish, bottom-living fishes, confined mostly to shallow warm seas. The body is short and stout, the head wide, and the eyes lie on top of the head. A large mouth is well equipped with teeth, reflecting a very predatory lifestyle. Many species are cryptically colored and match their background. Some species, such as the Oyster toadfish (*Opsanus tau*) of the eastern American seaboard, exhibit parental care. The large eggs are laid in a protected spot and guarded by the male until they hatch. The Venomous toadfish (*Thalassophryne maculosa*) lives only along the Caribbean coast of South America at depths of 30 to 60cm (1–2ft). As its common name implies, it can cause serious damage to unwary bathers or fishers. It is said to have the most highly developed stinging apparatus among fishes. Spines in the dorsal fin and on the gill covers are hollow and linked to venom sacs. The fish lies buried in the sand, usually with only its eyes protruding. When trodden on, the spines act like hypodermic needles and inject poison into the offending foot.

The Plainfin midshipmen (*Porichthys notatus*) from the Pacific and the Atlantic midshipmen (*Porichthys plectrodon*) from the Atlantic burrow during the day. Each species has a pattern of light organs along its sides that are used in courtship. A wide range of sounds, from whistles to grunts and growls, are produced by swimbladder-drumming muscles during courtship.

Some of the species of the genus *Halophryne*, from the Indo-Pacific, can live in freshwater and are sometimes sold in aquarium shops. Poisonous spines on the gill covers, however, should make the aquarist think twice before buying these cryptic predatory fishes.

Anglerfishes

ORDER LOPHIIFORMES

The anglerfishes can be divided conveniently into three groups: the first containing the goosefishes or monkfishes; the second, the frogfishes, handfishes, gapers, and batfishes; and the third, the deepsea Ceratioid anglerfishes. Nearly all of the approximately 310 species are characterized by having the first dorsal-fin spine placed on the tip of the snout and modified as a fishing apparatus.

The first two groups contain mostly bottom-dwelling forms, while the last is made up of mostly pelagic species living at mesopelagic, bathypelagic, and abyssal depths. Except for rare incursions into mouths of rivers by a few members of the group, all are confined to marine habitats.

The goosefishes or monkfishes (family Lophiidae), about 25 species, are large, depressed, bottom-living fishes. Their heads and mouths are enormous. The Common goosefish (*Lophius piscatorius*) is widely distributed along the European coast from the Barents Sea south to North Africa, to the Mediterranean and Black Sea, and extends from the surface down to 1,000 m (3,280 feet). It can reach a length of 1. m (5ft) and is a voracious predator. Normally this anglerfish feeds from the bottom – hiding among rocks and vegetation and partially buried in sand or mud, its body outline broken up by a series of irregular fleshy skin flaps, it wriggles its brightly colored bait and thereby attracts prey to its gaping mouth. Often, however, this species, along with other members of the family, comes up to the surface to snatch resting geese, ducks, cormorant, gulls, and other seabirds (hence the name "goosefish"). More than one report has described an anglerfish found dead having choked on a seabird a bit too large to swallow. Their voracity is further evinced by reports of lophiids caught in trawls with a belly full of fish, having been unable to resist gorging themselves on their fellow captives. The Goosefish is relatively uncritical about its diet and consumes any food item that comes close.

This and other members of the family, including members of the closely related genera *Lophiodes*, *Lophiomus*, and *Sladenia*, typically move

Above *The coloring of the Warty frogfish* (Antennarius maculatus) *varies enormously according to habitat. It lives on rocky reefs in the Indo-Pacific region and hunts using a large, fish-shaped lure.*

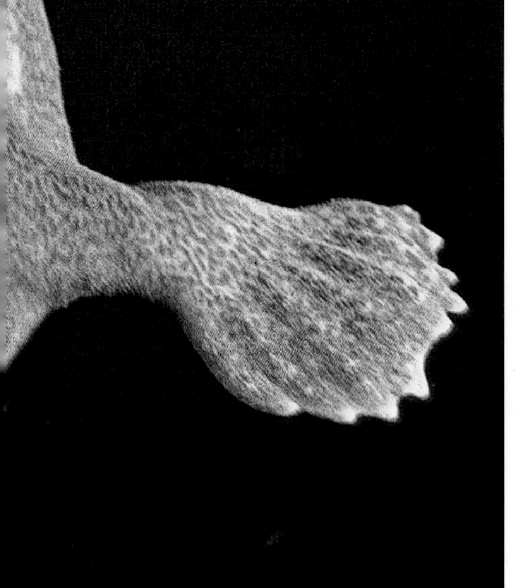

Left *A diver swimming near a frogfish in the Pacific Ocean. In the open, their distinctive shape makes them easy to spot; however, against reefs, their cryptic coloration is a highly effective camouflage.*

Below *An aggressive* Lophius piscatorius, *a member of the goosefish family. This fish is widely sold under the name of Monkfish – perhaps unsurprisingly, minus the head (which accounts for half the body length).*

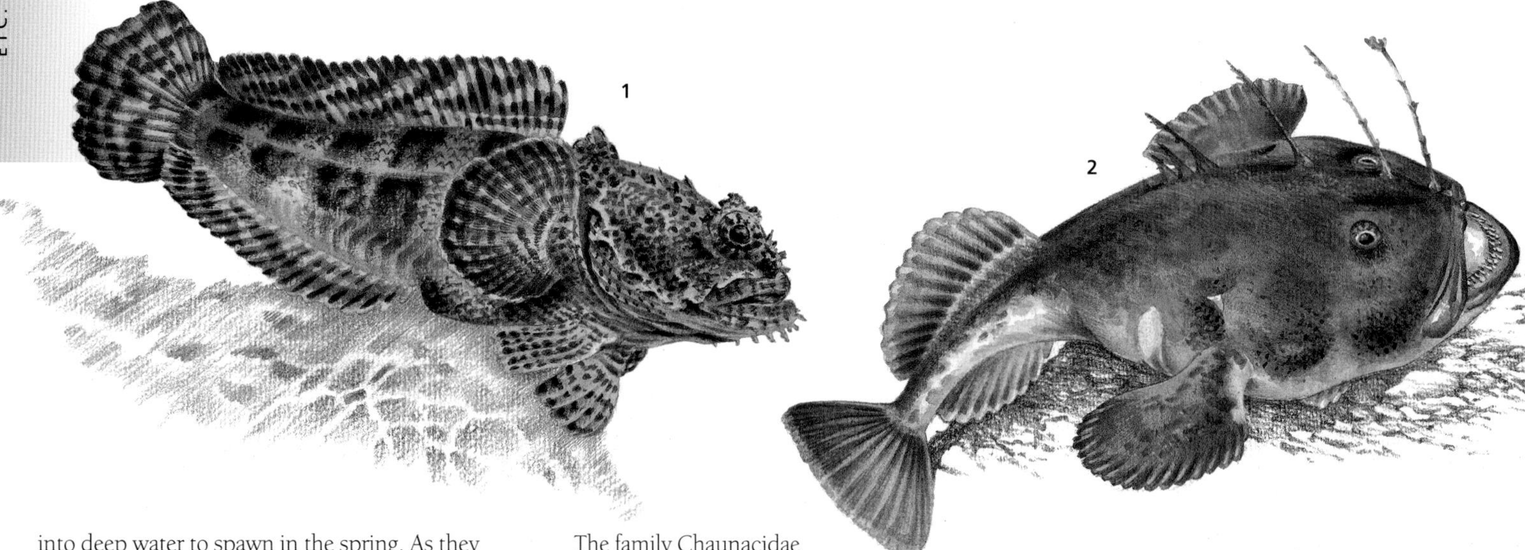

into deep water to spawn in the spring. As they are spawned, the eggs emerge from the female embedded within a remarkable ribbonlike mucous sheath or veil up to 9m (30ft) long and 60cm (2ft) wide. The buoyant veil brings the eggs to the surface where they hatch, providing the young developing larvae with a rich food resource of plankton. As they mature, the larvae eventually settle back down to the bottom.

Although often described as "flabby and revolting," the appearance of the Goosefish belies the nature of its flesh, which is delicately flavored and popular in Europe, where it is sold as Monkfish. The flesh of the tail is white and somewhat reminiscent of scampi in texture. Unfortunately, however, *Lophius* and its close allies have been overfished over much of their ranges and are seriously threatened with depletion.

The frogfishes, family Antennariidae, rarely grow longer than about 30cm (12in). They are either flattened from side to side or more nearly globular. Highly cryptic forms, nearly all live camouflaged on the bottom among sea grasses, rocks, and coral. One species, however, the Sargassum frogfish (*Histrio histrio*) lives in warm waters around the world, clinging to floating clumps of Sargassum weed. Like those of other members of the family, its pectoral and pelvic fins are muscular and elongate, resembling arms and legs, allowing it to "crawl" around among drifting vegetation. The lure of the Sargassum frogfish is not well developed, little more than a tiny slender filament, but in bottom-living, closely related forms, such as *Antennarius* spp., the fishing apparatus is strikingly sophisticated in design. In many species the device is long, half the length of the body or more, and equipped with a fleshy terminal bait, often brightly colored. The baits of some species appear to mimic small aquatic creatures, such as worms, shrimps, and other crustaceans. The Warty frogfish (*Antennarius maculatus*) of the western tropical Pacific, has a bait resembling a tiny fish. Not only is the mimicry remarkably fishlike, but the anglerfish also moves the bait through the water in such a way that it looks just like a swimming fish.

The family Chaunacidae is represented by two genera, *Chaunax* and *Bathychaunax*, but the number of species is unknown. While much of the scientific literature indicates a single widely distributed species, some ichthyologists estimate as many as 14 species. Living on the bottom at depths of 90 to more than 2,000m (300–6,600ft), the group contains large-mouthed, loose-skinned, globose forms that attain a maximum length of about 35cm (13.8in). All forms are pink to deep-reddish-orange.

The batfishes (family Ogcocephalidae) are highly depressed and flattened fishes found around the world in most tropical and subtropical seas. Viewed from above, they are either triangular or round in shape, bearing a narrow elongate tail behind. The mouth is terminal, the eyes are large and placed dorsally, and the armlike pectoral fins are directed backward. The pelvic fins lie far forward beneath the throat and, like many other benthic anglerfishes, are used in tandem with the muscular pectoral fins for "walking" along the sea floor. The reduced gill openings are located near the hind end of the body, just before the pectoral fins. Most species are inhabitants of deep water (down to about 2,500m; 8,200ft) and all spend the day on the bottom luring prey. The majority of batfishes are rather dull in coloration, a light gray to brown, but the Polka-dot batfish (*Ogcocephalus radiatus*) of the Bahamas and Gulf of Mexico is covered with yellow or reddish orange patches, the tip of the tail is black, and the belly is a bright coppery red. The Circular batfish (*Halieutaea fitzsimonsi*), a South African species, has been recorded in freshwater up the Tugela River in Natal, rather far from the coast. Batfishes in general, and the Shortnose batfish (*Ogcocephalus nasutus*) in particular, perhaps the most common species of the family in the western North Atlantic, are known to eat a wide variety of snails, in addition to some polychaete worms and crustaceans. One wonders how snails could possibly be attracted to wriggling baits such as those displayed by these anglerfishes.

The most bizarre members of this superorder, indeed of all animals, are the Deep-sea Ceratioid anglerfishes. The goosefishes, frogfishes, and their close allies have the first dorsal-fin spine placed far forwards on the tip of the snout, where, under precise muscular control, it is moved around like a fishing rod, with a fleshy appendage at the tip serving as bait. While this system works well in shallow, sunlit seas, it is useless in the pitch-black, lightless depths where ceratioids dwell. But in these derived inhabitants of the deepsea, the fleshy bait has been replaced by a bioluminescent organ, the light of which is produced by a tight cluster of millions of symbiotic bacteria. Among the approximately 160 species of ceratioids, some, like the oneirodids the Short-rod angler (*Microlophichthys microlophus*) and *Tyrannophryne pugnax*, have an very short fishing apparatus, not much more than a bulbous light organ attached to the snout, but in others, like members of the genus *Gigantactis*, the equipment can be over five times the length of the fish. In members of the genus *Lasiognathus*, the bioluminescent bait is even accompanied by a series of sharp bony hooks.

In *Thaumatichthys*, a rare ceratioid known from

Above *Representative species of anglerfishes and toadfishes:* ***1*** *Oyster toadfish* (Opsanus tau), *family Batrachoididae. This species, which has venomous dorsal spines, lives along the Atlantic seaboard of the USA south of Massachusetts and in the West Indies;* ***2*** *Goosefish* (Lophius americanus), *family Lophiidae. The enormous, upward-facing jaw of this species is equipped with slender, curved teeth;* ***3*** *Longlure frogfish* (Antennarius multiocellatus), *family Antennariidae;* ***4*** Halieutichthys aculeatus, *a species of batfish, family Ogcocephalidae;* ***5*** *Anglerfish (genus* Melanocetus) *family Melanocetidae;* ***6*** *Anglerfish (genus* Linophryne), *family Linophrynidae – the large hyoid barbel indicates that this is a female;* ***7*** *Anglerfish (genus* Gigantactis), *family Gigantactinidae, with a long, whiplike lure on the end of its snout.*

around 30 adult specimens, the fishing apparatus penetrates through the roof of the mouth so that the bait hangs just behind large fanglike teeth. But perhaps least explicable is *Neoceratias spinifer* in which the fishing equipment has been lost altogether – no one knows how this fish catches food.

Deepsea ceratioid anglerfishes also differ from their shallow-water relatives in having an extreme sexual dimorphism and a unique mode of reproduction in which the males are dwarfed and attach themselves (either temporarily or permanently) to the bodies of relatively huge females. In some species attachment is followed by tissue fusion and, eventually, by a connection of the circulatory systems so that the male becomes permanently dependent on the female for blood-transported nutrients, while the host female becomes a kind of self-fertilizing hermaphrodite.

Although the ceratioids are found worldwide, from high Arctic latitudes to the Southern Ocean, they are distributed patchily, preferring the more highly productive waters. Generally they are found at greater depths in the tropics than at the poles, but all these generalizations can be affected by local factors. Some are relatively shallow-living, for example, *Oneirodes carlsbergi*, which is most commonly found between about 300 and 400m, but most live far below this depth, averaging between 1,000 (3,300ft) and 2,500m (8,200ft), some extending down to at least 3,700m (12,140ft). Most live pelagically far off the bottom, but there is ample evidence that some, for example the Two-rod anglerfish (*Diceratias bispinosus*) and *Bufoceratias wedli*, are closely associated with the bottom; the extremely flattened members of the genus *Thaumatichthys* must certainly lead fully benthic lifestyles. In a most surprising discovery, a species of *Gigantactis* was filmed from a submersible swimming upside-down along the bottom and dragging its bait along the substrate, apparently trolling for benthic-living organisms. KEB/TP

SPECIAL FEATURE

SEXUAL PARASITES

Reproductive modes among deepsea ceratioid anglerfishes

AMONG THE MOST BIZARRE AND INTRIGUING of all animals, deepsea ceratioid anglerfishes, which constitute by far the most speciose group of vertebrates below oceanic depths of about 300m (900ft), differ remarkably from all other living organisms in a variety of ways. Most strikingly, they display an extreme sexual dimorphism and a unique mode of reproduction, in which the males are dwarfed and attach themselves either temporarily or permanently to the bodies of relatively gigantic females. The males of some families are adults at a surprisingly small size. For example, those of some members of the family Linophrynidae are mature at body lengths of only 8–10mm (0.3–0.4in), making them strong contenders for the title of "world's smallest vertebrate." On the other hand, the females of some species grow quite large: females of *Thaumatichthys* and *Gigantactis* are represented by specimens of 30–40cm (12–16in); the record body length for females of the genus *Himantolophus* is 46.5cm (18in); and females of *Ceratias*, by far the largest known of all the ceratioids, grow to at least 77cm (30in).

As well as being smaller than females by an order of magnitude, the males lack a luring apparatus, and those of most species are equipped with large well-developed eyes and relatively huge nostrils. It is hypothesized that the males find females in the dark abyssal depths by using a combination of vision, hence the big eyes, and by homing in on a female-emitted, species-specific smell. As the males mature from larvae to adults the normal jaw teeth are lost, but they are replaced by a set of pincerlike denticles at the anterior tips of the jaws for grasping and holding fast to a prospective mate. Once found, the males bite the female – attachment is usually on her belly, but it could be almost anywhere: on her side, on top of her head, her face, her lip, on one of her fins, even on the bioluminescent bait at the tip of her fishing apparatus. In most ceratioids, the males attach only for a short time, until spawning takes place, and then they let go to begin the search again for another female. But, in a few groups (only eight genera in four of the eleven recognized ceratioid families: Caulophrynidae, Ceratiidae, Neoceratiidae, and Linophrynidae), attachment is followed by fusion of the skin of male and female and, eventually, by a connection of the circulatory systems so that the male becomes permanently dependent on the female for blood-transported nutrients, while the host female becomes a kind of self-fertilizing hermaphrodite. This so-called "sexual parasitism," an approach to reproduction that is unique in all the world to these few groups of deepsea anglerfishes, usually results in a female with a single attached mate, but some females acquire two or three, and in very rare cases, even seven or eight.

The earliest recorded capture of a deepsea anglerfish took place in 1833, when a large female was washed ashore on the southwest coast of Greenland, poorly preserved and partially eaten by birds. It was officially described in 1837 as *Himantolophus groenlandicus* by the Danish zoologist Johannes Reinhardt (1776–1845). Although hundreds more specimens were found in subsequent decades, they were not studied closely, and so it escaped biologists' notice that they were all female. The question of where the males were would only be raised, and solved, in the next century.

In 1922, the Icelandic fisheries biologist Bjarni Saemundsson (1867–1940), examining a specimen collected by the recent "round-the-world" deepsea expeditions of the Danish research vessel *Dana*, was startled to find two little fishes hanging by their snouts to the belly of a large female deepsea anglerfish identified as *Ceratias holboelli*. Not recognizing them as dwarfed males, he described them as the young of the same species. This view was corrected just three years later, when the British researcher Charles Tate Regan (1878–1943) dissected a small fish attached to another female *C. holboelli* and concluded that it was a parasitic male. Regan wrote that the male fish is "merely an appendage of the female, and entirely dependent on her for nutrition, ... so perfect and complete is the union of husband and wife that one may almost be sure that their genital glands ripen simultaneously, and it is not too fanciful to think that the female may possibly be able to control the seminal discharge of the male and ensure that it takes place at the right time for fertilization of her eggs."

Sexual parasitism in ceratioid anglerfishes is now common scientific knowledge, yet there is still much about this remarkable reproductive mode that is not understood. For example, the physiological mechanisms that allow for sexual parasitism, which have intriguing and potentially significant biomedical relevance, have never been studied. Two especially important questions come to mind: With circulatory systems fused and female fluids diluting those of males, how do males manage the endocrine control that is required for sperm production? How are normal immuno-responses suppressed to enable tissue fusion between males and females? These and many other related questions remain for some future researcher to resolve. KEB/TP

Left *Unmetamorphosed anglerfishes are encased in a protective gelatinous skin. This is a female, as shown by the beginning of the illicium (fishing lure) above the eye.*

Right *At her command, two parasitic males are fused to the body of a female anglerfish* (Haplophryne mollis).

Silversides, Killifishes, and Ricefishes

Fishes that fly, fishes that give birth to live young, fishes used to control mosquitoes, and fishes that spawn to the cycles of the moon: these are some of the unusual lifestyles found in the series Atherinomorpha. Some of its members are extremely well known, thanks to their wide use in experimental studies of embryo development and their adaptability to aquarium conditions.

The series comprises some 1,290 species of minute to medium-sized fishes distributed worldwide in temperate and tropical regions, and inhabiting freshwater, brackish water, and seawater. This account recognizes three orders: the silversides, killifishes, and ricefishes and their allies.

Silversides

ORDER ATHERINIFORMES

Silversides are characterized by having a silvery lateral band at midbody, hence their common name, but this character is found in many other groups of fishes. Most silversides are narrow bodied and elongate, although some are relatively deep bodied, such as species of the rainbowfish genus *Glossolepis* of New Guinea, which are used as food fish. Larval atheriniforms are characterized by a single row of melanophores on the dorsal margin.

Many atherinomorphs have a prolonged development time, with fertilized eggs taking one week or more to hatch, as opposed to the more usual time in teleosts of from one to two days. The grunions, in particular the Californis grunion (*Leuresthes tenuis*), a species of silverside found on the West Coast of North America from southern California to Baja California, Mexico, is well known because of its spawning behavior, which is correlated with the lunar cycle. Grunion spawn during spring tide; the fertilized eggs are stranded during low tide, and hatching is stimulated by the waters of the returning high tide two weeks later.

Silversides, such as species of the North American genus *Menidia* found in coastal and gulf drainages, are used as bait fish. Another common name for silversides is smelts, though they are not related to the true smelts of the salmoniform family Osmeridae.

Killifishes

ORDER CYPRINODONTIFORMES

The killifishes are probably best known to the general public in the form of the livebearing fishes in the family Poeciliidae, which are extremely popular in the aquarium hobby. Included among them are the guppy – undoubtedly one of the most commonly kept fishes – and the mosquitofish, which consumes mosquito larvae and pupae and is used throughout the world as a natural mosquito-control agent. Poeciliids are also of great interest to biologists because there exist populations of

Below *The Celebes rainbowfish* (Marosatherina ladigesi) *is a small silverside from freshwater habitats around the Indonesian island of Sulawesi. Its limited distribution and collection for the aquarium trade have led to it being classified as Vulnerable by the IUCN.*

FACTFILE

SILVERSIDES, KILLIFISHES, ETC.

Series: Atherinomorpha

Superorder: Acanthopterygii

About 1,290 species in about 170 genera, 20 families, and 3 orders.

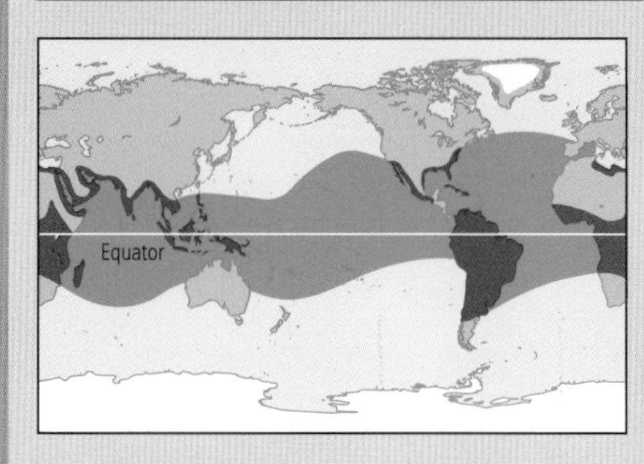

SILVERSIDES Order Atheriniformes

290 species in 49 genera and 6 families. Worldwide, freshwater and seawater. **Length** maximum 60cm (24in). Families: Atherinidae; Atherinopsidae, including the grunions (genus *Leuresthes*); Atherionidae; Melanotaeniidae; Notocheiridae; Phallostethidae. **Conservation status**: Six species Critically Endangered, including the Lake Wanam rainbowfish (*Glossolepis wanamensis*) and Glass blue-eye (*Kiunga ballochi*) of Papua New Guinea.

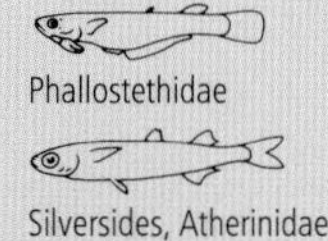
Phallostethidae

Rainbowfishes Melanotaeniidae

Silversides, Atherinidae

KILLIFISHES Order Cyprinodontiformes

800 species in 84 genera and 9 families. Pantropical and north temperate regions in freshwater and brackish water. **Length** maximum 30cm (12in). Families: Anablepidae, including foureyes (genus *Anableps*); Aplocheilidae; Cyprinodontidae; Fundulidae; Goodeidae; Poeciliidae, including guppy (*Poecilia reticulata*), mosquito fish (*Gambusia affinis*); Profundulidae; Rivulidae; Valenciidae. **Conservation status**: Eighteen species are Critically Endangered, including the Monterrey platy (*Xiphophorus couchianus*) of Mexico and the Leon Springs pupfish (*Cyprinodon bovinus*) of Texas, USA.

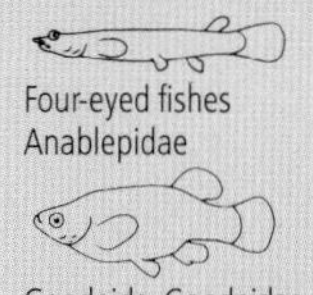
Four-eyed fishes Anablepidae

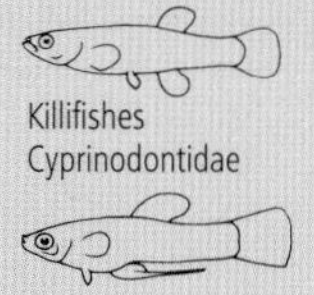
Killifishes Cyprinodontidae

Goodeids, Goodeidae

Poeciliidae

RICEFISHES AND ALLIES Order Beloniformes

200 species in about 37 genera and 5 families. Worldwide, freshwater and marine water. **Length** maximum 1m (3.3ft). Families: flying fishes (family Exocoetidae), halfbeaks (family Hemiramphidae), needlefishes (family Belonidae), ricefishes (family Adrianichthyidae), sauries (family Scombersocidae). **Conservation status**: Two species are Critically Endangered, the Duck-billed buntingi (*Adrianichthys kruyti*) and Popta's buntingi (*Xenopoecilus poptae*), both of Indonesia. Two further species are classed as Endangered, and 8 as Vulnerable.

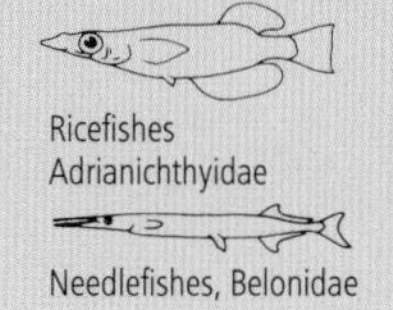
Ricefishes Adrianichthyidae

Flying fishes, Exocoetidae

Needlefishes, Belonidae

Sauries, Scomberesocidae

Above The best-known example of a fish that synchronizes its spawning behavior with lunar cycles is the California grunion (Leuresthes tenuis) of the North American Pacific coast. This marine species moves inshore and spawns in the sand of beaches at night, following the highest of the spring tides after both the new and full moons. As the waves roll in, the fish come ashore and deposit and fertilize their eggs in the wet sand near the top of the high-tide mark. The eggs are normally hidden about 8cm (3in) below the surface. The eggs develop in the sand, awaiting the next set of spring tides, some 12–14 days later. Then the eggs hatch and the young are washed into the sea.

Left Found throughout Central Africa, the Lyretail panchax (Aphyosemion australe) is an oviparous, nonannual killifish species that hangs its eggs on vegetation. Male killifishes, such as that shown here, have more flamboyant finnage and coloration than females.

some species that are composed entirely, or almost entirely, of females (see Life Without Males).

Other killifishes or killies in the Old and New World tropical families Aplocheilidae and Rivulidae, respectively, are also popular aquarium fishes, as well as pest-control agents. Their popularity as aquarium fishes is no doubt due in part to their bright and beautiful coloration, as well as to their hardy nature, which is so renowned that hobbyists often exchange fishes around the world by mailing them, wrapped only in a plastic bag with a little water and air and shipped in an insulated container. Included in these two tropical families are the annual killifishes, which are so named because adults rarely live longer than one rainy season, at the end of which time they spawn, leaving fertilized eggs in the drying muddy substrate. The eggs spend the dry season buried in the mud, lying quiescent until the rains return the following season. When the rains begin again the eggs are stimulated and hatch, and the cycle is repeated.

North American killifishes are less brightly colored, as are most temperate fishes, but no less well known, at least to biologists. The Mummichog (*Fundulus heteroclitus*) a species found in brackish water from Canada to the southern USA is widely used in experimental embryological studies. Its biology is probably better known than that of any other species of bony fish.

Life Without Males

The Amazon molly (*Poecilia "formosa"*) does not live in the Amazon Basin, but in more northerly, marine waters off Texas and Mexico. The common name of this species denotes the mythical race of woman warriors, and alludes to the fact that, like them, it lives in a society almost entirely devoid of males. In particular, *P. formosa* reproduces by gynogenesis, a form of unisexuality. Sperm is supplied by the males of two closely related species (*P. latipinna* and *P. mexicana*), but its sole function is to trigger cell division in the eggs (a process known as embryogenesis). Because none of the the males' sperm enters the egg and fertilizes it, the paternal genome makes no contribution to the genetic makeup of the next generation, which is, like the one before it, exclusively female. Moreover, as with all unisexual species, Amazon mollies are genetically identical, being descended from a single, founding female.

Some of the most spectacular cyprinodontiforms are those of the genus *Anableps*, found from southern Mexico to northern South America. These are the largest members of the order, some growing to over 30cm (12in). They are most well known for the characteristic from which their common name of foureyes (Spanish *cuatro ojos*) is derived; each eye is divided horizontally into two sections, and there are separate upper and lower corneas and retinas. *Anableps* species are usually found just below the surface of the water, and seen from above only by the tops of their eyes, which protrude above the surface. The upper eyes are used for vision above the water, while the lower eyes are used for vision below.

Killifishes of the families Poeciliidae (subfamilies: Poeciliinae and Goodeinae) and Anablepidae have species that are viviparous (i.e. there is internal fertilization of females by males, and females give birth to live young). Males of these viviparous species have anal fins modified for sperm transfer – in the Poeciliinae and Anablepidae the first few anal rays are usually more elongate and elaborate than those in the rest of the fin and are modified into a gonopodium. In the Goodeinae, males have an anal fin or "notch" known as the spermatopodium. At one time it was thought that all viviparous killifishes formed a natural group – that is, that they were more closely related to each other than any was to a group of oviparous or egg-laying killifishes. Yet this is not the case, with egg-laying killifishes often judged to be more closely related to a particular viviparous group. This insight into killifish relationships has brought a more general awareness that viviparity, although characterized by many complex anatomical and behavioral modifications, is a way of life that has arisen several times within the evolution of killifishes.

Ricefishes and Allies

ORDER BELONIFORMES

The order Beloniformes comprises two groups, the ricefishes (Adrianichthyoidei) and the halfbeaks, flying fishes, needlefishes, and sauries (Belonoidei or Exocoetoidei).

All ricefishes are contained in a single family. They are so called because they were discovered in Oriental rice paddies. The scientific name of the common ricefish genus *Oryzias* is derived directly from the generic name of rice plants, *Oryza*. Ricefishes are common in freshwater and brackish water from the Indian subcontinent throughout coastal Southeast Asia into China, Japan, and along the Indo-Australian archipelago as far as Sulawesi. The Medaka (*Oryzias latipes*) is a well-known model organism in experimental biology.

Halfbeaks are freshwater and marine fishes characterized by an elongate lower jaw and a short upper jaw, hence "half a beak." Most halfbeaks are oviparous, but some, such as the Indo-Australian Wrestling halfbeak (*Dermogenys pusilla*) have internal fertilization and are viviparous.

Species in the family of flying fishes do not exhibit true flight, as the name implies. They have expanded pectoral (and sometimes pelvic) fin rays that allow them to glide for several seconds after they propel themselves above the water surface.

In the needlefish family, species have an elongate upper as well as lower jaw; they are more or less fully beaked. The common name is a reference to the extremely sharp, needlelike teeth in the jaws. Most of the cosmopolitan temperate and tropical needlefishes are marine, whereas some, such as *Potamorhapsis guianensis* of the Amazon, live in freshwater. Needlefishes are characterized by having greenish-colored bones and also often muscle tissue. This does not, however, prevent them from being used as a food fish.

Sauries are commercially among the most important beloniform fishes. the Pacific saury (*Cololabis saira*), found in both the eastern and western Pacific, is an important species in fisheries in Japan. The scientific name *Scomberesox*, the type genus of the family, is a composite of *Scomber*, a name for mackerels, and *Esox*, the name for pikes and pickerels. Apparently, sauries seemed to early researchers to have features of those two distantly related groups – five to seven finlets behind the dorsal and anal fins being reminiscent of the mackerels, while the moderate-sized jaws with strong teeth resembled those of pikes and pickerels. LP

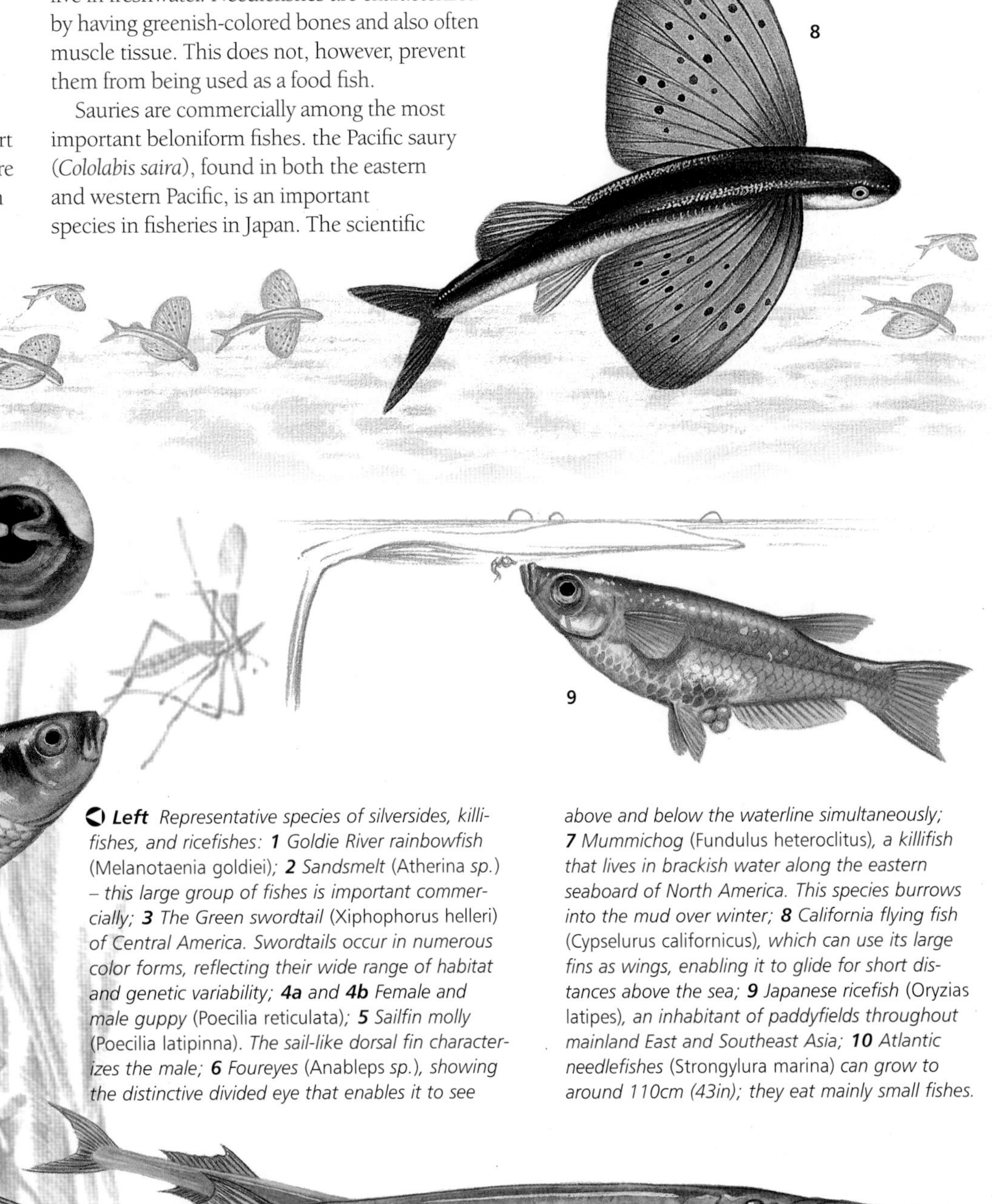

◀ ***Left** Representative species of silversides, killifishes, and ricefishes:* ***1*** *Goldie River rainbowfish* (Melanotaenia goldiei)*;* ***2*** *Sandsmelt* (Atherina sp.) *– this large group of fishes is important commercially;* ***3*** *The Green swordtail* (Xiphophorus helleri) *of Central America. Swordtails occur in numerous color forms, reflecting their wide range of habitat and genetic variability;* ***4a*** *and* ***4b*** *Female and male guppy* (Poecilia reticulata)*;* ***5*** *Sailfin molly* (Poecilia latipinna)*. The sail-like dorsal fin characterizes the male;* ***6*** *Foureyes* (Anableps sp.)*, showing the distinctive divided eye that enables it to see above and below the waterline simultaneously;* ***7*** *Mummichog* (Fundulus heteroclitus)*, a killifish that lives in brackish water along the eastern seaboard of North America. This species burrows into the mud over winter;* ***8*** *California flying fish* (Cypselurus californicus)*, which can use its large fins as wings, enabling it to glide for short distances above the sea;* ***9*** *Japanese ricefish* (Oryzias latipes)*, an inhabitant of paddyfields throughout mainland East and Southeast Asia;* ***10*** *Atlantic needlefishes* (Strongylura marina) *can grow to around 110cm (43in); they eat mainly small fishes.*

Perchlike Fishes

No other order of fishes approaches that of the perchlike fishes in the number of species and the variety of form, structure, and ecology. Indeed, the Perciformes constitute the largest of all vertebrate orders, containing about 150 families with more than 9,300 species. This figure represents almost 40 percent of all fishes.

Classification of the perchlike fishes continues to be the subject of much contention. Whether they form a natural assemblage is debatable; at present the perciforms are ill-defined and lack a single specialized character (or combination of characters) derived from a common ancestor to define the group (i.e. they are not monophyletic).

The earliest fossil record dates to the Upper Cretaceous (96–65 million years ago). In common with other orders in the larger assemblage known as the spiny-finned fishes, the perciforms have spines in both the anterior part of the dorsal fin (or separately, in front of the soft-rayed dorsal fin) and that of the anal fin. Spines are also present in the pelvic fins.

The large majority of perchlike fishes are marine shore fishes. Only around one-fifth of species – including notably the perches themselves (family Percidae) and most cichlids (family Cichlidae) – inhabit freshwater environments.

Left Combtooth blennies are among the smallest perchlike fishes, and some have evolved sophisticated survival strategies. As well as living in crevices, the Red Sea mimic blenny (Ecsenius gravieri) has the same coloration as the venomous Blackline fangblenny (Meiacanthus nigrolineatus), thus avoiding being preyed upon. This phenomenon is known as Batesian mimicry.

Widespread and Diverse

PERCIFORM FAMILIES

The most "typical" members of the perciforms, in terms of their morphology, are the species in the **perch family**. The perch body is typically deep and slender; the two dorsal fins are separate; the pelvics are near the "throat" and the operculum ends in a sharp, spinelike point. They are adapted to northern hemisphere temperatures; warm winters retard the maturing of sperm and eggs.

The European perch is a sedentary species that prefers lakes, canals, and slow-flowing rivers. The ruffe, or pope, of Europe and southern England is a bottom-feeding species, frequenting canals, lakes, and the lower reaches of rivers. Confusingly, it has contiguous dorsal fins.

The zander, or pikeperch, is a native of eastern Europe, but has been widely introduced to other parts of Europe as a sport fish. This predatory species, which takes roach, perch, and sticklebacks, is prized by anglers and valued as a food fish. The North American zander (or walleye) occurs naturally in wide, shallow rivers and lakes.

The North American darters are the most speciose group of percids, with about 145 species. The common name is derived from their habit of darting between stones, as they are bottom-dwelling fishes that lack swimbladders. While many species of darter are brightly colored, often in red and green, the Eastern sand-darter is an inconspicuous translucent species, which buries itself in sandy stream beds with only the eyes and snout protruding.

Like that of all other teleosts, the percid skeleton is basically bone, with some cartilage, although the skeletons of the perciform families of **louvars**

Left Cruising the clear waters of a coral reef, a beautifully marked Coral hind (Cephalopholis miniata) opens its mouth wide to feed. This predatory species occasioanally takes crustaceans, but mainly eats the small Sea goldie (Pseudanthias squamipinnis), a co-member of its own family (Serranidae).

(Luvaridae) and **ragfishes** (Icosteidae) are largely cartilaginous. The louvar, the only species in its family, may be related to the mackerels and tunas. It lives in tropical seas, grows to about 1.8m (6ft) and has a tapering, pinkish colored body. Its pectoral fins are sickle shaped; the pelvics are minute; its dorsal and anal fins are long, low, and set far back on the body. The louvar feeds on jellyfishes; its intestine is very long, with numerous internal projections that increase the absorbent surface area of the gut. The significance of the cartilaginous skeleton remains a mystery.

The ragfish is also the only species in its family. The name comes from its almost boneless appearance, which makes it look like a bundle of rags that has been dropped on the floor. The ragfish grows to some 2.1m (7ft). It is chocolate brown in color, the body shape is elliptical, and it lacks scales, spines, and pelvic fins. The distribution is the northeastern and mid-Pacific, from Japan to Alaska and southern California, where specimens are captured occasionally in trawls at 18–366m (60–1,200ft). It eats fishes and squid, and is itself fed on by sperm whales.

The large spotted **groupers** in the family Serranidae are voracious predators. Sometimes they are found with black, irregular lumps, either lying in the body cavity or bound by tissue to the viscera: these lumps are mummified sharp-tailed eels. Each eel is swallowed by the grouper and in its death throes punctures the gut; it gets squeezed into the body cavity, where it becomes mummified.

The Queensland grouper, a native of Australian seas, may weigh up to half a tonne and is another sea bass with a hearty appetite. This fish has been known to stalk pearl and shell divers, much as a cat stalks a mouse, a habit that has led to unfounded stories of divers being swallowed by groupers.

The Serranidae family also contains much smaller and highly colorful species, such as the colorful anthias, or fairy basslets (genus *Pseudanthias*), that abound on Indo-Pacific reefs. The huge, swarming schools containing hundreds or even thousands of bright red fishes attract any sport diver, and has led to these fishes probably being among the most photographed in the world. Apart from that, they, and many related species, are also popular with aquarium hobbyists.

Above *The boldly striated Oriental sweetlips* (Plectorhinchus orientalis) *gathers in small groups in the shelter of seaward coral reefs, hovering motionless in midwater. At the head of this shoal is another perch-like species, the Bluestripe snapper* (Lutjanus kasmira).

FACTFILE

PERCHLIKE FISHES

Series: Percomorpha

Superorder: Acanthopterygii.

Order: Perciformes

More than 9,300 species in about 1,500 genera and about 150 families.

DISTRIBUTION Worldwide in both marine water and freshwater.

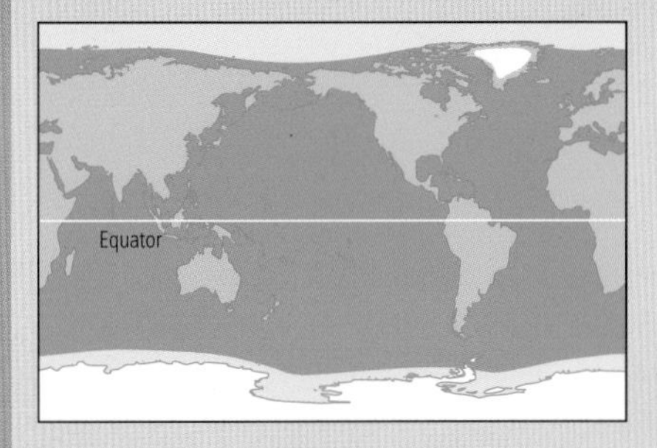

SIZE Length 1cm (0.4in)–5m (16ft); weight up to 900kg (2,000lb)

See Key Families of Perchlike Fishes ▷

Species in the families of **grunts** (Haemulidae) and **drums or croakers** (Sciaenidae) are pretty, tropical marine fishes that have earned their name from the noise they produce. In the grunts, the sound is produced by grinding together well-developed pharyngeal teeth. The drums' noise is caused by muscles vibrating the swimbladder,

not always attached directly thereto but running from either side of the abdomen to a central tendon situated above the swimbladder. Rapid twitches of the muscles vibrate the swimbladder walls which have a complex structure and act as a resonator to amplify the drumming sound. A swimbladder is absent in the drum genus *Menticirrhus,* so this fish produces only a weak noise by grinding its teeth. Both families contain several commercially important food fishes, and are fished for in many parts of the world.

Barracudas (family Sphyraenidae) are another group of perciforms reported to attack divers. They are tropical marine fishes, which in some areas, especially the West Indies, are more feared than sharks. The body is elongate and powerful; the jaws are armed with sharp, daggerlike teeth. Barracudas eat other fish and seemingly herd shoals, making the food easier to catch. Large individuals tend to be solitary, but younger barracudas aggregate in shoals. The barracuda makes very good eating but is notorious for being sporadically poisonous owing to the accumulation of toxin (ciguatera) acquired from the herbivorous fish on which it feeds, who for their part have accumulated the toxins from consumption of seasonal algae (dinoflagellates).

Mackerels, tunas, and bonitos or scombrids (family Scombridae) are also delicious perciforms. The scombrids are mainly schooling fishes of the open seas, cruising at speeds of up to 48km/h (30mph). Their bodies are highly streamlined, terminating in a large lunate caudal fin. Some scombrids have slots on the dorsal surface of the body, in which the spiny dorsal fin fits, thus reducing water resistance. Behind the dorsal and anal fins are a series of finlets, the number of which varies according to species. In all species the scales are either very reduced or absent.

The Common mackerel is found on both sides of the North Atlantic. On the European side it ranges from the Mediterranean to Ireland. It is a pelagic fish, which in summer forms enormous shoals at the water's surface near coasts to feed on small crustaceans and other plankton. In winter the shoals disband and move to deeper water,

How Tuna Fish Keep Warm

Tunas differ from other scombrids and most other teleosts in their ability to retain metabolic heat via a countercurrent heat-exchange system that operates in the muscles and gills. Red muscle occurs in large proportions in tunas. It contains many blood vessels, so the muscle cells are supplied with oxygen- and carbohydrate-enriched blood, enabling them to utilize highly efficient, aerobic metabolism. Aerobic metabolism uses up oxygen and frees energy to drive the muscle and as heat, which is retained in the body by the heat-exchange system. White muscle, found in large proportions in all other fish, has a very poor blood supply and carbohydrate is metabolized anaerobically. which liberates just enough energy to drive the muscle.

Fishes normally lose heat through their gills during respiration, but the tunas' countercurrent heat-exchange system ensures that the metabolic heat is returned to the body. The advantage to the tunas is twofold; the muscles operate at a higher temperature, helping the fish to achieve high speeds and allowing it to range further north. BB

where the fish approach a state of hibernation.

The skipjack tuna is a cosmopolitan marine species that owes its name to its habit of "skipping" over the surface of the water in pursuit of smaller fish.

Billfishes (families Xiphiidae and Istiophoridae), which includes swordfish, sailfishes, spearfishes, and marlins, are all fast-swimming fishes closely related to scombrids (see Mystery of the Swordfish). They include some of the world's most popular marine sport fishes. Several billfish species are known to be migratory, possibly to follow food. Billfishes are fish eaters; they erect the dorsal fin to prevent prey escaping. They can also use the bill as a club to maim their victims as they rush through a school of fishes.

Sailfishes undergo a remarkable change during their larval development. Larvae of about 9mm (0.3in) have both jaws equally produced and armed with conical teeth; the edge of the head above the eye has a series of short bristles; there are two long pointed spines at the back of the head; the dorsal fin is a long, low fringe and the pelvic fins are represented by short buds. At 6cm (2.4in) they begin to resemble the adult: the upper jaw elongates, the teeth disappear, the dorsal fin differentiates into two fins, the spines at the back of the head become reduced and the bristles disappear. Young swordfish also undergo a similar series of changes.

Left A school of Sawtooth barracudas (Sphyraena putnamae). *The two parallel rows of sharp teeth in the barracuda's upper and lower jaws are used to slash and tear pieces off their prey; barracudas do not have a wide enough gape to swallow large fish whole.*

Below The spinecheek anemonefish (Premnas biaculeatus) *lives in lagoons and around reefs and is commonly associated with the sea anemone species* Entacmaea quadricolor. *Members of this family (Pomacentridae) are mainly from the Indo-Pacific region.*

Several of the most spectacular and colorful fishes of tropical seas are perciforms. **Butterflyfishes** (family Chaetodontidae) are distributed worldwide in warm waters around coral reefs. Currently, some 115 species are known. Most are very brightly colored, but often with intricate patterns camouflaging the eye, making it difficult for a potential predator to distinguish the head from the tail. Many species have a dark vertical bar that runs through the eye, further disguising it, and to add to the confusion many species also have an eye spot near the caudal fin. Butterflyfishes delude would-be predators by swimming slowly backward. Once the predator lunges at the eye spot, the butterflyfish darts forward, leaving its attacker confused. The Indo-Pacific butterflyfish, also known as the forcepsfish or longnose butterflyfish, is so called because its snout is very long and used like a pair of forceps to reach deep into reef crevices.

The majority of butterflyfishes are obligate coral feeders that are totally dependent on healthy reefs. In many areas where coral death has occurred through direct human impact or by temperature-induced coral bleaching, the number of butterflyfishes has fallen dramatically.

The great beauty of butterflyfishes makes them attractive also to aquarium hobbyists, but – with relatively few exceptions – they are very difficult to keep in captivity, and are best left to highly experienced and skilled aquarium keepers.

Closely related to, and in the past commonly lumped with, the butterflyfishes family are **angelfishes**. Presently placed in their own family, Pomacanthidae, angelfishes are distinguished from butterflyfishes by their larger, rather rectangular bodies and heavy spines at the base of the gill cover. There are more than 80 species in the family, ranging from tiny dwarf angelfishes, only 8–15cm (3–6in) long, to spectacular giants like the Queen angelfish of the Caribbean, reaching nearly 0.5m (1.6ft). Many are sought-after aquarium fishes, that also do quite well under human care, since most are much less specialist in their feeding requirements than are the butterflyfishes. Several species are quite satisfactory food and occur in most tropical fish markets, but they are rarely of much economic importance.

The twenty-eight species of **clownfishes (anemonefishes)** and **damselfishes** belonging to the Pomacentridae are small, brightly colored fishes of warm shallow seas. Clownfishes live in association with large sea anemones. The relationship is intimate: the fish remains inside the anemone when it withdraws its tentacles. The clownfishes benefit by being protected from predators, so they never stray far from their anemones, but the latter can exist happily without the clownfishes. The anemones' sting-cells are lethal to nonsymbiont fishes, but the clownfishes' mucus coat is considerably thicker than in related species and it

appears to lack the chemical components that triggers the sting-cells. The beautiful color patterns of clownfishes and their fascinating behavior make them some of the most sought-after aquarium fishes. They reproduce easily in captivity and commercial farming of the most popular species is now well established.

Surgeonfishes (family Acanthuridae) are another family of colorful reef-dwelling fishes, of which there are about 72 species. The name alludes to a razor-sharp, lancetlike spine on either side of the caudal peduncle. In most species the spines lie in a groove and are erected when the fish is disturbed or excited. The spines are a formidable weapon, inflicting slash wounds on the victim, as the surgeonfish lashes its tail from side to side. Surgeonfishes normally travel solitary or in small groups, using their small incisorlike teeth to scrape plants and animals from reefs and rocks. Under special circumstances, however, some species, such as the spectacular powder-blue s urgeonfish, gathers in large foraging schools, where the individual is better protected from predators. A few species of surgeonfishes feed on plankton rather than algae. In parts of the Indo-Pacific, surgeonfishes are considered tasty food fish, but the offending caudal peduncle is cut off prior to sale. Several of the smaller and more colorful species are traded for the aquarium hobby.

Members of the family of **stargazers** (Uranoscopidae) are distributed widely in warm seas and earn their name from their eyes, which are set on top of the head so that they appear to be staring at the sky. Stargazers have electric organs situated just behind the eyes, which deliver a shock of up to 50 volts, enough to stun small fish – which are then eaten. The European stargazer is common in the Mediterranean and Black Sea; it grows up to 35cm (14in) and has flaps of tissue in the mouth that resemble worms, tempting potential prey to approach. Predators are deterred from eating the stargazer by grooved spines situated above each pectoral fin. At the base of the spine is a poison gland; as the spine inflicts a wound, poison is trickled into it via the groove. Stargazers are usually

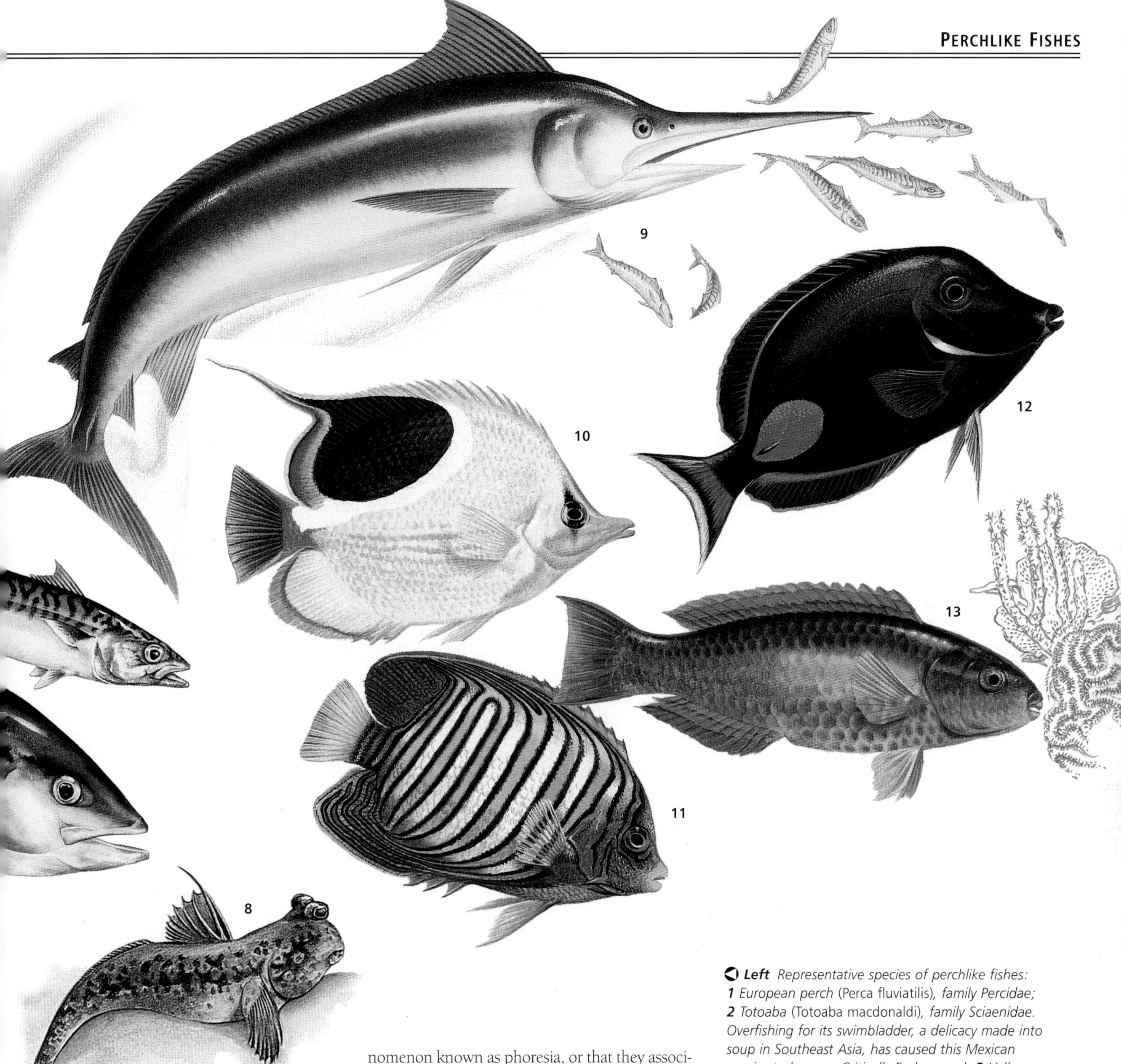

found buried in the sand, with only the eyes and snout tip protruding.

A few perciforms sometimes form unusual relationships with other vertebrates and invertebrates and even with floating objects. **Remoras or sharksuckers** (family Echeneidae) are slim fishes usually associated with sharks, large fishes, and occasionally turtles. Their dorsal fin is modified into a sucking disk, the rim of which is raised, and the platelike fin rays can be adjusted to create a strong vacuum between the disk and a remora's chosen partner. It is unknown what benefit such an association is to either the remoras or the animals to which they attach. It has been suggested that remoras are simply "hitching a ride," a phenomenon known as phoresia, or that they associate with sharks to feed on the scraps they can snatch from the shark's meal. Remoras have been observed entering the mouths of manta rays, large sharks, and billfishes, and it has been speculated that they may fulfill a role similar to that of cleaner fish. However, there is no documented evidence of remoras undertaking cleaning duties.

Despite their usual attachment to sharks, remoras are competent swimmers and often leave the "host" to forage. When free swimming in a group, remoras arrange themselves with the largest on top, smallest at the bottom, reminiscent of a stack of plates. The group swim in a circular fashion; it seems remoras do not like to swim unaccompanied. Ancient legend recounts that remoras can impede the progress of sailing vessels, even stop them. The remora is also reported to have

Left *Representative species of perchlike fishes:* ***1*** *European perch* (Perca fluviatilis), *family Percidae;* ***2*** *Totoaba* (Totoaba macdonaldi), *family Sciaenidae. Overfishing for its swimbladder, a delicacy made into soup in Southeast Asia, has caused this Mexican species to become Critically Endangered;* ***3*** *Yellow labidochromis* (Labidochromis caeruleus), *a mouthbrooding cichlid from Lake Malawi in East Africa; family Cichlidae;* ***4*** *Crevalle jack* (Caranx hippos), *an Eastern Atlantic species, family Carangidae;* ***5*** *Yellowfin tuna* (Thunnus albacares), *an important commercial food fish, family Scombridae;* ***6*** *Atlantic mackerel* (Scomber scombrus), *family Scombridae;* ***7*** *Dwarf pygmy goby* (Pandaka pygmaea), *family Gobiidae. At just 1.5cm in length, this species from the Philippines is the smallest freshwater fish in the world;* ***8*** *Mudskipper* (Periophthalmus *sp.*), *an inhabitant of mangrove swamps, family Gobiidae;* ***9*** *Atlantic blue marlin* (Makaira nigricans), *family Istiophoridae;* ***10*** *Saddle butterflyfish* (Chaetodon ephippium)*; family Chaetodontidae.* ***11*** *Regal angelfish* (Pygoplites diacanthus), *a popular fish in the aquarium trade, family Pomacanthidae.* ***12*** *Achilles tang* (Acanthurus achilles)*; family Acanthuridae.* ***13*** *Princess parrotfish* (Scarus taeniopterus)*; family Scaridae.*

magic powers, and a potion including one was supposed to delay legal proceedings, arrest aging in women, and slow down the course of love.

Pilotfishes in the family Carangidae also associate with sharks and rays. It was thought that the pilotfishes guided the sharks to their prey and in return received protection from their enemies by their proximity to such a formidable companion. Sharks and rays are in fact seeking food, and although pilotfishes gain from the hunting efforts of sharks, they never lead the foray.

The young of another carangid, the common Horse mackerel, shelter in the bell of the sombrero jellyfish (genus *Cotylorhiza*). Why these small fishes do not get stung is unexplained but possibly the absence of glutathione (an amino acid that stimulates release of sting cells) in their mucus coats protects them.

Members of two other families also associate with jellyfishes. Young butterfishes (Stromateidae) are laterally compressed fish that lack pelvic fins and shelter under the protection of the Portuguese man-of-war (genus *Physalia*). The closely related **driftfishes** (Nomeidae) are known as "man-o'-war" fishes and are distinguished from the preceding family by the presence of pelvic fins. Again, it is unknown how these fish gain immunity from the stinging cells of this jellyfish.

Among **gobies** (family Gobiidae), fishes in the genus *Evermannichthys* habitually live inside sponges. The bodies of these little fishes are slender and nearly cylindrical, allowing them easy access to the larger orifices on the sponge's surface. Scales are either absent or poorly developed, but along the lower posterior line of the sides are two series of large, well-separated scales, the edges of which are produced into long spines. A further series of four spined scales is situated in the middle line, behind the anal fin. It is thought that these structures are used by the fish for climbing up the inner surfaces of the sponge cavities.

Many Indo-Pacific genera of gobies, for example *Amblyeleotris* and *Cryptocentrus*, are commensal with digging snapping shrimps (genus *Alpheus*). The goby is usually found at the burrow entrance, while the snapping shrimp busily excavates it. When danger threatens, the goby dives into the burrow; this also alerts the shrimp, which follows the fish inside. The snapping shrimp will not emerge until the goby is once again on sentry duty at the burrow entrance.

The mudskippers, of the genus *Periophthalmus* (family Gobiidae), are found in tropical Africa, Asia, Australia, and Oceania, and spend a great part of their time walking or "skipping" about mangrove roots at low tide. During these periods the branchial chamber is filled with water and oxygen exchange continues over the gills. When the oxygen in this water is exhausted the mudskippers replace it with oxygenated water from a nearby puddle. The mudskippers can also respire through the skin (cutaneously) and have a highly vascular mouth and pharynx through which gaseous exchange can take place; they are therefore often seen sitting with their mouths gaping.

A number of species of **wrasse** (genus *Labroides*; family Labridae) have an unusual cleaning relationship with other fishes: they remove ectoparasites and clean wounds or debris. There are some 500 species in the wrasse family, usually nonschooling, brilliantly colored, and found on reefs in all tropical and temperate marine waters. The wrasses have well-developed "incisor" teeth that protrude like a pair of forceps from a protractile mouth; in some noncleaning species these teeth are used for removing the fins and eyes of other fishes.

Cleaner wrasse are small, brightly colored reef dwellers that occupy a specific area, the "cleaner station." Their diet is mostly parasitic organisms on the bodies and gills of fishes. The association between cleaner and customer is not permanent. Fishes requiring cleaning congregate at the cleaning stations and follow a specific behavior pattern that invites the cleaner to get to work. The customer allows the cleaner to move all over the body, including such sensitive areas as the eyes and mouth, and even to enter the branchial cavity to remove parasites from the gills. The cleaners benefit by immunity from predation during the cleaning and presumably at other times, since many of the customers are predators on fishes the size of these wrasses.

In North Atlantic waters there are also species of wrasse that indulge in cleaning behavior on other fishes. The control of parasites in salmon farms is increasingly done with the help of specimens of goldsinny and young ballan wrasse, corkwing wrasse and related species, rather than with biocides. Many wrasse species are eaten, but since most are relatively small and bony, they are more likely to end up in a fish soup than to be cooked by themselves. In Asian waters there are some species that are highly rated as food fish. The prize wrasse in this respect is undoubtedly the Napoleonfish or humphead wrasse, which has been severely overfished in many areas because of the tremendous demand for large specimens. It can grow to a stunning 230cm (7.5ft), weighing up to nearly 200kg (440lb), but large specimens are getting increasingly rare.

Parrotfishes family (Scaridae) is closely related to the wrasses and many species secrete a mucous nightshirt that surrounds the whole body. This mucous cocoon may take up to half an hour to secrete and as long for the fish to release itself from. Interestingly, this cocoon is not secreted every night but only under certain conditions, the causal factors of which are a mystery. It seems that the mucous cocoon may be a protective device, preventing odors from the parrotfishes reaching predatory fish, such as moray eels.

Species in the **leaffish** family (Nandidae) rely on crypsis (pretending to be something else) to

THE MYSTERY OF THE SWORDFISH

The swordfish is the sole representative of its family (Xiphiidae). It is a solitary fish, and may weigh up to 675kg (1,000lb). The snout is produced into a powerful, flattened sword. Swordfish live in all tropical and subtropical oceans but will enter temperate waters, occasionally straying as far north as Iceland.

The sword has a coat of small denticles similar to those found on sharks. Its function is unknown, but suggestions include a weapon (i.e. the swordfish strikes a shoal of fishes with lateral movements and then devours the mutilated victims) and as extreme streamlining, with the snout acting as a cutwater.

There are numerous accounts of large fish attacking boats, but often there is no attempt to discriminate between swordfish, spearfish, and sailfish, all of which have similar habits. There is no doubt that a swordfish could pierce the bottom of a boat and have the sword snap off in its struggles to withdraw it. In the Natural History Museum, London there is a sample of timber that a swordfish snout has penetrated to a depth of 56cm (22in). It is also reported that the wooden sailing ship HMS *Dreadnought* sprang a leak on a voyage from Ceylon (Sri Lanka) to London. Examination of the hull revealed a 2.5cm (1in) hole punched through the copper sheathing, which was reputed to have been made by a swordfish. Periodically swords are found in whale blubber. Whether these attacks on ships and whales are deliberate is unclear. The most likely explanation is that when a swordfish, which can travel at speeds up to 100km/h (60mph), encounters a boat or whale it finds it impossible to change course in time, and a collision becomes inevitable. BB

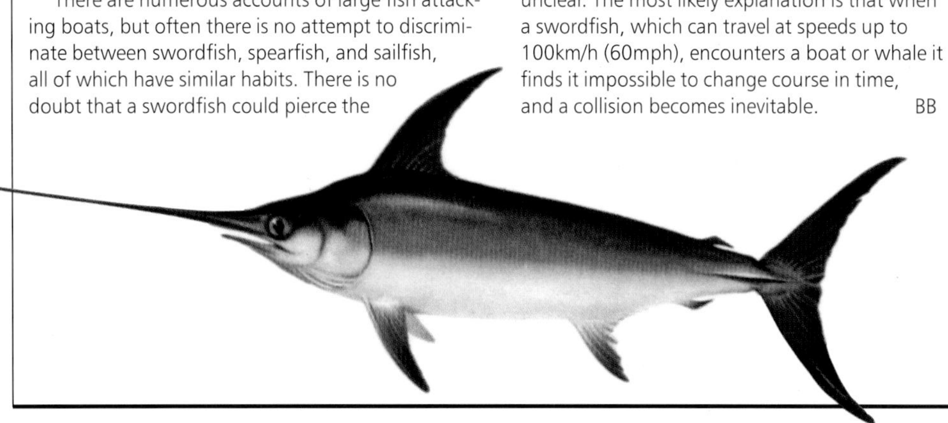

Right *Parrotfishes derive their name from the fact that their jaw teeth are fused together and resemble a parrot's beak. Like birds, they are also characterized by vibrant colors, a feature amply displayed by this* Scarus *sp. parrotfish sleeping on a reef off Borneo.*

catch their food. They live in tropical freshwaters in Africa, Southeast Asia, and South America. Southeast Asian leaffishes are very perchlike – none mimics leaves and the body is only slightly compressed. The most common leaffish of this area is *Badis badis* (sometimes placed in a family of its own, Badidae), which has a large number of different color forms and lives in streams of India and Indochina. The most spectacular leaffishes are those found in South America. They are deep-bodied fish with soft dorsal fin rays: they closely resemble floating leaves in both contours and marks – even a "stalk" is present, protruding from the lower jaw. Leaffish usually hide beneath rocks or in crevices, where they look like a dead leaf, then dart out to capture prey. The most famous leaffish, *Monocirrhus polyacanthus*, lives in the Amazon and Rio Negro basins of South America. Its body is leaf-shaped and tapered toward the snout, with an anterior barbel mimicking a leaf stalk. The fish reaches about 10cm (4in) in length and is a mottled brown, similar to dead leaves. It drifts with the current; on approaching a potential meal, the leaffish bursts into action and, assisted by its large protrusile mouth, engulfs fishes up to half its size.

The perciform family with the largest numbers of species is that of **cichlids** (Cichlidae). Even a conservative estimate indicates that there are at least 1,300 species, and some scientists believe the final count may reach around 5,000. To put this in context, only about 3,000 species of freshwater fishes have so far been recorded from the entire continent of South America. Cichlids are distributed widely in freshwaters of Central and South America, Africa, Syria, Iran, Madagascar, southern India, and Sri Lanka. Over half of the presently known cichlid species are found in Africa, especially the Great Lakes (Victoria, Malawi, and Tanganyika), each of which boasts 100 or more endemic species. The high number of endemisms and apparently rapid speciation make the cichlid fauna of these lakes interesting examples in evolutionary theory. Lake Malawi alone may hold about 1,000 species, of which the vast majority are endemic. However, the taxonomy is controversial and normal definitions of what constitutes a species are often difficult to apply to cichlids. Cichlids are characterized by a single nostril on each side of the head and two lateral lines on each side of the body. The pharyngeal bone is triangular, lying on the floor of the "throat;" its function is to break up food against a hard pad at the base of the skull, and it is of diagnostic importance in the identification of species.

Cichlids have evolved all kinds of dentition that allow them to cope with a varied diet. Vegetarians have bands of small, notched teeth in the jaws, sometimes with an outer chisel-like series for cutting weed or scraping algae off rocks. Fish-eating species have large mouths armed with strong, pointed teeth for securing struggling fish. The

SHOOTING DOWN INSECTS

Six species of archer fishes (*Toxotes* spp.) occur naturally in both freshwater and salt water from India and Malaysia to northern Australia. Their remarkable hunting technique involves spitting droplets of water at their prey. The archer fish has a groove in the roof of the mouth. The tongue is thin and free at the front but thick and muscular with a midline fleshy protuberance at the back. Once the tongue and protuberance are pressed against the roof of the mouth, the groove becomes a narrow tube. The thin, free end of the tongue acts as a valve. When the archer fish spots an insect, the tongue is pressed against the roof of the mouth, the gill covers are jerked shut, and the tip of the tongue is flicked, shooting out drops of water. Archer fish are able to compensate for refraction of light by placing the body vertically below the prey. A fully grown fish can shoot down an insect from up to 1.5m (5ft), whereas babies can shoot only a few centimeters without loss of accuracy. Experiments with adult archer fish suggest, however, that their aim may be quite haphazard and the "downing" of an insect owes more to sheer firepower than to sharp shooting. BB

Above *The Wolf-eel* (Anarrhichthys ocellatus) *lies in wait in rock crevices or among kelp beds for its prey (smaller fish, crustaceans, mollusks, and sea urchins). All the five species in the family to which it belongs (Anarhichadidae) are equipped with large, conical canine teeth at the front of the jaw and strong molars at the rear, with which they can deliver a powerful bite. Wolffishes inhabit the far northern waters of the Atlantic and Pacific and grow up to 2.5m (80in).*

Right *Native to small streams in the Orinoco river basin in South America, the Ram cichlid* (Mikrogeophagus ramirezi) *is a small, brightly colored fish that is very popular with aquarists. It spawns on the substrate, and the adults guard their eggs and fry.*

Below *The Lace or Pearl gourami* (Trichogaster leeri; *family Osphronemidae) of Southeast Asia is one of the "labyrinth fishes," so named for the accessory breathing organs on either side of their heads that enable them to breathe atmospheric air.*

mollusk-eating varieties have strong, blunt pharyngeal teeth to grind up mollusks, although in some species the lateral jaw teeth are modified, enabling the fish to remove the snail from its shell before swallowing it. Finally, in some species the dentition is greatly reduced and deeply embedded in the gums of a very distensible mouth. These species (paedophores) feed almost entirely on the eggs and young of mouthbrooding cichlids, which they force the parent to "cough up."

Cichlids are very popular aquarium fishes and some species, such as freshwater angelfish and discus fishes, count, like goldfish, among the truly domesticated fishes, of which a huge number of stunning color breeds have been produced.

Gouramis (Osphronemidae), **climbing gouramis** (Anabantidae), **kissing gouramis** (Helostomatidae), and **snakeheads** (Channidae) are four more perciform families renowned for their ability to breathe atmospheric air. The first three families are also called labyrinth fishes for their labyrinth-like accessory breathing organs. These breathing organs are located at the top of each gill chamber; they are hollow and formed from highly vascular skin lining the gill chambers. As the fish grows, the organs become more convoluted, increasing the surface area available for respiration. These fishes rely on atmospheric air for survival and quickly suffocate if denied access to the surface of the water.

One of the most spectacular labyrinth fishes is the Siamese fighting fish. This species is customarily brightly colored in aquaria, but has a rather unexciting, brownish-red coloration in the wild. This is another example of selective breeding being used to produce appealing, domesticated forms of fishes – similar to the enhancement by breeding of certain features in other pet animals, such as dogs and cats.

The **snakehead** family is related to the gouramis but contains long cylindrical fishes with flattened, rather reptilian-looking heads. These fish inhabit rivers, ponds, and stagnant marsh pools in Southeast Asia. The various species of snakehead differ in the extent to which each has developed the habit of breathing air. The accessory breathing organs are simpler than those found in anabantoids, consisting of a pair of cavities lined with a vascular, thickened, and puckered membrane. These lunglike reservoirs are not derived from the branchial chambers but are pouches of the pharynx.

Snakeheads are also able to move overland, but do so by a rowing motion of the pectoral fins. During prolonged periods of drought, snakeheads survive by burying themselves in mud and estivating; in hot, dry weather they become torpid. Since several species are popular food fishes that are often sold in food markets while still alive, there is a constant risk of snakeheads being introduced into areas where they do not occur naturally. Many are large, voracious predators, and where accidental introductions have taken place, the snakeheads have inflicted terrible damage on the local fauna. It is therefore extremely important that those handling live snakeheads are aware of the dangers and avoid spreading them to new areas. At the time of writing, 13 US states have banned ownership of live snakeheads. KEB/SAF

Key Families of Perchlike Fishes

The order Perciformes, with over 9,300 species in about 1,500 genera and over 150 families, is the largest of all fish orders – indeed, of all vertebrate orders.

Sea basses or groupers
Family Serranidae

Primarily marine worldwide. 449 species in 62 genera including: Black sea bass (*Centropristis striata*); Coney (*Cephalopholis fulva*); Queensland grouper (*Epinephelus lanceolatus*); Black grouper (*Mycteroperca bonaci*); hamlets (genus *Hypoplectrus*); Swallowtail seaperch (*Anthias anthias*); Sixstripe soapfish (*Grammistes sexlineatus*); Kelp bass (*Paralabrax clathratus*); Tattler (*Serranus phoebe*); Atlantic creolefish (*Paranthias furcifer*); Sand perch (*Diplectrum formosum*).

Perches
Family Percidae

Freshwater worldwide. 162 species in 10 genera including: European perch (*Perca fluviatilis*); Yellow perch (*Perca flavescens*); ruffe or pope (*Gymnocephalus cernuus*); North American darters (genera *Ammocrypta, Crystallaria, Etheostoma, Percina*) including Eastern sand-darter (*Ammocrypta pellucida*), Crystal sand-darter (*Crystallaria asprella*), and Slackwater darter (*Etheostoma boschungi*); pikeperches or zanders (genus *Sander*), including zander (*S. lucioperca*), North American zander or walleye (*S. vitreus*), sauger (*S. canadensis*).

Drums or croakers
Family Sciaenidae

Marine and freshwater worldwide. 270 species in 70 genera including: Freshwater drum (*Aplodinotus grunniens*); Silver croaker (*Bairdiella chrysoura*); White weakfish (*Atractoscion nobilis*); Silver seatrout (*Cynoscion nothus*); White croaker (*Genyonemus lineatus*); Black drum (*Pogonias cromis*); Queenfish (*Seriphus politus*); Polla drum (*Umbrina xanti*).

Wrasses
Family Labridae

Marine waters worldwide. 500 species in 60 genera including: goldskinny (*Ctenolabrus rupestris*); Ballan wrasse (*Labrus bergylta*); Corkwing wrasse (*Symphodus melops*); Tautog (*Tautoga onitis*); Napoleonfish or Humpback wrasse (*Cheilinus undulatus*); Spanish hogfish (*Bodianus rufus*); California sheephead (*Semicossyphus pulcher*); Señorita (*Oxyjulis californica*); Slippery dick (*Halichoeres bivittatus*); Bluehead (*Thalassoma bifasciatum*); African clown wrasse (*Coris formosa*); Birdmouth wrasse (*Gomphosus coereleus*); Harlequin tuskfish (*Choerodon fasciatus*).

Butterflyfishes
Family Chaetodontidae

Worldwide, in coral reef environments. 114 species in 10 genera including: Spotfin butterflyfish (*Chaetodon ocellatus*); Forceps fish (*Forcipiger longirostris*); bannerfishes (genus *Heniochus*), including Horned bannerfish (*H. varius*); Blacknosed butterflyfish (*Johnrandallia nigrirostris*); Longnosed butterflyfish (*Prognathodes aculeatus*); Western talma (*Chelmonops curiosus*).

Angelfishes
Family Pomacanthidae

Worldwide, in tropical waters. 74 species in 9 genera including: dwarf angelfishes (genus *Centropyge*), including Coral beauty (*C. bispinosus*) and Cherubfish (*C. argi*); Rock beauty (*Holacanthus tricolor*); Queen angelfish (*H. ciliaris*); Regal angelfish (*Pygoplites diacanthus*); Griffis angelfish (*Apolemichthys griffisi*); Blackstriped angelfish (*Genicanthus lamarck*).

Cichlids
Family Cichlidae

Worldwide, exclusively freshwater. c.1,300 species in 105 genera, including: discus (genus *Symphysodon*); Freshwater angelfish (*Pterophyllum scalare*); Longfin tilapia (*Oreochromis macrochir*); *Julidochromis marlieri*; Jewel fish (*Hemichromis bimaculatus*); Egyptian mouthbrooder (*Pseudocrenilabrus multicolor*); Malawi eye-biter (*Haplochromis compressiceps*); Keyhole cichlid (*Aequidens maronii*); Ramirez's dwarf cichlid (*Mikrogeophagus ramirezi*); Jack Dempsey (*Cichlasoma octofasciatum*); *Uaru amphicanthoides*; Mango tilapia (*Sarotherodon galilaeus*); Oscar (*Astronotus ocellatus*).

Combtooth blennies
Family Blenniidae

Mostly marine, occasionally brackish, rarely freshwater; worldwide distribution. 345 species in 53 genera including: Midas blenny (*Ecsenius midas*); Bicolor blenny (*E. bicolor*); Redlip blenny (*Ophioblennius atlanticus*); Ocellated dragonet (*Synchiropus ocellatus*); Jewel blenny (*Salarias fasciatus*); Striped blenny (*Chasmodes*

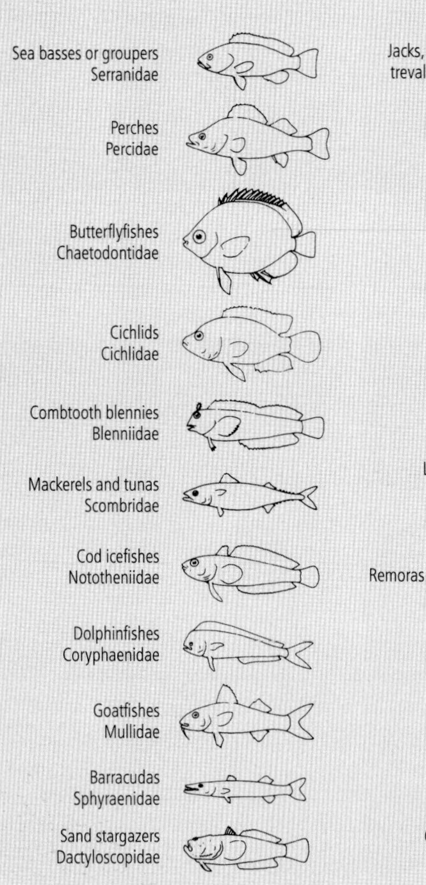

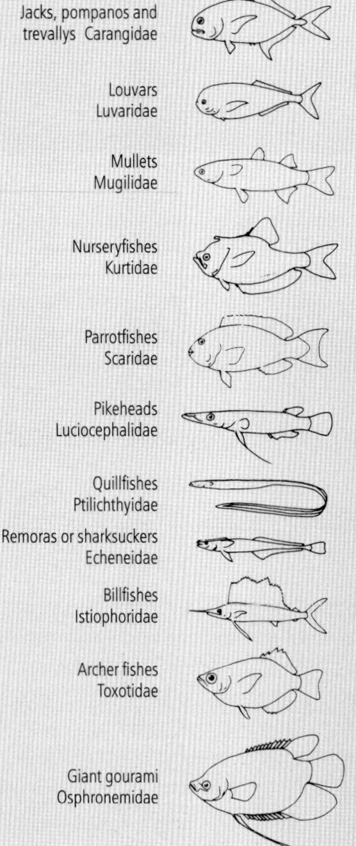

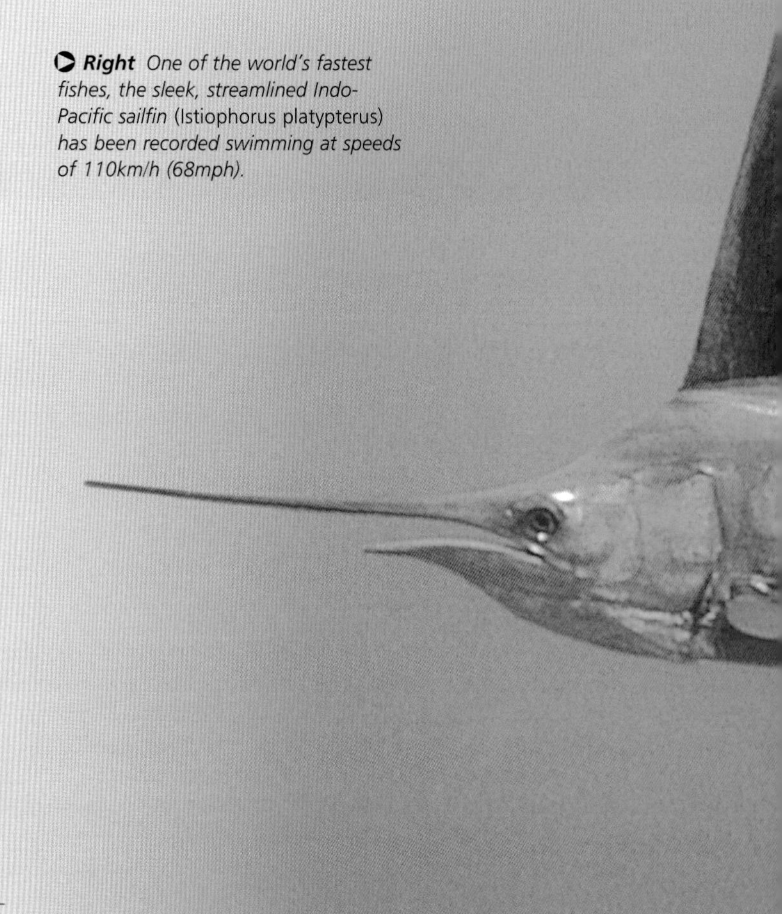

Right *One of the world's fastest fishes, the sleek, streamlined Indo-Pacific sailfin* (Istiophorus platypterus) *has been recorded swimming at speeds of 110km/h (68mph).*

bosquianus); Barred blenny (*Hypleurochilus bermudensis*); Feather blenny (*Hypsoblennius hentzi*); Molly miller (*Scartella cristata*); Disco blenny (*Meiacanthus smithi*); Forktail blenny (*M. atrodorsalis*); Muzzled blenny (*Omobranchus punctatus*).

Gobies
Family Gobiidae

c.1,875 species in 212 genera. Mostly tropical marine and brackish environments. Species and genera include: Bluecheek goby (*Valenciennea strigata*); Giant goby (*Gobius cobitis*); mudskippers (genus *Periophthalmus*), including Barred mudskipper (*P. argentilineatus*); Lemon goby (*Gobiodon citrinus*); Yellow prawn-goby (*Cryptocentrus cinctus*); Catalina or Blue-banded goby (*Lythrypnus dalli*); Golden-banded goby (*Brachygobius doriae*); Blackeye goby (*Rhinogobiops nicholsii*); Clown goby (*Microgobius gulosus*); Bridled goby (*Coryphopterus glaucofraenum*); Frillfin goby (*Bathygobius soporator*); Highfin goby (*Gobionellus oceanicus*); Spotted fringefin goby (*Eviota albolineata*); Transparent goby (*Aphia minuta*); Tusked goby (*Risor ruber*).

Mackerels, tunas, and bonitos
Family Scombridae

49 species in 15 genera. Tropical and temperate seas worldwide. Species include: Atlantic mackerel (*Scomber scombrus*); Horse mackerel (*Trachurus trachurus*); Skipjack tuna (*Katsuwonus pelamis*); Little tunny (*Euthynnus alletteratus*); Atlantic bonito (*Sarda sarda*); Albacore (*Thunnus alalunga*); Yellowfin tuna (*T. albacares*); Cero (*Scomberomorus regalis*); Spanish mackerel (*S. maculatus*); Plain bonito (*Orcynopsis unicolor*).

Other families, genera, and species include:

glassfishes (family Chandidae), including the Indian glassperch (*Parambassis ranga*); **dottybacks** and **eelblennies** (family Pseudochromidae), including Orchid dottyback (*Pseudochromis fridmani*), Carpet eel blenny (*Congrogadus subducens*); **prettyfins** or **longfins** (family Plesiopidae), including comet (*Calloplesiops altivelis*); **jawfishes** (family Opistognathidae); **sunfishes** (family Centrarchidae), including bluegill (*Lepomis macrochirus*); **louvar** (*Luvarus imperialis*) sole species of Luvaridae; **ragfish** (*Icosteus aenigmaticus*) sole species of Icosteidae; **cardinal fishes** (family Apogonidae), including Banggai cardinalfish (*Pterapogon kauderni*); **remoras** or **sharksuckers** (family Echeneidae); **dolphinfishes** (family Coryphaenidae, genus *Coryphaena*); **jacks, pompanos,** and **trevallys** (family Carangidae), including pilotfish (*Naucrates ductor*); **snappers** (family Lutjanidae); **grunts** (family Haemulidae), including sweetlips (genus *Plectorhinchus*), porkfish (*Anisotremus virginicus*); **threadfins** (family Polynemidae, genera *Polynemus, Eleutheronema*); **goatfishes** (family Mullidae), including Red mullet (*Mullus surmuletus*); **moonfishes** (family Monodactylidae); **archerfishes** (family Toxotidae, genus *Toxotes*); **leaffishes** (family Nandidae); **hawkfishes** (family Cirrhitidae); **damselfishes** (family Pomacentridae), including clownfishes (genera *Amphiprion, Premnas*); **parrotfishes** (family Scaridae); **eelpouts** (family Zoarcidae); **wolffishes** (family Anarhichadidae); **weeverfishes** (family Trachinidae); **stargazers** (family Uranoscopidae), including European stargazer (*Uranoscopus scaber*); **dragonets** (family Callionymidae); **rabbitfishes** (family Siganidae); **surgeonfishes** (family Acanthuridae), including Powder-blue surgeonfish (*Acanthurus leucosternon*); **barracudas** (family Sphyraenidae); **billfishes** (family Istiophoridae), including sailfishes (genus *Istiophorus*); **swordfish** (*Xiphias gladius*, family Xiphiidae); **gouramies** (family Osphronemidae), including Siamese fighting fish (*Betta splendens*); **snakeheads** (family Channidae).

About 269 species of perchlike fishes are threatened, with 5 Extinct in the Wild, including *Haplochromis lividus* and *Platytaeniodus degeni*; 55 species are Critically Endangered, including the Warsaw grouper (*Epinephelus nigritus*), totoaba (*Totoaba macdonaldi*), and Dwarf pygmy goby (*Pandaka pygmaea*); 28 species are Endangered, including the Bluestripe darter (*Percina cymatotaenia*); and 136 species are Vulnerable, including the Mexican darter (*Etheostoma pottsi*), Bigeye tuna (*Thunnus obesus*), and the Humpback wrasse .

Flatfishes

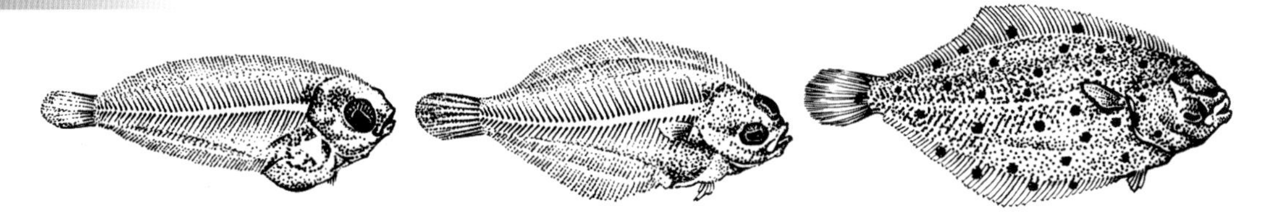

Above and right *Flatfishes start life as normal-shaped fish with an eye on each side and a horizontal mouth. As the larva grows, ABOVE, one eye migrates to the other side of the head and the mouth twists until the adult comes to lie permanently on one side. The consequent deformation in the adult can be seen in this image of a Leopard flounder* (Bothus pantherinus) *RIGHT from Tonga in the Indo-Pacific Ocean.*

AS THE COMMON NAME FOR THE ORDER indicates, flatfishes are noted for their flattened body shape, with eyes present on only one side of the body. Furthermore, the scientific order name, Pleuronectiformes, derives from the Greek for "side swimmers." Unique among fishes in being asymmetrical, flatfishes are believed to have evolved from a generalized symmetrical percoid (sea bass) body pattern in a fish that habitually rested on its side.

There are about 570 pleuronectid species, divided into 11 families. The members of the most primitive family – the Psettodidae – have rather perchlike pectoral and pelvic fins; only the eyes and long dorsal fin distinguish them from the seaperch, suggesting that flatfish evolved from perchlike ancestors.

All adult flatfishes are bottom living but their eggs, which contain oil droplets, float at or near the sea surface. The larvae take a few days to hatch; the fish appear symmetrical, with an eye on each side of the head and a ventrally situated mouth, further suggestive of their perchlike ancestry. When about 1cm (0.5in) long, a metamorphosis occurs that has profound effects on the symmetry of the skull and the whole fish. The changes are initiated when one eye migrates across the head to lie alongside the other, its passage being assisted by resorption of the cartilaginous bar of skull separating them. The nostril simultaneously migrates to the eyed or colored side. Except in the psettodids, the mouth also twists into the same plane as the eyes. The eye that migrates often characterizes particular families. Members of families such as the Scophthalmidae and the Bothidae are called "lefteye flounders" because their right eye usually migrates, so the uppermost, colored side is the left one. Pleuronectidae are right-eyed flounders because ultimately the right side is uppermost. In the psettodids, equal numbers are found lying on either side.

While these radical changes are taking place, the little fish sinks to the sea bottom. Flatfish do not have a swimbladder, so they remain lying at or near the bottom, on their blind side. The body shape of adult flatfishes is quite variable – the European turbot and its relatives are nearly as

FACTFILE

FLATFISHES

Series: Percomorpha

Superorder: Acanthopterygii

Order: Pleuronectiformes

About 570 species in about 123 genera and 11 families. Species and families include: **scaldfishes** (family Bothidae), including Peacock flounder or Plate fish (*Bothus lunatus*), European scaldfish (*Arnoglossus laterna*); **turbot** or **lefteye flounders** (family Scophthalmidae), including European turbot (*Psetta maxima*), megrim (*Lepidorhombus whiffiagonis*); **pleuronectids** or **right-eye flounders** (family Pleuronectidae), including plaice (*Pleuronectes platessa*), European flounder (*Platichthys flesus*), Lemon sole (*Microstomus kitt*), halibut (*Hippoglossus hippoglossus*); **psettodids** (family Psettodidae); **soles** (family Soleidae), including European sole (*Solea solea*); **tongue fishes** or **tongue soles** (family Cynoglossidae); **American soles** (family Achiridae), including **Drab** or **Freshwater sole** (*Achirus achirus*).

DISTRIBUTION Worldwide in both marine waters and freshwaters.

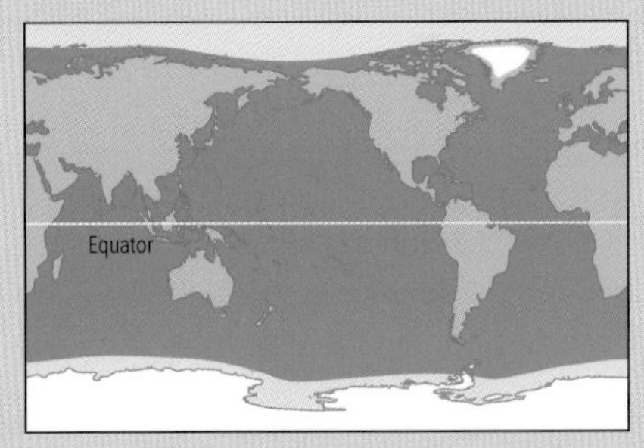

SIZE Length 4.5cm (1.8in) – 2.5m (8ft); weight 2g (0.07oz) –316kg (697lb)

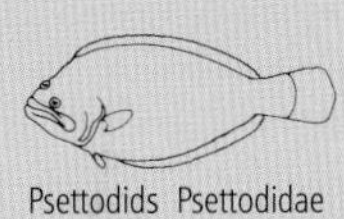

CONSERVATION STATUS The halibut is classed as Endangered and the Yellowtail flounder (*Pleuronectes ferrugineus*) is classed as Vulnerable.

broad as long, whereas the tongue soles (family Cynoglossidae) are long and narrow. Frequently, flatfish bury themselves, by flicking sand or by wriggling movements of the body, leaving just their eyes and upper operculum (gill cover) exposed. There is a special channel that connects the gill cavities. Water is pumped from the mouth over both sets of gills, but the expired water from the gills on the buried side is diverted through the channel and expired from the exposed side.

Many flatfish are predominantly brown on the colored side, although they often have spots and blotches of orange thus enabling them to blend with the substrate. The pleuronectids, however, are masters of disguise among fish as they can change their color to match the substrate. When placed on a chequered board some species can reproduce the squares with reasonable accuracy. All flatfish are carnivorous but their methods of catching prey are quite diverse. Scaldfishes, members of the left-eyed family Bothidae, are daytime hunters that feed on other fish. They swim actively after their prey and have very acute vision. Species of soles (family Soleidae) and tongue soles (Cynoglossidae) hunt at night for mollusks and polychaete worms, which they locate by smell. These families of flatfish both have innervated filamentous tubercles instead of scales on the blind side of the head, which probably enhance their sense of smell. The pleuronectids are intermediate: some, like the halibut, actively prey on fishes, and others, like plaice, hunt polychaete worms and crustacea, relying on smell and visual acuity to locate their prey.

The majority of flatfishes are marine, but a few species can live in seawater or freshwater. The European flounder frequently migrates up rivers to feed and is found up to 65km (40 miles) inland in the summer, returning to spawn in the sea in the fall. The American flatfish, the Drab sole (*Achirus achirus*), is a freshwater species, often kept by aquarists. It has a large surface-area-to-weight ratio and can suspend itself by surface tension at the water surface. It can also attach itself to rocks or the sides of aquaria by creating a vacuum between the underside of its body and the substrate.

There is no obvious difference between the sexes in most species of flatfishes, although among the scaldfishes the male often has some filamentous dorsal and pelvic fin rays or other visible sexual dimorphism.

Many flatfishes, such as Dover sole, flounder, and halibut, are highly esteemed as food fishes, and some have considerable commercial importance. The structure of flatfish is very convenient for cooks. They cook quickly and evenly. They are easily filleted, and the bones are rarely troublesome. BB/KEB/SAF

*◐ **Left** Juvenile Starry flounders* (Platichthys stellatus) *camouflaged on the seabed off the New England coast, USA. The young and adults of this species travel up rivers, sometimes as far as 120km (75 miles).*

Triggerfishes and Allies

Fascinated by their bizarre forms and traits, the 1st-century Roman author Pliny the Elder included pufferfishes and oceanic sunfishes in his 37-volume encyclopedia Naturalis Historiae. *Tetraodontiforms now make up 5 percent of the world's tropical marine fishes and remain one of the most specialized groups of teleost fishes.*

Tetraodontiformes are an order of mostly marine fishes that have the teeth fused into a beak. Among their number are poisonous fishes, inflatable fishes, and one of the largest oceanic teleosts. None of them have scales; instead, they are covered either with spines or with skin so thick that little can penetrate it.

Triggerfishes (Balistidae) are named for the interlocking triggerlike mechanism of their first and second dorsal fin spines; the small second spine must be released before the larger first spine can be depressed. Triggers, with their bony scales, have an easily recognized overall appearance, with their opposite and almost symmetrical dorsal and anal fins actively undulating as the major propulsion mechanism. Many triggers have striking color patterns and inhabit coral reefs. The filefishes (Monacanthidae) are rather similar to triggerfishes, but have very small, rough, scales and the dorsal spines are much further forward than in the triggerfishes. Filefishes have extremely small mouths and feed by picking up small invertebrates. Many have an expandable dewlap between the chin and the anal fin.

The boxfishes and cowfishes in the family Ostraciidae have been described as bony cuboid boxes with holes for the mouth, eyes, fins, and the vent. Some species also have two small, hornlike processes over the eyes, hence the common name cowfish. The rigid outside skeleton (exoskeleton) is formed by fused bony scutes. Boxfishes are slow-swimming, brightly colored fishes of shallow tropical seas. In case their armor should be thought inadequate against predators, boxfishes can also secrete a virulent toxin if molested. There are about 33 species, some of which can grow up to 60cm (2ft); most are shorter than 30cm (1ft).

The pufferfishes or blowfishes (Tetraodontidae) derive their common names from their ability to inflate the body with water (or air if lifted above water) as a defense tactic. In the inflated state they

Above Ocean sunfish (Mola mola) *grow from a juvenile length of just 0.6cm (0.25in) to a length of 3m (9.8ft) when fully developed. California sea lions usually prey upon the juveniles.*

Below This night-foraging Black-blotched porcupinefish (Diodon liturosus) *is displaying the distinctive defensive form taken by pufferfishes and porcupinefishes. When inflated like this – either with water or air – the lateral spikes of the scutes stick out.*

FACTFILE

TRIGGERFISHES AND ALLIES

Series: Percomorpha

Superorder: Acanthopterygii

Order: Tetraodontiformes

About 340 species in about 100 genera and 9 families: **spikefishes** (Triacanthodidae); **triplespines** (Triacanthidae); **triggerfishes** (Balistidae), including Picassofish or humuhumu (*Rhinecanthus aculeatus*); **filefishes** (Monacanthidae); **boxfishes** or **trunkfishes** (Ostraciidae), including cowfishes (genus *Lactoria*); **pufferfishes** or **blowfishes** (Tetraodontidae), including fugu (genus *Takifugu*) and sharpnose puffers or tobies (genus *Canthigaster*); **porcupinefishes** (Diodontidae); **molas** or **oceanic sunfishes** (Molidae), including Ocean sunfish (*Mola mola*); **Three-toothed puffers** or **pursefishes** (Triodontidae).

DISTRIBUTION Worldwide in tropical and temperate waters.

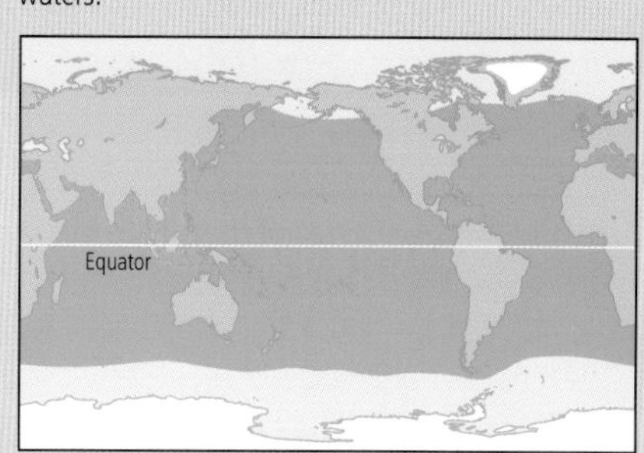

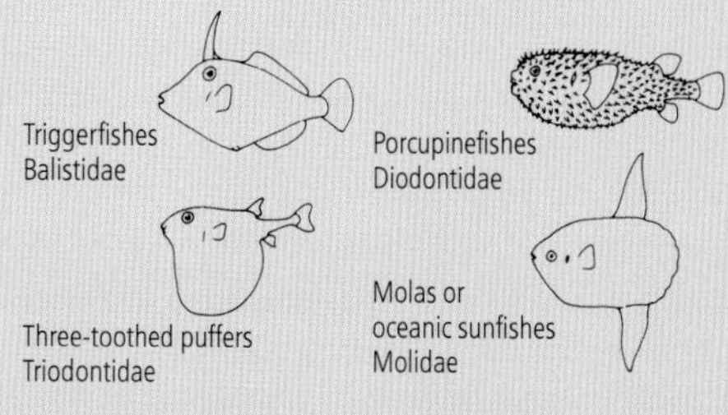

SIZE Length 2.5cm–3m (1in–9.8ft); **weight** maximum 2,300kg (5,000lb).

CONSERVATION STATUS 3 species are classed as Vulnerable, including the Queen triggerfish (*Balistes vetula*).

Above *Among the most highly prized of the aquarium fishes, the lone Clown triggerfish* (Balistoides conspicillum) *inhabits Indo-Pacific waters. Its bright yellow mouth is thought to deter predators.*

are literally balls that look both inedible and are extremely difficult for a predator to grasp. Untypical of the order, several pufferfishes actually live in freshwater.

The Freshwater pufferfish is a striking black and yellow species widespread throughout the Zaïre system and some other West African rivers. This species and some of the other African freshwater species are occasionally kept in aquaria, but they are aggressive inhabitants. All pufferfishes are very poisonous but, despite that, some are valued as a delicacy. Some species are eaten as *fugu,* particularly in Japan. The fishes are prepared by specially trained cooks,to avoid any possibility of the toxic parts being eaten or contaminating the flesh. The lethal poison, tetraodotoxin, is found in the fish's gut, liver, ovary, and skin. Serious poisoning has resulted from ill-prepared *fugu*.

The oceanic sunfishes (Molidae) are the giants of the order. The Ocean sunfish is the largest species, probably weighing up to 2,300kg (5,000lb). Seen from the side, this brownish-blue fish is nearly circular, with the caudal fin reduced to a mere skin-covered fringe, but with the dorsal and anal fins produced into "oars" used for locomotion. It is most often seen lying on its side at the surface, allegedly basking but probably dying. A rare film of a young specimen alive shows that it swims rapidly in an upright position by vigorous sculls of the expanded fins. The fish's diet consists of jellyfishes, crustaceans, mollusks, and zooplankton. Below the scaleless skin is a very thick layer of tough gristle. Although not common, the Ocean sunfish lives worldwide in tropical and subtropical waters. BB/KEB/SAF

Above *The strange-looking Longhorn cowfish* (Lactoria cornuta) *has no known sexual dimorphism and feeds on benthic invertebrates by blowing jets of water at the sandy substrate. Adults are solitary, while juveniles are often found in small groups.*

Seahorses and Allies

UNUSUAL AND VERY DISTINCTIVE IN SHAPE, it is not surprising that many people find it hard to believe that seahorses are a type of fish. With their upright posture, horselike head, and strong prehensile tail, they certainly present an unfamiliar picture. However, seahorses are only the better known members of a large, diverse group of fishes within the order Syngnathiformes.

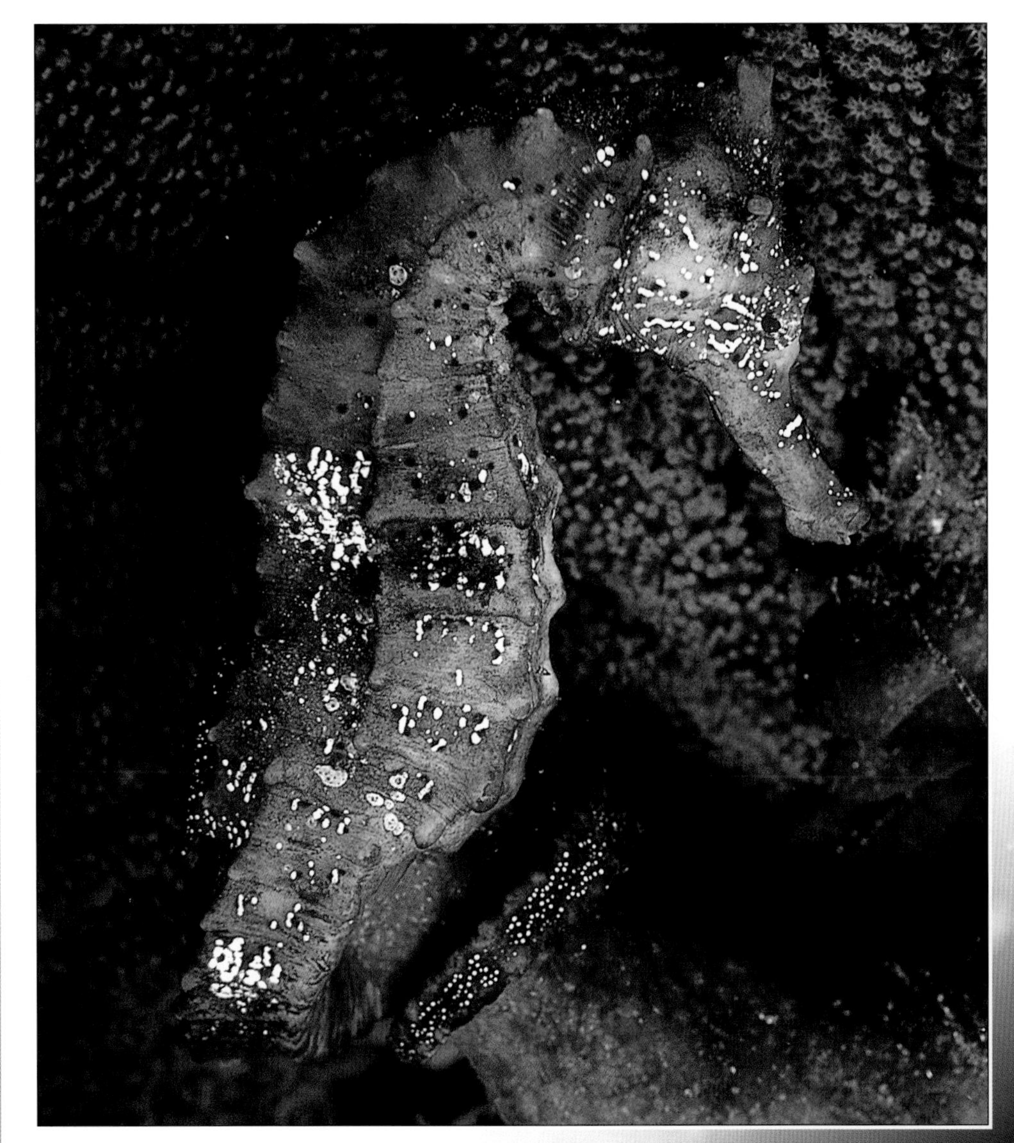

As well as the seahorses, the other members of the order Syngnathiformes – pipefishes, trumpetfishes, cornetfishes, snipefishes, and shrimpfishes – are an almost entirely marine group. Only a few pipefish species live permanently in freshwater. Important unifying characters of these fishes are long snouts with a small terminal mouth, the elongation of the first few vertebrae (in the shrimpfish, the first six vertebrae form over three-quarters of the length of the vertebral column), and the structure of the first dorsal fin, which, when present, consists not of fin rays but of prolonged processes associated with the vertebrae.

FACTFILE

SEAHORSES AND ALLIES

Series: Percomorpha

Superorder: Acanthopterygii.

Order Syngnathiformes

About 241 species in 60 genera and 6 families: **seahorses** and **pipefishes** (family Syngnathidae), including seahorses (genus *Hippocampus*); **ghost pipefishes** (family Solenostomidae); **trumpetfishes** (family Aulostomidae); **cornetfishes** (family Fistulariidae); **snipefishes** (family Macrorhamphosidae); **shrimpfishes** (family Centriscidae).

DISTRIBUTION Worldwide in tropical and temperate seas; some in brackish and freshwater.

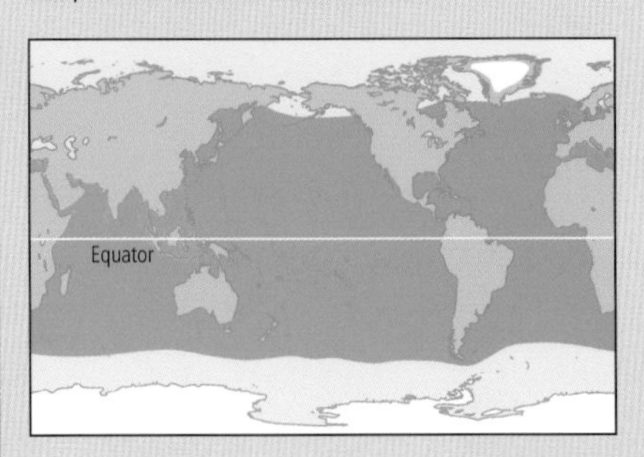

SIZE Length 2cm (0.8in)–1.8m (5.9ft).

Shrimpfishes Centriscidae

Seahorses and pipefishes Syngnathidae

CONSERVATION STATUS At least 27 species are threatened, including the River pipefish (*Syngnathus watermeyeri*), which is Critically Endangered and the Cape seahorse (*Hippocampus capensis*), which is Endangered; the rest are classed as Vulnerable (20 are seahorses).

The shrimpfishes have an extremely compressed body entirely enclosed in thin bony sheets. Only the downturned posterior part of the body is free, allowing tail-fin movement for locomotion. They live in shallow warm seas, sometimes among sea urchin spines, where they shelter for protection. The deep-bodied snipefish lives in deeper water and is covered with prickly denticles and a row of scutes on the chest. Apart from the lack of parental care, little is known of their reproductive behavior.

The pipefishes, however, show a remarkable series of reproductive adaptations. The simplest strategy, in the subfamily of nerophiine pipefishes, is for the eggs to be loosely attached to the

Above *One of three species in the family Aulostomidae, the quirky, slow-moving Chinese trumpetfish* (Aulostomus chinensis) *uses stealth and camouflage to prey on small fishes and shrimps.*

Left *Measuring up to 30cm (12in) long, the Pacific seahorse* (Hippocampus ingens) *is one of the world's largest seahorse species. Seahorses use their prehensile tails to cling on to plants, corals, or sponges; here, the Pacific seahorse is on a Red gorgonian coral.*

Below *In the warm southern oceans lives the Harlequin ghost pipefish* (Solenostomus paradoxus)*. The larger female (pictured) has capelike pelvic fins that form a pouch. Within this, the eggs are attached to short filaments.*

abdomen of the male. A more elaborate condition is present in some syngnathines where the eggs are individually embedded in spongy tissue covering the male's ventral plates. Further protection in other groups is provided by the development of lateral plates partially enclosing the eggs. In all cases the male carries the eggs.

The seahorses – which are merely pipefishes with the head at right angles to the body, a prehensile tail, and a dorsal fin adapted for locomotion – exhibit the ultimate in egg protection. The trend seen in the development of protective plates is continued until a full pouch (or marsupium) is formed, with a single postanal opening. The female has an ovipositor by which the eggs are placed in the male's pouch until the pouch is full. Apparently, this simple act is not always done without mishap, and some eggs may be lost.

Hatching time varies with temperature, the young leaving the pouch between 4 and 6 weeks after the eggs were deposited. In some larger species, the male helps the young to escape by rubbing his abdomen against a rock, in others there are vigorous muscular spasms, which may expel the young with considerable velocity. After "birth" the male flushes out his brood sac by expansion and contraction to expel egg remains and general debris to prepare for the next breeding season. This may occur relatively soon, and three broods a year are not unknown.

Over the last several years, increasing focus has been put on the conservation status of seahorses. Destruction of marine seagrass habitats, which are important to most seahorse species, is continuing at an alarming rate worldwide. Added to this is the commercial fishing of seahorses for the Chinese traditional medicine trade, which is thought to exacerbate the problem in some areas. Small-scale fishing also takes place for the curio and aquarium trade. As of May 2004, all seahorse species have had some limited protection through CITES trade regulations, and there are efforts in many countries, including China, to increase farming. Hopefully, habitat destruction will also slow down, helping to protect these marvelous fishes.

Apart from small numbers for the aquarium trade, the other groups of Syngnathiformes have little economic value. KEB/SAF

Other Spiny-finned Fishes

The remaining orders of spiny-finned fishes span a huge variety of forms and ecologies. Groups represented range from the familiar sticklebacks to the exotic and highly diverse families of the order Scorpaeniformes.

Squirrelfishes and Allies

ORDER BERYCIFORMES

Beryciformes (with five families) are large-headed marine fishes that live in temperate and tropical oceans, many in deep water and cave habitats. Most families have rough scales, a characteristic alluded to in the common name of the Monocentridae – pinecone fishes. This family comprises just four species in two genera (*Monocentrus* and *Cleidopus*). These rounded little fishes live in small schools in the Indo-Pacific Ocean. The body is covered with irregular bony plates and the soft-rayed dorsal fin is preceded by a few large, alternately angled spines. The pelvic spines

Above *The Pinecone or Pineapple fish* (Cleidopus gloriamaris), *which inhabits the western Pacific, is covered in a mosaic-like pattern of rough scales. It lives in caves or around rocky ledges and coral reefs.*

Right *The lugubrious appearance of the bottom-dwelling John Dory* (Zeus faber) *deters some people from eating it, even though it is an excellent food fish. It can grow to a length of 66cm (26in).*

Other Spiny-finned Fishes

Squirrelfishes and Allies
Order: Beryciformes

Temperate to tropical oceans. Length maximum 60cm (2ft). About 130 species in 18 genera and 5 families, including: **squirrel-** and **soldierfishes** (Holocentridae); **alfonsinos** (Berycidae); **flashlight fishes** (Anomalopidae); **pinecone fishes** (Monocentridae), including *Monocentrus japonicus*.

Dories and Allies
Order: Zeiformes

Oceans in mid- to deep water. Length maximum about 1m (3.3ft). About 39 species in at least 20 genera and 6 families, including: **dories** (genus *Zeus*; family Zeidae), including John Dory (*Z. faber*); boarfish (*Capros aper*).

Pricklefishes, Whalefishes, and Allies
Order: Stephanoberyciformes

Oceanic, all oceans except Arctic and Mediterranean. Length maximum 40cm (1.3ft). About 86 species in 28 genera and 9 families, including: **bigscales** and **ridgeheads** (Melamphaidae); **mirapinnids** (Mirapinnidae), including hairyfish (*Mirapinna esau*); **flabby whalefishes** (Cetomimidae).

Swamp eels and Allies, or Synbranchids
Order: Synbranchiformes

Mainly freshwater (a few occasionally in brackish waters) of C and S America, Africa, Asia, Indo-Australian archipelago, and NW Australia. Length maximum 1.5m (5ft). About 87 species in 12 genera and 3 families, including: **swamp eels** (Synbranchidae), including Marbled swamp eel (*Synbranchus marmoratus*), Rice eel (*Monopterus albus*); **spiny eels** (Mastacembelidae), including the Lesser spiny eel (*Macrognathus aculeatus*); **chaudhuriids** (Chaudhuriidae), including *Nagaichthys filipes*. The Blind swamp eel (*Ophisternon infernale*) is Endangered.

Sticklebacks
Order: Gasterosteiformes

Worldwide in marine, brackish, and freshwater. Length maximum 20cm (8in). About 216 species in 11 genera and 5 families including: **sticklebacks** (Gasterosteidae), including Brook stickleback (*Culea inconstans*), Fifteen-spined stickleback (*Spinachia spinachia*), Four-spined stickleback (*Apeltes quadracus*), Nine-spined stickleback (*Pungitius pungitius*), Three-spined stickleback (*Gasterosteus aculeatus*); **seamoths** (Pegasidae); **Armored stickleback** (*Indostomus paradoxus*; only species in the family Indostomidae); **tubesnouts** (Aulorhynchidae), including the Tube-snout (*Aulorhynchus flavidus*) and Tubenose (*Aulichthys japonicus*). *Pungitius hellenicus* is Critically Endangered and *Pegasus laternarius* is Vulnerable.

Mailcheeked fishes
Order: Scorpaeniformes

Worldwide in seawater and freshwater (although distribution more disjunct in fresh waters). Length maximum 2m (6.5ft). Nearly 1,300 species in about 266 genera and 25 families, including: **flying gurnards** (Dactylopteridae); **scorpionfishes** (Scorpaenidae), including lionfish (*Pterois volitans*), Atlantic redfish (*Sebastes marinus*), stonefishes (genus *Synanceia*), including Indo-Pacific stonefish (*S. verucosa*); **sea robins** or **gurnards** (Triglidae); **sablefishes** (Anoplopomidae); **sculpins** (Cottidae), including bullhead (*Cottus gobio*), Shorthorn sculpin (*Myoxocephalus scorpius*); **armored sea robins** and **pogges** (Agonidae); **lumpfishes** (Cyclopteridae); **Australian prowfishes** (Pataecidae). 10 species are threatened – 3 are Critically Endangered, including the Pygmy sculpin (*Cottus paulus*); 2 are Endangered, including the Acadian redfish (*Sebastes fasciatus*); and 5 are Vulnerable, including the St. Helena deepwater scorpionfish (*Pontinus nigropunctatus*). The Utah Lake sculpin (*Cottus echinatus*) is now Extinct.

Squirrelfishes and allies
Pinecone fishes
Monocentridae

Dories and allies
Dories
Zeidae

Pricklefishes, Whalefishes and Allies
Bigscales and ridgeheads
Melamphaidae

Swamp eels and Allies
Mastacembelidae

Sticklebacks
Sticklebacks
Gasterosteidae

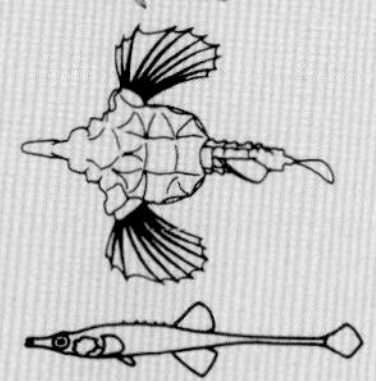

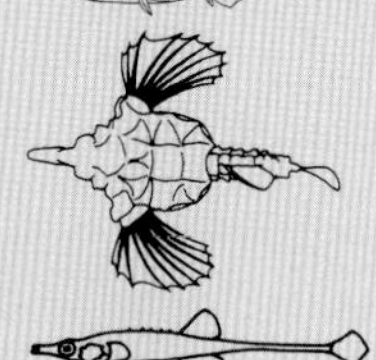

Seamoths
Pegasidae

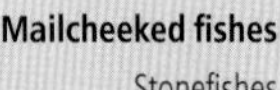

Armored stickleback
Indostomidae

Mailcheeked fishes
Stonefishes
Synanceiidae

Australian prowfishes
Pataecidae

FACTFILE

OTHER SPINY-FINNED FISHES

Series: Percomorpha

Superorder: Acanthopterygii.

Orders: Beryciformes, Zeiformes, Stephanoberyciformes, Synbranchiformes, Gasterosteiformes, Scorpaeniformes

At least 1,858 species in about 355 genera, 53 families and 6 orders.

Equator

are massive and erectile. Although pinecone fishes do not grow to more than 23cm (9in) long, they are commercially viable in Japan, where they are eaten. They have two small luminous organs on each side of the lower jaw with colonies of bacteria providing the light. The blue-green light can be "switched" off and on by closing and opening the mouth.

Luminous organs are also present under the eye in the nocturnal flashlight fishes of the family Anomalopidae. Each organ is a flat white color in daylight but at night glows with a blue-green light. The luminous organs blink on and off when functioning and are controlled in some species by rotating the entire gland, and in others by covering it with a eyelidlike membrane. The light is turned on and off in repeated patterns, typically with some 10 seconds light, and 5 seconds, darkness. The fishes probably have three main advantages from this: first, the light helps the fishes to keep together in the otherwise total darkness; second, the light probably attracts planktonic crustaceans that the fishes feed on; and finally, the light confuses potential predators so they miss their prey. The flashlight fishes live on reefs in the Red Sea, Indo-Pacific, and the Eastern Pacific Ocean, in schools from 20 to 50 specimens, normally in depths below 30 meters (c. 100ft). Apart from a very limited demand for the aquarium trade, they are of no commercial importance although, if caught, the luminous gland is removed and used as bait in subsistence fisheries.

The family Berycidae contains mid- to deepwater, large-eyed species with compressed bodies, commonly known as alfonsinos. Most species are red or pink, and are therefore sometimes sold under the confusing names of "redfish" or "red bream." The flesh is excellent and fetches a good price at market. The two very similar species, the Splendid alfonsino (*Beryx splendens*) and the Alfonsino (*B. decadactylus*), are found together circumglobally in tropical and subtropical seas.

The best known of all members of Beryciformes are the squirrel- and soldierfishes in the family Holocentridae. Most of the 66 known species are twilight- and night-active coral-reef inhabitants. During the day they stay in caves and crevices on the reef, where they stand out with their bright red color patterns and are easily spotted by sport divers. They are hardy aquarium fishes, but too voracious to be really popular. Being fairly small and very bony fishes, most species are not in great demand as food, although they are found in fish markets worldwide.

Dories and Allies

ORDER ZEIFORMES

The order Zeiformes comprises deep-bodied, extremely compressed fishes. The most familiar species is the John Dory, which has pronounced protrusile jaws and feeds on small fishes and crustaceans. The origin of its common name is a subject of much debate. Some argue that it derives from the French *jaune d'orée* (with a yellow edge) in allusion to the yellowish color of the body. In some countries, its common name is the vernacular for "St. Peter's Fish" (*Saint-Pierre* in French, *pez de San Pedro* in Spanish). This name alludes to the single dark blotch on each side, which are thought to represent the thumb and forefinger prints of St. Peter, who allegedly took tribute money from its mouth. The same distinction has, however, been accorded to the haddock in other countries (even though the biblical incident that is the source of the story took place in the Sea of Galilee, which is freshwater). The John Dory is found in the eastern Atlantic and Mediterranean, and the same or possibly distinct, related species also live in the Indian Ocean and the Pacific. The fish yields excellent, bone-free fillets of white flesh, and is particularly highly prized in the Mediterranean and Australia.

Pricklefishes, Whalefishes, and Allies

ORDER STEPHANOBERYCIFORMES

The order Stephanoberyciformes (of which some families are sometimes placed in a separate order, Cetomimiformes) is poorly known, but contains fascinating oceanic deepwater fishes. The bigscale fishes or ridgeheads of the family Melamphaidae, of which there are some 33 species, are especially notable. They are small (maximum 15cm/6in), subcylindrical fishes with large heads, blunt short snouts and long, abruptly narrowed caudal peduncles. The scales are typically very large and clearly distinguishable. Another member of the order is the hairyfish of the family Mirapinnidae, characterized by a short, hairlike pile that covers the body. It also has two halves of caudal fin overlapping and large, winglike pelvic fins. The species is thought to spend most of its time in very deep water, but is so far known from a single 5.5cm (2.2in) specimen caught at the surface north of the Azores.

Swamp Eels and Allies

ORDER SYNBRANCHIFORMES

The order Synbranchiformes consist of eel-like freshwater fishes. The swamp eels (family Synbranchidae) occur in South and Central America, West Africa, southern Asia, and Australia. Their common name derives from their shape, and from the fact that they often live in poorly oxygenated waters. Instead of having a gill opening on each side of the head there is a single, common slit on the underside. In some species the gill chamber is divided internally into two chambers by a tissue-dividing wall (septum). Often the gill chamber is distensible and, by filling it with water, the fish can "breathe" while traveling overland. Species in stagnant water absorb atmospheric air either through a modified section of the gut, well provided with blood vessels, or via lunglike chambers extending from the branchial cavity. In both cases the air is taken in through the mouth.

At least one swamp eel species, the Marbled swamp eel (*Synbranchus marmoratus*), can burrow in the mud and estivate much like the lungfish, so avoiding droughts. Pectoral and pelvic fins are absent, and the dorsal and anal fins are reduced to mere ridges of skin without fin rays.

The Rice eel from Asia has colonized irrigation ditches in paddyfields. They grow to 1m (3.3ft) long and are an important source of food, not least because they can stay alive, hence fresh, for a long time if kept moist. The male makes a bubble nest, where the female lays the eggs. They and the newly hatched young are guarded by the male.

The family Mastacembelidae, or spiny eels, are also greatly valued in some areas as food fishes. More colorful species are kept as aquarium fishes. Spiny eels live in a wide variety of habitats, ranging from the clear East African inland seas to quite swampy areas. Many are air breathers, and utilize this ability to survive in poorly aerated water or mud. Like many other species, the Lesser spiny eel spends the daytime in a mud burrow, excavated by rocking and wriggling movements that submerge it at a constant rate, leaving only the tips of the nostrils protruding.

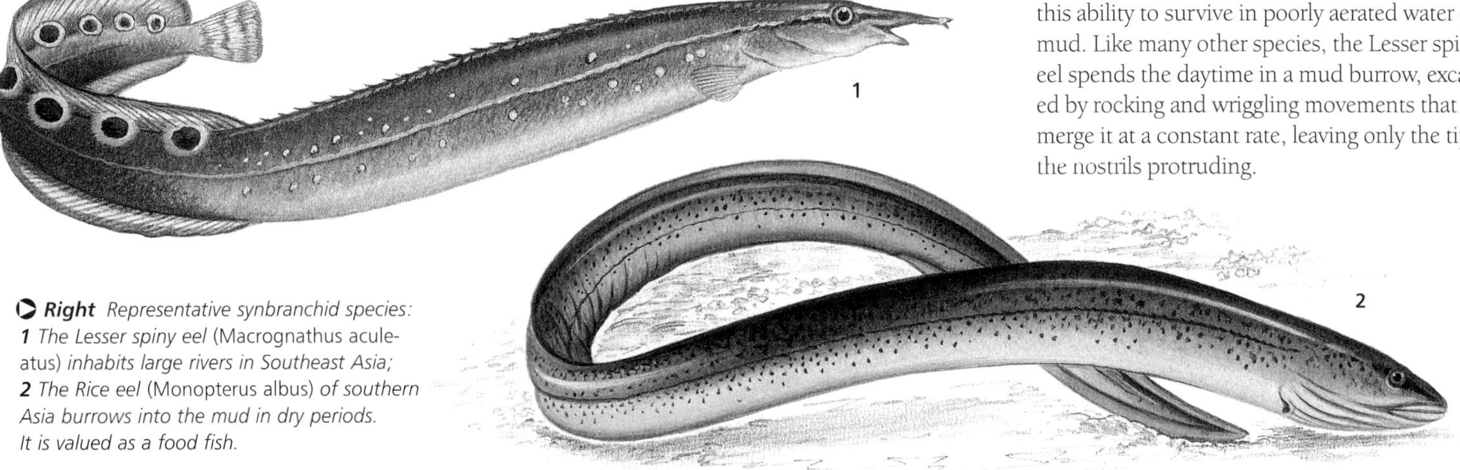

Right *Representative synbranchid species:* **1** *The Lesser spiny eel* (Macrognathus aculeatus) *inhabits large rivers in Southeast Asia;* **2** *The Rice eel* (Monopterus albus) *of southern Asia burrows into the mud in dry periods. It is valued as a food fish.*

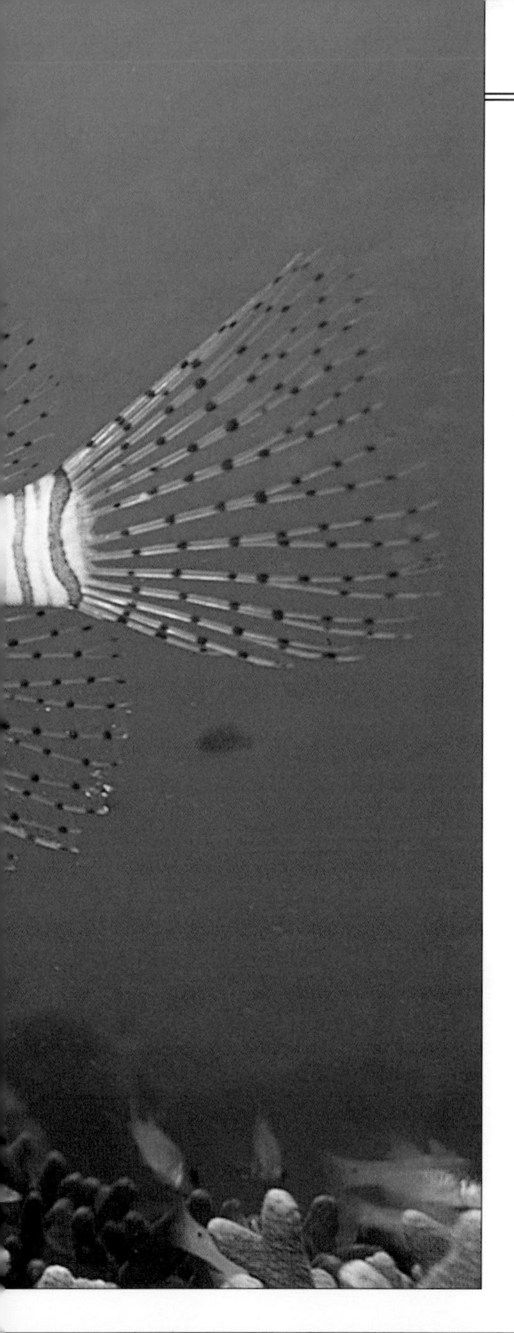

◐ *Left The Red lionfish* (Pterois volitans), *a member of the order Scorpaeniformes, is a formidable predator. Hunting at night, it uses its fanned-out pectoral fins to corner small prey, which it stuns with its venomous dorsal spines before swallowing them whole.*

◑ *Right The body of the Short dragonfish* (Eurypegasus draconis), *a species of sea moth, is protected by bony plates with a reticulated pattern that give the fish excellent camouflage against rocky substrates. It is widespread in oceans from the Red Sea east to the central Pacific.*

Sticklebacks

ORDER GASTEROSTEIFORMES

The most commonly known fishes of the order Gasterosteiformes are the sticklebacks. The "humble" stickleback is found in most freshwaters, brackish waters, and sometimes coastal waters throughout Eurasia and northern America. It is highly variable in form, and some authors consider there to be a large species complex rather than a single species, *Gasterosteus aculeatus*. Although mostly "three-spined," two- and four-spined individuals occur. There are populations in Canada that never have pelvic fins. Freshwater forms usually have fewer bony scutes than brackish or marine forms. Generally, sticklebacks never grow to more than 7.5cm (3in) long, but in lakes in the Queen Charlotte Islands, off the coast of British Columbia, there are dark pigmented forms that grow to 20cm (8in) long.

The Fifteen-spined stickleback is a solitary marine species, living in marine water on the European Atlantic coast. The Nine-spined stickleback (which usually has 10 spines) has at most only a few small scutes. It is nearly as widespread as the Three-spined but rarely enters brackish or saline waters. Two genera are confined to northern America: the Brook stickleback (usually with 5 spines), which is fairly widespread and may even be found in coastal waters of reduced salinity in the north; and the Four-spined stickleback, which lives only in the northeastern part of northern America. Despite the presence of protective spines on the back and in the pelvic fins, sticklebacks form an important part of the food chain and are eaten by larger fish, birds, and otters.

The tubesnouts are primitive relations of the stickleback found in the cooler waters either side of the North Pacific. The American species known as the Tube-snout, which resembles the European Fifteen-spined stickleback, lives in huge shoals. No nest is built but the female lays sticky eggs along the stipe of the giant kelp (large seaweed), which is first bent over and then glued down. The male defends the eggs. Its Japanese relative, the Tubenose, is poorly known, but is reputed to lay its eggs inside a seasquirt (*Cynthia*).

Related to sticklebacks, but superficially very different, are the sea moths (Pegasidae). They are small, tropical marine fishes with a body encased in bony scutes and variously developed snouts. There are only five species, in two genera. Because of their unusual shape, and the fact that the shape is maintained when the fish is gutted and dried, they are often sold as curios in the Far East. Apart from the fact that they are bottom-living egglayers and eat small invertebrates such as worms and crustaceans, little is known about their biology.

In Burma, Cambodia, western Malaysia, and Thailand, an unusual species known as the Armored stickleback is found. It lives in freshwater swamps with soft bottoms and dense vegetation. The systematic position of this species has been much discussed. Superficially the species

THE RED BREAST OF THE THREE-SPINED STICKLEBACK

It is probably for its breeding behavior that the Three-spined stickleback is best known. The male, in his breeding dress of red breast and bright blue body, is well known to children as a prized "tiddler" to be caught in spring. The male builds a roughly spherical nest from strands of water plants stuck together by secretions from his kidneys. The choice of site varies and, oddly, there is some evidence that males with more bony scutes prefer a sandy locality, while those with fewer plates a muddier one. The males' bright colors serve both to advertise the nest site to the female and to warn other males to keep away. An attracted female is courted, shown the nest, and, if she approves, lays her eggs there. More than one female may be induced to lay. The fertilized eggs are guarded by the male, who fans them and removes diseased eggs. During the parental phase, the male changes to an inconspicuous dark livery. As the eggs hatch, the male progressively destroys the nest and, a few days after hatching, the young are left on their own. KEB resembles the above-mentioned tubesnouts, but various morphological evidence suggests that it could have its closest relatives among the Syngnathiformes. Currently, it is generally thought to be most closely related to sea moths.

Mailcheeked Fishes

Order Scorpaeniformes

Mailcheeked fishes, or Scorpaeniformes, are an order of predominantly shallow-water marine fish, but the order also contains some notable deep-water species, such as the Atlantic redfish that is found off the North-Atlantic coasts at 100–1,000m (330–3,300ft) depth. Like the redfish, many other Scorpaeniformes species are also predominantly reddish in color. Otherwise, they have a general appearance rather similar to that of the Perciformes – spiny first dorsal fin, ctenoid scales, spines on the head, and pelvic fins well forward). The order contains about 25 families, including the sea robins or gurnards, the fascinatingly beautiful coral-reef lionfishes, the extremely poisonous stonefishes of Australasia, the sculpins, and the armored sea robins and pogges.

Some sculpins, species of the family Cottidae, are common in freshwater. The large head of *Cottus gobio* has given it the name bullhead in the United Kingdom. The vast majority of the approximately 300 sculpin species known are, however, marine. The greatest diversity occur along the North Pacific coastline, but there are also many species in Atlantic coastal areas. Most live in shallow areas, but a few are known from waters as deep as 2,000m (6,600ft). Although some species are eaten, mainly in soup, hardly any have much commercial value. In Greenland, large specimens of the Shorthorn sculpin (*Myoxocephalus scorpius,* which grows to 60cm/2ft) are cooked in a similar way to cod, and are eaten with pleasure.

The flying gurnards (family Dactylopteridae) are not gurnards and neither do they fly. The myth that they are capable of flight arises from their greatly expanded, colorful, fanlike pectoral fins. The purpose of these fins may well be to frighten away potential predators. The "flying" gurnards are heavily built, bottom-living fish with a heavy, bony skull. They spend most of their time "walking" along sandy bottoms, using their transformed, fingerlike pelvic fins to probe for crustaceans, mollusks, and worms that hide in the substrate, as is known also in the true gurnards or sea robins (family Triglidae).

Most scorpionfishes have venomous fin spines that can inflict nasty wounds on humans, as many fishermen will know from experience. In some species the venom can have lethal effects. This is particularly true for the Indo-Pacific stonefish, which possesses the strongest venom and most sophisticated venom apparatus that is known from any fish. Attached to the 9th through 13th dorsal fin spine are large venom glands, which at the slightest pressure squirt venom through a slit in the fin spine into the victim who carelessly handles or accidentally steps on the well camouflaged, bottom-living fish. Stings cause immediate burning pains, often followed by inflammation of lymph glands, breathing difficulties, vomiting, and muscular spasms. Deaths have occurred, and even in mild cases, the recovery period may last several months. KEB/SAF

Below *The venom in the dorsal fin spines of the Reef stonefish* (Synanceia verrucosa) *is highly toxic and has occasionally caused human fatalities. Even so, the fish still falls prey to the larger sharks and rays.*

Oarfishes and Allies

K*NOWN AS "KING OF THE HERRINGS" BY northern Europeans, the oarfish* (Regalecus glesne) *and other Lampridiformes have found a place in folklore as harbingers of abundance or paucity of food fish. Previously placed within the series Percomorpha, the Lamprifomes are now considered to occupy their own superorder, Lampridiomorpha. Lampridiform fish are scarce, spectacular in color, weirdly shaped, and cause great excitement when washed ashore.*

The oarfish is a silvery, ribbonlike fish, with a red dorsal fin the length of its body. The head profile resembles that of a horse. The anterior rays of the dorsal fin are elongated like a mane and the scarlet pelvic fins are very long, with bladelike expansions at the end (hence its common name). The caudal fin is reduced to a few streamers. This spectacular fish, which grows to over 9m (30ft) long, is thought to have been the origin of many sea-serpent stories, since they often mention a red mane.

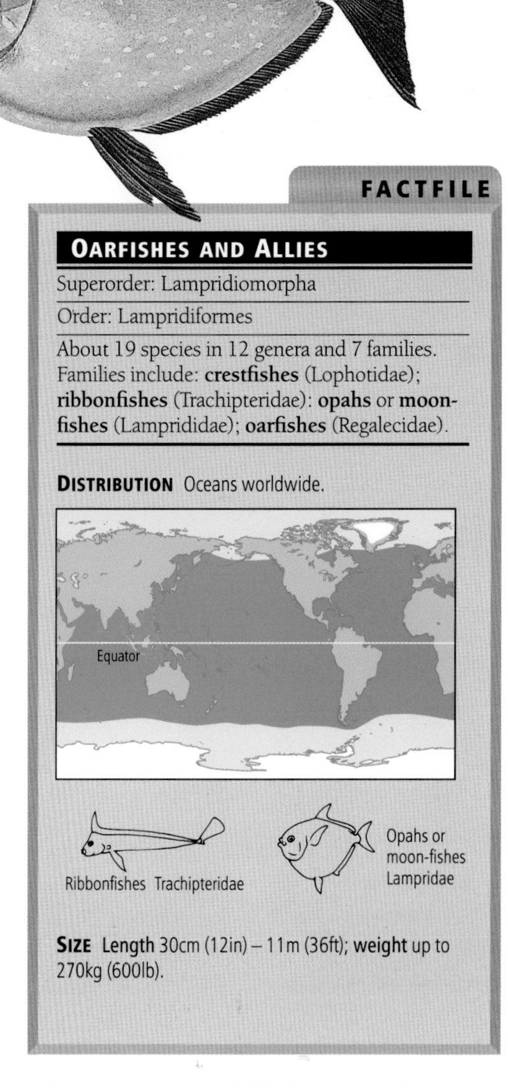

Above 1 *The extraordinary oarfish* (Regalecus glesne), *the longest of all fishes, is rarely seen;* **2** *The Opah or Spotted moonfish* (Lampris guttatus) *is sometimes taken as by-catch in longline fishing for tuna.*

FACTFILE

OARFISHES AND ALLIES

Superorder: Lampridiomorpha

Order: Lampridiformes

About 19 species in 12 genera and 7 families. Families include: **crestfishes** (Lophotidae); **ribbonfishes** (Trachipteridae): **opahs** or **moonfishes** (Lamprididae); **oarfishes** (Regalecidae).

DISTRIBUTION Oceans worldwide.

SIZE Length 30cm (12in) – 11m (36ft); **weight** up to 270kg (600lb).

The oarfish, one of only two species in the family Regalecidae, is found worldwide and probably lives in moderately deep water when healthy (those at the surface are usually moribund). Almost nothing is known of its biology. Examples caught with a full stomach reveal a diet of small crustaceans. Largely on theoretical grounds, it has been convincingly reasoned that the oarfish swims at an angle of 45°, with the long red mane streaming out horizontally. The blades of the pelvic fins bear a large number of chemoreceptor cells. The pelvic fins are now regarded as chemical probes that are held out in front of the animal so that the fish can detect its prey before it reaches it, and organize its respiratory cycle for maximum efficiency in sucking in the small crustaceans. Locomotion is probably by undulations of the dorsal fin, producing an upright attitude and relatively sedate passage through the water.

As regards body shape, there are two distinct groups of Lampridiformes. The Lophotidae, Stylephoridae, Regalecidae, and Trachipteridae have ribbon-shaped bodies; propulsion is effected largely by the dorsal fin and the skin has prickles, or tubercles, the function of which is to lessen water drag. The other group contains the family Veliferidae and the opahs or moonfishes (Lamprididae). The body is deep and propulsion is by enlarged, winglike pectoral fins powered by powerful red muscles attached to an enlarged shoulder girdle.

The coloration of the opahs is extraordinary. Their back is azure, which merges into silver on the belly. The sides have white spots and the whole is overlaid with a salmon-pink iridescence that fades quickly after death. The fins are bright scarlet. Despite being toothless and of a seemingly cumbersome shape, they feed on midwater fishes and squid, which testifies to the swimming efficiency of the winglike pectoral fins. There is one worldwide species, which reaches 1.5m (5ft) in length and over 90kg (200lb) in weight.

The ribbonlike lophotids are known as crest fishes because the bones of the top of the head project forward over the eyes and in front of the mouth, like a dorsal keel. The expansion of the bones is used as a base to attach some of the muscles that work the locomotory dorsal fin. Apart from their surrealistic shape, they also possess an ink sac that lies close to the intestine and discharges via the cloaca. Very few undamaged specimens have been found, but they grow to over 1.2m (4ft) long. Their distribution is uncertain, but they have been found off Japan and South Africa.

The ribbonfishes (family Trachipteridae) have the silvery body and fins of their relatives as a color pattern. Some add dark spots and bars to this pattern. As with many of the long-bodied lampridiforms, the swimbladder is reduced and the skeleton correspondingly lightened, thus achieving neutral buoyancy. When adult, the caudal fin consists of a few rays of the upper lobe, elongated and turned upwards at right angles to the body. The ribbonfishes' lifestyle is thought to be like that of other members of the order. KEB/SAF

Bichirs, Coelacanths, and Lungfishes

THESE THREE VERY DISTINCT GROUPS OF fishes are grouped together here solely on the grounds that all are anachronistic – in other words, they all appear to belong to past times, rather than the present. The popular contradiction in terms "living fossils" may be an appealing cliché to describe them, but does not really advance an understanding of their relationships.

Fossil coelacanths first appeared in rocks of the Devonian period (417–354 million years ago) and continued to occur until about 70 million years ago when, along with the dinosaurs, they disappeared from the fossil record. Since the coelacanth was rediscovered in 1938 and again in 1952, around two hundred have been transferred to scientific institutions but, although its anatomy has been detailed, little is known of its biology and its taxonomic relationships are still an unresolved controversy. Indeed, in a symposium volume published by the California Academy of Sciences in 1979, there are several contradictory papers, each advocating different groups of fishes as the closest relatives of the coelacanth.

Above *True to its common name, the Ornate bichir* (Polypterus ornatipinnis) *has an intricate marbling on its skin. This species is found throughout the basin of the Congo River in Central Africa and in Lake Tanganyika.*

Bichirs

POLYPTERIDAE

The bichirs are endemic to the freshwaters of Africa. The family contains just two genera, the true bichirs with at least ten species in one genus (*Polypterus*) and the Rope or Reedfish – the sole representative of its genus (*Erpetoichthys*). To which other major group of fishes the bichirs are related is a source of much debate. Almost every group has been suggested at one time or another over the last century.

Bichirs are primitive-looking fishes. They have thick, diamond-shaped scales that articulate by a "peg and socket" joint, gular (throat) plates, and an upper jaw that is fixed to the skull. As a group, they retain a surprising number of primitive features, ranging from the tail skeleton, to the possession of two lungs (see below). Yet their fossil record is scanty, the oldest known remains coming from deposits of the Cretaceous period (142–65 million years ago). All the fossils are in Africa, largely within the present area of distribution.

True bichir species owe their generic name (*Polypterus*, from ancient Greek, means "many-fins") to the row of small finlets on the back, each consisting of a stout spine supporting a series of rays. This arrangement is sometimes referred to as a "flag and pole" system – hence the group's alternative name, flagfishes. The pectoral fin has a

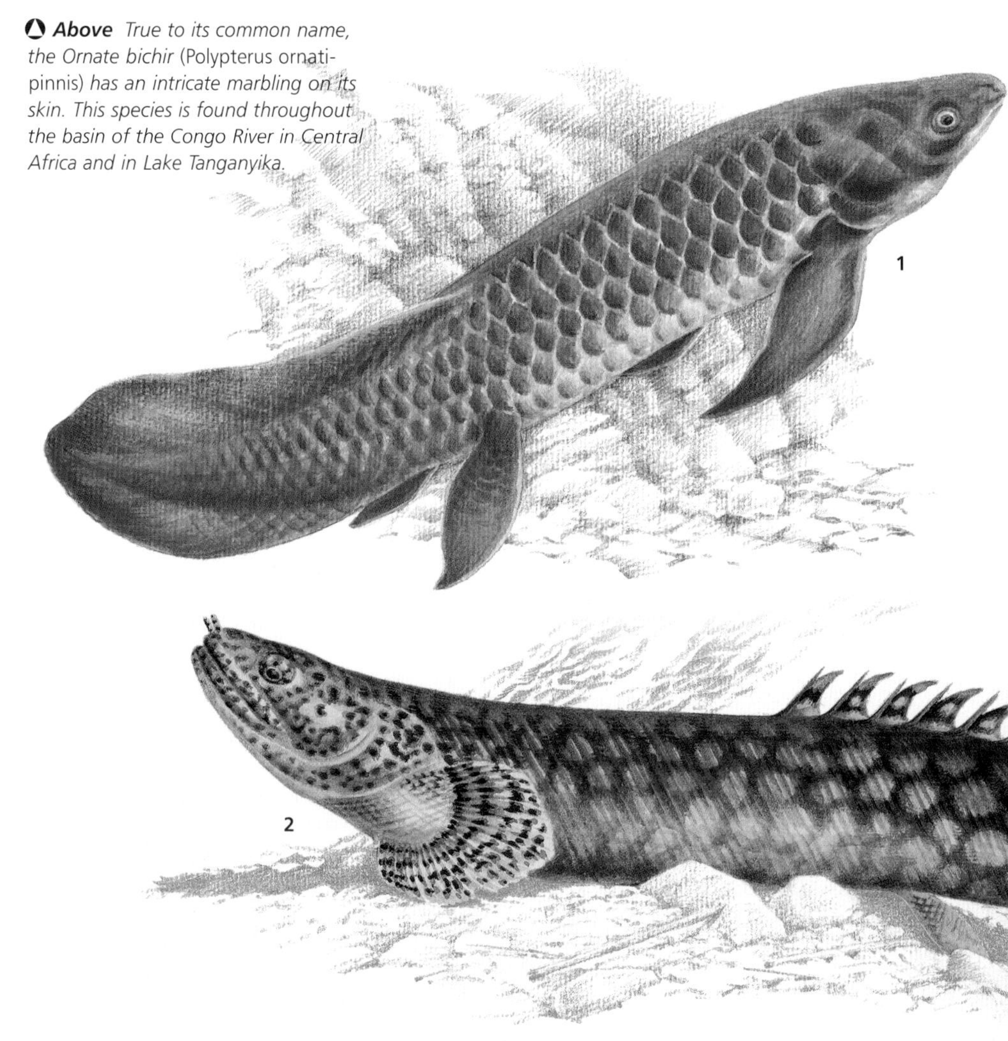

stout, scale-covered base and, although the tail is symmetrical in appearance, its internal structure retains the primitive, upturned, heterocercal condition in which the upper lobe skeleton is longer than the lower one, as in sharks.

Bichirs live in sluggish freshwaters and their swimbladders have a highly vascularized lining that is used as a lung, enabling them to live in poorly oxygenated conditions. Shallow swampy waters, such as those produced when rivers overflow, are preferred for spawning, during which several hundred eggs are laid among vegetation and abandoned. These eggs usually hatch in less than a week, yielding young bichirs with well-formed external gills, another primitive characteristic.

Bichirs are largely nocturnal fishes, not known for an active lifestyle. At night they feed on smaller fishes, amphibians, or large aquatic invertebrates. Within Africa, they are confined to the tropical regions in drainages emptying into the Atlantic Ocean or the Mediterranean. Living bichirs generally grow no longer than 75cm (2.5ft), although the Congo bichir (*Polypterus endlicheri congicus*) can attain a maximum length close to 1m (3.2ft). However, to judge from the size of the scales, some fossil species might have been twice as long.

The Reedfish is a slender, eel-like version of the true bichirs. It lacks pelvic fins and the subsidiary rays on the isolated spines. A much smaller species than most bichirs (it can grow only to around 40cm/16in), it lives exclusively in reedy areas in coastal regions of West Africa near the Gulf of Guinea. It appears to eat mostly aquatic invertebrates.

Coelacanths

COELACANTHIDAE

Much of the biology of the coelacanth has to be inferred from its anatomy and catching records, since no specimens have been kept alive for long. Apart from the mysterious first catch (see In Search of Old Fourlegs), all subsequent coelacanths had, until recently, been caught off the islands of Grand Comore and Anjouan at depths of 70–400m (230–1,300ft). Most had been caught during the first few months of the year in the monsoon season. These two islands are composed of highly absorbent volcanic rock and it has been argued, with some possible corroboration from its kidney structure, that the coelacanth lives in areas where fresh rainwater leaks out into the sea. In 2000, the range of this species (*Latimeria chalumnae*) was dramatically expanded, with the direct observation of some live specimens within the Sodwana Bay Jesser Canyon of the St. Lucia Marine Reserve off the northern Kwazulu-Natal coast of South Africa.

The coelacanth has the build of a lurking predator, and fish remains have been found in the stomach. It has very large eggs: 20 about the size of a tennis ball (around 9cm/3.5in across, making them the largest fish eggs known) were found in a 1.6m (5.3ft) long female. The only known embryos, which still had yolk sacs and were therefore probably not close to birth, were more than 32cm (12.5in) long and it has been estimated that the gestation period is well over a year. The females are larger than the males and can live for at least 11 years.

Although *L. chalumnae* and its recently (1997) discovered relative, *L. menadoensis*, are superficially similar to the last known fossilized coelacanths, there are some interesting anatomical differences. In the present-day coelacanths, unlike their Cretaceous forebears, the swimbladder is nonfunctional and is filled with fat. The fish, however, retain the triple tail, the lobed paired fins, a skull with a hinge in it, and the rough cosmoid scales of their fossil ancestors.

It might perhaps be supposed that the discovery

FACTFILE

BICHIRS, COELACANTHS, & LUNGFISHES

Orders: Polypteriformes, Coelacanthiformes, Ceratodontiformes, Lepidosireniformes

ORDER POLYPTERIFORMES

BICHIRS OR POLYPTERIDS
Family Polypteridae
10 species in 2 genera. Freshwaters in Africa. **Length** maximum c. 1m (3.3ft), but most smaller. Species and genera include: bichirs (*Polypterus* spp.) and **Reedfish** or **Ropefish** (*Erpetoichthys calabaricus*).

ORDER COELACANTHIFORMES

COELACANTHS
Family Coelacanthidae
2 species in 1 genus: *Latimeria chalumnae* and *L. menadoensis*. Oceans off Comoro Islands (between Madagascar and Africa), Eastern Cape and Kwazulu-Natal (South Africa), NW Sulawesi (Indonesia). **Length** maximum 1.8m (5.9ft). **Conservation status**: *L. chalumnae* is classed as Critically Endangered.

SUPERORDER CERATODONTIMORPHA
ORDERS CERATODONTIFORMES, LEPIDOSIRENIFORMES

LUNGFISHES Families Ceratodontidae, Lepidosirenidae and Protopteridae
6 species in 3 genera. Freshwaters of Brazil, Paraguay, Africa, and Queensland (Australia). **Length** maximum 2m (6.5ft). Species and genera include: **Australian lungfish** (*Neoceratodus forsteri*), **South American lungfish** (*Lepidosiren paradoxa*), and **African lungfish** (*Protopterus* spp.).

Australian lungfishes
Family Ceratodontidae

South American lungfishes
Family Lepidosirenidae

African lungfishes
Family Protopteridae

Left *Representative species of lungfishes, bichirs, and coelacanths:* ***1*** *Australian lungfish* (Neoceratodus forsteri), *the most stout-bodied of the lungfishes;* ***2*** *Short-finned bichir* (Polypterus palmas)*;* ***3*** *Coelacanth* (Latimeria chalumnae) *– once the sole coelacanth species, a Southeast Asian relative has now been found.*

of such "anachronistic" organisms would resolve many disputes about evolution, yet, in truth, these extraordinary fishes only generate more questions: Why, for example, have some anatomical changes occurred but not others? Why, if they have kidney similarities with freshwater fishes, do they live in the sea? How long do the embryos take to develop? How are the eggs fertilized internally? Does cannibalism between embryos exist, as has been suggested? Are populations really as widely scattered as they appear to be? And how can one explain the huge geographical gap that exists between the two species?

Lungfishes

LEPIDOSIRENIDAE, PROTOPTERIDAE, CERATODONTIDAE

The three genera and six species of lungfishes are now confined to the freshwaters of the Amazon, West and Central Africa, and the Mary and Burnett Rivers in the far southeast of Queensland in Australia. Their fossils, by contrast, are found in rocks from Greenland to Antarctica and Australia. As with some other ancient fishes, it is known that they represent an important organizational advance somewhere in the evolution from aquatic vertebrates to land-living vertebrates, but exactly how they fit in is still a matter of some uncertainty.

Living lungfishes are grouped into three families: the South American lungfish (*Lepidosiren paradoxa*) has its own family, the Lepidosirenidae, while the African *Protopterus* (4 species) form the Protopteridae. The Australian species, *Neoceratodus forsteri*, is the sole representative of its family, the Ceratodontidae.

Neoceratodus retains more of the primitive features of its Devonian ancestors than do its relatives. It has large, overlapping scales and lobed, paired fins fringed with rays that retain the primitive characteristic of being more numerous than their supporting bones. The dorsal, caudal, and anal fins are all joined up. Unlike its relatives, however, the Australian lungfish cannot survive

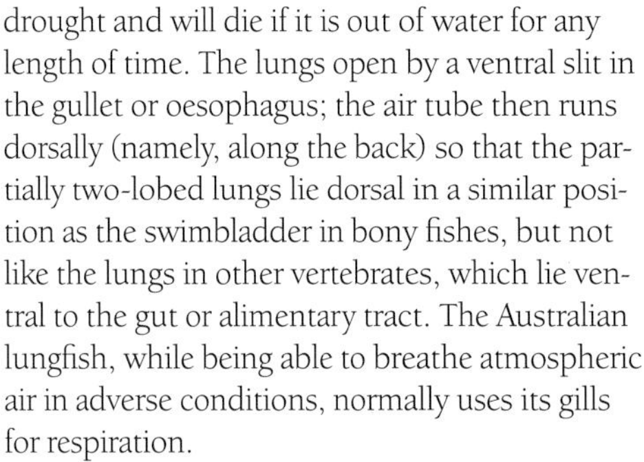

drought and will die if it is out of water for any length of time. The lungs open by a ventral slit in the gullet or oesophagus; the air tube then runs dorsally (namely, along the back) so that the partially two-lobed lungs lie dorsal in a similar position as the swimbladder in bony fishes, but not like the lungs in other vertebrates, which lie ventral to the gut or alimentary tract. The Australian lungfish, while being able to breathe atmospheric air in adverse conditions, normally uses its gills for respiration.

Lungfishes are carnivores, and when fully grown (about 1.5m, 5ft) eat large invertebrates, frogs, and small fish. The teeth consist of a pair of sharp plates in each jaw, which work in a shearing action.

Reproduction in the Australian lungfish has been observed in shallow water in August when, after a rudimentary courtship consisting of the male nudging the female, the eggs are scattered

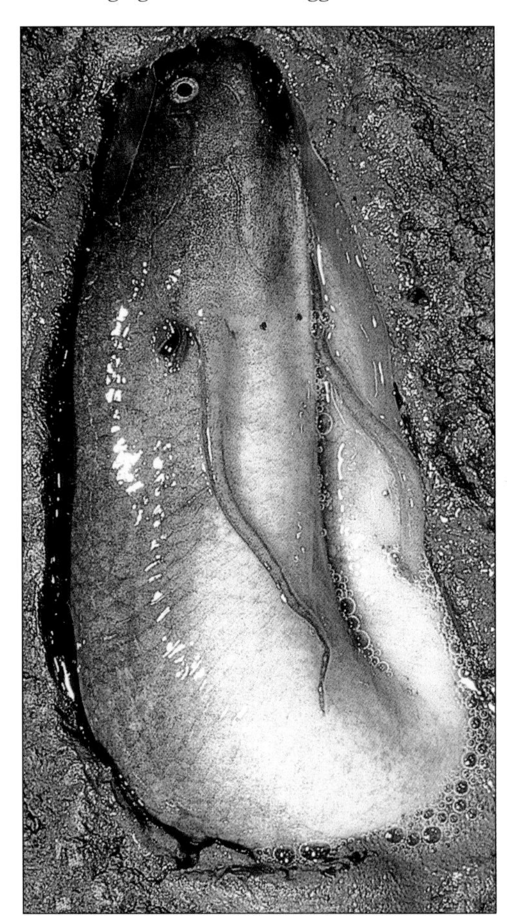

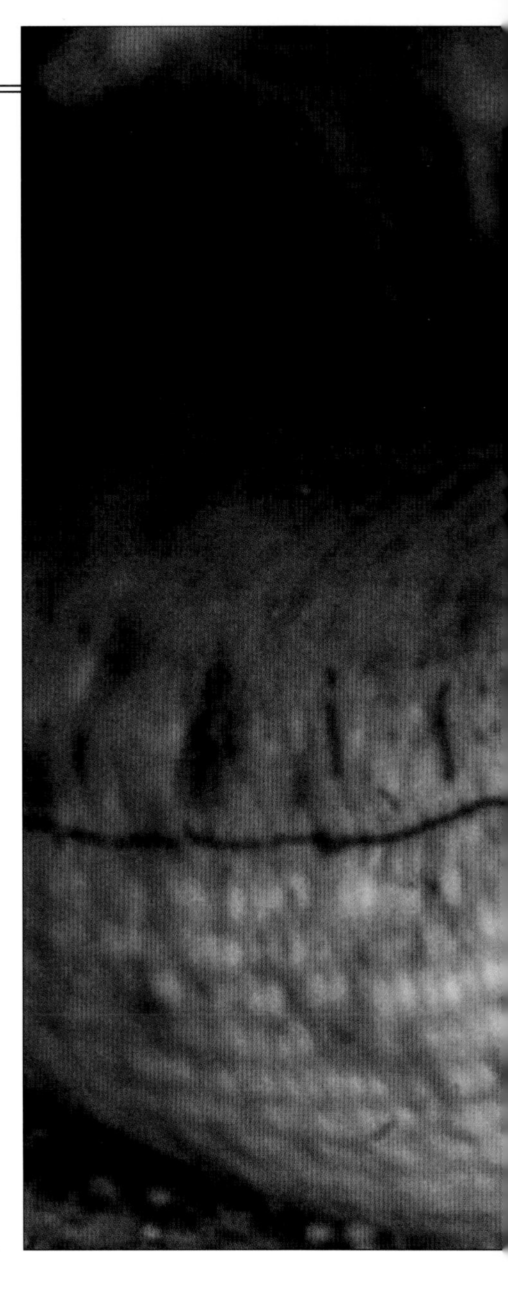

over a small area of dense weed and fertilized. There is no indication of any further parental concern. Although native to two small river systems, the interest in this rare species is such that stocks have been transplanted into other Australian rivers. At least three of the introductions have been successful and the fishes have bred.

The biology of the African species is better known. At least two, the African lungfish (*Protopterus annectens*) from west and southern Africa and the Spotted lungfish (*P. dolloi*) from the Zaïre basin, are known to survive drought by a type of summer hibernation (estivation) in cocoons. The widespread East African species, the Ethiopian or Speckle-bellied lungfish (*P. aethiopicus*) is thought to be capable of estivating, but rarely does so in the wild as its waters are less likely to dry up. The same applies to *P. amphibius*.

African lungfish are more elongate than their Australian relatives. Small scales cover the body and the paired fins are long and threadlike. They are aggressive predators and, in the case of the Ethiopian lungfish, can grow to more than 2m (6.5ft) long. The lungs are paired and lie in a ventral position, as in terrestrial vertebrates – and in

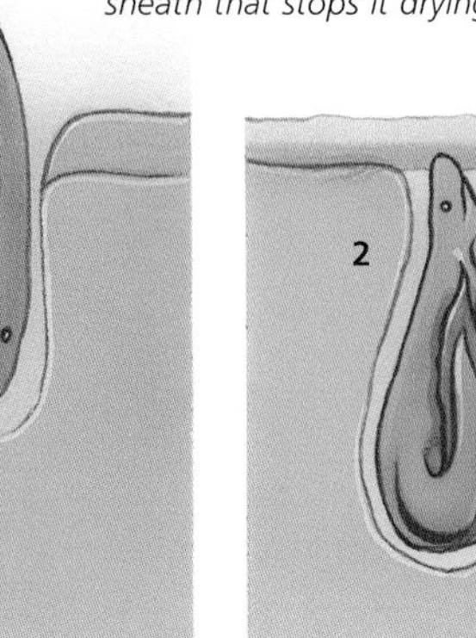

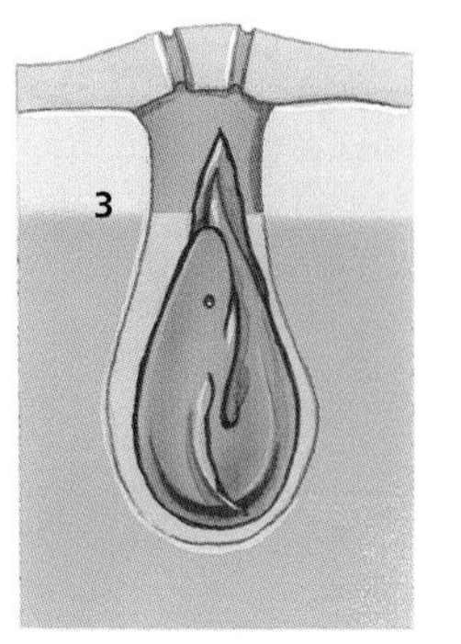

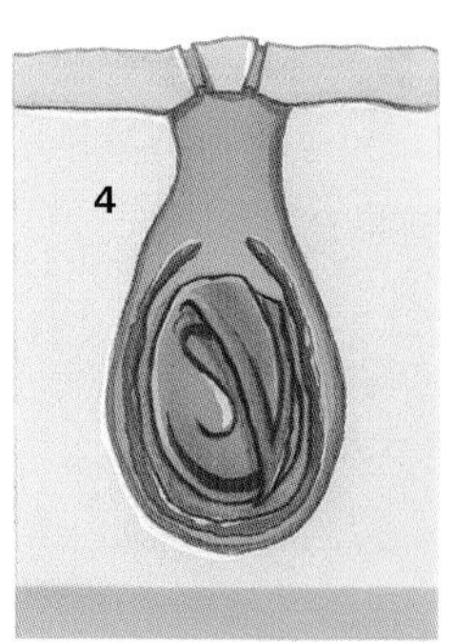

Below *Estivation in the African lungfish.* ***1*** *As the waters evaporate, the fish burrows into the mud;* ***2*** *It then turns back on itself and* ***3*** *settles in the base of the burrow with its tail wrapped around itself;* ***4*** *Finally, it secretes a mucous sheath that stops it drying out.*

Above *Inhabiting swampy floodplains with oxygen-poor waters in the basins of the Amazon and Paraná rivers, the South American lungfish* (Lepidosiren paradoxa) *satisfies its oxygen requirements by breathing air at the surface.*

sharp contrast to their Australian relative. The use of lungs in breathing requires a four-chambered heart, so, unlike bony fish, the lungfishes' auricles (upper heart chambers) and ventricles (lower heart chambers) are divided functionally by a partition so that blood is circulated to the lungs as a bypass from the normal body and gill circulation – as happens in mammals, humans included.

All the African lungfishes build nests. In the Ethiopian lungfish, it often takes the form of a deep hole dug by the male, which then guards the newly hatched young for about two months. As well as driving away would-be predators, the male also aerates the water in the nest. The nest of the Spotted lungfish is much more elaborate and has an underwater entrance. The terminal brood chamber in which the eggs develop may be in swampy ground and may be open at the top. Aeration vents are also built into it by the male.

The larvae of African (and South American) lungfish have external gills, the degree of development varying with the amount of oxygen in the water. At metamorphosis (i.e. the change from the juvenile to the adult form), the external gills are usually resorbed and the lung and gill respiration takes over. Occasionally, however, vestigial (tiny) external gills remain throughout life.

Protopterus annectens lives in swamps and rivulets that are prone to drying out, often for months on end. In order to survive such harsh conditions, the fish burrows into a tube that it excavates in the soft mud as the water level falls. Using its mouth and general body pressure, the lungfish widens out the bottom of the tube until it can turn around. As the water drops below the opening of the tube the fish then closes the mouth of the tube with a plug of porous mud, curls up in the lower chamber and secretes large quantities of a special mucus that hardens to form an encasing cocoon, with only an opening for the mouth. The cocoon retains moisture, while the porous mud plug allows breathing. During estivation, as in hibernation, the metabolic rate is greatly reduced and the basic energy needed for survival comes from the breakdown of muscle tissue. In this state, lungfishes have been known to survive four years of drought, although, normally, an incarceration of only a few months would be necessary. Eventually, when the rains return and the river floods again, the waters dissolve the cocoon and the fish emerges.

The South American lungfish (*Lepidosiren paradoxa*) looks similar to the African species, with an eel-like body and feelerlike pectoral fins, but with fleshier-based, wider pelvic fins. It also possesses the general characteristics of a cartilaginous vertebral column and a common opening for the waste products, eggs, and sperm.

Lepidosiren is able to estivate, but its refuge is much simpler than that of its African relatives and no cocoon is produced. An elaborate nest is made in the breeding season, during which time the pelvic fins of the male bear a large number of blood-rich filaments. The function of these filaments is unknown, but hypotheses have included the release of oxygen into the water of the nest, or that the filaments act as supplementary gills to reduce the number of visits to the surface while the fish is guarding its young. KEB/JD

In Search of "Old Fourlegs"

The discovery of the Coelacanth

It is universally accepted in biology that the absence of records does not imply certainty of extinction. The coelacanths represent perfect examples of this.

On a hot summer's day in 1938 Captain Goosen's boat, *Nerine*, docked at East London in South Africa. At that time, Marjorie Courtenay-Latimer was the curator at the East London Museum, and local skippers were accustomed to her frequent visits to the port to obtain fish specimens for the museum. At 10.30 a.m. on 22 December she was telephoned and told that the *Nerine* had returned with some specimens for her. Among the catch was a large blue fish with flipperlike fins and a triple tail, which she had never seen before. After several attempts she found a taxi driver willing to take her and her 1.5m (5ft) long, oily, smelly prize back to the museum. There, after searching through reference books, the nearest she could come to identifying the fish was as "a lung fish gone barmy."

Realizing that her find was important, she tried to contact Dr J. L. B. Smith (whose name will be forever linked with the fish), the ichthyologist at Rhodes University, Grahamstown. It was then mid-summer in South Africa and temperatures were high. How, therefore, was this important scientific find to be preserved? The local mortuary refused to have the corpse in its cold store, so finally, a local taxidermist, Mr R. Centre, although admitting inexperience in fish-stuffing, agreed to help. He wrapped the body in cloth, soaked it in formalin and placed it in a makeshift ichthyosarcophagus ("fish coffin").

On 26 December there was still no reply from Smith. An examination of the body showed that the formalin had not penetrated and that the internal organs were rotting. Pragmatism dictated that the decaying parts should be thrown away and what could be preserved should be preserved.

On 3 January 1939 a telegram arrived from J. L. B. Smith. It read "most important preserve skeleton and gills = fish described." The ensuing search of local rubbish heaps failed to find the discarded organs. Parallel disasters now revealed themselves. The early photographs taken of the fresh fish had been spoiled, while the museum trustees, not thinking the fish to be important, had ordered the skin to be mounted before, on 16 February, Smith finally arrived. He stared at the mounted skin and said: "I always knew, somewhere, or somehow, a primitive fish of this nature would appear."

He described the fish as *Latimeria chalumnae*, in recognition of Marjorie Courtenay-Latimer and the Chalumna River, off which the fish had been caught.

Why had this conspicuous, spiny-scaled fish remained unnoticed for so long? Its large eye and lurking predator shape seemed to suggest that it did not normally live off East London. Unless the only living specimen had been caught, there must be more – but where?

The hunt was long and extremely exciting. It took 14 years, involved a great deal of work and almost entailed commandeering the South African Prime Minister's private aircraft. For all the details the reader is referred to J. L. B. Smith's book, *Old Fourlegs: The Story of the Coelacanth* (London: Longman, Green, 1956), as well as Keith Thomson's similarly titled book, *Living Fossil: The Story of the Coelacanth* (Hutchinson Radius, 1991).

Just before Christmas 1952 (note the date again), Smith received a telegram from Captain Eric Hunt in the Comoro Islands. It read: "repeat cable just received have five-foot specimen

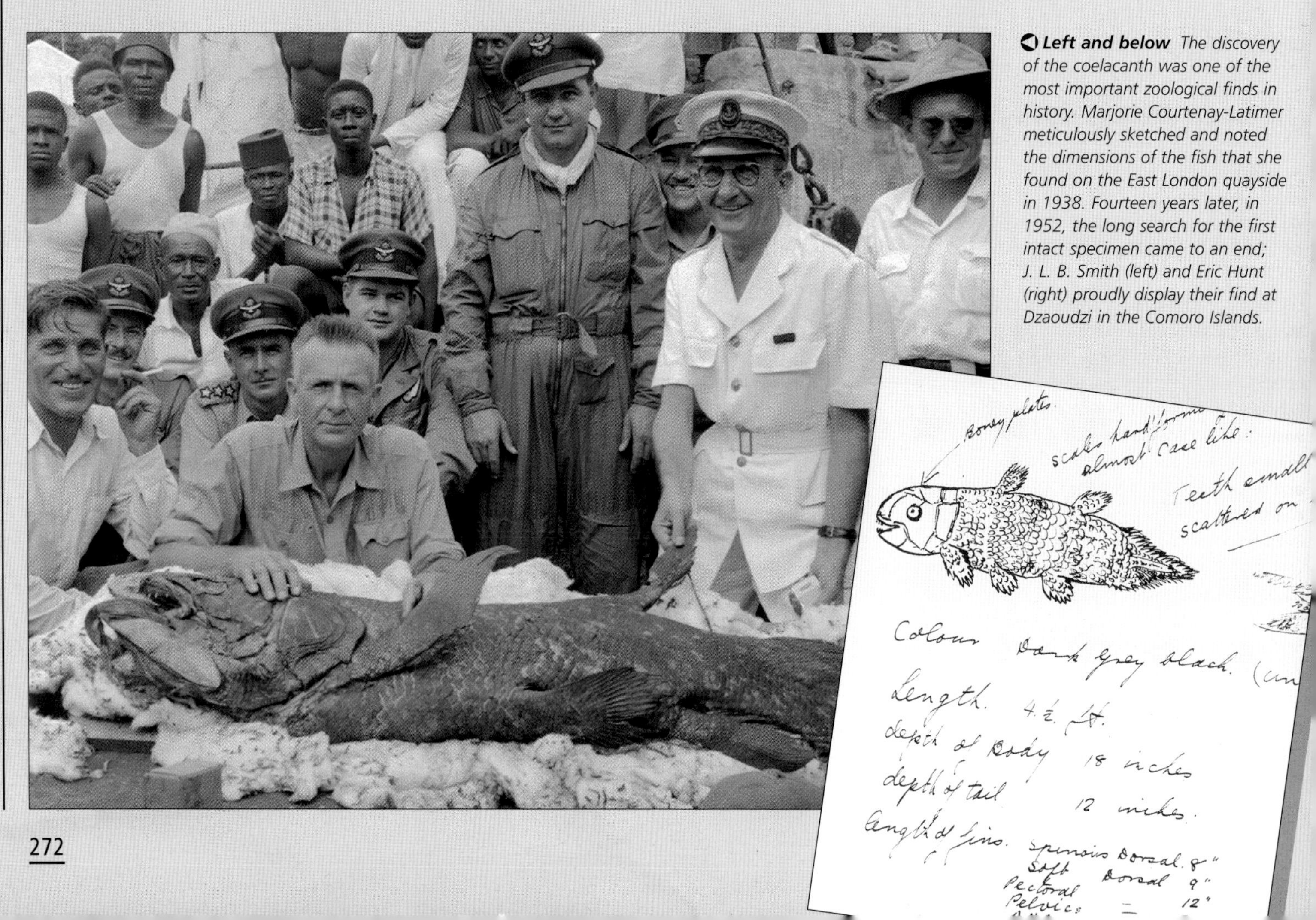

Left and below *The discovery of the coelacanth was one of the most important zoological finds in history. Marjorie Courtenay-Latimer meticulously sketched and noted the dimensions of the fish that she found on the East London quayside in 1938. Fourteen years later, in 1952, the long search for the first intact specimen came to an end; J. L. B. Smith (left) and Eric Hunt (right) proudly display their find at Dzaoudzi in the Comoro Islands.*

Above *"Old Fourlegs" is something of a misnomer, since the coelacanth swims conventionally and does not, as Smith speculated, propel itself along the seafloor with its prominent pelvic and anal fins. This specimen was sighted off the Comoros.*

Right *Arnaz Erdmann swimming with a* Latimeria menadoensis. *Fishermen in Sulawesi had long referred to their indigenous coelacanth as* Rajah Laut, *or "King of the Sea."*

coelacanth injected formalin here killed 20th advise reply hunt Dzaoudzi."

Although this discovery was very exciting to scientists, the inhabitants of the Comoro Islands (where Dzaoudzi is located) were unmoved. They were familiar with this fish, had dubbed it *Gombessa* (meaning "taboo" in reference to its foul taste), and thought it a worthless catch, although the coarse scales could be used to roughen bicycle tyre inner-tubes when mending punctures.

All subsequent findings of coelacanths were made off Comoro waters, making this the center of coelacanth distribution for many years. Then, in September 1997, Mark and Arnaz Erdmann, biologists honeymooning in Sulawesi, Indonesia (some 10,000 km/6,200 miles away from the Comoros), came across a fish in a local market that dramatically updated the coelacanth story and destroyed the Comoro Islands' exclusive claim as the only place on Earth where the coelacanth was found.

The Sulawesi fish was definitely a coelacanth, but of a slightly different type. This was confirmed ten months later when a live specimen was landed. These coelacanths were different in color to the Comoro ones (brown with golden-colored flecks, instead of bluish with pinkish-white flecks). DNA studies corroborated the difference, and the new coelacanth was named *Latimeria menadoensis*.

On 28 October 2000, divers in Jesser Canyon in Sodwana Bay, in a marine reserve off the northern Kwazulu-Natal coast of South Africa, spotted three live coelacanths at a depth of 104m (340ft). One month later, three further specimens were filmed in the same area. These finds suggested that the coelacanth might be more widespread in the western Indian Ocean than was once thought.

Diving at such depths can be extremely dangerous and, in this case, ended in tragedy for one of the divers, who never regained consciousness following a blackout after surfacing. KEB/JD

Sharks

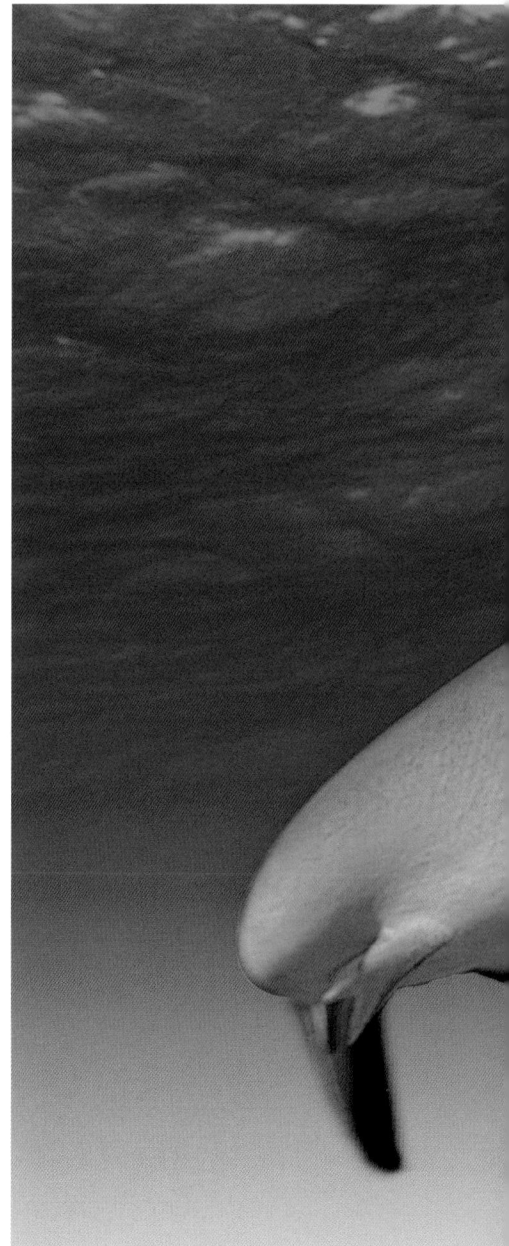

THE TALES OF ANCIENT MARINERS AND modern media hyperbole have given most people the idea that sharks are savage predators, but this is true only of a minority of species. A group of fishes that have survived for around 400 million years, the elasmobranchs – sharks, skates, and rays – have five or more gill slits on either side of their heads and a cartilaginous skeleton. These characteristics distinguish them from most other fishes, which have a single gill cover on each side of the head and bony skeletons. Sharks have exceptional sensory organs, and some give birth to live young.

With few natural predators, many sharks grow slowly, mature late, and have very few young. The recent increase in affluence in Southeast Asia has led to a huge demand for shark fins to supply the shark fin soup market. Sharks are now being fished at a much greater rate than they can reproduce and some species are threatened with extinction if this trend continues.

Sophisticated Hunters

TEETH AND SENSORY SYSTEMS

A notable feature of sharks is their teeth. In the large highly predatory sharks, these teeth are large and razor sharp, used for shearing and shredding their prey into bite-sized pieces. When biting, the shearing is often aided by body motions such as rotating the body or rapid shaking of the head. Those preying on fishes have long, thin teeth to help them catch and hold on to struggling and slippery fish. Sharks that feed on the bottom have flattened teeth for crushing the shells of their mollusk and crustacean prey. At any one time most sharks may have several rows of teeth in their mouths. Only the first row or two are actively used for feeding, the remaining rows are replacement teeth in various stages of formation, with the newest teeth being at the rear. As a tooth in the functioning row breaks or is worn down, it falls out and a replacement tooth moves forward in a sort of conveyor-belt system. The largest species, the Basking shark and the Whale shark, have minute teeth that play no role in feeding; rather, they feed in a similar way to the baleen whales, filtering the water for plankton. Basking sharks have modified gill rakers; Whale sharks have spongy tissue supported by the gill arches and can also swallow small shoaling fishes.

Sharks find their prey through a number of sensory systems. Many have better eyesight than

Above *Not all sharks are large. The Australian marbled catshark (*Atelomycterus macleayi*) only grows to around 60cm (2ft). Catsharks are named for their elliptical, catlike eyes. They are also characterized by two dorsal fins set far back.*

FACTFILE

SHARKS

Class: Chondrichthyes

Orders: Chlamydoselachiformes, Hexanchiformes, Heterodontiformes, Orectolobiformes, Scyliorhiniformes, Triakiformes, Odontaspidiformes, Isuriformes, Carcharhiniformes, Squaliformes, Squatiniformes, Pristiophoriformes.

At least 370 species in at least 74 genera, 21 families and 12 orders (other taxonomic accounts, e.g. Nelson 1994 (3rd edn.) list only 8 orders).

DISTRIBUTION Worldwide in tropical, temperate and polar oceans at all depths.

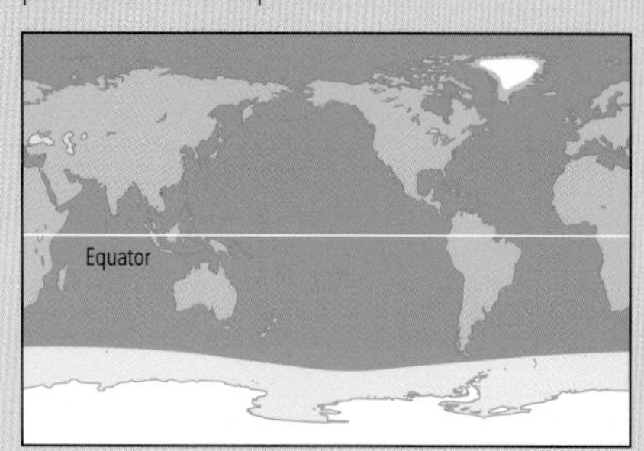

SIZE Length 15cm–12m (6 inches–40ft); weight 1–12,000kg (1–26,500lb).

See The 12 Orders of Sharks ▷

was once believed and unlike most bony fishes, are able to control the size of the pupil. Sharks that hunt in dim light or the dark have a tapetum that reflects the light so that it stimulates the retina a second time; in the dark, light reflected from these eyes makes them shine like those of a cat. Many sharks have a nictitating membrane, which serves as a protective eyelid. When the fish gets close to its prey, it closes this membrane and switches over to other sensors, particularly its Ampullae de Lorenzini (a series of pits around the snout). Great white sharks do not have nictitating membranes but roll their eyes backward for protection while striking prey. The Ampullae de Lorenzini are sensitive to other stimuli, but their use as electroreceptors is of prime importance. Using these electroreceptors, sharks are capable of picking up an impulse of just one-millionth of a volt, which is less than the electrical charge

Above *Millions of years of evolution have given sharks the beautifully streamlined shape and powerful musculature that make them such efficient hunters. In addition, the snout is equipped with highly sensitive scent detectors. This Grey reef shark* (Carcharhinus amblyrhynchos) *exemplifies such adaptations.*

Above right *Triangular, sawlike teeth make the Great white shark* (Carcharodon carcharias) *a fearsome predator. These serrated teeth are designed to tear chunks of flesh from the flanks of large prey, as the shark shakes its head from side to side.*

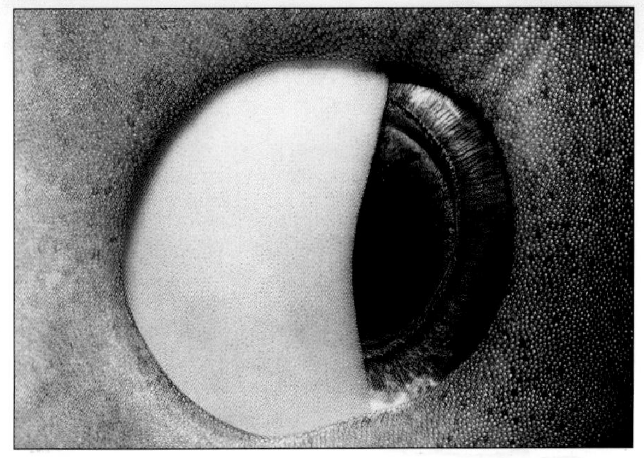

Right *A nictitating membrane closing over the eye of a Tiger shark* (Galeocerdo cuvier). *This requiem shark species eats a wide variety of prey, including other sharks and sea snakes, and the membrane helps protect its eyes during an attack.*

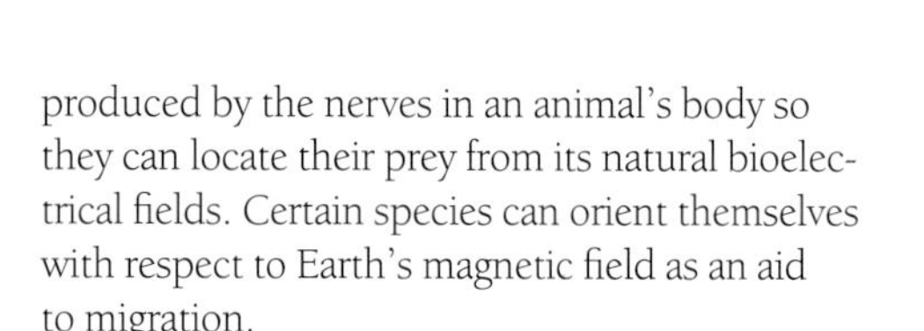

produced by the nerves in an animal's body so they can locate their prey from its natural bioelectrical fields. Certain species can orient themselves with respect to Earth's magnetic field as an aid to migration.

In common with all other fishes, sharks have a lateral-line system – a series of sensors along both sides of the body that pick up pressure waves caused by the movements of another animal or even by the shark itself approaching a stationary object. Some species have sensory barbels around their mouth that taste the seabed for prey. Sharks have a keen sense of smell and can detect one part of blood in a million parts of seawater.

Frilled Shark

ORDER CHLAMYDOSELACHIFORMES

The most primitive living shark is the Frilled shark (*Chlamydoselachus anguineus*), sole representative of this order. It has broad-based, tricuspid (trident-like) teeth, which are otherwise found only in fossil sharks. Its common name is derived from its long, floppy gill flaps, forming a frill around the head. Its primitive and unique characteristics place it in its own family. First discovered off Japan on Sagami Bay in the 1880s, deep trawling has shown that it lives at depths of 300–600m (1,000–2,000ft) over a wide area off the coasts of Australia, Chile, California, Europe, and South Africa (where specimens may possibly be a separate *Chlamydoselachus* species). It grows to 2m (6ft) long and has a thin, eel-like body and feeds on small fishes swallowed whole. The female develops eggs in her body, producing 6–12 young per litter (i.e. the species is viviparous).

Six- and Seven-gilled Sharks

ORDER HEXANCHIFORMES

The six- and seven-gilled sharks (also known as the cow sharks) are so named because they have developed one or two extra sets of gill slits. They prefer cold water, and in the tropics live in deep water. No nictitating membrane is present in these species. They reach 4.5m (15ft) in length and feed on other fishes. The upper-jaw teeth are long and tapered, while the lower-jaw teeth are short and wide with unique, strong, multiple serrations. They also develop eggs internally and produce up to 40 young.

Catsharks and False Catsharks

ORDER SCYLIORHINIFORMES

The catsharks and false catsharks comprise about 18 genera and 87 species. Mostly living in cold or deep waters, they are found worldwide and many have mottled patterns and do not lose this pigmentation pattern as they mature. Living on or near to the seabed, they feed on mollusks, crustaceans, and bottom-dwelling fishes. Some species have sensory barbels that help them locate their prey.

Most species are small, maturing at about 1–1.5m (3–4.5ft) but some are larger, such as the False catshark, which may reach 3m (9ft). When threatened, the aptly named swell sharks swallow water or air into their stomachs, enlarging their girth to three or four times its normal size.

Smooth Dogfish Sharks

ORDER TRIAKIFORMES

The smooth dogfish sharks (triakoids) live in the shallows of tropical, subtropical, and temperate seas. They are moderately large sharks, reaching about 2m (6ft) in length. Despite the fact that they are bottom dwellers, feeding on mollusks, crustaceans, and fishes, they do not lie or crawl on the seabed. Most species have modified crushing and grinding teeth.

The Leopard shark from the eastern North Pacific has a beautiful color pattern, with dark gray to black spots on a silvery background, which makes it popular for public aquariums.

Virtually all species make long migrations, spending winters in the tropics and migrating

Left *Representative species of sharks:* **1** *Frilled catfish* (Chlamydoselachus angineus; *sole member of the order Chlamydoselachiformes*), *the most primitive form of shark;* **2** *Necklace carpetshark* (Parascyllium variolatum; *order Orectolobiformes*); **3** *Whale shark* (Rhincodon typus; *order Orectolobiformes*) *– this Vulnerable species is the world's largest fish;* **4** *Cuban ribbontail catshark* (Eridacnis barbouri; *order Scyliorhiniformes*); **5** *Leopard shark* (Triakis semifasciata; *order Triakiformes*), *caught both commercially and as a game fish in the Eastern Pacific;* **6** *Goblin shark* (Mitsukurina owstoni; *order Odontaspidiformes*) *– this species has specialized jaws that can suddenly project far forward to catch prey;* **7** *Great white shark* (Carcharodon carcharias; *order Isuriformes*); **8** *Thintail thresher* (Alopias vulpinus; *order Isuriformes*), *widespread in tropical and temperate waters;* **9** *Longnosed sawshark* (Pristiophorus cirratus; *order Pristiophoriformes*) *– this species occurs only off southern Australia;* **10** *Ornate angelshark* (Squatina tergocellata; *order Squatiniformes*), *from the Eastern Indian Ocean.*

to temperate waters in the summer. Evidence suggests that these migrations are regulated by water temperature and this, in turn, affects where the sharks spawn. Females of these species develop embryos in the uterus and deliver 10–20 young at a time. Although these sharks are considered harmless to humans there is at least one authenticated instance of the Leopard shark attacking a man in northern California.

Horn or Port Jackson Sharks

ORDER HETERODONTIFORMES

The horn sharks live in the Indian and Pacific oceans. Sluggish, bottom dwellers up to 1.65m (5.4ft) long, they lie, sometimes in groups, under kelp beds or on shallow rocky reefs and occasionally sandy patches during the day and disperse to feed at night when their prey are most active. The genus name, *Heterodontus* (Greek for "different teeth"), alludes to the fact that they have pointed front teeth and rear teeth fused molarlike into crushing plates – ideal for holding, breaking, and then crushing shelled mollusks and crustaceans. Stockily built, species of this order have prominent brow ridges above their eyes, which give them the appearance of having horns – hence the name horn sharks. Oviparous, they lay eggs, which are unique in having a screw-shaped spiral form. The female forces these into crevices between rocks or pieces of coral and each egg case contains one pup (baby shark).

Orectoloboids

ORDER ORECTOLOBIFORMES

The orectolobiformes are closely related families of sharks that are tropical to subtropical fishes. They are found mostly in the Indo-Pacific, while two species are found in the Atlantic Ocean.

In size the orectolobiformes range from the epaulette sharks – 1m (3ft) long – to the Whale shark, which has been reliably reported as reaching over 12m (40ft) in length and is the world's largest fish. In most species the young are born with spotted or banded patterns, which fade as the animals' mature.

Most species lay eggs (oviparous), but a few develop them internally (viviparous). Usually less than 12 young are produced per litter.

All species except the Whale shark are bottom dwellers, the skeletons of the pectoral fins are modified so that they can use the fins for walking on the seabed; even when disturbed many species will crawl away rather than swim.

Most orectolobiformes feed on mollusks and crustaceans and have teeth for crushing and grinding. All have sensory barbels around the mouth. The carpet sharks eat mainly fishes, so they have long, thin teeth. The Whale shark is a filter feeder, having its gill arches specially modified to filter out the planktonic organisms and small animals that it feeds on such as krill, squid, anchovies, sardines, and mackerel. Its redundant teeth are

Above *Off the coast of Costa Rica, Central America, a pack of Whitetip reef sharks* (Triaenodon obesus) *gather to hunt. These small, slim sharks are sluggish during the day, but become active at night. They often live in lagoons and on coral reefs.*

Below *Distinct features of horn sharks, such as the Port Jackson shark* (Heterodontus portusjacksoni), *include hornlike ridges above the eyes, pointed front teeth, and crushing rear-teeth plates, and a bottom-dwelling habit.*

Above The Spotted wobbegong (*Orectobolus maculatus*) *has a flattened head and distinctive coral-shaped skin flaps around the edge of its snout. It inhabits shallow coastal waters off southern Australia.*

minute. The Whale shark's vast bulk requires constant fuel, so it swims continuously, filtering the ocean for its food. Found worldwide, in all tropical, subtropical, and warm temperate seas, when first described by Smith, the generic name used was *Rhineodon* but in a later descriptions he used both *Rhincodon* and *Rhineodon*. *Rhincodon* is now the accepted version; some researchers feel that it should be placed in a separate family.

Sand Tiger, False Sand Tiger, and Goblin Sharks

ORDER ODONTASPIDIFORMES

The sand tiger, false sand tiger, and goblin sharks are all fairly large, reaching 3–3.5m (10–12ft) in length. The sand tiger sharks (5 species, with a new one possibly found off Columbia's Malpelo Islands) are found worldwide in shallow temperate and tropical waters. Largely fish eaters, they have long, thin teeth that protrude from the mouth to give a ferocious, snaggle-toothed appearance. This, combined with their docile nature, has made them favorites in public aquariums. Known as the Tiger Shark in North America, Grey Nurse Shark in Australia, and Spotted Raggedtooth Shark in South Africa, this is one species whose scientific name has changed from *Carcharias taurus* to *Eugomphodus taurus* to *Odontaspis taurus* and is now back to *Carcharias taurus*.

Reproduction in this order is viviparous – embryos develop in the uterus – but as with many shark species, there is intrauterine cannibalism. The female begins with 6–8 embryos per uterus but as they grow, the largest in each oviduct devours its siblings, embryos, and other unfertilized eggs; only two young are eventually born.

The False sand tiger shark, or Crocodile shark, lives in deeper waters off China and East and West Africa. The Goblin shark is perhaps the most bizarre-looking of all living sharks. Projecting from its "forehead" is a flattened, spade-shaped, horn-like growth whose function is unknown, the mouth can extend forward under this horn or retract under the eye. The Japanese fishermen that first caught it called it *Tenguzame*, which means "Goblin shark." Like the Frilled shark, it was originally caught in the 1890s in Japan's Sagami Bay. Since then, it has been caught worldwide at depths of 300m (1,000ft) and more. However, apart from its length, 4.3m (14ft), little is known about this species, although DNA studies suggest that it became specialized early in its evolution. While alive, Goblin sharks are a translucent pinkish-white colour, but become very dark brown shortly after death.

Requiem Sharks

ORDER CARCHARHINIFORMES

The requiem sharks are probably the largest group of living sharks, with about 100 species in 10 genera. In body shape and behavior they are the "typical" shark people think of. They reach 3.5m (11.5ft) in length and live in all tropical and temperate seas.

The Bull shark lives in tropical to subtropical coastal waters worldwide and enters freshwater for lengthy periods. It has been reported 3,700km (2,300 miles) up the River Amazon, 2,900km (1,800 miles) up the River Mississippi, more than 1,000km (620 miles) from the ocean in the River Zambezi, and in Lake Nicaragua. Sharks from these freshwaters were once thought never to enter the ocean and thus to be distinct species and were named after the waters that they were found in, for example *Carcharhinus nicaraguensis*.

Above *Lesser spotted dogfish embryos in their egg capsules (mermaid's purses). Sharks have three modes of egg development. Most are viviparous, with eggs developing and hatching in the womb. In some viviparous species, the eggs develop in the womb connected to the wall by a yolk "placenta." These two types give birth to fully developed young. The third type (represented by the dogfish) is oviparous, with the female attaching fertilized eggs to vegetation.*

The 12 Orders of Sharks

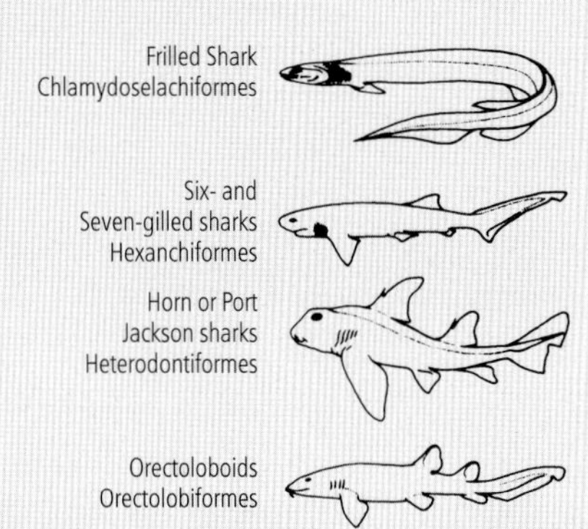

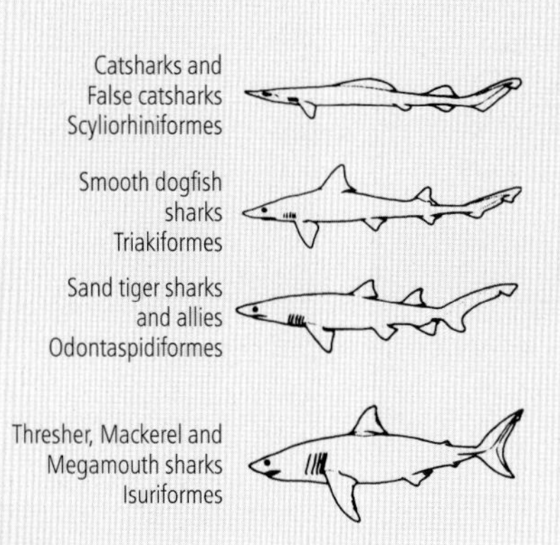

Requiem sharks
Carcharhinoidiformes

Spiny dogfish
and Allies
Squaliformes

Angel sharks
Squatiniformes

Sawsharks
Pristiophoriformes

Frilled Shark
Order: Chlamydoselachiformes

Lives off coasts of California, Chile, Europe, S Africa, Japan, Australia. 1 family – Chlamydoselachidae – containing the sole species *Chlamydoselachus anguineus*. One South African species may be a separate subspecies.

Six- and Seven-gilled sharks or hexanchoids
Order: Hexanchiformes

Worldwide in cold marine waters at shallow to moderate depths. 1 family – Hexanchidae – containing 5 species in 3 genera including: Bluntnose sixgill shark (*Hexanchus griseus*), Bigeye sixgill shark (*Hexanchus nakamurai*), Broadnose sevengill shark (*Notorynchus cepedianus*), Sharpnose sevengill shark (*Heptranchias perlo*). The Bluntnose sixgill shark is presently classed as Vulnerable.

Horn or Port Jackson sharks
Order: Heterodontiformes

Tropical Western Indian and Eastern and Western Pacific Oceans. 1 family – Heterodontidae – containing 8 species of the genus *Heterodontus*, including: Horn shark (*H. francisci*), Crested bullhead shark (*H. galeatus*), Japanese bullhead shark (*H. japonicus*), Port Jackson shark (*H. portusjacksoni*), Zebra bullhead shark (*H. zebra*).

Orectoloboids
Order: Orectolobiformes

Indo-Pacific Ocean; 2 species in Atlantic Ocean. About 31 species in 13 genera and 7 families, including: banded catsharks and epaulette sharks (family Hemiscyllidae), including epaulette sharks (genus *Hemiscyllium*); carpet sharks (family Orectolobidae), including the Spotted wobbegong (*Orectolobus maculatus*); collared carpet sharks (family Parascyllidae), including the Necklace carpetshark (*Parascyllium variolatum*) and the Barbelthroat carpetshark (*Cirrhoscyllium expolitum*); nurse sharks (family Ginglymostomatidae), including the Nurse shark (*Ginglymostoma cirratum*) and the Tawny nurse shark (*Nebrius ferrugineus*); Whale shark (*Rhincodon typus*, sole species of the family Rhincodontidae). The Whale shark and the Bluegray carpet shark (*Heteroscyllium colcloughi*) are both classed as Vulnerable.

Catsharks and False catsharks
Order: Scyliorhiniformes

Worldwide in cold or deep marine water. About 87 species in at least 15 genera and 3 families including: catsharks (family Scyliorhinidae), including Lesser spotted dogfish (*Scyliorhinus canicula*), swell sharks (genus *Cephaloscyllium*); False catshark (*Pseudotriakis microdon*; family Pseudotriakidae), 2 species, one on both sides of the North Atlantic and one in the Western Pacific; finback catsharks (family Proscylliidae), including Cuban ribbontail catshark (*Eridacnis barbouri*), Graceful catshark (*Proscyllium habereri*), and the Slender smooth-hound (*Gollum attenuatus*). Four species are listed as Lower Risk/Near Threatened, including the Pyjama shark (*Poroderma africanum*).

Smooth dogfish sharks
Order: Triakiformes

Worldwide in tropical, subtropical, and temperate seas. 1 family – Triakidae – containing about 30 species in 9 genera, including: Whiskery shark (*Furgaleus macki*) of the eastern Indian Ocean; Longnose houndshark (*Iago garricki*) of the western Pacific; Leopard shark (*Triakis semifasciata*) of the eastern North Pacific, the Tope (Soupfin) shark (*Galeorhinus galeus*) and the Sailback houndshark (*Gogolia filewoodi*). The Whitefin tope shark (*Hemitriakis leucoperiptera*) is Endangered, while 3 species are Vulnerable, including the Tope shark, and 4 species are classed as Lower Risk/Conservation Dependent and 3 as Lower Risk/Near Threatened.

Goblin sharks and allies
Order: Odontaspidiformes

Worldwide in tropical and temperate seas. 7 species in 4 genera and 3 families, including: False sand tiger shark or Crocodile shark (*Pseudocarcharias kamoharrai*; sole species of the family Pseudocarchariidae); Goblin shark (*Mitsukurina owstoni*; sole species of the family Scapanorhynchidae); sand tiger sharks (family Odontaspididae), including Sand tiger shark (*Carcharias taurus*) and Bigeye sand tiger (*Odontaspis noronhai*). The Sand tiger shark is Vulnerable and the Crocodile shark is Lower Risk/Near Threatened.

Thresher sharks and allies
Order: Isuriformes

Worldwide in tropical and temperate seas. 10 species in 6 genera and 3 families including: mackerel sharks (family Isuridae), including Basking shark (*Cetorhinus maximus*), Great white shark or man-eater or White shark (*Carcharodon carcharias*), Shortfin mako shark (*Isurus oxyrinchus*), Porbeagle shark (*Lamna nasus*), Salmon shark (*Lamna ditropis*); Megamouth shark (*Megachasma pelagios*; Megachasmidae); thresher sharks (family Alopiidae), including Bigeye thresher (*Alopias superciliosus*). Both the Basking shark and the Great white shark are Vulnerable, while the Shortfin mako and the Porbeagle shark are Lower Risk/Near Threatened.

Requiem sharks
Order: Carcharhiniformes

Worldwide in tropical and temperate seas. 1 family – Carcharhinidae – containing about 100 species in 10 genera including: Blacktip reef shark (*Carcharhinus melanopterus*), Bull shark (*C. leucas*), Silvertip shark (*C. albimarginatus*), Oceanic whitetip shark (*C. longimanus*), hammerhead sharks (genus *Sphyrna*), Great hammerhead shark (*S. mokarran*), Scalloped hammerhead (*S. lewini*), Tiger shark (*Galeocerdo cuvier*), and Blue shark (*Prionace glauca*). The Ganges shark (*Glyphis gangeticus*) and the Pondicherry shark (*C. hemiodon*), along with several new species that are awaiting description, are Critically Endangered; 2 species are Endangered: the Borneo shark (*Carcharhinus borneensis*) and the Speartooth shark (*Glyphis glyphis*), and 2 are Vulnerable: the Smoothtooth blacktip (*Carcharhinus leiodon*) and the Sharptooth lemon shark (*Negaprion acutidens*).

Spiny dogfish and allies
Order: Squaliformes

Worldwide in cold or deep seas. 1 family – Squalidae – containing about 70 species in around 12 genera including: bramble sharks (genus *Echinorhinus*), Bramble shark (*E. brucus*), Prickly shark (*E. cookei*), cigar or cookiecutter sharks (genera *Isistius*, *Squaliolus*) including the Cookiecutter shark (*Isistius brasiliensis*), sleeper sharks, including the Pocket shark (*Mollisquama parini*), Common Spiny dogfish (*Squalus acanthias*). The Spiny dogfish is classed as Lower Risk/Near Threatened.

Angel sharks
Order: Squatiniformes

Worldwide in tropical and temperate seas. 1 family – Squatinidae – containing about 10 species of the genus *Squatina*, including: Angelshark (*S. squatina*), Hidden angelshark (*S. occulta*), Australian angelshark (*S. australis*), African angelshark (*S. africana*), and the Angular angelshark (*S. guggenheim*). The Hidden angelshark (*S. occulta*) is Endangered and 2 species are Vulnerable – the Angelshark and the Angular angelshark.

Sawsharks
Order: Pristiophoriformes

Bahamas, off coast of S Africa, western Pacific Ocean from Japan to Australia. 1 family – Pristiophoridae – containing 5 species in 2 genera: Sixgill sawshark (*Pliotrema warreni*), Longnose sawshark (*Pristiophorus cirratus*), Shortnose sawshark (*P. nudipinnis*), Japanese sawshark (*P. japonicus*), Bahamas sawshark (*P. schroederi*).

Note: Nelson 1994 and the Fishbase database (www.fishbase.org) list only 8 orders of sharks. The Scyliorhiniformes and the Triakiformes are subsumed within the Carcharhiniformes, and the Frilled shark within the Hexanchiformes, while the Odontaspidiformes and Isuriformes merge to form a new order, the Lamniformes.

Right *A Scalloped hammerhead shark* (Sphyrna lewini) *being cleaned by two Angelfishes. Widespread throughout the world's warmer oceans, hammerheads are often caught in commercial fishing, either intentionally (their fins are highly prized) or as bycatch.*

All *Carcharhinus* species are widespread, and in summer some migrate long distances into temperate waters. They have a metallic gray or brown dorsal coloration. Some, however, have the edges of their fins tipped with white or black, hence the names, Silvertip, Whitetip, and Blacktip sharks.

The largest requiem shark, the Tiger shark, reaches over 6m (20ft) in length and is unquestionably one of the most dangerous of all sharks. As a scavenger it will swallow anything that it can get down its throat – including shoes, cans, birds, and human body parts. Juvenile Tiger sharks have dark bands on a silvery gray background, a coloration from which their name derives, but these bands fade with age.

The hammerhead sharks are so called because of the large lateral expansions of their heads with their eyes set on the ends. Except for their unique heads, which define the genus (or, possibly, two genera), they are typical requiem sharks. It has been mooted that their hammer-shaped heads help streamline their bodies, or give them a better field of vision, but more research suggests that the elongate head contains extra electrodetectors, the Ampullae de Lorenzini.

Hammerhead sharks are often seen moving their heads from side to side like a metal detector above sandy bottoms and then burrowing into the sand to grab hidden fishes – mostly stingrays; they also make regular migrations following Earth's magnetic field. The Great hammerhead shark is the largest, growing to more than 5m (15ft) long while the Scalloped hammerhead is the one most seen by divers.

Spiny Dogfish and Allies

ORDER SQUALIFORMES

The spiny dogfish sharks are cold-water forms, worldwide in distribution. All develop eggs viviparously, producing about 12 young per litter. In size they range from less than 30cm (1ft) to over 6m (20ft). Many, especially the deepwater species, feed on squid and octopus.

In the North Atlantic, the Common Spiny dogfish – also known as the Pickled Dogfish or Spurdog – is an important food fish. Tens of millions are caught every year and stocks are collapsing. Spiny dogfish rarely exceed 1m (3ft) in length, travel in schools and migrate long distances, moving into Arctic waters each summer. Each dorsal fin is preceded by a spine, which has venom-producing tissue at its base. The venom is painful to humans, but not fatal.

Many deepwater species, especially in the genus *Etmopterus*, have light-producing organs along the sides of their body, possibly attracting their deepwater squid prey as well as providing camouflage through counter-illumination. Their large eyes are very sensitive at low light levels.

The small but very thin cigar sharks (especially genus *Isistius*) are equipped with greatly elongated teeth in their lower jaw. They swim up to a larger animal (a fish, squid, or even a cetacean), bite it, and then, with a twist of the body, cut out a perfectly circular piece of flesh from their prey. This feeding technique has given rise to their alternative common name – cookiecutter sharks.

Sleeper sharks, the giants among spiny dogfish sharks, are the only sharks that permanently inhabit Arctic waters, often under the ice. They feed on seals and fishes, and are thought to be the only sharks with flesh that is poisonous to both humans and dogs.

The bramble sharks are unusual in having extremely large, flat dermal denticles ("teeth") widely dispersed over the skin, giving them a "brambly" appearance. There are probably two species, one in the Atlantic (*Echinorhinus brucus*) and one in the Pacific (*Echinorhinus cookei*). Although large, over 2.7m (9ft) long, their skeleton is not calcified, so it is extremely soft.

Thresher, Mackerel, and Megamouth Sharks

ORDER ISURIFORMES

The thresher, mackerel, and megamouth sharks are among the largest sharks in the world and are found in tropical and temperate seas.

The thresher sharks derive their name from the extremely long, thin upper lobe of their caudal fin, which may be as long as the rest of their body. Swimming into a school of small fishes, they are believed to use their tail like a whip, thrashing it among the school, killing or stunning the fishes, which are then eaten. They grow to some 6m (20ft) and give birth to only a few young, but the pups of the largest species are about 1.5m (5ft) long.

An exciting shark discovery was that of the aptly named Megamouth, which was first captured off Hawaii in November 1976. Over 5m (16ft) long, it has now been found off Japan, Indonesia, the Philippines, the USA, Brazil, and Senegal. The Megamouth shark is a filter feeder

that migrates vertically, spending the day at depth and rising to about 12m (40ft) to feed at night. It has been speculated that the lining of the roof of its mouth may attract prey with bioluminescence. Researchers think that it is distantly related to the Basking shark, but is sufficiently different to be considered separately. The Megamouth is preyed upon by the Cookiecutter shark (*Isistius brasiliensis*); it is also feared that, with the growth in deep-water fishing, this species will increasingly be taken as bycatch.

The family of mackerel sharks contains some of the best-known sharks, including the Porbeagle, Mako, Basking, and the unjustifiably notorious Great white sharks. They are large and most live in all tropical and temperate seas. The Basking shark can reach 10m (33ft) but lives only in temperate seas. All have an unusual caudal fin, with the lobes being nearly equal in length, caudal keels on either side of the tail, and are relatively fast swimmers; most species are fish eaters. Some species are known for breaching – making spectacular leaps into the air. The reason for this behavior is not known, but it has been suggested that it is an attempt to dislodge skin parasites. The Basking shark is known to have collided with boats during such leaps. Most, if not all, species are homoiothermic – that is, they maintain their body temperature above that of their surroundings.

Reaching over 6m (20ft) in length, the Mako shark is one of the fastest fish in the world and has been recorded swimming at speeds over 95km/h (60mph).

The world's most notorious shark by far is the Great white shark – also called the White pointer, Blue pointer, Man-eater, or simply the White shark. It is the species most often cited in references to shark attacks on humans, although many of these should be ascribed to the Tiger shark and the Bull shark (see Shark Attacks). Feeding primarily on marine mammals (the only shark to do so), its broad, serrated teeth are capable of

Above *Strips of Whale shark meat drying in the Philippines. For millions of people who depend on subsistence fishing, sharks are a plentiful and cheap source of protein. Exploitation for their meat, fins, and other byproducts has endangered many species.*

Below *The huge-mouthed Basking shark* (Cetorhinus maximus) *of the mackerel shark order inhabits temperate waters, feeding on plankton near the surface. Its gill rakers are well developed, while its teeth are vestigial. Despite its great size and terrifying appearance, it poses no threat to humans.*

biting large chunks of flesh from whales, seals, and sea lions. Known to reach 6.7m (20ft) in length, its average length is 4.5m (15ft). Reproduction is viviparous and the developing embryos are then nourished by eating unfertilized eggs (oophagy). Like many sharks, the Great White shark is countershaded; only the belly is white, the dorsal surface being blue-gray to gray-brown or bronze.

At around 2.7m (9ft) long, the Porbeagle and the Salmon shark (sometimes known as the Pacific porbeagle) are the smallest members of the mackerel shark family. They live in the Atlantic and Pacific oceans, respectively.

The Basking shark is second in size to the Whale shark. Commonly 10m (33ft) long, it is a filter feeder. Its teeth are minute, and modified gill rakers are used for sieving the plankton. Its liver yields vast amounts of oil and the fish has been the subject of local fisheries in the North Atlantic. The Basking shark's name comes from its behavior of swimming and resting at the surface.

Angelsharks

ORDER SQUATINIFORMES

The angelsharks are extremely distinctive in appearance; being very flat, they are considered to be more closely related to the skates and rays than to the more "typical" sharks. They grow to more than 1.8m (6ft) in length, and there are about 12–18 species in the genus *Squatina*, found in all tropical to temperate seas. An anterior lobe of the pectoral fins extends in front of their gill slits. They have long, thin teeth, and lie camouflaged in fairly shallow water, waiting for prey to swim by, before swiftly lunging out and capturing it in their highly protrusible jaws. Although normally lethargic, they can move very rapidly when catching prey. The angelsharks are viviparous, producing about 10 young per litter.

Sawsharks

ORDER PRISTIOPHORIFORMES

Having long, flat, bladelike snouts edged with teeth of varying sizes, sawsharks bear a striking resemblance to sawfishes; however, they are true sharks. They are quite rare and grow to a length of about 1.8m (6ft). One species, the Sixgill sawshark (*Pliotrema warreni*), has an extra set of gills. The genus *Pristiophorus* has seven species, most of which inhabit the western Pacific and Southwestern Indian Ocean, but also including one Atlantic species found in deep water off the Bahamas, Cuba, and Florida. They have a pair of long thin barbels under the sawlike rostrum, which help them find their mollusk and crustacean prey on the seafloor. Their teeth are flat and broad for crushing and their "saws" appear to be used only for defense. Sawsharks are viviparous, producing 3–22 young per litter with the saw-teeth folded back at birth, thus preventing injury to the mother. JJ/GDi

SHARK ATTACKS

The most famous characteristic of sharks is their alleged propensity to kill and eat humans. But very few shark species have been implicated in unprovoked attacks. The idea that sharks are waiting for people to enter the water so that they can attack them is false. To put it in perspective, every year more people are killed by beestings, 10 times more people are killed by lightning, and thousands more are killed by other people or in automobile accidents than by sharks. The International Shark Attack File authenticates 70–100 unprovoked shark attacks annually, with 5–15 fatalities.

The Great white shark has the worst reputation for attacking humans. Yet Great whites feed mainly on sea mammals. The shark usually surprises its prey with one huge bite, and then retreats to allow its victim to die. For this reason, many humans survive the attack of a Great white shark if they are saved before being consumed. Death, however, may result from massive blood loss or damage to organs.

Attacks by other species of sharks are for feeding. Statistically, the next most dangerous sharks are the Tiger, Bull, and Sand tiger shark. Divers dive unmolested among shoals of 300 Scalloped hammerhead sharks, although there is thought to be a small risk from Great Hammerhead sharks. All predatory sharks feed most actively at dawn and dusk and most sharks will scavenge when the opportunity occurs. Their sensors home in on the smell of dead fishes being carried by spear-fishermen, as well as the vibrations from dying, speared, or hooked fishes and fishes that are already under attack by other predators, so many attacks are linked to these events. Similarly, the vibrations given by bathers splashing on the surface will attract sharks. To any predatory fish, any other fish on the surface would appear to be one that is in trouble and therefore easy prey. Divers that are submerged are at much less risk than divers or swimmers on the surface.

Despite their proven sensitivity to blood, sharks are heavily attracted to decaying flesh; divers and fishermen chum for sharks with fish oil, dead fishes, frozen fishes, or horsemeat. Problems linked to this occur when bathers and surfers are in the water in areas near to where abattoirs discharge their waste or fishermen clean their catch. In recent years, divers have ventured near or into bait-balls – tightly knit shoals of fishes that are already under attack by sharks or dolphins – so it is not surprising that some of them have been bitten in error. Similarly, changes in current patterns sometimes bring shoals of fishes close inshore – aerial shots taken off Florida showed bathers frolicking among such shoals but oblivious to the sharks that were feeding on them. Another common fallacy is that sharks will not attack in the presence of dolphins.

Every year more and more divers enter the water and many of them go out of their way to get close to sharks. This activity should increase the chances of a shark attack but it does not happen. The reverse is true and more and more people realize that if they are sensible about where they go and what they do, the chance of such an attack is rare.

A great deal of research has gone into preventing shark attacks. Various chemicals (including detergents and a skin secretion produced by the Red Sea Moses sole), striped wetsuits, all-enclosing bags, air bubbles, methods of slowing one's heartbeat, and electric fields have all been tried. Most measures are useless, while others will deter some species of sharks but attract others. Netting works well for all sizes of sharks off popular bathing beaches but often kills many of them. Heavy chain-mail suits are useful when hand-feeding sharks up to 2m (6ft) long, but the only sure way of avoiding attack by the larger predatory sharks is to be inside a sturdy cage. JJ/GDi

Below *A Great white shark inspects a caged diver off Australia. Attacks by this species may often be a case of mistaken identity – a swimmer or surfer, seen from below, looks much like a seal.*

Skates, Rays, and Sawfishes

THE MERE MENTION OF RAYS TO A MARINER conjures up images of gigantic breaching devilfishes, venomous stingrays, electric rays, or numbfishes, which can stun the unwary. All these reputations reflect truth and are based on anatomical adaptations, but what these relatives of sharks are really doing is responding defensively to human attempts to catch them. Skates and rays are edible and consumed by most fishing cultures.

Skates and rays are found worldwide; some species commute into brackish water and 18 species of the three genera of subfamily Potamotrygoninae live in freshwater rivers with Atlantic drainage in South America. Some species of sawfishes are known to swim up rivers, even as far as Lake Nicaragua.

The order is most closely related to the angelsharks and sawsharks. All have their pectoral fins extending well in front of the gill arches and fused to the sides of the head, and gill slits underneath the body. Mostly sedentary, feeding on mollusks, crustaceans, and small fishes, they have large openings (spiracles) positioned clear of the bottom that take in water clear of sand and silt and then pump it over the gills and out through the ventral gill slits.

Extraordinary Forms

SKATE, RAY, AND SAWFISH FAMILIES

Members of the sawfish family are easily recognized by the "saw" that emerges from the front of the body, a similar protuberance to that of sawsharks. It is used both to capture food and as a defensive weapon. Swimming into a school of fishes, sawfishes rapidly slash their saw back and forth, stunning or killing fishes in the school, then consuming any immobile individuals. The saw is also used to dig for prey hidden in the substrate.

The number of teeth on the saw varies with the species. The teeth within the jaw are short and flattened for crushing the shells of crustaceans and mollusks. Including the saw, the Green sawfish can measure over 7m (24ft). The Freshwater sawfish lives mostly in rivers and lakes but occasionally enters the sea. Female sawfishes develop their eggs internally (viviparous). Living in waters that are heavily fished and subject to pollution, most sawfish species, such as the Common sawfish, are in danger of extinction.

Right *Golden cownose rays* (Rhinoptera steindachneri) *make an eerie sight as they gather en masse off the Galápagos Islands. Shoals may contain hundreds of individuals.*

FACTFILE

Skates and Rays

Class: Chondrichthyes

Order: Rajiformes

Over 465 species in 62 genera and 12 families.

Distribution Worldwide in tropical, subtropical, and temperate waters.

Size Length 30cm–7.3m (1–24ft); width 10cm–6.7m (4 inches–22ft).

See Skate, Ray, and Sawfish families ▷

Species in the guitarfish and shovelnose ray families look like sawfishes without saws, but have larger pectoral fins. They range in length from about 75cm (2.5ft) for the Atlantic guitarfish to about 3m (10ft) in the Whitespotted guitarfish. The Skarkfin or Bowmouth guitarfish has an unusual mouth that undulates, rising and falling like waves. They feed in shallow water on mollusks and crustaceans and have flattened crushing teeth. All species develop their eggs internally.

Members of both *Rhinobatos* and *Rhynchobatus* have a fairly long front extension (rostrum), thus giving the front part of the body, the disk, a heart-shaped appearance. The remaining genera (*Rhina*, *Platyrhina*, *Zapteryx*, and *Platyrhinoides*) have much shorter rostra and the disk looks round. Many species have enlarged dermal denticles, often called thorns, on their dorsal surface.

All members of the electric ray and Lesser electric ray families produce electricity. Most live in shallow water although some occur at great depths. Slow swimmers, spending most of their time on the seabed, they feed on fishes and invertebrates, which they capture by stunning them with electric shocks. Their electric organs, composed of modified muscle tissue, are between the pectoral fins and the head, one on each side. The electric shock from these organs can reach 220 volts and is also used in defense. In ancient Greece and Rome, the shocks of the species *Torpedo nobiliana* were used as a treatment for maladies such as headache.

The Lesser electric ray grows to about 30cm (1ft), while the Atlantic torpedo ray grows to over 2m (6ft) long. Their disks are round, the tail short and stubby in most species, and the eyes usually small. In some species, e.g. the Blind Numbfishes from deep water off New Zealand, the eyes are very poor, the fish use electroreceptors to see. The skin is entirely free of scales in all species and is often beautifully marked. For reproduction, eggs develop internally until hatching.

The largest group of skates and rays is the Rajidae family, with over 250 species. They are found worldwide in cool waters – even in the tropics they are common in deep, cold water at depths greater than 2,100m (7,000ft). Always closely associated with the seabed where they hide in the sand or mud with only their eyes and spiracles protruding, they feed mostly upon mollusks and crustaceans, although they occasionally catch fishes. The smallest species, the Little skate, which is common off the Atlantic coast of North America, reaches only about 50cm (20in) long. The Big skate from the Pacific coast of North America has been recorded at over 2.5m (8ft).

The enlarged pectoral fins of species in this family and their fairly long snout give the disk a diamond shape. The enlarged pectoral fins of stingrays, eagle rays, and manta rays as well as skates are often called wings; and from their graceful up-and-down movements in swimming, it is easy to see why. In some skates the pelvic fins have been greatly enlarged and elongated and they can use these to "walk" over the seabed. The tails of skates have weak electric organs; long, thin, and covered with strong, sharp thorns, they are used as a defense weapon. Most species have

Skate, Ray and Sawfish Families

Eagle rays and Mantas
Family Myliobatidae

Worldwide in tropical and subtropical oceans. About 42 species in 7 genera including: **Spotted eagle ray** *(Aetobatus narinari)*, **Bat ray** *(Myliobatis californica)*, **Manta ray** *(Manta birostris)*.

Electric rays
Family Torpedinidae

Tropical and subtropical oceans, mainly in shallow water. 14 species in 2 genera including: **Atlantic torpedo ray** *(Torpedo nobiliana)*, **Marbled electric ray** *(T. sinuspersici)*.

Lesser electric rays and Allies
Family Narcinidae

Tropical and subtropical oceans, mainly in shallow water. About 24 species in 9 genera including: **Lesser electric ray** *(Narcine brasiliensis)*, **Blind numbfish** *(Typhlonarke aysoni)*.

Guitarfishes
Family Rhinobatidae

Tropical and temperate areas of the Atlantic and Indo-Pacific oceans. About 45 species in 7 genera including: **Whitespotted guitarfish** *(Rhynchobatus djiddensis)*, **Atlantic guitarfish** *(Rhinobatos lentiginosus)*. Conservation status: **Brazilian guitarfish** *(Rhinobatos horkeli)* Critically Endangered.

Sawfishes
Family Pristidae

Tropical/subtropical oceans in coastal, estuarine, and freshwater. c. 6 species and 2 genera, including: **Common sawfish** *(Pristis pristis)*, **Freshwater sawfish** *(P. microdon)*, **Green sawfish** *(P. zjisron)*, **Narrow sawfish** *(Anoxypristis cuspidata)*. Conservation status: 2 species Critically Endangered – **Largetooth sawfish** *(P. perotteti)* and Common sawfish.

Skates
Family Rajidae

Worldwide, mostly in deepsea waters but a few species are found in shallow, inshore waters. Over 250 species in at least 18 genera including: **Big skate** *(Raja binoculata)*, **Little skate** *(R. erinacea)*. Conservation status: 2 species are Endangered, including the **Blue** or **Common skate** *(Dipturus batis)*, 1 is Vulnerable and 3 are Lower Risk/Near Threatened.

Stingrays
Family Dasyatidae

Tropical and subtropical oceans. About 70 species in c. 9 genera. 3 genera in S America inhabit fresh water. Species include: **Atlantic stingray** *(Dasyatis sabina)*, **Southern stingray** *(D. americana)*, **Manzana ray** *(Paratrygon aireba)*. Conservation status: 5 species Endangered, including the **Pincushion ray** *(Urogymnus ukpam)*, and 3 species are Vulnerable.

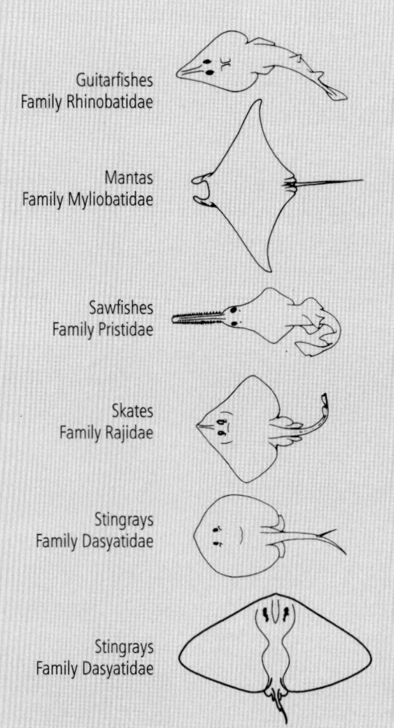

Deepwater stingray
Family Plesiobatidae

Mozambique, Hawaii, and (perhaps) S China. Sole species *Plesiobatis daviesi*.

Sixgill stingrays
Family Hexatrygonidae

Off S Africa, W Pacific, from Hong Kong to Japan. 6 species in 1 genus, including: **Sixgill stingray** *(Hexatrygon bickelli)*.

Round stingrays
Family Urolophidae

W Atlantic, E Indian, and Pacific Oceans. About 40 species in 3 genera, including: **Spotted round ray** *(Urobatis maculatus)*.

Butterfly rays
Family Gymnuridae

Atlantic, Indian, and Pacific Oceans; some species enter estuaries. About 14 species in 2 genera, including: **California butterfly ray** *(Gymnura marmorata)*.

Shovelnose and Shark rays
Family Rhinidae

Indo-West Pacific Ocean. About 6 species in 2 genera, including: **White-spotted shovelnose ray** *(Rhynchobatus djiddensis)*, **Shark ray** *(Rhina ancylostoma)*.

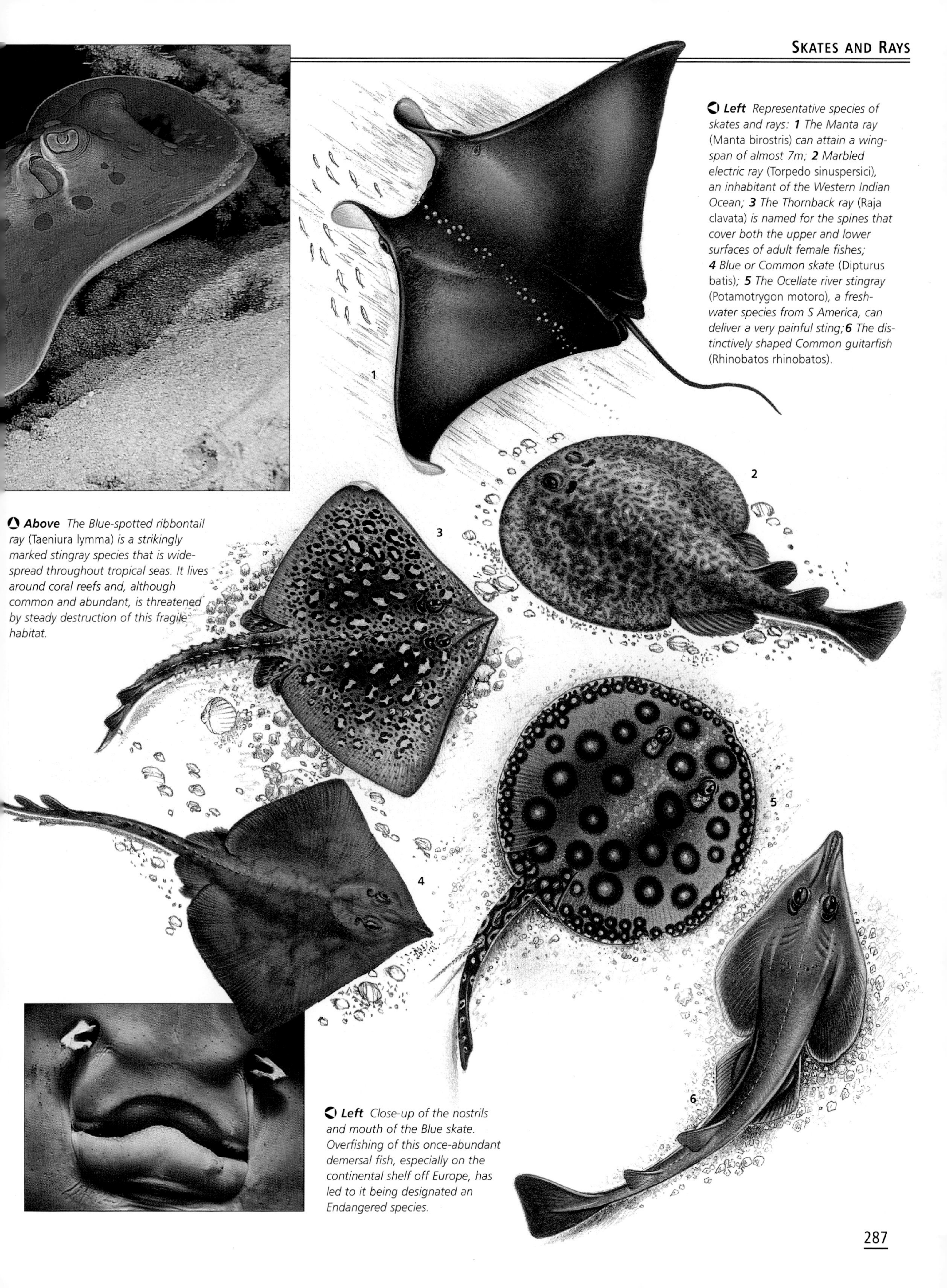

Left Representative species of skates and rays: **1** The Manta ray (Manta birostris) can attain a wingspan of almost 7m; **2** Marbled electric ray (Torpedo sinuspersici), an inhabitant of the Western Indian Ocean; **3** The Thornback ray (Raja clavata) is named for the spines that cover both the upper and lower surfaces of adult female fishes; **4** Blue or Common skate (Dipturus batis); **5** The Ocellate river stingray (Potamotrygon motoro), a freshwater species from S America, can deliver a very painful sting; **6** The distinctively shaped Common guitarfish (Rhinobatos rhinobatos).

Above The Blue-spotted ribbontail ray (Taeniura lymma) is a strikingly marked stingray species that is widespread throughout tropical seas. It lives around coral reefs and, although common and abundant, is threatened by steady destruction of this fragile habitat.

Left Close-up of the nostrils and mouth of the Blue skate. Overfishing of this once-abundant demersal fish, especially on the continental shelf off Europe, has led to it being designated an Endangered species.

spines or thornlike structures or bucklers on the dorsal surface of the body.

Typical skates (Rajidae) have two dorsal fins on the tail, Arynchobatidae have one, and Anacanthobatidae none. Skate egg cases are leathery oblong capsules with stiff pointed horns at the corners, often washed up on beaches, they are called mermaid's purses. Like sharks, skates and rays live longer and produce fewer young than most fishes, so they are particularly vulnerable to overfishing. Heavily fished for their tasty wings, populations are decreasing rapidly in many parts of the world.

Stingrays are named for the one or more spines on the dorsal side of their tail. They are found worldwide in warm tropical and subtropical waters, some migrating into temperate waters in the summer. They spend a lot of time camouflaged on the seabed, often partially covered by sand. They can swim rapidly when disturbed, or in pursuit of fish. They also eat mollusks and crustaceans, and have flat, crushing teeth. The disk can be diamond shaped or almost round, causing a distinction between the round stingrays and the square stingrays. Two genera are called butterfly rays (*Gymnura* and *Aetoplatea*) with reference to

Above *Exquisitely patterned and hydrodynamically streamlined, a Spotted eagle ray* (Aetobatus narinari) *glides through the waters of the Caribbean.*

their wide wings and short, stubby tails.

The spines on stingrays' tails have venom sacs and are used for defense, the fish striking out with their tails if stepped on. Each spine has angled barbs, which allow easy penetration, but make it very difficult to remove. Sharks, especially hammerheads, prey on stingrays and are seemingly immune. The venom is painful but rarely fatal to humans. Like most fish venoms, it is a high-molecular-weight protein that is easily broken down by heat; the wound should be immersed in water that is as hot as one can stand for 60–90 minutes. When wading in shallow water, one should shuffle rather than lift one's feet .

The Atlantic stingray measures only about 30cm (1ft) across its disk. The largest species is the Indo-Pacific Smooth stingray, which has a total length of over 4.5m (15ft) and a disk width of over 2m (7ft). Several deaths have been attributed to this species in Australia. Even very large specimens lie in shallow water and impale swimmers in the chest or abdomen with their large spine, which can be 30cm (1ft) long.

Although most species are marine, some South American genera live only in freshwater (e.g. *Potamotrygon* and *Paratrygon*). Their osmoregulatory physiology has adapted so completely to freshwater that they rapidly die in salt water, their upper limit of salt tolerance being about 50 percent that of seawater. They have an almost perfectly circular disk and are all beautifully marked with spots and bars. Lying in shallow water, covered by mud, they are feared for their poisonous spines. Like all other stingrays, the females develop their eggs internally, with up to twelve young per litter.

The three genera (*Myliobatus*, *Aetobatus,* and *Rhinoptera*) of eagle rays – named for their large pectoral wings – are found worldwide in tropical and subtropical seas. Their wingspan can reach 2.4m (8ft). They have no frontal protrusion (rostrum), giving them a pug-nosed appearance. Their whiplike tail may be more than twice the length of their disk, with one or more spines at its base. The wings taper and are pointed at the tips, and both the eyes and spiracles are large. The teeth are fused into large crushing plates. They eat shellfish, which they find by squirting water from the mouth and blowing away the sand. Eagle rays can swim fast enough to breach and glide through the air.

The manta rays are the giants of the skates and rays and have one of the largest brains among fishes. Although stingless, they are closely related to the stingrays. The manta, whose name comes from the Spanish word for blanket or cloak, reaches a wingspan of over 6.7m (22ft), while the smallest juveniles have wingspans of little more than 1m (40 in). The wingspan of mobulas (devil rays) is 1–3m (3–10ft). The mouth of mantas is located at the front of the head and the teeth are only in the lower jaw, while that of mobulas is situated slightly beneath the head and the teeth are in both jaws. Mobulas are primarily pelagic.

Giant filter feeders, mantas have two extensions of the pectoral fins called cephalic lobes that project forward from the head and are used to funnel plankton and fish fry into the mouth. These lobes resemble horns – hence the name devilfish, although this name also alludes to their habit of rubbing against anchor lines to remove skin parasites, causing the anchors to drag. Usually solitary except when breeding, they can be pelagic but prefer areas near land. When food is plentiful they stay inshore around a single reef. As with Basking sharks, when plankton rises to the surface in the afternoon, they feed at the surface. Like eagle rays, they can swim at great speed and make spectacular breaches, possibly to dislodge parasites. Reproduction is ovoviviparous, the one or two young hatch from eggs inside the mother and are nourished with a milky fluid from projections in the uterus until born alive. Mantas have been fished commercially in the Philippines and the Sea of Cortez and are often caught in drift-nets. JJ/GDi

Chimaeras

Left *It is easy to see how the Ratfish* (Chimaera monstrosa) *earned its common name.*

A LARGE, BLUNT HEAD, AN ERECTABLE spine in front of the first dorsal fin, and a gill cover over the four gills that leaves a single opening are some characteristics of the chimaeras. Named for the she-monster of Greek mythology, which had a lion's head, a goat's body, and a serpent's tail, these unusual relations of the sharks are also known as ratfish or rabbitfish; one of the generic names (Hydrolagus) *literally means "water* (hydros) *rabbit* (lagus)*".*

Chimaeras live in cold water, often at great depths – some have been recorded as deep as 2,400m (8,000ft). They are poor swimmers. Instead of using powerful side-to-side body movements like most fishes, especially sharks, they swim by flapping their pectoral fins, which makes them bob up and down in a clumsy fashion. Chimaeras usually keep close to the seafloor and have been observed motionless on the bottom, perched on the tips of their fins.

Like all cartilaginous fishes, the males have pelvic claspers to introduce sperm into the female. However, male chimaeras also have a second pair of retractable claspers in front of the pelvic fins that are probably used to hold the female during mating. Mature males also have a clasper on the forehead. Long-nosed and plow-nosed chimaeras have elongated, fleshy, and flexible head projections (rostra), which are covered with electrical and chemical receptors.

Chimaeras are durophagous – a term meaning that they habitually feed on items that are hard. Their teeth are fused together to form three crushing plates, one in the lower jaw and two in the upper jaw, which are used to crush the shells of their food, mollusks, crustaceans, and a few small fishes. Their large eyes are adaptations to low ambient light.

Young chimaeras are covered by short, stout dermal denticles (minute teeth), which are lost as they mature, except in long-nosed chimaeras, which keep some of them for life. All chimaeras lay fairly large eggs, 15–25cm (6–10in) with a hard leathery shell that hatch in 6–8 months.

Most sharks take in water through the mouth to pass over the gills but chimaeras take in water through large nostrils connected to special channels, which direct the water to the gills. Unlike sharks, chimaeras have their gills in a common chamber protected by a flap (operculum).

The flesh of chimaeras is eaten mostly in Australia, New Zealand, and China, although the fillets are best pre-soaked in freshwater to remove the slight taste of ammonia. In the past, chimaera liver was used as a source of machine oil. JJ/GDi

Below *The Plownose chimaera or Ghost shark* (Callorhincus milii) *occurs in the southwestern Pacific Ocean, off New Zealand and Southern Australia. It uses its distinctive plow to search out its main food item, shellfish.*

FACTFILE

CHIMAERAS

Class: Chondrichthyes

Order: Chimaeriformes

31 species in 6 genera and 3 families.

Distribution Subarctic, subantarctic, temperate and tropical waters.

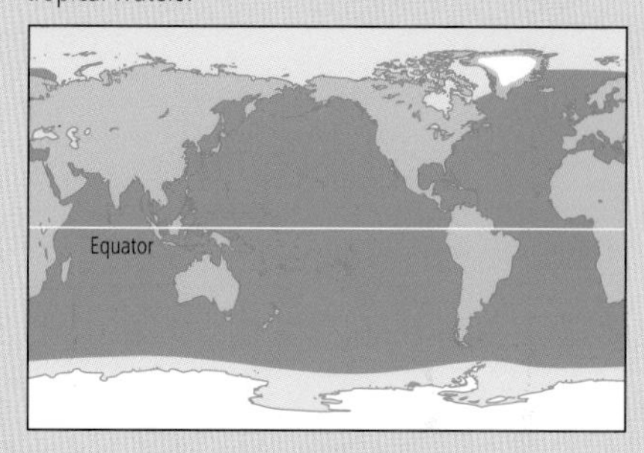

Size Length 60–90cm (2–3ft) in most species to 1.8–2.2m (6–8ft) in *Chimaera monstrosa* and *Hydrolagus purpurescens*.

SHORT-NOSED CHIMAERAS
Family Chimaeridae
21 species in 2 genera, *Chimaera* and *Hydrolagus*. Atlantic, Pacific, and Indian oceans.

LONG-NOSED CHIMAERAS
Family Rhinochimaeridae
6 species in 3 genera, *Rhinochimaera*, *Harriotta*, and *Neoharriotta*. Scattered localities worldwide in temperate and tropical seas.

PLOW-NOSED CHIMAERAS OR ELEPHANT FISHES
Family Callorhinchidae
4 species of the genus *Callorhinchus*. Southern Hemisphere.

Glossary

FISHES

Adaptation features of an animal that adjust it to its environment. Adaptations may be genetic, i.e. produced by evolution and hence not alterable within the animal's lifetime, or they may be phenotypic, i.e. produced by adjustment on the behalf of the individual and may be reversible within its lifetime.

Adipose fin a fatty fin behind the rayed DORSAL FIN, normally rayless (exceptionally provided with a spine or pseudorays in some catfish).

Adult a fully developed and mature individual, capable of breeding but not necessarily doing so until social and/or ecological conditions allow.

Air bladder see SWIMBLADDER.

Algae very primitive plants, e.g. epilithic algae, algae growing on liths (i.e. stones).

Ammocoetes the larval stage of the lamprey.

Anadromous of fishes that run up from the sea to spawn in freshwater.

Benthic the bottom layer of the marine environment.

Biotic community a naturally occurring group of plants and animals in the same environment.

Breaching leaping clear of the water.

Brood sac or pouch a protective device made from fins or plates of one or other parent fish in which the fertilized eggs are placed to hatch in safety.

Cartilage gristle.

Caudal fin the tail fin.

Caudal peduncle a narrowing of the body in front of the caudal fin.

Cerebellum a part of the brain.

Cilia tiny hairlike protrusions.

Class a taxonomic level. The main levels (in descending order) are Phylum, Class, Order, Family, Genus, Species.

Cogener a member of the same genus.

Colonial living together in a colony.

Colony a group of animals gathered together for breeding.

Conspecific a member of the same species.

Cryptic camouflaged and difficult to see.

Ctenoid scales of "advanced" fishes that have a comblike posterior edge, thereby giving a rough feeling.

Cutaneous respiration breathing through the skin.

Cycloid scales with a smooth posterior (exposed) edge.

Denticle literally a "small tooth;" used of dermal denticles, i.e. toothlike scales (all denticles are dermal in origin).

Diatoms small planktonic plants with silicaceous tests (shells).

Dimorphism the existence of two distinctive forms.

Disjunct distribution geographical distribution of taxons that is marked by gaps. Many factors may cause it.

Display any relatively conspicuous pattern of behavior that conveys specific information to others, usually to members of the same species; often associated with courtship but also e.g. threat displays.

Dorsal fin the fin on the back.

Ecology the study of plants and animals in relation to their natural environmental setting. Each species may be said to occupy a distinctive ecological NICHE.

Endostyle a complex hairy (ciliated) groove that forms part of the feeding mechanism of the larval lamprey.

Epigean living on the surface. See also HYPOGEAN.

Esca the luminous lure at the end of the ILLICIUM (the fishing rod) of the anglerfishes.

Family either a group of closely related species or a pair of animals and their offspring.

Feces excrement from the digestive system passed out through the anus.

Fin in fishes the equivalent of a leg, arm, or wing.

Fin girdles bony internal supports for paired fins.

Ganoid scales a primitive type of thick scale.

Gape the width of the open mouth.

Genus (plural genera) a taxonomic division superior to species and subordinate to family (see TAXONOMY).

Gills the primary respiratory organs of fish. Basically a vascularized series of slits in the PHARYNX allowing water to pass and effect gas exchange. The gills are the bars that separate the gill slits.

Gill slits the slits between the gills that allow water through.

Gular plates bony plates lying in the skin of the "throat" between the two halves of the lower jaw in many primitive and a few living bony fishes.

Heterocercal a tail shape in which the upper lobe is longer than the lower and into which the upturned backbone continues for a short distance.

Hypogean living below the surface of the ground, e.g. in caves.

Illicium a modified dorsal fin ray in anglerfishes, which is mobile and acts as a lure to attract prey.

Introduced of a species that has been brought from lands where it lives naturally to lands where it has not previously lived. Some introductions are natural but some are made on purpose for biological control, farming, or other economic reasons.

Invertebrate animals lacking backbones, e.g. insects, crustacea, coelenterates, worms of all varieties, echinoderms etc.

Krill small shrimplike marine crustaceans that are an important food for certain species of seabirds, whales, and fishes.

Lamellae platelike serial structures (e.g. gill lamellae) usually of an absorbent or semipermeable nature.

Larva a pre-adult form unlike its parent in appearance.

Lateral-line organs pressure-sensitive organs lying in a perforated canal along the side of the fish and on the head.

Marine living in the sea.

Maxillary bone the posterior bone of the upper jaw. Tooth-bearing in primitive fish, it acts as a lever to protrude the tooth-bearing anterior bone (premaxilla) in advanced fish.

Metamorphosis a dramatic change of shape during the course of ontogeny (growing up). Usually occurs where the adult condition is assumed.

Mollusk a shellfish.

Monotypic the sole member of its genus.

Natural selection the process whereby individuals with the most appropriate ADAPTATIONS are more successful than other individuals, and hence survive to produce more offspring. To the extent that the successful traits are heritable (genetic) they will therefore spread in the population.

Neoteny a condition in which a species becomes sexually mature and breeds while still in the larval body form, i.e. the ancestral adult body stage is never reached.

Niche the position of a species within the community, defined in terms of all aspects of its lifestyle (e.g. food, competitors, predators, and other resource requirements).

Olfactory sac the sac below the nostrils containing the olfactory organ.

Opercular bones the series of bones including the operculum (gill flap) and its supports.

Operculum the correct name for the bone forming the gill flap.

Order a level of taxonomic ranking. See CLASS and Introduction.

Osmosis the tendency for ions to flow through a semipermeable membrane from the side with the greatest concentration to the side with the least. Thus, in the sea, fish fluids pick up ions and have to get rid of them, while retention of vital ions is essential in freshwater.

Oviparous egg-laying.

Ovipositor a tube by which eggs are inserted into small openings and cracks.

Ovoviviparity the retention of eggs and hatching within the body of the mother.

Pelagic the upper part of the open sea, above the BENTHIC zone

Pelvic girdle the bones forming the support for the pelvic fins.

Perianal organ an organ around the anus.

pH a measure of the acidity or alkalinity of water: pH7 is neutral; the lower the number the more acid the water, and vice versa.

Pharyngeal teeth teeth borne on modified bones of the gill arches in the "throat" of the fish.

Pharynx that part of the alimentary tract that has the gill arches.

Photophore an organ emitting light.

Piscivore fish eater.

Plankton very small organisms and larvae that drift largely passively in the water.

Predator an animal that forages for live prey; hence "anti-predator behavior" describes the evasive actions of prey.

Predator an animal that forages for live prey; hence "anti-predator behavior" describes the evasive actions of the prey.

Prehensile capable of being bent and/or moved.

Rostrum snout.

Scale a small flat plate forming part of the external covering of a fish; hence deciduous scale, a scale that easily falls off the fish.

Scutes bony plates on or in the skin of a fish.

Solitary living on its own, as opposed to social or group-living in lifestyle.

Spawning the laying and fertilizing of eggs, sometimes done in a spawning ground.

Specialist an animal whose lifestyle involves highly specialized strategems, e.g. feeding with one technique on a particular food.

Species a taxonomic division subordinate to genus and superior to subspecies. In general, a species is a group of animals similar in structure and which are able to breed and produce viable offspring. See TAXONOMY.

Spiracle a now largely relict GILL SLIT lying in front of the more functional gill slits.

Subcutaneous canal a canal passing beneath the skin.

Subfamily a division of a FAMILY.

Suborder a subdivision of an order.

Subspecies a recognizable subpopulation of a single species, typically with a distinct geographical distribution.

Swimbladder or air bladder A gas- or air-filled bladder lying between the gut and the backbone. The swimbladder may be open via a duct to the PHARYNX so that changes of pressure can be accommodated by exhalation or inhalation of atmospheric air. If closed, gas is secreted or excreted by special glands. Its main function is buoyancy but it can also be used in some species for respiration, sound reception, or sound production.

Taxonomy the science of classifying organisms. It is very convenient to group together animals which share common features and are thought to have common descent. Each individual is thus a member of a series of ever-broader categories (individual–species–genus–family–order–class–phylum) and each of these divisions can be further divided where it is convenient (e.g. subspecies, superfamily, or infraorder). The SPECIES is a convenient unit in that it links animals according to an obvious criterion, namely that they interbreed successfully. However, the unit on which NATURAL SELECTION operates is the individual: it is by the differential reproductive success of individuals bearing different characteristics that evolutionary change proceeds.

Teleosts a group of fishes, defined by particular characters. The most familiar fishes are almost all teleosts.

Temperate zone an area of climatic zones in mid-latitude, warmer than the northerly areas, but cooler than subtropical areas.

Territory an area defended from intruders by an individual or group. Originally the term was used where ranges were exclusive and obviously defended at their borders. A more general definition of territoriality allows some overlap between neighbors by defining territoriality as a system of spacing wherein home ranges do not overlap randomly – that is, the location of one individual or group's home range influences that of others.

Tropics strictly, an area lying between 22.5°N and 22.50°S. Often, because of local geography, animals' habitats do not match this area precisely.

Tubercles small keratinized protrusions of unknown and doubtless different functions, which are either permanent or seasonally or irregularly present on the skin of fishes.

Type the species on which the definition of a genus depends.

Vascularized possessed of many small, usually thin-walled, blood vessels.

Velum a hood around the mouth of larval lampreys (ammocoetes) that is a feeding adaptation.

Ventral on the lower or bottom side or surface: thus ventral or abdominal glands occur on the underside of the abdomen.

Vertebrate an animal with a backbone. primitively consisting of rigidly articulating bones.

Vestigial a characteristic with little or no contemporary use, but derived from one which was useful and well developed in an ancestral form.

Villi small hairlike processes that often have an absorptive function.

Viviparous producing live offspring from within the body of the mother.

Vomerine teeth teeth carried on the vomer, a median bone near the roof of the mouth.

Weberian apparatus a modification of the anterior few vertebrae in ostariophysan fishes (carps, catfish, characins, etc) that transmit sound waves as compression impulses from the SWIMBLADDER to the inner ear, thereby enabling the fish to hear.

AQUATIC INVERTEBRATES

Abdomen a group of up to 10 similar segments, situated behind the THORAX of crustaceans and insects, which in the former group may possess appendages.

Abyss, abyssal the part, or concerning the part, of the ocean, including the ocean floor, that extends downward from some 4,000m (13,000ft).

Acoelomate having no COELOM (main body cavity).

Acrorhagi groups of NEMATOCYSTS.

Adaptation a characteristic that enhances an organism's chances of survival in the environment in which it lives, in comparison with the chances of a similar organism lacking the same characteristic.

Adult a fully developed and mature individual capable of breeding, but not necessarily doing so until social and/or ecological conditions allow.

Americ having a body not divided into SEGMENTS.

Amphipod a CRUSTACEAN of the invertebrate order amphipoda. Includes many freshwater and marine shrimps.

Ampulla a small contractile fluid reservoir associated with the tube-feet of some echinoderms.

Anabiosis suspended animation, with a low metabolic rate enabling an animal to survive adverse environmental conditions, particularly desiccation.

Ancestral stock a group of animals usually showing primitive characteristics, which is believed to have given rise to later more specialized forms.

Anisogamous of reproduction, involving gametes of the same species that are unalike in size or in form.

Antennae the first pair of head appendages of uniramians and the second pair of crustaceans.

Antennules the first pair of head appendages of crustaceans.

Apophysis an outgrowth or process on an organ or bone.

Aquatic associated with, or living in water.

Arthropod an invertebrate, such as an insect, spider, or crustacean, which is TRIPLOBLASTIC and COELOMATE, with a chitinous, jointed EXOSKELETON, paired, jointed limbs, and a lack of NEPHRIDIA and CILIA; in some classifications, a member of the phylum Arthropoda; here includes the members of the phyla Crustacea, Chelicerata, Uniramia, Tardigrada, Pentastomida, Onychophora.

Arachnid a member of the Arachnida, a class of the phylum Chelicerata, which includes spiders, scorpions, mites, and ticks.

Asexual reproduction reproduction that does not include fertilization (exchange of GAMETES) or MEIOSIS. See BINARY FISSION: BUDDING; GEMMULE; PARTHENOGENESIS.

Asymmetrical having no plane of symmetry, e.g. an animal of indeterminate shape that cannot be divided into two halves that are mirror images.

Atrium the volume enclosed by the tentacles of an endoproct; also the chamber through which the water current passes before leaving the body of a sea squirt or lancelet.

Autrotrophy the synthesis, by an organism, of its own organic constituents from inorganic material, i.e. independent of organic sources; autotrophic organisms may synthesize food phototrophically (e.g. green plants) or chemotrophically (e.g. bacteria) via inorganic oxidations.

Axon a long process of a nerve cell, normally conducting impulses away from nerve cell body.

Axopodium a stiff filament or pseudopodium which radiates outward from the body of a heliozoan or radiolarian.

Bacterium member of a division of unicellular or multicellular microscopic PROKARYOTIC organisms, lacking CHLOROPHYLL. Distinct from both plants and animals, rod-like, spherical or spiral in shape, occasionally forming a mycelium.

Benthic associated with the bottom of seas or lakes.

Bilateral symmetry a bilaterally symmetrical animal can be halved in one plane only to give two halves which are mirror images of each other. Most multicellular animals are bilaterally symmetrical, a form of symmetry generally associated with a mobile, free-living lifestyle.

Binary fission a form of ASEXUAL REPRODUCTION of a cell in which the nucleus divides, and then the CYTOPLASM divides into two approximately equal parts.

Biomass a measure of the abundance of a life form in terms of its mass.

Bipectinate comblike, of structures with two branches, particularly the OSPHRADIUM and/or CTENIDIUM in some mollusks.

Biramous of those ARTHROPODS (e.g. crustaceans) with forked ("two-branched") appendages.

Bivalve a shell or protective covering composed of two parts hinged together and which usually encases the body; also, a member of the molluskan class Bivalvia, which includes most bivalved animals.

Bothria long, narrow grooves of weak muscularity present in Pseudophyllidea (an order of tapeworm); form an efficient sucking organ.

Branchial hearts contractile hearts near the base of each CTENIDIUM in certain mollusks.

Brood sac a thoracic pouch of certain crustaceans into which fertilized eggs are deposited and where they develop.

Buccal mass a muscular structure surrounding the RADULA, horny jaw, and ODONTOPHORE of a mollusk (not a bivalve).

Budding a form of ASEXUAL REPRODUCTION in which a new individual develops as a direct growth from the parent's body.

Calcareous composed of, or containing, calcium carbonate as in the spicules of certain sponges or the shells of mollusks.

Cambrian a geological period some 543–490 million years ago, also the oldest system of rocks in which fossils can be used for dating, containing the first shelled fossil remains.

Carapace the dorsal shield of the exoskeleton covering part of the body (mainly anterior) of most crustaceans; particularly large in e.g. crabs, it protects the animal from both predation and water loss.

Carboniferous a geological period some 354–290 million years ago.

Carnivore an animal that feeds on other animals.

Catabolic of processes involving breakdown of complex organic molecules by living organisms, typically animals, resulting in the liberation of energy.

Caudal relating to the tail or to the rearmost SEGMENT of an invertebrate.

Cecum a blindly ending branch of the gut or other hollow organ.

Cellulose the tough, fibrous fundamental constituent of the cell walls of all green plants and some algae and fungi.

Cephalization development of the head during evolution; different organisms show different degrees of cephalization, generally according to their "level of evolution."

Cephalopod a member of an order of mollusks including such marine invertebrates as squids, octopuses, and cuttlefishes.

Cephalothorax the fusion of head and anterior thoracic segments in certain crustaceans to form a single body region that may be covered by a protective CARAPACE.

Cerata (sing. ceras) projections on the back of some shell-less sea slugs, often brightly colored, which may bear NEMATOCYSTS from cnidarians and may act as secondary respiratory organs.

Cercaria a swimming larval form of flukes; produced asexually by REDIA larvae while parasitic in snails; cercaria infects a new final or intermediate host via food or the skin.

Chaetae the chitinous bristles characteristic of annelid worms.

Chela the pincerlike tip of limbs in some arthropods.

Chelicera one of the first pair of appendages behind the mouth of a chelicerate.

Chelicerate a member of the phylum Chelicerata. Chelicerates possess CHELICERA and include scorpions, spiders, and horseshoe crabs.

Chitin a complex nitrogen-containing polysaccharide which forms a material of considerable mechanical strength and resistance to chemicals; forms the external "shell" or CUTICLE of arthropods.

Chloroplast a small granule (plastid) present in cells and containing the green pigment chlorophyll; site of PHOTOSYNTHESIS.

Chordate a member or characteristic of the phylum Chordata, animals that possess a NOTOCHORD.

Chromatophore a cell with pigment in its CYTOPLASM.

Cilia (sing. cilium) the only differences between FLAGELLA and cilia are in the former's greater length and the greater number of cilia found on a cell; flagella measure up to 22μm, cilia up to 10μm. Ciliary feeders feed by filtering minute organisms from a current of water drawn through or toward the animal by CILIA.

Ciliated having a number of cilia on a surface that beat in a coordinated rhythm; ciliary action is a common method of moving fluids within an animal body or over body surfaces, employed in CILIARY FEEDING, and a common means of locomotion in microscopic and small animals. The ciliated ciliates are a major class (Ciliata) of protozoans.

Cirri (sing. cirrus) in barnacles, paired thoracic feeding appendages; in protozoans, short, spinelike projections in tufts called CILIA; in annelids, broad flattened projections situated dorsally on segments; in flukes and some turbellarian flatworms the cirrus is the male copulatory apparatus.

Class a rank used in the classification of organisms; consists of a number of similar ORDERS (in some cases only one order may be distinguished); similar classes are grouped into a PHYLUM.

Cleavage see RADIAL CLEAVAGE: SPIRAL CLEAVAGE.

Clitellum the saddlelike region of earthworms, which is prominent in sexually mature worms.

Coelom the main body cavity of many TRIPLOBLASTIC animals, situated in the middle layer of cells, or MESODERM, and lined by EPITHELIUM. In many organisms the coelom contains the internal organs of the body and plays an important part in collecting excretions, which are removed via NEPHRIDIA or coelomoducts.

Coelomate having a COELOM.

Coelomocyte a free cell in the coelom of some invertebrates, which appears to be involved with the excretion of waste material, wound healing, and regeneration.

Colony an organism consisting of a number of individual members in a permanent colonial association.

Commensalism a relationship between members of different species in which one species benefits from the relationship, often by access to food; the other species neither benefits nor is harmed.

Community a naturally occurring group of different organisms inhabiting a common environment, sometimes named for one of its members, e.g. the *Donax* community of sandy beaches named for a genus of bivalve mollusks.

Compound eyes the type of eyes possessed by most crustaceans and insects, composed of many long, cylindrical units (ommatidia), each of which is capable of light reception and image formation.

Conjugation the union of GAMETES or two cells (in certain bacteria); or the process of sexual reproduction in most ciliates.

Convergent evolution the evolution of two organisms with some increasingly similar characteristics but different ancestry.

Copepod a small marine CRUSTACEAN of the invertebrate order Copepoda.

Copulation the process by which internal fertilization is accomplished; the transfer of sperm from one member of a species to another via specialized organs.

Corona the characteristic ciliated, wheel-like organ at the anterior end of rotifers.

Coxa the basal segment of an arthropod appendage, which joins the limb to the body.

Cretaceous a geological period extending from some 144 to 65 million years ago.

Cross-fertilization the fusion of male and female GAMETES produced by different individuals of the same species.

Crustaceans members of a class within the phylum Arthropoda typified by five pairs of legs, two pairs of antennae, head and thorax joined, and calcareous deposits in the exoskeleton, e.g. crayfishes, crabs, shrimps.

Cryptobiosis a form of suspended animation enabling an organism to survive adverse environmental conditions.

Ctenidium one of the pair of gills within the MANTLE CAVITY of some mollusks.

Cuticle the external layer covering certain multicellular animals (e.g. arthropods), formed from a collagen-like protein or CHITIN, which is secreted by the EPIDERMIS. The cuticle acts as a physiological barrier between the animal and its external environment, may reduce water loss, acts as a barrier to the entry of microorganisms, and in arthropods acts as an EXOSKELETON.

Cyst a thick-walled protective membrane enclosing a cell, larva, or organism.

Cytoplasm all the living matter of a cell excluding the nucleus.

Dactylozooid the specialized defensive polyp of colonial hydrozoans.

Desiccation loss of water, or drying out.

Detritivore an animal that feeds on dead or decaying organic matter.

Detritus organic debris derived from decomposing organisms that provides a food source for a large number of organisms.

Deuterostome a member of a major branch of multicellular animals (the others are PROTOSTOMES). The mouth is formed as a secondary opening, and the original embryonic blastopore becomes the anus. The embryo undergoes radial cleavage, the body cavity (ENTEROCOEL) arises as a pouch from the ENDODERM, and the central nervous system is dorsal.

Devonian a geological period from some 417 to 354 million years ago.

Dextral of spirally coiled gastropod shells in which, as is usual, the whorls rise to the right and the aperture is on the right where the shell is viewed from the side.

Diatom a single-celled alga, a component of the PHYTOPLANKTON.

Dimorphism the presence of two distinct forms, e.g. in color or size, in a species or population.

Dioecious having separate sexes.

Diploblastic a multicellular animal having a body composed of two distinct cellular layers, the ECTODERM and ENDODERM.

Dispersal the movement of individuals away from their previous home range, often as they approach maturity.

Display a relatively conspicuous pattern of behavior that conveys specific information to others, usually involving visual elements.

Dinoflagellate a unicellular organism of the PHYTOPLANKTON characterized by the possession of two FLAGELLA, one directed posteriorly, the other lying at right angles to the posterior flagellum.

Diverticulum (plural -ae) a blind-ending tube forming a side branch of a cavity or passage.

DNA deoxyribonucleic acid, a complex molecule, found almost exclusively in chromosomes of plants and animals, whose "double helix" structure contains the hereditary information necessary for an organism to replicate itself.

Dorsal situated at or related to, the back of an animal, i.e. the side that is generally directed upwards.

Dorso-ventral a plane running from the top to the bottom of an animal, as in a dorso-ventrally flattened horseshoe crab or sea slater.

Ecology the study of animal and plant communities in relation to each other and their natural surroundings.

Ecosystem an intricate community of organisms within a particular environment, interacting with one another and with the environment in which they live.

Ectoderm (is) the superficial or outer germ layer of a multicellular embryo, which develops mainly into the skin, nervous tissue, and excretory organs.

Ectoparasite a parasite that lives on the outside of its host and may be permanently attached or come into contact with the host only when feeding or reproducing.

Elongate relatively long in comparison with width.

Endemic confined to a given region, such as an island or country.

Endocuticle the inner layer of the crustacean CUTICLE, which is composed of CHITIN.

Endoderm (is) the innermost of the three germ layers in the early embryo of most animals, developing into, for example, in jellyfishes the lining of the

ENTERON or, in many animals, the GUT lining.

Endoparasite a parasite that lives permanently within its host's tissues (except for some reproductive or larval stages). Often there are primary and SECONDARY HOSTS for different stages of the life cycle. Endoparasites are typically highly specialized.

Endoskeleton an internal skeleton, as in echinoderms and vertebrates.

Enterocoelom a COELOM that is thought to have arisen from cavities in the sacs of the MESODERM of the embryo.

Enteron the body cavity of cnidarians, which is lined with ENDODERM and opens to the exterior via a single opening, the mouth.

Epibenthic living on the seabed between the low-water mark and some 200m (670ft) depth.

Epicuticle the outer layer of the crustacean CUTICLE, a thin, nonchitinous protective layer.

Epidermis the outer tissue layer of the epithelium.

Epithelium a sheet or tube of cells lining cavities and vessels and covering exposed body surfaces.

Epizoic a sedentary animal that is attached to the exterior of another animal but is not parasitic, i.e. is epizoic.

Esophagus part of the foregut of certain invertebrates connecting the pharynx with the stomach or crop and concerned with the passage of food along the gut.

Eukaryote a cell, or organism possessing cells, in which the nuclear material is separated from the CYTOPLASM by a nuclear membrane and the genetic material is borne on a number of chromosomes consisting of DNA and protein; the unit of structure in all organisms except bacteria and blue-green algae.

Exoskeleton the skeleton covering the outside of the body, or situated in the skin.

Extracellular digestion digestion of food within an organism but not within its constituent cells.

Family a rank used in the classification of organisms, consisting of a number of similar GENERA (or sometimes only one). Similar families are grouped into an ORDER. In zoological classifications the name of the family usually ends in -idae.

Fibril a small fiber, or subdivision of a fiber; used as contractile ORGANELLES in protozoans.

Filamentous a type of structure, e.g. a crustacean GILL, in which the branches are threadlike, but not sub-branched, and are arranged in several series along the central axis.

Filter feeding a form of SUSPENSION FEEDING in which food particles are extracted from the surrounding water by filtering. Filtering requires the setting up of a water current usually by means of CILIA, with mucus being used to trap particles and sometimes to filter them from the surrounding water.

Fission see BINARY FISSION.

Flagellum (plural flagella) a fine, long thread, moving in a lashing or undulating fashion, projecting from a cell.

Flame cell the hollow, cup-shaped cell lying at the inner end of a protonephridium, important in the excretory system of some invertebrates. The inner end bears FLAGELLA, whose beating causes body fluids to enter the NEPHRIDIUM.

Fragmentation a form of SEXUAL REPRODUCTION in which an organism produces eggs in SEGMENTS of its body, which then break off and themselves split after leaving the host body, allowing the eggs to develop eventually into new organisms.

Free-living having an independent lifestyle, not directly dependent on another organism for survival.

Funnel part of the molluskan "foot" in cephalopods responsible for respiratory currents to the CTENIDIA and for jet propulsion.

Gamete a female (ovum) or male (spermatozoan) reproductive cell whose nucleus and often CYTOPLASM fuses with another gamete, so constituting fertilization.

Gametocyte a cell that undergoes MEIOSIS to form GAMETES; an oocyte forms an ovum (female gamete) and a spermatocyte forms a spermatozoan (male gamete).

Ganglion a small discrete collection of nervous tissue containing numerous cell bodies. The nervous system of most invertebrates consists largely of such ganglia connected by nerve cords, which may be concentrated into a cerebral ganglion constituting the "brain."

Gastrozooid a type of individual POLYP in colonial hydrozoans that captures and ingests prey.

Gemmule a mass of sponge cells that acts as a resting stage under adverse conditions and is composed of amoebocytes surrounded by two membranes in which SPICULES are embedded.

Gene the unit of the material of inheritance, a short length of chromosome and the set of characters that it influences in a particular way.

Generalist an unspecialized animal, not adapted to a particular niche; may be present in a variety of habitats.

Genus a rank used in classifying organisms, consisting of similar SPECIES (in some cases only one species). Similar genera are grouped into FAMILIES.

Gill the respiratory organ of aquatic animals.

Gill book a type of gill, possessed by e.g. horseshoe crabs, formed by the five posterior pairs of appendages on the OPISTHOSOMA.

Gizzard part of the alimentary canal where food is broken up, preceding main digestion. In crustaceans its walls bear hard "teeth."

Gonoduct the duct through which sperm and eggs are released into the surrounding water.

Gut the alimentary canal; a tube concerned with the digestion and absorption of food. In most animals there are two openings (cnidarians and flatworms have only one) – the mouth into which food is taken and the anus from which material is ejected.

Hemal system a tubular system of undecided function present in echinoderms.

Hemocoel the major secondary body cavity of arthropods and mollusks, which is filled with blood. Unlike the COELOM, it does not communicate with the exterior and does not contain germ cells. However, body organs lie within or are suspended in the hemocoel. It functions in the transport and storage of many essential materials.

Herbivore animal that feeds on plants.

Hermaphrodite an animal producing both male and female GAMETES; among unisexual animals, hermaphrodites may occur as aberrations.

Hermatypic of corals, reef-building corals with commensal zooanthellae.

Heterotrophic heterotrophic organisms are unable to synthesize their own food substances from inorganic material, therefore they require a supply of organic material as a food source. They include all animals, all fungi, most bacteria, and a few flowering plants.

Hibernation dormancy in winter.

Holoplankton organisms in which the whole life cycle is spent in the PLANKTON.

Host see INTERMEDIATE HOST; PRIMARY HOST; SECONDARY HOST.

Hybrid a plant or animal resulting from a cross between parents that are genetically different, usually from two different species.

Hydrostatic skeleton a fluid-filled cavity enclosed by a body wall that acts as a skeleton against which the muscles can act.

Hyperparasite (verb hyperparasitize) an organism which is a parasite upon another parasite.

Infusariiform a larval stage of mesozoans produced in members of the order Dicyemida by the hermaphrodite RHOMBOGEN generation, and in the order Orthonectida by free-living males and females and reinfecting the host.

Inorganic material material not derived from living or dead animals or plants, carbon atoms being absent from the molecular structure.

Intermediate host an organism that plays host to parasitic larvae before they mature sexually in the final or definitive host.

Interstitial living in the spaces between SUBSTRATE particles.

Intracellular digestion digestion of food within the cell.

Introduction a species that has settled in lands where it does not occur naturally as a result of human activities.

Invertebrate an animal that is not a member of the subphylum vertebrata of the Chordata, i.e. it lacks a skull surrounding a well-developed brain and does not have a skeleton of bone or cartilage.

Isogamy (adjective isogamous) a condition in which the GAMETES produced by a species are similar, i.e. not differentiated into male and female.

Jurassic a geological period that extended from about 206 to 144 million years ago.

Keratin a tough, fibrous protein rich in sulfur, the outer layer of the CUTICLE of nematode worms is keratinized.

Kinety in ciliate protozoans, a row of kinetosomes and FIBRILS; from kinetosomes arise CILIA, the fibrils linking each kinetosome in a longitudinal row.

Kingdom the uppermost rank of classification dividing bacteria and blue-green algae, algae, plants, fungi, protista, and animals into their respective kingdoms.

Krill shrimplike CRUSTACEANS of the genera Euphausia, Meganyctiphanes etc. occurring in very great numbers in polar seas, particularly of Antarctica, where they form the principal prey of baleen whales.

Lacunae minute spaces in invertebrate tissue containing fluid.

Larva a general term for a distinct pre-adult form into which most invertebrates hatch from the egg and which may develop directly into adult form or into another larval form.

Lymph an intercellular body fluid drained by lymph vessels; contains all the constituents of blood plasma except protein, and varying numbers of cells.

Macronucleus one of two nuclei present in ciliate protozoans.

Macrophagous diet of pieces that are large relative to the size of animal; feeding usually occurs at intervals

Madreporite a delicate, perforated sieve plate through which seawater may be drawn into the WATER VASCULAR SYSTEM of echinoderms: may be internal (e.g. sea cucumbers) or prominent external convex disk (starfishes).

Malpighian tubule/gland a tubular excretory gland that opens into the front of the hindgut of insects, arachnids, myriapods, and water bears.

Mandible the paired appendages behind the mouth of crustaceans and uniramians, used in biting and chewing, and having grinding and biting surfaces.

Mantle a fold of skin covering all or part of the body of mollusks; its outer edge secretes the shell.

Mantle cavity the cavity between the body and MANTLE of a mollusk, containing the feeding and/or respiratory organs.

Maxilla paired head appendages of crustaceans and uniramians, which are located behind the MANDIBLES on the fifth segment. They act as accessory feeding appendages.

Maxilliped the first one, two, or three pairs of thoracic limbs of malocostracan crustaceans, which have turned forward and become adapted as accessory feeding appendages rather than being involved in locomotion.

Maxillule paired head appendages of crustaceans and uniramians, which are located on segment six behind the MAXILLAE. They also function in the manipulation of food.

Medusa the free-swimming sexual stage of the cnidarian life cycle, produced by the asexual BUDDING POLYPS.

Megalopa a postlarval stage of brachyuran crustaceans in which, unlike the adult (e.g. crab), the abdomen is large, unflexed, and bears the full number of appendages.

Meiosis cell division whereby the DNA complement is halved in the daughter cells. Compare MITOSIS.

Meroplankton organisms passing part of their life cycle in the PLANKTON, usually the larval forms of BENTHIC animals. Compare HOLOPLANKTON.

Merozoite a stage in the life cycle of some parasitic protozoans that enters red blood corpuscles of the host.

Mesenchyme embryonic connective tissue consisting of scattered, irregularly branching cells in a jellylike matrix; gives rise to connective tissue, bone, cartilage, and blood.

Mesoderm the cell layer of TRIPLOBLASTIC animals that develops into tissues lying between the ENDODERM and ECTODERM.

Mesogloea the layer of jellylike material between the ECTODERM and ENDODERM of cnidarians, such as jellyfishes etc.

Mesozoic a geological era ranging from 248 to 65 million years ago, comprising the TRIASSIC, JURASSIC, and CRETACEOUS systems.

Metabolic rate the rate at which the chemical processes within an organism take place.

Metameric having many similar SEGMENTS constituting the body.

Metamorphosis the period of rapid transformation of an animal from larval to adult form, often involving destruction of larval tissues and major changes in morphology.

Metazoan an animal, as in the vast majority of invertebrates, whose body consists of many cells in contrast to PROTOZOANS, which are unicellular; a member of the subkingdom Metazoa.

Microfilaria the larval form of filaroid nematode worm parasites found in the SECONDARY HOST, usually mosquitoes.

Microflora microscopic bacteria occurring in the soil.

Microhabitat the particular parts of the habitat that are encountered by an individual in the course of its activities.

Micronucleus one of two nuclei found in the protozoan ciliates, the smaller micronucleus provides the gametes during conjugation. See MACRONUCLEUS.

Microphagous diet a diet of pieces of food that are minute relative to the animal's own size; feeding occurs continually.

Microtubule a very small long, hollow cylindrical vessel conveying liquids within a cell.

Miracidium a ciliated larva of flukes that emerges from eggs passed out with the feces of the vertebrate host and parasitizes snails, where it reproduces asexually.

Mitosis cell division in which daughter cells replicate exactly the chromosome pattern of the parent cell, unlike meiosis.

Molt periodic shedding of the arthropod EXOSKELETON. Possession of a hardened exoskeleton prevents continuous growth until the adult stage is reached. Molting occurs under hormonal control after the secretion of a new and larger CUTICLE. An increase in size occurs during the short period prior to the hardening of the new cuticle, involving water or air uptake into the internal spaces. New tissue then grows into these spaces after the hardening of the new cuticle, i.e. between molts.

Monoblastic organisms having a single cell layer (e.g. sponges).

Monophyletic descended from a common ancestor. Some scientists hold a monophyletic view of arthropods, while others recognize several phyla of jointed-limbed invertebrates with separate evolutionary origins.

Morphology the structure and shape of an organism.

Multicellular composed of a large number of cells.

Myotome a block of muscle, one of a series along the body of a lancelet, sea squirt larva, or vertebrate.

Natural selection the mechanism of evolutionary change suggested by Charles Darwin, whereby organisms with characteristics that enhance the chance of survival in the environment in which they live are more likely to survive and produce more offspring with the same characteristics than organisms without those characteristics or with other less advantageous characteristics.

Nauplius the first larval stage of some crustaceans, which is divided into three segments, each possessing a pair of jointed limbs that develop into the adult's two pairs of antennae and the mandibles. The nauplius uses its limbs in feeding and locomotion.

Nekton aquatic organisms, such as fish, which, unlike the smaller PLANKTON, can maintain their position in the water column and move against local currents.

Nematocyst the characteristic stinging ORGANELLE of cnidarians (e.g. jellyfishes) located particularly on the tentacles. A short process at one end of the ovoid cell (cnidoblast) containing the nematocyst acts as a trigger opening the lidlike OPERCULUM. Water entering the cnidoblast swells the nematocyst, a long threadlike tube coiled up inside. The nematocyst discharges, ensnaring prey in its barbed coils or releasing poison down the tube into the victim.

Nematogen the first and subsequent early generations of certain mesozoans (order Dicyemida), parasites of immature cephalopods.

Nephridiopore the pore by which a NEPHRIDIUM opens into the external environment.

Nephridium an excretory tubule opening to the exterior via a pore (nephridiopore). The inner end of the tubule may be blind, ending in FLAME CELLS or it may open into the COELOM via a ciliated funnel.

Nerve cord a solid strand of nervous tissue forming part of the central nervous system of invertebrates

Niche the position of a species within the community, defined in terms of all aspects of its lifestyle.

Nocturnal awake and active by night, particularly of animals that hunt for food by night.

Notochord a row of vacuolated cells forming a skeletal rod lying lengthwise between the central nervous system and gut of all CHORDATES.

Oligomeric/ous having a few segments constituting the body.

Omnivore an animal that feeds on both plant and animal tissue.

Operculate the condition of gastropods having an OPERCULUM.

Operculum a lidlike structure; the calcareous plate on the top surface of the foot of some gastropods, serving to close the aperture when the animal withdraws into the shell.

Opisthosoma the posterior body region of chelicerates, which may be segmented in primitive forms but generally has the segments fused.

Order a group used in classifying organisms, consisting of a number of similar FAMILIES (sometimes only one family). Similar orders are grouped into a CLASS.

Ordovician a geological period extending from about 490 to 443 million years ago.

Organ part of an animal or plant that forms a structural and functional unit, e.g. spore, lung.

Organelle a persistent structure forming part of a cell, with a specialized function within it analogous to an ORGAN within the whole organism.

Organic material material derived from living or dead animals and plants, molecules making up the organisms being based on carbon, the other principle elements being oxygen and hydrogen.

Osmotic pressure and regulation osmotic pressure is the force that tends to move water in an osmotic system, i.e. the pressure exerted by a more concentrated solution on one of a lower concentration. The body fluids of a freshwater animal will exert an osmotic pressure on the surrounding aqueous medium, causing water to enter the animal. Osmoregulation is the maintenance of the internal body fluids at a different osmotic pressure from that of the external aqueous environment.

Osphradium a patch of sensory EPITHELIUM located on gill membranes of mollusks.

Paleozoic a geological era ranging from 543–248 million years ago, comprising the CAMBRIAN, ORDOVICIAN, and SILURIAN systems in the older or lower Paleozoic sub-era and the DEVONIAN, CARBONIFEROUS, and PERMIAN systems in the newer or Upper Paleozoic sub-era.

Palp an appendage, usually near the mouth, which may be sensory, aid in feeding, or be used in locomotion.

Papilla a small protuberance ("little nipple") above a surface

Parapodium one of a pair of appendages extending from the sides of the segments of polychaete worms.

Parthenogenesis the development of a new individual from an unfertilized egg. It occurs when rapid colonization is important under adverse environmental conditions, or when there is an absence or only a small number of males in the population.

Pathogen an agent which causes disease, always parasitic.

Pedicellariae minute, pincerlike grooming and defensive structures on the body of sea urchins and starfishes.

Pedipalp an appendage borne on the third prosomal segment of chelicerates, sensory or prehensile in horseshoe crabs, adapted for seizing prey in scorpions, and sensory or used by the male in reproduction in spiders.

Peduncle a narrow part supporting a longer part, e.g. the muscular stalk by which the body of an endoproct is attached to the SUBSTRATE.

Pelagic of organisms or lifestyles in the water column, as opposed to the bottom SUBSTRATE.

Pentamerism the fivefold RADIAL SYMMETRY typical of echinoderms.

Pericardial cavity the cavity within the body containing the heart. In vertebrates a hemococtic space, which is an expanded part of the blood system, supplying blood to the heart.

Peristalsis rhythmic waves of contraction passing along tubular organs, particularly the GUT, produced by a layer of smooth muscle.

Permian a geological period from 290 to 248 million years ago, marking the end of the PALEOZOIC era.

Pharynx part of the alimentary tract or gut behind the mouth, often muscular.

Pheromone a chemical substance that when released by an animal influences the behavior or development of other individuals of the same species.

Photosynthesis the synthesis of organic compounds, primarily sugars, from carbon dioxide and water using sunlight as a source of energy and chlorophyll or some other related pigment for trapping the light energy.

Phyletic concerning evolutionary descent.

Phylogeny the evolutionary history or ancestry of a group of organisms.

Phylum a major group used in the classification of animals. Consists of one or more CLASSES. Several (sometimes one) phyla make up a KINGDOM.

Physiology the study of the processes that occur within living organisms.

Phytoplankton microscopic algae that are suspended in surface waters of seas and lakes where there is sufficient light for PHOTOSYNTHESIS to take place.

Pinnate of tentacles, GILLS, resembling a feather or compound leaf in structure, with similar parts arranged either side of a central axis.

Pinnule a jointed appendage present in large numbers on the arms of crinoids giving a featherlike appearance, hence the name feather star.

Plankton drifting or swimming animals and plants, many minute or microscopic, which live freely in the water and are borne by water currents owing to their limited powers of locomotion.

Planula the free-swimming ciliated larva of cnidarians (jellyfishes and allies).

Plasma the fluid medium of the blood in which highly specialized cells are suspended; mainly water, containing a variety of dissolved substances that are transported from one part of the body to another.

Plasmodium the asexual stage of orthonectid mesozoans, resembling the protozoan plasmodium, which divide repeatedly by FISSION, filling the hosts' tissue spaces.

Pneumostome the aperture to the lunglike MANTLE CAVITY of pulmonates.

Polyp the stage, the most important in the life cycle of most cnidarians, in which the body is typically tubular or cylindrical, the oral end bearing the mouth and tentacles and the opposite end being attached to the SUBSTRATE.

Polysaccharide a carbohydrate produced by a combination of many simple sugar or monosaccharide molecules, e.g. starch and CELLULOSE.

Primary host the main host of a parasite in which the adult parasite or the sexually mature form is present.

Proboscis a tubular organ that may be extended from the mouth of many invertebrates such as moths and butterflies; in ribbon worms, the proboscis can be everted.

Proglottides (sing. proglottis) the segments that make up the "body" of a tapeworm. When mature, each proglottis will contain at least one set of reproductive organs.

Prokaryote cell having, or organism made of cells having, genetic material in the form of simple filaments of DNA, not separated from the CYTOPLASM by a nuclear membrane (cf. EUKARYOTE). Bacteria and blue-green algae have cells of this type.

Prosoma the anterior body region of CHELICERATES, composed of eight SEGMENTS, analogous to the head and THORAX of other arthropods, or the CEPHALOTHORAX of chelicerates. The segments are generally fused and are distinguishable only in the embryo.

Prostomium the anterior nonsegmental region of annelid worms, bearing the eyes, ANTENNAE, and a pair of PALPS; comparable to the head of other phyla.

Protein a complex organic compound composed of numerous amino acids joined together by peptide linkages, forming one or more folded chains. The sequence of amino acids is peculiar to a particular protein.

Protoconch the first shell of a gastropod which is laid down by the larva.

Protonephridium a type of excretory organ in which the tubule usually ends in a FLAME CELL.

Protostome a member of one major branch of the multicellular animals (the other, complementary, branch is the DEUTEROSTOMES). The mouth is formed from the embryonic blastopore, the embryo undergoes SPIRAL CLEAVAGE, the body cavity is formed by the MESODERM splitting into two, and the central nervous system is ventral.

Protozoan an animal-like organism of the phylum Protozoa, Kingdom Protista, differing from animals in consisting of one cell only, but resembling them and plants, and differing from bacteria in having at least one well-defined nucleus.

Pseudocoel now called a blastocoelom, a fluid-filled body cavity between the body wall and the gut of some groups, such as nematode worms, which arises from the blastocoel, an embryonic cavity that appears in the early stages of development.

Pseudopodium (plural: -a) a temporary projection of the cell when the fluid endoplasm flows forward inside the stiffer ectoplasm. Occurs during locomotion and feeding.

Pseudotrachea a branched TUBULE resulting from intuckings of the CUTICLE of certain terrestrial isopods, which acts as a specialized respiratory surface. Pseudotracheae resemble the TRACHEAE of uniramians and certain arachnids, although they have evolved independently.

Pulmonate having a lung, e.g. certain snails and slugs.

Pygidium the terminal, nonsegmental region of some invertebrates that bears the anus.

Radial symmetry a form of symmetry in which the body consists of a central axis around which similar parts are arranged symmetrically.

Radula the "toothed tongue" of mollusks, a horny strip with ridges or "teeth" on its surface which rasp food. Absent in members of the class Bivalvia.

Ray a radial division of an echinoderm, e.g. a starfish "arm."

Redia a larval type produced asexually by a previous larval stage of flukes (Trematoda). Lives parasitically in snails and reproduces asexually, giving rise to more rediae or to CERCARIAE.

Reticulopodium a type of PSEUDO-PODIUM characteristic of the foraminiferans; reticulopodia are threadlike, branched, and interconnected.

Rhombogen an hermaphrodite form of dicyenid mesozoan derived from a NEMATOGEN when the cephalopod host has reached maturity. It resembles the nematogen morphologically and gives rise to INFUSARIIFORM larvae.

Rostrum the anterior plate of the crustacean CARAPACE, present in malacostracans, which extends toward the head and ends in a point.

Sclerotization hardening of the arthropod CUTICLE by TANNING.

Scolex the head region of a tapeworm, which attaches to the wall of the host's gut by suckers and/or hooks.

Secondary host the host in which the larval or resting stages of a parasite are present.

Sedentary sedentary organisms, or stages in the life cycle of certain organisms, are permanently attached to a SUBSTRATE; as opposed to FREE-LIVING.

Segment a repeating unit of the body that has a structure fundamentally similar to other segments, although certain segments may be grouped together into TAGMATA to perform certain functions, as in the head, THORAX, or ABDOMEN.

Segmentation the repetition of a pattern of segments along the length of the body, or along an appendage. The similarity between different segments of an animal may be imperfect, particularly the segments forming the head.

Septum a portion dividing a tissue or organ into a number of compartments.

Seta (plural setae) a bristlelike projection on the invertebrate EPIDERMIS.

Sexual reproduction reproduction involving MEIOSIS and fertilization, usually fusion of two GAMETES, one female and one male. See also COPULATION, CONJUGATION, FRAGMENTATION.

Siliceous composed of or containing silicate, as in the skeleton of glass sponges (see SPICULES).

Silurian a geological era, 443–417 million years ago.

Sinistral of gastropod shells, with whorls rising to the left and not as usual to the right (compare DEXTRAL).

Sinus a space or cavity in an animal's body.

Siphon a tube through which water enters and/or leaves a cavity within the body of an animal, e.g. in mollusks and in sea squirts.

Solitary a lifestyle in which an organism exists by itself and not in permanent association with others of the same species.

Specialist an organism having special adaptations to a particular habitat or mode of life; its range of habitats or variety of modes of life may thus be limited and, as a result, its evolutionary flexibility also.

Speciation the origin of species, the diverging of two like organisms into different forms, resulting in new species.

Species a taxonomic rank, the lowest commonly used; reproductively an isolated group of interbreeding organisms. Similar species make up a GENUS.

Spermatophore a package of sperm produced by males, usually of species in which fertilization is internal but does not involve direct COPULATION.

Spermatotheca an organ, usually one of a pair, in a female or hermaphrodite that receives and stores sperm from the male.

Spicule a mineral secretion (calcium carbonate or silica) of sponges that forms part of the skeleton of most species and whose structure is of importance in sponge classification.

Spiral cleavage a form of embryonic division that occurs in PROTOSTOMES; in the spiral arrangement of cells, any one cell is located between the two cells above or below it. In all other many-celled animals, i.e. DEUTEROSTOMES, there is radial CLEAVAGE.

Spore a single-celled or multicelled reproductive body that becomes detached from its parent and gives rise directly or indirectly to a new individual.

Sporocyst a saclike body formed by the MIRACIDIUM larva of a blood fluke while within the snail, the intermediate host; produces numerous CERCARIAE, over 3,000 per day from a single sporocyst.

Sporozoite SPORE produced in certain protozoans, which then develops into gametes.

Statocyst the balancing organ of a number of invertebrates consisting of a vesicle containing granules of sand or calcium carbonate. These granules move within the vesicle and stimulate sensory cells as the animal moves, so providing information on its position in relation to gravity.

Stolon the tubular structure of colonial cnidarian POLYPS that anchors them to the SUBSTRATE and from which the polyps arise.

Stomodeum a region of unfolding ECTODERM from which derive the mouth cavity and foregut in many invertebrates.

Strobila the "body" of a tapeworm, consisting of a string of segments, through which food is absorbed from the gut of the host.

Stylet a small, sharp appendage, for example in water bears, used to pierce plant cells.

Suborder members of an ORDER forming a group of organisms that differ in some way from the other members but also resemble them in many characteristics.

Substrate the surface or sediment on or in which an organism lives.

Superfamily a division containing a number of families or a single family differing in some way from other families that are included in the same ORDER.

Suspension feeding a feeding mechanism in which small organisms and other matter suspended in the water are removed and consumed.

Symbiosis a close and mutually beneficial relationship between individuals of two species.

Synapse the site at which one nerve cell is connected to another.

Tagmata (sing. tagma) functional body regions of arthropods and annelids consisting of a number of segments; e.g. the head, THORAX, and ABDOMEN of crustaceans.

Tanning hardening of the arthropod CUTICLE achieved by the cross-linking of the protein chains by arthoquinones, involving also polyphenol and polyphenoloxidase catalysts.

Taxon a taxonomic grouping of organisms or the name applied to it.

Taxonomy the study of the classification of organisms according to resemblances and differences.

Tegumental gland a gland below the EPIDERMIS of the crustacean cuticle. Ducts from the glands convey the constituents of the EPICUTICLE to the cuticle surface during molting, when the new epicuticle is formed.

Telson the posterior segment of the arthropod abdomen which is present only embryonically in insects. In certain crustaceans the telson is flattened to form a tail fin, which is used in swimming.

Terrestrial associated with, or living on the earth or ground.

Tertiary the geological period of time from the end of the CRETACEOUS era 65 million years ago to the present time, divided into a number of epochs, the last 1.8 million years sometimes distinguished as the Quaternary Period.

Test an external covering or "shell" of an invertebrate, especially sea squirts (tunicates), sea urchins etc; is in fact an internal skeleton just below the EPIDERMIS.

Thorax the segmented body region of insects and crustaceans that lies behind the head and which typically bears locomotory appendages. Up to 11 segments are present in crustaceans but only 3 in insects.

Tissue a region consisting mainly of cells of the same sort and performing the same function, associated in large numbers and bound together by cell walls (plants) or by intercellular material (animals).

Torsion the process of twisting of the body in the larval stage of gastropods.

Trachea a cuticle-lined respiratory tubule of uniramians and certain arachnids that is involved in gas exchange. Tracheae open to the exterior via a spiracle, which can often be sealed to reduce desiccation. The tracheae are branched and ramify into the tissues, they end in thin-walled, blind-ending tracheoles within the cells.

Triassic a geological period extending from 248 to 206 million years ago, marking the beginning of the MESOZOIC era.

Trichocyst rodlike or oval ORGANELLE in the ECTODERM of protozoans, which may discharge a long thread on contact with prey.

Trochophore an oval or pear-shaped, free-swimming, planktonic larval form of organisms from different phyla, including segmented worms and mollusks.

Tube feet or podia hollow, extensive appendages of echinoderins connected to the WATER VASCULAR SYSTEM that may have suckers, or serve as stiltlike limbs, or be ciliated to waft food particles toward the mouth.

Tubule long hollow cylinder within a cell, normally for conveying or holding liquids.

Tunicin, tunic a form of CELLULOSE, the main constituent in the fibrous matrix forming the tunic or TEST of sea squirts.

Unicellular an organism composed of only a single cell.

Uniramian a member of the phylum Uniramia, which includes the insects, centipedes and millipedes. They possess a single pair of antennae and mandibles. The appendages are basically unbranched or UNIRAMOUS in contrast to those of crustaceans.

Uniramous condition describing an arthropod limb that is not branched and is present in insects and myripods (hence the Uniramia) and chelicerates. Some crustacean limbs are secondarily uniramous where one of the branches of the BIRAMOUS limb has been lost.

Upwelling an upward movement of ocean currents, resulting from convection, causing an upward movement of nutrients and hence an increase in plankton populations.

Urogenital tract ducts and tubules common to the genital and urinary systems voiding via a common aperture.

Uropod flattened extension of the sixth abdominal appendage of malacostracan crustaceans, which together with the flattened TELSON form a tail fin used in swimming.

Vacuole a fluid-filled space within the CYTOPLASM of a cell, bounded by a membrane.

Valve in bivalves, one half of the two-valved shell.

Vascular containing vessels that conduct fluid – in animals usually blood – as in the vascularized MANTLE CAVITY of pulmonate snails.

Veliger a free-swimming larval form of mollusks possessing a VELUM; develops from a TROCHOPHORE; foot, mantle, shell, and other adult organs are present.

Velum the veil-like ciliated lobe of the VELIGER larva, used in swimming; also the inward-projecting margin of the umbrella in most hydrozoan medusae.

Ventral situated at, or related to, the lower bottom side or surface.

Vermiform a wormlike larval stage of dicyenid mesozoans formed within the axial cells of the NEMATOGEN generation, or generally meaning "wormlike."

Vertebrate an organism that belongs to the subphylum Vertebrata (Craniata) of the phylum Chordata; differs from other chordates and invertebrates in having a skull that surrounds a well-developed brain, and a skeleton of cartilage and bone.

Water vascular system or **ambulacral system** a system of canals and appendages of the body wall that is unique to echinoderms, derived from the coelom and used, for example, in locomotion in starfishes.

Zoea a planktonic larval form of some decapod crustaceans that possess a segmented THORAX, a CARAPACE, and at least three pairs of BIRAMOUS thoracic appendages. In contrast to the antennal propulsion of the NAUPLIUS, these thoracic appendages are used in locomotion. The abdominal pleopods appear but are not functional until the postlarval stage.

Zoochlorella a symbiotic green alga of the Chloroplyceae, which occurs in the amoebocytes of certain freshwater sponges, the gastrodermal cells of some hydra species and the jellylike connective tissue (parenchyme) of certain turbellarian flatworms.

Zooid a member of a colony of animals that are joined together; may be specialized for certain functions.

Zoospore a motile spore that swims by means of a flagellum, is produced by some unicellular animals and algae, and is a means of ASEXUAL REPRODUCTION.

Zooplankton small or minute animals that live freely in the water column of seas and lakes, consisting of adult PELAGIC animals or the larval forms of pelagic and some BENTHIC animals; most are motile, but the water movements determine their position in the water column.

Zygote a fertilized ovum before it undergoes cleavage.

Bibliography

The following list of titles indicates key reference works used in the preparation of this volume, and those recommended for further reading.

Adouette, A., Balavoine, G., Lartillot, N., Lespinet, O., Prud'homme, B. and de Rosa, R. (2000) The new animal phylogeny: Reliability and implications. In: *PNAS* 97, 9 (25 April), pp. 4453–4456.

Alexander, R. McN. (1979) *The Invertebrates*. Cambridge University Press, Cambridge.

Allen, T. B. (2001) *Shark Attacks: Their Causes and Avoidance*. Lyons Press, New York.

Amlacher, E (1970) *A Textbook of Fish Diseases*. T.F.H. Publications, Inc., Neptune City, New Jersey.

Anon (1986) *Sharks: Silent Hunters of the Deep*. Reader's Digest.

Banister, K. and Campbell, A. (eds.) (1985) *The Encyclopedia of Underwater Life*. George Allen & Unwin.

Barnes, R. S. K., Calow, P. and Olive, P. J. W. (1993) *The Invertebrates – a New Synthesis* (2nd edition). Blackwell Scientific Publications, Oxford.

Barrington, E. J. W. (1982) *The Invertebrate Structure and Function*, Van Nostrand Reinhold, New York.

Bliss, D. E. (ed.) (1982) *The Biology of the Crustacea, vols 1–10*. Academic Press, London and New York.

Bond, C. E. (1979) *Biology of Fishes*. Saunders College Publishing.

Bonner, W. N. and Berry, R. J. (eds.) (1981) *Ecology in the Antarctic*. Academic Press, London.

Bright, M. (2002) *Sharks*. Natural History Museum, London.

Brusca, R.C. and Brusca, P. J. (2002) *Invertebrates* (2nd edition) Sinauer Associates, Inc., Sunderland, Massachusetts.

Campbell, A. C. (1982) *The Hamlyn Guide to the Flora and Fauna of the Mediterranean Sea*, Hamlyn, London.

Campbell, A. C. (1984) *The Country Life Guide to the Sea Shores and Shallow Seas of Britain and Europe*, Country Life Books, London.

Compagno, L. J. V. (1989) *FAO Species Catalogue Vol. 4: Sharks of the World. Part 1: Hexanchiformes to Lamniformes*. FAO, Rome, Italy.

Compagno, L. J. V. (1984) *FAO Species Catalogue Vol. 4: Sharks of the World. Part 2: Carcharhiniformes*. FAO, Rome, Italy,

Corbera, J., Sabetés, A. and García-Rubies, A. (1996) *Peces de Mar de la Península Ibérica*. Planeta, S.A., Madrid.

Dawes, J., (1991) *Livebearing Fishes: A Guide to Their Aquarium Care, Biology and Classification*. Blandford, London.

Dawes, J., Lim, L. L., and Cheong, L. (eds.) (1999) *The Dragon Fish*. Kingdom Books.

Ellis, R. (1989) *The Book of Sharks*. Grosset and Dunlap.

Fretter, V. and Graham, A. (1976) *A Functional Anatomy of Invertebrates*, Academic Press, London, New York, San Francisco.

Helfman, G. S. , Collette B. B., and Facey, D. E. (1997) *The Diversity of Fishes*. Blackwell Scientific Publications, Oxford.

Hennemann, R. M. (2001) *Sharks & Rays - Elasomobranch Guide of the World*. IKAN Unterwasserarchiv.

Herman, L. M (1980) *Cetacean Behavior: Mechanisms and Functions*. John Wiley & Sons, Chichester.

Hieronimus, H. (2002) *All Rainbows and Related Families*.Verlag A.C.S. GmbH.

Kedera, H., Igarashi, T., Kuroiwa N., Maeda, H., Mitani, S., Mori, F., and Yamasaki, K., (1994) *Jurassic Fishes*. T.F.H. Publications, Inc., Neptune City, New Jersey.

Kempkes, M. and Schäfer. F. (1998) *All Livebearers and Halfbeaks: Guppys, Platys, Mollys* (Verlag A.C.S. GmbH, 1998)

Lawrence, J. (1987) *A Functional Biology of Echinoderms*.Croom Helm, London and Sydney.

Lever, C. (1996) *Naturalized Fishes of the World*. Academic Press

Mann, A. J. and Williams W. D. (1982) *Textbook of Zoology: Invertebrates*, Macmillan, London.

Meffe, G. K. and Snelson, F. F., Jr. (eds.) (1989) *Ecology and Evolution of Livebearing Fishes (Poeciliidae)*. Prentice Hall.

Moyle, P. B. and Cech, J. J. (Jr.) (2000) *Fishes: An Introduction to Ichthyology* (4th edition) Prentice-Hall, Inc.

National Audubon Field Guide to Fishes (North America) (2nd edition, 2002). Alfred A. Knopf, New York.

National Audubon Field Guide to North American Seashore Creatures (1981). Alfred A. Knopf, New York.

Nelson, J. R. (1994) *Fishes of the World* (3rd edition). John Wiley and Sons, Inc., New York and Chichester.

Neilsen, C. (1995) *Animal Evolution: Interrelationships of the Living Phyla* Oxford University Press, Oxford.

Neilsen, C. (1998) Origin and evolution of animal life cycles. In: *Biological Reviews*, 73, pp. 125–155.

Ono, R. D., et al. (1983) *Vanishing Fishes of North America*. Stone Wall Press, Inc.

Page, L. M. and Burr, B. M. (1991) *A Field Guide to Freshwater Fishes (North America, North of Mexico)*. Peterson Field Guide Series, Houghton Mifflin Co., Boston.

Paxton, J. R. and Eschmeyer, W. N. (1998) *Encyclopedia of Fishes* (2nd edition). Academic Press, New York and London.

Quinn, J. R. (1992) *Piranhas – Fact and Fiction*. T.F.H. Publications, Inc., Neptune City, New Jersey.

Ross, R. A. and Schäfer, F. (2000) *Freshwater Rays*. Verlag A.C.S. GmbH.

Schäfer, F. (1997) *All Labyrinths – Bettas, Gouramis, Snakeheads, Nandids*. Verlag A.C.S. GmbH.

Scheel, J. J. (1998) *Atlas of Killifishes of the Old World*. T.F.H. Publications, Inc., Neptune City, New Jersey.

Schraml, E. (1998) *African Cichlids: I–Malawi Mbuna*. Verlag A.C.S. GmbH.

Spotte, S. (1992) *Captive Seawater Fishes (Science and Technology.* John Wiley and Sons, Inc., New York and Chichester.

Stevens, J. (ed) (1987) *Sharks*. Facts On File, New York.

Taylor, L. R. (ed.) (1997) *Sharks & Rays - The Ultimate Guide to Underwater Predators*. Harper Collins Publishers.

Thompson, T. E. (1976) *Biology of the Opisthobranch Molluscs 1*, Ray Society, London.

Thompson, T. E. and Brown, G. H. (1984) *Biology of the Opisthobranch Molluscs 2*. Ray Society, London.

Thompson, K. S. (1991) *Living Fossil – The Story of the Coelacanth*. Hutchinson Radius.

Warner, G. F. (1977) *The Biology of Crabs*, Elek, London.

Watson, R. (1999) *Salmon, Trout and Charr of the World – A Fisherman's Natural History*. Simon Hall Press.

Wishnath, L. (1993) *Atlas of Livebearers of the World*. T.F.H. Publications, Inc., Neptune City, New Jersey.

Yonge, C. M. and Thompson, T. E. (1976) *Living Marine Molluscs*. Collins, London.

Index

Page numbers in **bold** indicate extensive coverage or feature treatment of a subject.

C

G

H

I

J

K

L

Picture Credits

Prelims: OSF: Daniel Cox; Doug Perrine

Ardea: 119, Kurt Amsler 245, Kev Deacon 246-7, Jean-Paul Ferrero 51, JM Labat 253, Ken Lucas 32b, 82, 226-7, P. Morris 160-1, 164-5, Mark Spencer 262, Ron & Valerie Taylor 261b 29, 261b, Ron Taylor 279t; Alissa Arp/ San Francisco State University: 110; Beverly Factor: 21; Biophotos Associates: 115; Bruce Coleman Collection: Franco Banfi 211b, Sven Halling 42, Malcolm Hey 35, 244t, C & S Hood 260, Pacific Stock 4/5, 234-5, Kim Taylor 252c; Coral Reef Research Institute : 38; Corbis: Hal Beral 46, Lester V Bergman 17, Jonathan Blair 145, Brandeon D. Cole 57, Mimmo Jodice 148/9, Jeffrey L. Rotman 155, Stuart Westmorland 78; Corbis Sygma: Thierry Prat 63; Dr G.L. Baron: 71b; Mark Erdmann: 273c; Frank Lane Picture Library: F. Bavendam/Minden Pictures 101, Susan Dewinsky 109, Foto Natura Stock 224-5, W.T. Miller 82 insert, Flip Nicklin/Minden Pictures 47b; Imagequestmarine.com: 142t, Peter Herring 190, 204-5, 208t, 208b, 209t; Natural Visions: Peter David 239, Heather Angel 67; Nature Picture Library: Dan Burton 182-3, Brandon Cole 142b, 152b, Georgette Douwma 32c, 244b, 256-7, 259t, Jeff Foott 241, Jurgen Freund 22, 282t, David Hall 284-5, Alan James 201, Reijo Juurinen 156-7, Avi Klapfer & Jeff Rotamn 254-5, 281, Conrad Maufe 275c, John Downer Productions 152bl, Fabio Liverani 49, 152t, Naturbild 187, Michael Pitts 6b, 141, Jeff Rotman 256, 275b, 278-9, 282b, 283, John Sparks 153, Sinclair Stammers 70; NHPA: A.N.T 195b, ANT Photo Library 266, 289, Pete Atkinson 34, 56b, Anthony Bannister 56t, Bill Coster 64, Daniel Heuclin 263, Image Quest 3D 37b, Scott Johnson 95t, B. Jones & M. Shimlock 32t, 33, 65, 72, 108, 138/9, 140, 211t, Lutra 188-9, 220-1, Trevor McDonald 23t, 58b, 286-7, Ashod Francis Papazian 265, Peter Parks 11, 121t, Tom & Theresa Stack 28t, MI Walker 25, Nobert Wu 28b, 37t, 238, 251, 264-5; Oxford Scientific Films: 3, 4, 10, 12/13, 20, 39, 74-75, 111b, 117t, 132/3, 160, 175, 178, 196t, Doug Allan 230, Kathie Atkinson 52b, 84, 104, 137t, Tobias Bernhard 86/7, 128b, 259b, 274-5, Waina Cheng 8, Paulo De Oliveira 170, 181, 279b, Mark Deeble & Victoria Stone 83b, Dr F. Ehrenstrom & L. Beyer 121b, 220t, 233, 235b, David Fleetham 2, 9, 50b, 52t, 79, 83t, 90/1, 123, 174, 176, 210, 247b, 261t, 278b, Stephen Foote 117b, Jeff Foote/Okapia 202t, 203t, 203b, David Fox 60b, Gary Gaugler 77, Max Gibbs 176/7, Lawrence Gould 177, Karen Gowlett-Holmes 7, 23b, 24, 27, 86t, 92, 94t, 102, 103, 118-9, 126-7, 126b, 126l, 128/9, 134/5, 134b, 258, Green Cape PTY Ltd 50t, Howard Hall 48/9, 167, Mark Hamblin 47t, Richard Hermann 98/9, 100, 122, Frank Huber 202b, Rodger Jackman 128t, 171, Paul Kay 30/1, 130b, Breck P. Kent 186, Richard Kirby 60t, Rudie Kuiter 85, 96/7, 101t, 131, 194, 195t, 213, Zig Leszczynski 180, Alastair MacEwen 18/9, Victoria A. McCornick 198/9, Prof H. Melhhorn/Okapia 75, Colin Milikins 48, 54/5, 166, Patrick Morris 191, Tammy Peluso 130t, Michael Pitss 172/3, Science Pictures Ltd 196b, Sue Scott 198, 231, Frithjof Skibbe 87, Gerard Soury 85b, 274, 288, Survival Anglia, Harold Taylor 58t, 92/3, Konrad Wothe 6t, Norbert Wu 94/5, 154, 252t; PA Photos: EPA 150; Photomax: Max Gibbs 204, 214, 216/7, 220b, 224, 232, 235t, 240, 241b, 252b, 268; Premaphotos Wildlife: Ken Preston-Mafham 41t, 68, 105, Dr Rod Preston-Mafham 134t; SAIAB: 272c, 272b; Seaphot: John Lythgoe 99; Seapics: Shedd Aquar/Ceisel 209b; Science Photo Library: 273t, Martin Dohrn 111t, Eye of Science 16, 69, Claude Nuridsany & Marie Perennou 271, David Scharf 71t, Andrew Syred 40, John Walsh 112; Still Pictures: Roland Birke 14/15; Welcome Trust Medical Photographic Library: Graham Budd 137b.

Diagrams by: Martin Anderson, Simon Driver